EVOLUTIONARY COMPUTATION

Techniques and Applications

EVOLUTIONARY COMPUTATION

Techniques and Applications

Edited by
Ashish M. Gujarathi, PhD
B. V. Babu, PhD

Apple Academic Press Inc.	Apple Academic Press Inc.
3333 Mistwell Crescent	9 Spinnaker Way
Oakville, ON L6L 0A2	Waretown, NJ 08758
Canada	USA

Library and Archives Canada Cataloguing in Publication

Evolutionary computation : techniques and applications / edited by Ashish M. Gujarathi, PhD, B.V. Babu, PhD.

Includes bibliographical references and index.
Issued in print and electronic formats.
ISBN 978-1-77188-336-8 (hardcover).--ISBN 978-1-77188-337-5 (pdf)
1. Evolutionary computation. 2. Genetic algorithms. 3. Software engineering. I. Gujarathi, Ashish M., author, editor II. Babu, B. V., author, editor

TA347.E96E96 2016	006.3'823	C2016-905450-0	C2016-905451-9

Library of Congress Cataloging-in-Publication Data

Names: Gujarathi, Ashish M., editor. | Babu, B. V., editor.
Title: Evolutionary computation : techniques and applications / editors, Ashish M. Gujarathi, PhD, B.V. Babu, PhD.
Other titles: Evolutionary computation (Gujarathi)
Description: Toronto ; Waretown, New Jersey : Apple Academic Press, [2017] |
Includes bibliographical references and index.
Identifiers: LCCN 2016035469 (print) | LCCN 2016037827 (ebook) | ISBN 9781771883368 (hardcover : acid-free paper) | ISBN 9781771883375 (ebook) | ISBN 9781771883375 (ebook)
Subjects: LCSH: Evolutionary computation. | System engineering. | Mathematical optimization.
Classification: LCC TA347.E96 E96 2017 (print) | LCC TA347.E96 (ebook) | DDC 006.3/823--dc23
LC record available at https://lccn.loc.gov/2016035469

Apple Academic Press also publishes its books in a variety of electronic formats. Some content that appears in print may not be available in electronic format. For information about Apple Academic Press products, visit our website at **www.appleacademicpress.com** and the CRC Press website at **www.crcpress.com**

CONTENTS

LIST OF CONTRIBUTORS

Robert T. F. Ah King
Department of Electrical and Electronic Engineering, University of Mauritius, Reduit 80837, Mauritius

S. B. D. V. P. S. Anauth
Department of Electrical and Electronic Engineering, University of Mauritius, Reduit 80837, Mauritius

B. V. Babu
Vice Chancellor, Galgotias University, Greater Noida, Uttar Pradesh, India, E-mail: profbvbabu@gmail.com, Tel: +91-12-04806849

Hema Banati
Dyal Singh College, Department of Computer Science, University of Delhi, Delhi, E-mail: banatihema@hotmail.com

U. Sabura Banu
Professor Department of Electronics and Instrumentation Engineering, BS Abdur Rahman University, Vandalur, Chennai – 600048, Tamilnadu, India

Nirupam Chakraborti
Department of Metallurgical and Materials Engineering, Indian Institute of Technology, Kharagpur, 712 302, West Bengal, India

Erik Cuevas
Departamento de Ciencias Computacionales, Universidad de Guadalajara, CUCEI, CU-TONALA, Av. Revolución 1500, Guadalajara, Jal, México

Wael Mohamed Fayek
Assistant Professor, Department of Electrical Engineering, Helwan University, Cairo, Egypt

Paulo Fazendeiro
Universityof Beira Interior, DepartmentofInformatics, Portugal
Instituto de Telecomunicações (IT), Portugal,E-mail: fazendeiro@ubi.pt

A. Garg
School of Mechanical and Aerospace Engineering, Nanyang Technological University, 50 Nanyang Avenue, 639798, Singapore

Ravindra D. Gudi
Professor, Department of Chemical Engineering, Indian Institute of Technology Bombay, Powai, Mumbai, India, E-mail: ravigudi@iitb.ac.in, Tel: +91-22-25767231

Ashish M. Gujarathi
Department of Petroleum and Chemical Engineering, College of Engineering, Sultan Qaboos University, P.O. Box 33, Al-Khod, Muscat-123, Sultanate of Oman, Phone: +968 2414 1320, E-mail: ashishgujrathi@gmail.com, ashishg@squ.edu.om

Vishal Jain
Department of Mechanical Engineering, Birla Institute of Technology and Science, BITS Pilani, K.K. Birla Goa Campus, Zuarinagar, Goa 403726, India

Manas Ranjan Kabat
Department of Computer Science and Engineering, Veer Surendra Sai University of Technology, Burla, India, E-mail: kabatmanas@gmail.com

Sanjay S. Kadam
Evolutionary Computing and Image Processing Group (ECIP), Center for Development of Advanced Computing (C-DAC), Savitribai Phule Pune University Campus, Ganeshkhind, Pune–411007, India

Saket Kansara
ESTECO Software India Pvt. Ltd., Pune, Maharashtra, India

Nikilesh Krishnakumar
Department of Mechanical Engineering, Birla Institute of Technology and Science, BITS Pilani, K.K. Birla Goa Campus, Zuarinagar, Goa 403726, India

G. N. Sashi Kumar
Scientific Officer, Computational Studies Section, Machine Dynamics Division, Bhabha Atomic Research Centre, Trombay, Mumbai, India–400 085, Tel.: +91-22-2559-3611; E-mail: gnsk@barc.gov.in

Nikos D. Lagaros
Assistant Professor, Institute of Structural Analysis and Antiseismic Research, School of Civil Engineering, National Technical University of Athens, 15780, Greece

O. P. Malik
Professor Emeritus, Department of Electrical and Computer Engineering, University of Calgary, Alberta, Canada

Shikha Mehta
Dyal Singh College, Department of Computer Science, University of Delhi, Delhi, E-mail: mehtshikha@gmail.com

Mohit Mishra
Department of Computer Science and Engineering, Indian Institute of Technology (Banaras Hindu University), Varanasi, India

Tamoghna Mitra
Thermal and Flow Engineering Laboratory, Åbo Akademi University, Biskopsgatan 8, FI-20500 Åbo, Finland

Kamran Morovati
Information Security Center of Excellence (ISCX), Faculty of Computer Science, University of New Brunswick, 550 Windsor St., Head Hall E128, Fredericton, NB, E3B 5A3, Canada

Iordanis A. Naziris
PhD Candidate, Laboratory of Building Construction and Building Physics, Department of Civil Engineering, Aristotle University of Thessaloniki, 54124, Greece

Diego Oliva
Dpto. Ingeniería del Software e Inteligencia Artificial, Facultad Informática, Universidad Complutense de Madrid, 28040 Madrid, Spain, E-mail: doliva @ucm.es; pajares@ucm.es

Ying Chuan Ong
Department of Chemical and Biomolecular Engineering, National University of Singapore, Engineering Drive 4, Singapore 117585, Republic of Singapore

Valentín Osuna-Enciso
Departamento de Ingenierías, CUTONALA, Universidad de Guadalajara, Sede Provisional Casa de la Cultura – Administración, Morelos #180, Tonalá, Jalisco 45400, México, E-mail: valentin.osuna@cutonala.udg.mx

Gonzalo Pajares
Dpto. Ingeniería del Software e Inteligencia Artificial, Facultad Informática, Universidad Complutense de Madrid, 28040 Madrid, Spain, E-mail: doliva @ucm.es; pajares@ucm.es

S. K. Pal
Senior Research Scientist, Scientific Analysis Group, Defence Research and Development
Organization, Ministry of Defence, Govt. of India, New Delhi, India

Millie Pant
Department of Applied Science and Engineering, Saharanpur Campus, IIT Roorkee, India

Kyriakos Papaioannou
Professor Emeritus, Laboratory of Building Construction and Building Physics, Department of Civil
Engineering, Aristotle University of Thessaloniki, 54124, Greece

Sumeet Parashar
ESTECO North America, Novi, MI 48375, USA

Manoj Kumar Patel
Department of Computer Science and Engineering, Veer Surendra Sai University of Technology, Burla,
India, E-mail: patel.mkp@gmail.com

Marco Pérez-Cisneros
Departamento de Ciencias Computacionales, Universidad de Guadalajara, CUCEI, CU-TONALA, Av.
Revolución 1500, Guadalajara, Jal, México

K. V. R. B. Prasad
Professor, Department of E.E.E., MITS, P.B. No: 14, Kadiri Road, Angallu (V), Madanapalle – 517325,
Chittoor District, Andhra Pradesh, India, E-mail: prasad_brahma@rediffmail.com

Paula Prata
University of Beira Interior, Department of Informatics, Portugal
Instituto de Telecomunicações (IT), Portugal, E-mail: pprata@di.ubi.pt

Manojkumar Ramteke
Department of Chemical Engineering, Indian Institute of Technology Delhi, Hauz Khas, New
Delhi–110 016, India, E-mail: ramtekemanoj@gmail.com, mcramteke@chemical.iitd.ac.in

G. P. Rangaiah
Department of Chemical and Biomolecular Engineering, National University of Singapore, Engineering
Drive 4, Singapore 117585, Republic of Singapore

Satya Prakash Sahoo
Department of Computer Science and Engineering, Veer Surendra Sai University of Technology, Burla,
India, E-mail: sahoo.satyaprakash@gmail.com

Abdus Samad
Department of Ocean Engineering, Indian Institute of Technology Madras, Chennai–600036, India

Henrik Saxén
Thermal and Flow Engineering Laboratory, Åbo Akademi University, Biskopsgatan 8, FI-20500 Åbo,
Finland

Munawar A. Shaik
Associate Professor, Department of Chemical Engineering, Indian Institute of Technology Delhi, Hauz
Khas, New Delhi, India, E-mail: munawar@iitd.ac.in, Tel: +91-11-26591038

Shivom Sharma
Department of Chemical and Biomolecular Engineering, National University of Singapore, Engineering
Drive 4, Singapore 117585, Republic of Singapore

Tarun Kumar Sharma
Amity School of Engineering & Technology, Amity University Rajasthan, Jaipur, India

Pravin M. Singru
Department of Mechanical Engineering, Birla Institute of Technology and Science, BITS Pilani, K.K. Birla Goa Campus, Zuarinagar, Goa 403726, India

K. Tai
School of Mechanical and Aerospace Engineering, Nanyang Technological University, 50 Nanyang Avenue, 639798, Singapore

Vibhu Trivedi
Department of Chemical Engineering, Indian Institute of Technology Delhi, Hauz Khas, New Delhi–110 016, India

V. Vijayaraghavan
School of Mechanical and Aerospace Engineering, Nanyang Technological University, 50 Nanyang Avenue, 639798, Singapore

R. V. Yampolskiy
Associate Professor, Department of Computer Engineering and Computer Science, University of Louisville, KY, USA

Daniel Zaldívar
Departamento de Ciencias Computacionales, Universidad de Guadalajara, CUCEI, CU-TONALA, Av. Revolución 1500, Guadalajara, Jal, México

LIST OF ABBREVIATIONS

ABC	Artificial bee colony
ACO	Ant colony optimization
AGA	Adaptive genetic algorithm
AGSAA	Adaptive genetic simulated annealing algorithm
AHP	Analytic hierarchy process
AI	Artificial intelligence
ANN	Artificial neural networks
APSS	Adaptive power system stabilizer
AR	Aspect ratio
ARNA-GA	Adaptive RNA-GA
ARX	Auto regressive model with eXogenous signal
ASF	Achievement scalarization function
ASP	Active server pages
AVR	Automatic voltage regulator
BF	Bacterial foraging
BFA	Bacterial foraging algorithm
BFOA	Bacterial foraging optimization algorithm
BioGP	Bi-objective genetic programming
BITS	Birla Institute of Technology and Science
BNNT	Boron nitride nanotubes
BT	Bi-level thresholding
BW	Bandwidth factor
CAD	Computer aided design
CEC	Congress on Evolutionary Computation
CFD	Computational fluid dynamics
CGA	Cockroach GA
CNT	Carbon nanotubes
COCOMO	Constructive cost model
CPSS	Conventional power system stabilizer
CTT	Constrained tree traversal
CV	Constraint violations
DE	Differential evolution
DFS	Depth first search

DLOC	Developed line of code
DM	Decision maker
DNA-HGA	DNA based hybrid genetic algorithm
DNM	Deep net meta-crawler
DRI	Directly reduced iron
DSSM	Delayed S-shaped model
EAs	Evolutionary algorithms
EC	Evolutionary computation
EMO	Electromagnetism-like optimization
EP	Evolutionary programming
ES	Evolutionary strategies
EXMP	Exponential model
FCCU	Fluid catalytic cracking unit
FEM	Finite element methods
FLC	Fuzzy logic control
FN	False negative
FP	False positive
FSI	Fire safety index
FTA	Fault tree analysis
GA	Genetic algorithm
GMV	Generalized minimum variance
GNFS	General number sieve algorithm
GP	Genetic programming
GPUs	Graphic processing units
H DNA-GA	Hybrid DNA-GA
HLGA	Hajela and Lin's genetic algorithm
HM	Harmony memory
HMCR	Harmony-memory consideration rate
HS	Harmony search
HSA	Harmony search algorithm
HSMA	Harmony search multi-thresholding algorithm
IA	Inverse-anticodon
IACC	International Advance Computing Conference
ICA	Imperialist competitive algorithm
ICIT	International Conference on Industrial Technology
ICMST	International Conference on Mining Science and Technology
IDE	Integrated DE

IET	Institute of Engineering and Technology
IJCNN	International Joint Conference on Neural Networks
IL	Intermediate locations
IT	Information technology
ITAE	Integral time absolute error
JG	Jumping genes
JPSO	Jumping PSO
JSP	Java server pages
KLOC	Kilo line of code
LCI	Lower confidence interval
LO	Local optimum
LOC	Lines of code
LPPE	Low density polyethylene
M DNA-GA	Modified DNA-GA
MACK	Multivariate adaptive cross-validating Kriging
MACO	Multiobjective ACO
MAPE	Mean absolute percentage error
MAXGEN	Maximum number of generations
MCDM	Multi-criteria decision making
MCN	Maximum cycle number
MD	Molecular dynamics
MDE	Modified differential evolution
ME	Methodology
MFR	Mixed flow reactor
MGGP	Multi-gene genetic programming
MGO	Molecular geometry optimization
MIMO	Multi input multi output
MINLP	Mixed-integer nonlinear programming
MM	Maximum-minimum
MNFE	Mean number of function evaluations
MODE	Multiobjective differential evolution
MOEAs	Multiobjective evolutionary algorithms
MOO	Multiobjective optimization
MOPs	Multiobjective optimization problems
MRP	Multiple routing paths
MSE	Mean square error
MSIGA	Modified self-adaptive immune GA
MT	Multilevel thresholding

MU	Monetary units
NACO	Niched ant colony optimization
NEMS	Nano electro mechanical systems
NFE	Number of function evaluations
NI	National Instrument
NLP	Nonlinear programming
NM	Normal-mutation
NMP	Non-minimum-phase
NNs	Neural networks
NP	Nondeterministic polynomial time
NP	Number of population size
NPGA	Niched-Pareto genetic algorithm
NPI	Node parent index
NSGA	Nondominated sorting genetic algorithm
NSGA-II	Elitist nondominated sorting genetic algorithm
NWeSP	Next generation web services practices
OBL	Opposition based learning
OF	Objective functions
PA	Phthalic anhydride
PA	Pole assignment
PAES	Pareto archived evolution strategy
PAR	Pitch adjusting rate
PESA	Pareto envelope based selection algorithm
PET	Polyethylene terephthalate
PIRGA	Protein inspired RNA-GA
PLCs	Programmable logic controllers
PM	Polynomial mutation
PM10	Particulate matter
POS	Pareto-optimal solutions
POWM	Power model
PS	Pole shift
PSNR	Peak signal-to-noise ratio
PSO	Particle swarm optimization
PSSs	Power system stabilizers
QS	Quadratic sieve
RBF	Radial basis functions
RGA	Relative gain array
RJGGA	Real jumping gene genetic algorithm

RMSE	Root mean square error
RN	Random number
RSM	Response surface models
RSS	Random search space
SA	Sensitivity analysis
SA	Simulated annealing
SBX	Simulated binary crossover
SE mean	Standard error of mean
SEDP	Software engineering design problems
SFLA	Shuffled Frog Leaping Algorithm
SGA	Simple genetic algorithm
SI	Swarm intelligence
SIMD	Single instruction multiple data
SOP	Single objective optimization problem
SOR	Steam over reactant
SPEA	Strength-Pareto evolutionary algorithm
SPMD	Single program multiple data
SPP	Scout production period
SR	Success rate
SRP	Single routing path
STD	Standard deviation
SWCNT	Single-walled carbon nanotube
TBP	True boiling point
TCR	Time constant regulator
TEM	Transmission electron microscopy
TN	True negative
TP	True positive
TSP	Traveling salesman problem
UBBPSO	Unified bare-bones particle swarm optimization
UCI	Upper confidence interval
VFSA	Very fast-simulated annealing
WNYIPW	Western New York Image Processing Workshop

PREFACE

Evolutionary computation has gained popularity in the recent past due to its several advantages over deterministic methods. Unlike deterministic methods, the evolutionary computation methods start with multiple solutions (both for single and multiobjective optimization (MOO) studies) so as to obtain wide range of initial population members. Evolutionary computation methods are also highly preferred to solve multiobjective optimization problems where conflicting objective functions are involved. The individual evolutionary computation method is expected to converge—to a single optimum solution (global solution) for single objective optimization and—to the Pareto optimal front for multiobjective optimization studies. As multiobjective optimization algorithm results in a set of solutions, the two goals are associated with each multiobjective optimization algorithm. The algorithm should converge to the true Pareto front and the algorithm should maintain a diverse set of solutions on the Pareto front. While achieving these two goals of MOO and dealing with two search spaces, the search for the true Pareto front in case of MOO study depends upon the following key issues, such as, number and type of decision variables (continuous, discontinuous) and nature of decision variable space; type of objective functions (minimization, maximization) and nature of objective space; nonlinearity and stiffness of model equations; type of constraints (equality and inequality); an ability of algorithm to handle the search spaces of objectives and decision variables. This book talks about recent advancement of evolutionary computation on both theory and applications.

This book is broadly organized in three sections. Part 1 contains 7 chapters on 'Theory and applications in engineering systems.' Part 2 has 6 chapters on 'Theory and applications of single objective optimization studies.' Part 3 has 8 chapters and is titled as 'Theory and applications of single and multiobjective optimization studies'

Chapter 1 gives introduction on theory of evolutionary algorithms, single and multiobjective optimization. This chapter also includes paragraph on organization of book. In chapter 2 the Bio-Mimetic adaptations of genetic algorithm and their applications to chemical engineering is discussed. An overview of use of surrogate-assisted evolutionary algorithms in various aspects of engineering optimization including single as well as multiobjective optimization evolutionary algorithms is given in chapter 3. Chapter 4 talks about an application of iron making using evolutionary algorithms. They discussed modern ironmaking process optimization aspects using blast furnace and rotary kiln operation. Soft computing techniques, such as Artificial Neural Networks, and evolutionary algorithm are also briefly discussed. Optimization of burden distribution in blast furnace and of production parameters, such as CO_2 emissions, and coal injection with respect to multiobjective optimization framework are performed. Chapter 5 represents theory and an application of harmony search optimization for multilevel thresholding in digital images. Chapter 6 talks about swarm intelligence in software engineering design problems. Chapter 7 discusses integrated gene programming approach in simulating the tensile strength characteristic of BNNTs based on aspect ratio, temperature and number of defects of nanoscale materials.

In Part 2, in chapter 8 an application of a nonlinear transformation for representing discrete variables as continuous variables is presented and discussed an alternate method for solving MINLP problems by converting them into equivalent NLP problems. In chapter 9, an application of optimal design of shell and tube heat exchanger using differential evolution algorithm is discussed. Chapter 10 represents the review of various evolutionary algorithms to solve the QoS-aware multicast routing problem. Chapter 11 focuses on the study of the effective parallelization of the canonical GA. Chapter 12 presents an efficient approach for populating deep web repositories using Shuffled Frog Leaping Algorithm (SFLA). Multi-loop fractional order PID controller is optimally tuned using Bat algorithm for quadruple tank process in chapter 13. A practical approach for multiobjective shape optimization using multiobjective ant colony optimization in presented chapter 14. In chapter 15, a review on nature-inspired computing techniques for integer optimization presented. Chapter 16 presents an application of genetic algorithms on adaptive power system stabilizer. Applications of evolutionary computation for solving the budget allocation problem of the fire safety upgrading of a group of buildings are discussed in chapter 17. Chapter 18 focuses on the development and comparative application of elitist multiobjective evolutionary algorithms (MOEAs) for voltage and reactive power optimization in power systems. In chapter 19 stochastic algorithms are used for solving several complex test problems and pooling problems. The authors have evaluated the performance of integrated DE (IDE), unified bare-bones particle swarm optimization (UBBPSO), modified simulated annealing and very fast-simulated annealing (VFSA) algorithms on pooling problems. The NSGA-II algorithm with SBX-A and SBX-LN crossover probability distributions are used to test several multiobjective functions in chapter 20. Finally in chapter 21 an application of evolutionary algorithms for malware detection and classification based on the dissimilarity of op-code frequency patterns is presented.

It is a great pleasure to acknowledge the valuable contribution from several authors across the globe. We wish to express our heartfelt gratitude and indebtedness to the managements of Sultan Qaboos University, Oman and Galgotias University, India.

We are greatly thankful to the editorial staff of Apple Academic Press for their constant encouragement and their plentiful efforts for editing the book.

—Ashish Madhukar Gujarathi, PhD
B. V. Babu, PhD

ABOUT THE EDITORS

Ashish M. Gujarathi, PhD
Assistant Professor, Petroleum and Chemical Engineering Department, College of Engineering, Sultan Qaboos University, Sultanate of Oman

Ashish M. Gujarathi, PhD, is currently an Assistant Professor in the Petroleum and Chemical Engineering Department of the College of Engineering at Sultan Qaboos University, Sultanate of Oman. He was formerly a Lecturer of Chemical Engineering at the Birla Institute of Technology and Science (BITS) in Pilani, India. Dr. Gujarathi has over 11 years of experience as a chemical engineer with diverse work experience comprising a blend of academic research, teaching, and industrial consultancy work. A prolific author with articles, book chapters, and conference proceedings to his credit, he is also an editorial board member of several journals, including the *Journal of Developmental Biology and Tissue Engineering* and the *International Open Access Journal of Biology and Computer Science* and has acted as a reviewer for several international journals as well as for books and conference proceedings. His research interests include reaction engineering; process design and synthesis; process modeling, simulation, and optimization; polymer science and engineering; petrochemicals; parametric estimation and optimization of major chemical processes; evolutionary computation; and biochemical engineering.

B. V. Babu, PhD
Vice Chancellor, Galgotias University, Greater Noida, India

B. V. Babu, PhD, is currently Vice Chancellor at Galgotias University in Greater Noida, India. An acknowledged researcher and renowned academician, Dr. Babu has 30 years of teaching, research, consultancy, and administrative experience. Formerly, he was the Pro Vice Chancellor of DIT University, Dehradun, and founding Director of the Institute of Engineering and Technology (IET) at J.K. Lakshmipat University, Jaipur. He is a member of many national and international academic and administrative committees and professional organizations. Professor Babu is a distinguished academician and an acknowledged researcher and is well known internationally for his algorithm MODE (Multi Objective Differential Evolution) and its improved variants. Overall he has over 235 research publications to his credit. He has published several books and has written chapters, invited editorials, and articles in various books, lecture notes, and international journals. He organized several international and national conferences, workshops, and seminars and also chaired several technical sessions. He has been an invited speaker and has delivered keynote addresses at various international conferences, seminars, and workshops. He is the recipient of CSIR's National Technology Day Award for recognition his research work as well as of many other awards. He is the life member/fellow of many professional bodies, such as IIChE, IE (I), ISTE, ICCE, IEA, SOM, ISSMO, IIIS, IAENG, SPE, ISTD, etc. He is editor-in-chief and editorial board member of several international and national scientific journals.

PART I

THEORY AND APPLICATIONS IN ENGINEERING SYSTEMS

CHAPTER 1

INTRODUCTION

ASHISH M. GUJARATHI[1] and B. V. BABU[2]

[1]*Department of Petroleum and Chemical Engineering, College of Engineering, Sultan Qaboos University, P.O. Box 33, Al-Khod, Muscat-123, Sultanate of Oman, Phone: +968 2414 1320, E-mail: ashishgujrathi@gmail.com, ashishg@squ.edu.om*

[2]*Professor of Chemical Engineering and Vice Chancellor, Galgotias University, Greater Noida, India, E-mail: profbvbabu@gmail.com*

CONTENTS

1.1   INTRODUCTION

An optimization problem involving more than one objective to be optimized is referred as multiobjective optimization problem (MOOP). The optimum solution corresponding to a single objective optimization problem refers to the optimal basic feasible solution (satisfying bounds of variables and the constraints). However, in case of multiobjective optimization, the optimum solution refers to a compromised (not necessarily the optimum with respect to any objective) set of multiple feasible solutions. In the most general form, the multiobjective optimization problem (with m objectives, n variables, p inequality constraints and q equality constraints) can be expressed as given by Eq. (1.1):

$$
\left.\begin{aligned}
&Min \quad f(x_i) = \left[f_1(x_i), f_2(x_i), \ldots, f_m(x_i) \right]^T \\
&\text{subject to} \\
&u_j(x_i) \geq 0, \quad j = 1, 2, \ldots, p \\
&v_k(x_i) = 0, \quad k = 1, 2, \ldots, q \\
&x_i^{(L)} \leq x_i \leq x_i^{(U)}, \quad i = 1, 2, \ldots n
\end{aligned}\right\} \tag{1.1}
$$

The optimization algorithms can be broadly classified into two categories, i.e., traditional or classical methods and the non-traditional or population based search algorithms. The traditional algorithms often start with a single point (guess value) and end up with a single point solution. The ideal out-come of a single objective optimization problem is a single global solu-tion. However, the outcome of gradient-based traditional algorithms largely depends on its control parameters such as the step size and the direction of search that are being used. In a complex and non-linear search space (as shown in Figure 1.1), which may involve multiple local and a single global solution, an inefficient local search algorithm may get trapped at local optimal solution. In contrast, evolutionary algorithms, which mimic nature's principle of survival of the fittest, start with multiple population points [7, 8, 16, 17, 19, 20]. Due to the strong genetic operators, evolutionary algorithms are found to achieve the global optimum in majority of industrial applica-tions for single objective optimization [2].

In case of MOO problems, the decision maker is always interested in obtaining a solution suitable to his/her design requirements, i.e., a single solution. But due to the multiobjective nature of the problem and the asso-ciated trade-off, the desired solution may vary as per the decision makers need and the choice. Thus providing multiple solutions rather than a single optimum solution (traditional multiobjective optimization algorithms give single optimum solution) would be an advantage to the decision maker, so that one can have a choice of selecting one from the several equally good

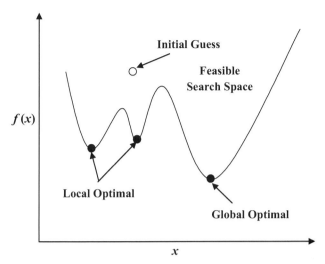

FIGURE 1.1   Local and global optimal solutions of a complex search space.

solutions from the Pareto front. The specialty of such solutions is that as we move from one solution to the other we gain in terms of one objective at the cost of loss in another objective involved in the study. Such a set of solutions is referred as the Pareto optimal set and the solutions in this set are nondominated with respect to each other.

1.1.1   DEFINITION OF DOMINANCE

A solution $x^{(1)}$ is said to dominate the other solution $x^{(2)}$, if both the following conditions 1 and 2 are true.

1. The solution $x^{(1)}$ is no worse than $x^{(2)}$ in all objectives, or $f_j\left(x^{(1)}\right) \not\triangleright f_j\left(x^{(2)}\right)$ for all $j=1,2,\dots M$;
2. The solution $x^{(1)}$ is strictly better than $x^{(2)}$ in at least one objective, or $f_j\left(x^{(1)}\right) \triangleleft f_j\left(x^{(2)}\right)$ for at least one $(j \in 1,2,\dots M)$.

If any of the two conditions is violated, the solution $x^{(1)}$ does not dominate the solution $x^{(2)}$ [7]. As multiobjective optimization algorithm results in a set of solutions, the following two goals are associated with each multiobjective optimization algorithm.

1. The algorithm should converge to the true Pareto front;
2. The algorithm should maintain a diverse set of solutions on the Pareto front.

In pursuit of achieving the convergence, the algorithm may lose diversity of solutions in the Pareto front. But it is worth to mention here that though both convergence and divergence issues are equally important in MOO study, the diverse set of solutions is meaningless if the algorithm did not converge to the true Pareto front. Thus, any efficient multiobjective optimization algorithm should first focus on achieving the convergence to the true Pareto front and then new solutions in the neighborhood of solutions on Pareto front may be searched to enhance the divergence. Though both the goals of achieving convergence and maintaining diverse set of solutions are important, focus should first be given in algorithm to attain the convergence.

Unlike single objective optimization, the MOO problems deal with two kinds of search space. Figure 1.2 shows the decision variable space and objective space with global and local Pareto fronts. The decision variables space is obtained by plotting decision variables and constraints within bounds. On evaluation, each point in the decision variable space generates an additional point in the objective space. Thus a given algorithm proceeds based on the comparison of objective function values from the objective space but perturbation of variables occurs in the decision variable space. Thus multiobjective optimization problems are more difficult to solve as compared to single objective optimization problems.

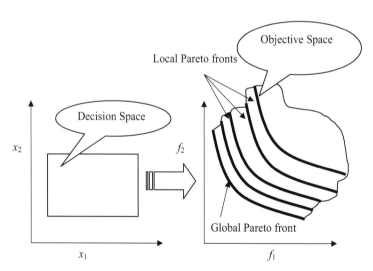

FIGURE 1.2 Decision space, objective space, local and global Pareto fronts involved in multiobjective optimization study.

In case of simple test problems (where there exists a direct relationship between the objective function and the decision variables), the cost of the objective function can easily be evaluated and then used in evolutionary algorithm. However, in case of industrial problems, the mathematical model needs to be evaluated first. Once the model is formulated, it needs to be integrated and simulated using suitable numerical technique. Judicious choice of numerical technique is made to solve the mathematical model. During the model based evaluation of Pareto front the decision variables are initialized randomly within the bounds. These decision variables are then converted to the specific input form of model. These input parameters are passed to the model. The model is integrated and simulated along the space coordinate and/or time domain. The objectives are evaluated from the output of the model. The termination criteria are checked, and if not terminated, the offspring is generated after applying the corresponding genetic operators. Selection is performed based on the objective function values of the parent and the offspring. The algorithm continues until the termination criteria are met.

While achieving these two goals of MOO and dealing with two search spaces, the search for the true Pareto front in case of MOO study depends upon the following key issues:

- number and type of decision variables (continuous, discontinuous) and nature of decision variable space;
- type of objective functions (minimization, maximization) and nature of objective space;
- nonlinearity and stiffness of model equations;
- type of constraints (equality, inequality);
- ability of algorithms to handle the search spaces of objectives and decision variables.

Out of the above-mentioned five aspects, first four are problem specific. However, Pareto front output also largely depends on the algorithm's ability to converge towards the true Pareto front and then produces a well diverse set of solutions. An inefficient algorithm may get trapped at local optimal nondominated set of solutions (Figure 1.2) or may result in a single point solution. Few of the reasons by which algorithm may result in local Pareto front or a single point solutions are:

- Algorithms may not produce a superior offspring which is nondominated with respect to other solutions in the current population;

- An inefficient selection scheme of algorithm may restrict a new solution to enter in the current population;
- In case of binary coded algorithms, accuracy of newly obtained solutions depends on the number of bits used in defining string;
- Of its inability to handle the complexity of a problem (i.e., multi-dimensional decision variable space).

Thus it is necessary to have an algorithm, which not only overcome above limitations but also results in a well diverse Pareto front and close to true Pareto front. The literature survey on several evolutionary algorithms shows a large demand towards developing new algorithms. The algorithm output also depends on the complexity of MOO problems. The industrial engineering problems and some of the test problems involve multi-dimensional decision variable space, multi-modal objective space with equality and inequality constraints. Some of the commonly observed trade-offs associated with process design decisions involved in industrial problems are described below.

1.1.2   INDUSTRIAL STYRENE REACTOR

Styrene is commercially produced from ethyl benzene [3, 21, 22]. The main reaction producing styrene is a reversible endothermic reaction. As per the Li-Chatelier's principle, for a reversible endothermic reaction, high temperature and low pressure favors the rate of forward reaction. But as the temperature is increased, other side products (due to thermal cracking), such as toluene, benzene, etc. are formed. Thus, at low temperature the yield and productivity are low, while selectivity is high. If the temperature of reactor is increased, the selectivity decreases (due to the formation of byproducts) but the yield increases. But the objectives of the process are to increase simultaneously the yield, selectivity and the productivity. If the decision variables such as temperature of ethyl benzene (feed), steam over reactant (SOR), operating pressure and initial ethyl benzene flow rate are used subject to a constraint on temperature, the optimization problem would become more complex and trade-off among the objectives would clearly be observed. The decision maker has to sacrifice for one of the objectives, while achieving the better value of another objective. Thus, in case of industrial styrene reactor there exits potential trade-off among the said objectives of simultaneous maximization of yield, productivity and the selectivity. Pareto fronts obtained for case-1 (i.e., simultaneous maximization of yield and the selectivity) using various strategies of Multiobjective Differential Evolution (MODE) algorithms (namely, MODE, MODE-III, Elitist MODE, Hybrid MODE and Trigonometric MODE) are shown in Figure 1.3a. The results of Figure 1.3a are re-plotted (for better clarity

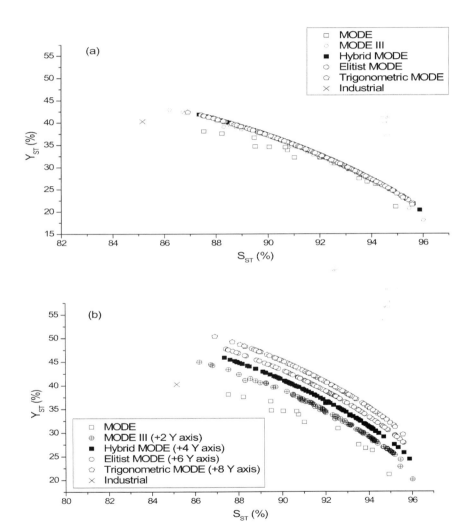

FIGURE 1.3 (a) Pareto fronts obtained for case-1 using strategies of MODE algorithms; (b) The results of figure (a) are re-plotted (for better clarity of Pareto fronts) with vertical shift in value of ordinate by +2 in MODE III, +4 in Hybrid MODE, +6 in Elitist MODE and +8 in Trigonometric MODE data points.

of Pareto fronts) in Figure 1.3b with vertical shift in value of ordinate by +2 in MODE III, +4 in Hybrid MODE, +6 in Elitist MODE and +8 in Trigonometric MODE data points. Detailed results on multiobjective optimization of styrene reactor using various strategies of MODE algorithms is available in our recent publications [9, 10, 14].

1.1.3 LOW DENSITY POLYETHYLENE (LPPE) TUBULAR REACTOR

Polyethylene is commercially produced by both high pressure (free-radical) and low pressure (ionic) addition ethylene polymerization processes. Two types of reactors (tubular and stirred autoclave) are essentially applied in the free-radical high-pressure polymerization processes. Ethylene free-radical polymerization is conducted in the presence of free-radical initiators, such as azo-compounds, peroxides, or oxygen at very high pressures (1300–3400 bars) and high temperatures (225–610 K). Under the reaction conditions employed in high-pressure processes, LDPE is produced as a result of short-chain branching formation. A commercial reactor consists of 3–5 reaction zones and several cooling zones [5]. The reactor includes a number of initiator side-feed points. The temperature and flow rate of each coolant stream entering a reaction/cooling zone is used to control the temperature profile in the reactor. A mixture of ethylene, a free-radical initiator system, and a solvent are injected at the entrance of reactor. Maximization of monomer conversion is one of the objectives to be considered during MOO of LDPE. While maximizing the conversion, the undesirable side chain concentration (sum of methyl, vinyl, and vinylidene) also increases. Thus, minimization of unwanted side products and maximization of monomer conversion gives rise to conflicting set of objectives. More meaningful and industrially important results can be generated if these sets of objectives are coupled with equality constraints on number average molecular weight. Figure 1.4 shows the Pareto optimal solutions for various end point constraints on the number-average molecular weight (i.e., $M_{N,f}$= 21, 900 ± 20; $M_{N,f}$= 21, 900 ± 200; $M_{N,f}$= 21, 900 ± 1100) using MODE III algorithm. The Pareto front obtained with an end point constraint, of $M_{N,f}$= 21, 900 ± 1100, covers a wide range of solutions as compared to the range of solutions obtained when a strict constraint on $M_{N,f}$ is used [13, 14].

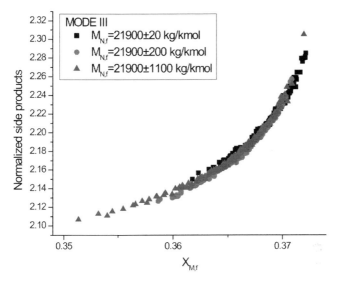

FIGURE 1.4 Pareto optimal solutions for various end point constraints on the number-average molecular weight using MODE III algorithm.

1.1.4 SUPPLY CHAIN AND PLANNING

The supply chain is basically the integrated network [6] among retailers, distributors, transporters, storage facilities and suppliers that participate in the sale, delivery and production of a particular product for the following purposes:

1. maximizing the overall profit generated;
2. increasing the competitiveness of the whole chain;
3. minimizing the system wide costs while satisfying the service level requirements;
4. matching the supply and demand profitably for products and services.

It is due to the above reasons that the supply chain optimization problem is considered as a multiobjective optimization problem (MOOP). The supply chain problem therefore has to be considered as a whole (system optimization) without placing the individual preferences of the individual objectives. The built up supply chain model should be capable of integrating all the entities so that the flow of information happens among the entities in order to meet the highly fluctuating demand of the market.

The important issues that drive the supply chain models and subsequently govern its design are:

1. inventory planning and management;
2. transportation and logistics management;
3. facilities location and layout design;
4. flow of information among the entities.

These four drivers represent the major flows associated with supply chain problem. In order to maximize overall profitability, it is not possible to get a unique solution that satisfies either all the criteria or the objectives. If all the objectives are satisfied then the solution obtained could be a non-Pareto optimal point. Hence in multiobjective optimization problem, we are interested in set of solutions (rather than a single solution), which are non-inferior with respect to each other and are part of Pareto optimal front. Simultaneous optimization of individual objectives is necessary without giving weightage to individual objectives. A goal programming approach to optimization would not result in a set of solutions and a compromised, but a single solution, would result in a single run. Evolutionary algorithms have shown a good potential to generate multiple equally good solutions for many engineering and test problems both for single and multiobjective optimization problems. Hence an attempt to solve such problems using newly developed evolutionary algorithms may result in a possibly better set of solutions.

Thus, to deal with above-mentioned problems there is a need towards development of new and efficient algorithms. To judge the robustness of newly developed algorithms, it needs to be tested on several benchmark test problems and then applied on industrial applications. This motivated us to design following objectives of the present study research.

Figures 1.5 and 1.6 show the Pareto fronts between total operating cost and the ratio of manufacturing cost to total operating cost obtained using MODE algorithm and NSGA-II [18]. Figure 1.6 also shows the effect of NP on Pareto front after 10 generations and comparison of results of MODE study with NSGA-II, where the results are taken from the literature [4, 14, 18]. Several real world problems are recently successfully solved using strategies of multiobjective differential evolution algorithms [11, 13, 15].

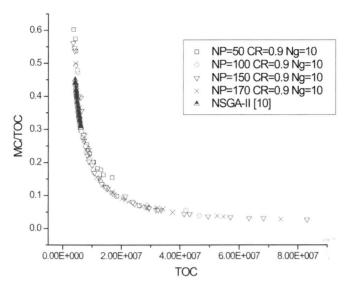

FIGURE 1.5 Comparison of Pareto fronts between TOC and MC/TOC using NSGA-II and MODE and effect of *NP* on Pareto front using MODE algorithm.

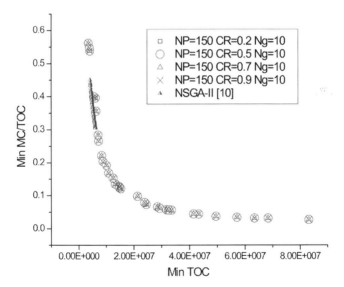

FIGURE 1.6 Pareto front between TOC and MC/TOC using MODE (at various values of CR) and NSGA-II.

1.2   ORGANIZATION OF BOOK

This book consists of 21 chapters on evolutionary algorithms, which includes both the theory and applications. The book is divided into three sections. Chapter 1–7, 8–14 and 15–22 are divided into Part 1, Part 2, and Part 3, respectively. Chapter 1 gives introduction on theory of evolutionary algorithms, single and multiobjective optimization. Detailed literature review is discussed both for theory on multiobjective optimization and applications. The applications discussed from literature includes, multiobjective optimization of styrene reactor, LDPE tubular reactor, and supply chain and planning. This chapter also includes paragraph on organization of book. Chapter 2 is written by Trivedi and Ramteke and they have discussed the biomemetic adaptations of genetic algorithm and their applications to chemical engineering. The typical applications include multiobjective optimization of parallel reactions in mixed flow reactors and multiobjective optimization of fluidized catalytic cracker unit. Brief working principles of various bio-mimetic adaptations of genetic algorithms (such as jumping gene adaptation of GA, Altruistic GA, GA Based on Biogenetic Law of Embryology, GAs Based on Biological Immune System, Lamarckian GAs, GAs based on Baldwin Effect, GAs based on DNA and RNA Computing, Cockroach GA, GAs with Heterosexual Crossover, etc.) are discussed. Chapter 3 written by Kannsara, Parashar and Samad covers an overview of use of surrogate-assisted evolutionary algorithms in various aspects of engineering optimization including single as well as multiobjective optimization evolutionary algorithms. They discussed how surrogate modeling could help during optimization and different surrogate modeling techniques. It also briefly describes how some existing approaches combine surrogate modeling and evolutionary optimization techniques to speed up engineering optimization procedure. Mitra, Saxén, and Chakraborti talks about an application of iron making using evolutionary algorithms in Chapter 4. They discussed modern ironmaking process optimization aspects using blast furnace and rotary kiln operation. Soft computing techniques, such as Artificial Neural Networks, Evolutionary Algorithm, etc., are briefly discussed. Optimization of burden distribution in blast furnace and of production parameters, such as CO_2 emissions, and coal injection with respect to multiobjective optimization framework are discussed. It is recommended to focus on the use of evolutionary techniques in control and decision making systems for the blast furnace. A need for development of faster evolutionary algorithms, which require less calculations before the optima are achieved is also presented. Chapter 5 by Diego Oliva et al. represents theory and an application of harmony search optimization for multilevel thresholding in digital images. Chapter 6 by Sharma and Pant talks about swarm intelligence in software engineering design problems. Their work focuses on concentrates on the recent model instigated by the scavenge behavior of honeybee swarm employed to solve optimization problems in Software Engineering Design. They have shown that Artificial Bee Colony greedy algorithm is analyzed on several problems such as parameter estimation of software reliability growth models, optimizing redundancy level in modular software system models and estimating the software cost parameters arising in the field of Software Engineering. They concluded that the proposed I-artificial bee colony *greedy* algorithm outperformed basic ABC in terms of solution quality as well as convergence rate for the considered Software design problems. Singru et al. in Chapter 7 discussed integrated gene programming approach in simulating the tensile strength characteristic of BNNTs based on aspect ratio, temperature and number of defects of nanoscale materials. In their study the predictions obtained from the proposed model are in good agreement with the actual results. The dominant process parameters and the hidden non-linear relationships are unveiled, which further validate the robustness of their proposed model.

In Chapter 8, Munawar and Gudi presented an application of a nonlinear transformation for representing discrete variables as continuous variables and discussed an alternate method for solving MINLP problems by converting them into equivalent NLP problems. A parameter based on rapid change in the objective function is used to aid in deciding when to switch from deterministic to stochastic solution. Selected examples from literature are used to illustrate the effectiveness of the hybrid evolutionary method in their study. In Chapter 9, Munawar and Babu discussed an application of optimal design of shell and tube heat exchanger using differential evolution algorithm. They carried out detailed study of Differential evolution strategies by varying the control parameters of DE (e.g., DR, F, and NP, etc.). Detailed algorithm specific study for optimal design of shell and tube heat exchanger is reported. In Chapter 10 Sahoo and Patel represented the review of various evolutionary algorithms to solve the QoS-aware multicast routing problem. Fazendeiro and Prata in Chapter 11 focused on the study of the effective parallelization of the canonical GA. They presented a complete characterization of the relative execution times of the atomic operators of the GA, varying the population cardinality and the genotype size. It is complemented with an analysis of the achieved speedups. The findings of the assessment of the parallelization potential at different granularity levels altogether with the analysis of data parallelism are reported. Chapter 12 written by Banati and Mehta presents an efficient approach for populating deep web repositories using Shuffled Frog Leaping Algorithm (SFLA). The work contributes a simple and effective rule based classifier to recognize the deep web search interfaces. Deep web semantic queries are evolved using SFLA to augment the retrieval of deep web URLs. Multi-loop fractional order PID controller is optimally tuned using Bat algorithm for quadruple tank process in Chapter 13 by Sabura Banu. Gain matrix was computed to measure the interaction of process at steady state. The stability of control loop pairing is analyzed by Niederlinski index. Their proposed controller is validated for servo, regulatory and servo-regulatory problems and the result shows that the scheme will result in a simple design of the multi-loop fractional order PID controller for quadruple tank process. Their study revealed that parameters optimized using Bat algorithm is better than parameters optimized using Genetic Algorithm and Particle Swarm Optimization technique.

Shashi Kumar presented a practical approach for multiobjective shape optimization using multiobjective ant colony optimization in Chapter 14. His study showed that optimizer has capability to search the domain outside the initial parametric space specified by designer. Cases studies in Chapter 14 have shown that shape optimization problem is a strong function of parameterization. The meshless solver minimizes on grid generation for every change in shape of the body. This study reveals that the optimization tools require the use of robust CFD solvers, meshless SLKNS solver along with Kinetic diffusion solver is one such combination of robust solver. In Chapter 15, the authors present a review on nature-inspired computing techniques for integer optimization. Chapter 16, by Fayek and Malik presents an application of genetic algorithms on adaptive power system stabilizer. Their chapter majorly focused on how to develop a controller that can draw the attention of the control engineer by attempting a compromise solution: use the GA for modeling the system dynamics, use the linear feedback controller (e.g., PS-Control) because of its simplicity and acceptance. They also discussed how to use a GA to represent the dynamic characteristics of the power system and real-time tests with APSS on a scaled physical model of a single-machine connected to an infinite-bus power system. Nazaris et al. discussed applications of evolutionary computation for solving the budget allocation problem of the fire safety upgrading of a group of buildings in Chapter

17. The proposed generic selection and resource allocation (S&RA) model was applied to the fire protection measures upgrade of the 20 monasteries of the Mount Athos, yielding feasible fund allocation solutions for different budget scenarios and resulted in optimal selection and resource allocation for network of buildings. Chapter 18 by Anauth and Ah King focuses on the development and comparative application of elitist multiobjective evolutionary algorithms (MOEAs) for voltage and reactive power optimization in power systems. A multiobjective Bacterial Foraging Optimization Algorithm (MOBFA) has been developed based on the natural foraging behavior of the Escherichia coli bacteria and its performance compared with two contemporary stochastic optimization techniques; an improved Strength Pareto Evolutionary Algorithm (SPEA2) and the Nondominated Sorting Genetic Algorithm-II (NSGA-II). The MOBFA was found to be a viable tool for handling constrained and conflicting multiobjective optimization problems and decision making. Ong, Sharma and Rangaiah (in Chapter 19) used stochastic algorithms for solving several complex test problems and pooling problems. They have evaluated the performance of integrated DE (IDE), unified bare-bones Particle swarm optimization (UBBPSO), modified simulated annealing and very fast-simulated annealing (VFSA) algorithms on pooling problems. They found that the performance of IDE was the better than other algorithms tested as it has close to perfect success rate (SR) and has relatively low minimum number of function evaluations (MNFE) for unconstrained problems. IDE was also the most reliable and efficient algorithm for the constrained problems tested in their study. The NSGA-II algorithm with SBX-A and SBX-LN crossover probability distributions are used to test 20 multiobjective functions by Prasad and Singru in Chapter 20. The NSGA-II algorithm with SBX-LN crossover probability distribution found better optimal solutions with good diversity for different multiobjective functions. Finally Chapter 21 by Morovati and Kadam presented application of evolutionary algorithms for malware detection and classification based on the dissimilarity of op-code frequency patterns extracted from their source codes. In binary classification, a new method of malware detection has been proposed by the authors. In multi-class classification, the Random Forest classifier yielded betters result in comparison to other classifiers. All the classifiers successfully achieved a very high accuracy, which confirms the strength and efficiency of the suggested malware detection method. They concluded that if ±2% is considered as an accuracy tolerance to compare classifiers together, all techniques used in their study have a good potential to be used as the malware detector.

1.3 CONCLUSIONS

The brief introduction on evolutionary multi-objective optimization is given in this chapter. The basic theory on comparison of deterministic methods and stochastic methods is discussed. Multiobjective optimization aspects are briefly discussed with respect to the decision variable- and objective-search space. Some of the recent applications in the field of evolutionary multiobjective optimization are also discussed in brief.

KEYWORDS

- deterministic methods
- evolutionary algorithms
- multiobjective optimization
- optimization
- stochastic methods

REFERENCES

1. Agrawal, N., Rangaiah, G. P., Ray, A. K., & Gupta, S. K., Multiobjective optimization of the operation of an industrial low-density polyethylene tubular reactor using genetic algorithm and its jumping gene adaptations. Industrial and Engineering Chemical Research, 2006, *45*, 3182–3199.
2. Angira, R. Evolutionary computation for optimization of selected nonlinear chemical processes. PhD Thesis, Birla Institute of Technology and Science (BITS), Pilani, India, 2005.
3. Babu, B. V., Chakole, P. G., & Mubeen, J. H. S. Multiobjective differential evolution (MODE) for optimization of adiabatic styrene reactor. Chemical Engineering Science, 2005, *60*, 4822–4837.
4. Babu, B. V. & Gujarathi, A. M. Multiobjective differential evolution (MODE) for optimization of supply chain planning and management, proceedings of IEEE Congress on Evolutionary Computation, CEC. 2007, 2732–2739.
5. Brandoline, Capiati, N. J., Farber, J. N., & Valles, E. M. Mathematical model for high-pressure tubular reactor for ethylene polymerization. Industrial and Engineering Chemistry Research, 1988, *27*, 784–790.
6. Chopra, S., & Meindl, P. Supply Chain Management: Strategy, Planning and operation: Pearson Education, Singapore, 2004.
7. Deb, K. Multiobjective optimization using evolutionary algorithms; John Wiley and Sons Limited, New York, 2001.

8. Goldberg, D. E. Genetic algorithms in search, optimization and machine learning. Addission-Wesley, Reading, MA, 1989.

9. Gujarathi, A. M., & Babu, B. V. Optimization of adiabatic styrene reactor: a hybrid multiobjective differential evolution (H-MODE) approach, Industrial and Engineering Chemistry Research 2009, *48(24)*, 11115–11132.

10. Gujarathi, A. M., & Babu, B. V. Multiobjective optimization of industrial styrene reactor: Adiabatic and pseudo-isothermal operation, Chemical Engineering Science 2010a, *65(6)*, 2009–2026.

11. Gujarathi, A. M., & Babu, B. V. Hybrid multiobjective differential evolution (H-MODE) for optimization of polyethylene terephthalate (PET) reactor, International Journal of Bio-Inspired Computation 2010b, *2(3–4)*, 213–221.

12. Gujarathi, A. M., & Babu, B. V. Multiobjective optimization of industrial processes using elitist multiobjective differential evolution (Elitist-MODE), Materials and Manufacturing Processes 2011, *26(3)*, 455–463.

13. Gujarathi, A. M., Motagamwala, A. H., & Babu, B. V. Multiobjective optimization of industrial naphtha cracker for production of ethylene and propylene, Materials and Manufacturing Processes 2013, *28(7)*, 803–810.

14. Gujarathi, A. M. Pareto optimal solutions in Process Design Decisions using Evolutionary Multiobjective optimization. PhD Thesis, Birla Institute of Technology and Science (BITS), Pilani, India, 2010.

15. Gujarathi, A. M., Sadaphal, A., & Bathe, G. A. Multiobjective Optimization of Solid State Fermentation Process, Materials and Manufacturing Processes 2015, *30(4)*, 511–519.

16. Gujarathi, A. M., & Babu, B. V. In: Handbook of Optimization, Zelinka, I. Snásel, V., & Abraham, A. (Eds.). Multiobjective Optimization of Low Density Polyethylene (LDPE) Tubular Reactor Using Strategies of Differential Evolution, Handbook of Optimization, Springer Berlin Heidelberg, 2013, pp. 615–639.

17. Onwubolu G. C., & Babu, B. V. New optimization techniques in engineering, Springer-Verlag, Heidelberg, Germany, 2004.

18. Pinto, E. G. Supply chain optimization using multi-objective evolutionary algorithm, technical report, available online at http://www.engr.psu.edu/ce/Divisions/Hydro/Reed/Education/CE%20563%20Projects/Pinto.pdf, as on 12th June,. 2007.

19. Rangaiah G. P. (Editor), Stochastic Global Optimization: Techniques and Applications in Chemical Engineering, Vol. 2 in the Advances in Process Systems Engineering, World Scientific, Singapore, 2010.

20. Rangaiah G. P., & Bonilla-Petriciolet A. (Eds.), Multiobjective Optimization in Chemical Engineering: Developments and Applications, John Wiley, 2013.

21. Sheel, J. G. P., & Crowe, C. M. Simulation and optimization of an existing ethyl benzene dehydrogenation reactor, Canadian Journal of Chemical Engineering, 1969, *47*, 183–187.

22. Yee, A. K. Y., Ray, A. K., & Rangiah, G. P. Multiobjective optimization of industrial styrene reactor. Computers and Chemical Engineering, 2003, *27*, 111–130.

CHAPTER 2

BIO-MIMETIC ADAPTATIONS OF GENETIC ALGORITHM AND THEIR APPLICATIONS TO CHEMICAL ENGINEERING

VIBHU TRIVEDI and MANOJKUMAR RAMTEKE

Department of Chemical Engineering, Indian Institute of Technology Delhi, Hauz Khas, New Delhi–110 016, India, E-mail: ramtekemanoj@gmail.com, mcramteke@chemical.iitd.ac.in

CONTENTS

2.1   INTRODUCTION

In our day-to-day life, we often face situations where we have to find the best combination of contradictory factors to get our tasks done. For instance, finding the best combination of traveling time and cost of a journey while planning a tour, finding the best combination of cost and comfort while purchasing a vehicle or finding the best combination of salary package and job satisfaction while searching for a job, etc., often involve the choices that are conflicting in nature. All such situations require satisfying multiple non-commensurate objectives simultaneously and thus lead to multi-objective optimization problems (MOPs). These situations frequently arise in the fields of science, engineering, management and economics. In our day-to-day life, we often cope with such situations using a qualitative decision making, however, finding an optimal solution of a complex industrial MOP, qualitatively, is not so easy and thus a detailed quantitative analysis is required. For quantitative assessment, a general mathematical formulation of an optimization problem is given as follows:

Objectives:

$$\min(\text{or}\max)I_i(x_1,x_2,.....,x_n), \qquad i=1,2,.....,m \qquad (1)$$

Constraints:

$$g_j(x_1,x_2,.....,x_n)\geq 0, \quad j=1,2,.....,r \qquad (2)$$

$$h_k(x_1,x_2,.....,x_n)=0, \quad k=1,2,.....,s \qquad (3)$$

Bounds:

$$x_l^L \leq x_l \leq x_l^U, \quad l=1,2,.....,n \ \text{T} \qquad (4)$$

where, $x_1 - x_n$ are n decision variables, $I_1 - I_m$ are m objective functions, $g_1 - g_r$ are r inequality constraints and $h_1 - h_s$ are s equality constraints. x_l^L and x_l^U are the lower and the upper limits of lth variable.

For $m = 1$ [single objective optimization problem (SOP)], the above formulation often leads to a single optimal solution whereas MOPs ($m \geq 2$) lead to a set of 'equally good' or 'nondominated' solutions. It is to be noted that, there exists no solution in the population having better values of all objectives than these nondominated solutions. These nondominated solutions provide a trade-off between all the objectives and commonly referred as Pareto optimal solutions. Often the operating solution from a set of non-dominated solutions is selected based on intuition or hand on experience of an underlying process and is a focus of current research.

Chemical engineering systems involve a concatenating of various units such as reactors, crackers, distillation columns, froth floatation cells, absorbers, crystallizers, etc. To maximize the profit, these units should be operated, planned and scheduled optimally. This often leads to the complex MOP formulations. However, for the sake of illustration, a simple MOP of parallel reactions [1] in a mixed flow reactor (MFR) is analyzed in detail. The reaction scheme is given as follows:

$$R \xrightarrow{k_0} S \qquad\qquad (r_s) = k_0; k_0 = 0.025$$
$$R \xrightarrow{k_1} T \text{ (desired)} \qquad (r_T) = k_1 C_R; k_1 = 0.2 \, \text{min}^{-1}$$
$$R \xrightarrow{k_2} U \qquad\qquad (r_U) = k_2 C_R^2; k_2 = 0.4 \, \text{liter mol}^{-1} \text{min}^{-1}$$

Here, T is the desired product whereas S and U are undesired products. The first reaction is of zero order, second is of first order and third is of second order. The rates of production of S, T and U are given in the above reaction scheme. The initial concentration of R is $C_{R0} = 1$ mol/min and the feed flow rate is $v = 100$ L/min (all data is taken from Levenspiel [2]). From an economic point of view, the design stage optimization of a given MFR requires to maximize the yield of the desired product (i.e., instantaneous fractional yield of T, θ(T/R)) while minimizing the reactor volume (V). These two objectives are conflicting in nature since the order of the desired reaction is lower than overall order of undesired reactions [2]. Thus, a two-objective optimization problem can be formulated as follows:

Objectives:

$$\max I_1(C_R) = \theta(\text{T/R}) = \frac{k_1 C_R}{k_0 + k_1 C_R + k_2 C_R^2} = \frac{0.2 C_R}{0.025 + 0.2 C_R + 0.4 C_R^2} \quad (5)$$

$$\min I_2\left(C_R\right)=V=v\frac{\left(C_{R0}-C_R\right)}{-r_R}=100\frac{\left(1-C_R\right)}{0.025+0.2C_R+0.4C_R^2} \qquad (6)$$

Constraints:

$$\theta(T/R)\le 1 \qquad (7)$$

Bounds:

$$0\le C_R\le 1 \qquad (8)$$

Here, C_R is the instantaneous concentration of R. The Pareto optimal solutions of this problem are shown in Figure 2.1 and the curve formed by these solutions is known as Pareto optimal front [3]. In general, such plots can be spaced in multiple dimensions for MOPs depending on the number of objectives present. Any solution which does not lie on the Pareto optimal front will be dominated by at least one nondominated solution in terms of both (or all) objectives. For example, solution point C is dominated whereas solution points A and B are nondominated (see Figure 2.1).

Until 1960 s, optimization studies were mostly concentrated on solving the SOPs and very often, the MOPs were converted to SOPs by combining

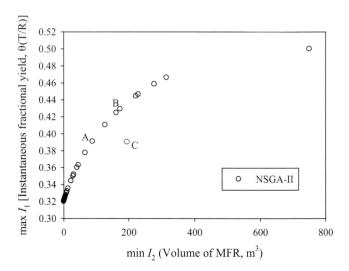

FIGURE 2.1 Pareto optimal plot for the MOP of parallel reactions in an MFR. Points A and B represent nondominated solutions whereas point C represents a dominated solution.

the multiple objectives to a single objective using different weights. In later decades, several techniques such as goal programming [4], Kuhn-Tucker multipliers [5], Pontryagin's minimum principle [6], the ε -constraint methods [7], the method of indifference functions [7], parametric approach [8], the utility function method [9], the lexicographic approach [9], etc., are developed to handle MOPs. All these techniques are termed as conventional techniques and were frequently used until 1990. These techniques were mostly derivative-based and often the quality of solutions was dependent on the choice of the weights selected. In the last two decades, the solving procedure of MOPs is completely revamped with the arrival of meta-heuristic algorithms. These techniques are stochastic in nature, derivative-free and often based on interesting themes from different areas of science. Most prominent among such techniques are genetic algorithm [10] (GA), evolutionary programming (EP) [11], differential evolution (DE) [12], particle swarm optimization (PSO) [13], ant colony optimization (ACO) [14], artificial bee colony algorithm (ABC) [15], etc. The focus of present study is GA, which is inspired from evolutionary biology. The detailed description of GA is given in the next section.

2.2   GENETIC ALGORITHM FOR MULTIOBJECTIVE OPTIMIZATION

Meta-heuristic algorithms based on evolution process are commonly referred as evolutionary algorithms. One of the most widely recognized evolutionary algorithms is GA. It is based on Darwin's theory of evolution of species. Several versions of GA have been developed over the years. These are listed in their chronological order in Table 2.1. Among these, NSGA-II is the most popular version and its binary-coded form is described next.

In NSGA-II, the variables are represented with binaries as the latter closely symbolize genes. Suppose a given MOP problem has n number of variables. Each of these variables is represented by a set of l_{str} number of binaries. Thus, n such sets of binaries are arranged in a string to constitute a chromosome comprising total l_{chrom} $(= n \times l_{str})$ number of binaries. In the initialization step, l_{chrom} number of binaries is generated (either 0 or 1) using a random number (RN). For this, an RN is generated using an RN generator and if it comes out to be in the range, $0 \leq RN < 0.5$, the corresponding binary is taken as '0' else it is taken as '1.' This procedure is repeated with the successive generation of RNs to assign l_{chrom} binaries in a chromosome

TABLE 2.1 Chronology of GA

S. No.	Year	Name	Reference	Evolutionary Operators	Type	Approach	Modification
1	1975	Genetic Algorithm (GA)	Holland [10]	Crossover, Mutation	Single-objective	Non-Pareto	Original version
2	1985	Vector Evaluated Genetic Algorithm (VEGA)	Schaffer [16]	Crossover, Mutation	Multiobjective	Non-Pareto	Selection based on individual objectives
3	1989	Simple Genetic Algorithm (SGA)	Goldberg [17]	Crossover, Mutation	Multiobjective	Pareto	Selection based on nondominated ranking
4	1992	Hajela and Lin's Genetic Algorithm (HLGA)	Hajela and Lin [18]	Crossover, Mutation	Multiobjective	Non-Pareto	Fitness assignment by a weighted sum method
5	1993	Multiobjective Genetic Algorithm (MOGA)	Fonseca and Fleming [19]	Crossover, Mutation	Multiobjective	Pareto	Selection based on nondominated ranking with some penalty on dominated individuals
6	1994	Niched-Pareto Genetic Algorithm (NPGA)	Horn et al. [20]	Crossover, Mutation	Multiobjective	Pareto	Tournament selection
7	1994	Nondominated Sorting Genetic Algorithm (NSGA)	Srinivas and Deb [21]	Crossover, Mutation	Multiobjective	Pareto	Modified ranking scheme by sorting the individuals into nondominated groups
8	1999	Strength Pareto Evolutionary Algorithm (SPEA)	Zitzler and Thiele [22]	Crossover, Mutation	Multiobjective	Pareto	Introduction of an elitist mechanism

TABLE 2.1 Continued

S. No.	Year	Name	Reference	Evolutionary Operators	Type	Approach	Modification
9	2000	Pareto Archived Evolution Strategy (PAES)	Knowles et al. [23]	Mutation	Multiobjective	Pareto	One parent produces one offspring, elitism similar to SPEA
10	2000	Pareto Envelope based Selection Algorithm (PESA)	Corne et al. [24]	Crossover, Mutation	Multiobjective	Pareto	Includes virtues of SPEA and PAES with slightly different elitism operation
11	2001	Pareto Envelope based Selection Algorithm-II (PESA-II)	Corne et al. [25]	Crossover, Mutation	Multiobjective	Pareto	Selection mechanism differs from PESA as it is based on hyperboxes (regions in objective space containing solution) rather than individual solutions
12	2002	Elitist Nondominated Sorting Genetic Algorithm (NSGA-II)	Deb et al. [26]	Crossover, Mutation	Multiobjective	Pareto	Introduction of crowding distance, Elitist mechanism does not use external memory but combines the best parents with best off-springs obtained
13	2007	ε-preferred NSGA-II	Sulflow et al. [27]	Crossover, Mutation	Multiobjective	Pareto	Two crossover operators are used, dealt with high dimension (up to 25 objective) problems using modified ranking
14	2013	Elitist Nondominated Sorting Genetic Algorithm for Many-objective Optimization	Jain and Deb [28]	Crossover, Mutation	Multiobjective	Pareto	Modified tournament selection and elitism to handle the problems with 4 or more objectives

and thereafter for a population of N_p chromosomes. A typical representation of a chromosome is shown in Figure 2.2 for $n = 2$ and $l_{str} = 3$.

The next step is to decode the binaries into real values. For this, the binaries of a given variable are first converted to a decimal number as follows:

$$\left(\text{decimal number}\right)_l = \left(\sum_{z=1}^{l_{str}} 2^{z-1} \times B_z\right)_l ; l \in [1, n] \tag{9}$$

where, B_z represents zth binary bit of lth variable. The corresponding decimal number is then mapped [8] between lower (x_l^L) and upper (x_l^U) limits of the lth variable to obtain its real value as follows:

$$x_l = x_l^L + \frac{x_l^U - x_l^L}{2^{l_{str}} - 1} \times \left(\text{decimal number}\right)_l ; l \in [1, n] \tag{10}$$

For example, if the variable is represented with $l_{str} = 3$, it can have only $2^{l_{str}}$ (= 8) discrete values in the search space starting from a lower limit, represented by [0, 0, 0] to the upper limit, represented by [1, 1, 1]. These permutations along with their respective real values for $x_l^L = 0$ and $x_l^U = 100$ are shown in Figure 2.3. It is to be noted that the higher length of binary strings leads to a better accuracy of mapping the search space.

After decoding, the real values of variables are used to evaluate fitness of a chromosome using fitness functions (F_m). Fitness function may differ from objective function (I_m). It is due to the fact that meta-heuristic codes are often written for either maximization or minimization and one has to convert the objectives in the compatible form. A popular method [8] of such conversion is:

$$\min F_m = \frac{1}{1 + \max I_m} \text{ or } \max F_m = \frac{1}{1 + \min I_m} \tag{11}$$

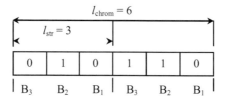

FIGURE 2.2 A typical chromosome ($n = 2$ and $l_{str} = 3$) used in NSGA-II.

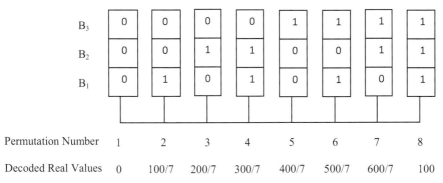

	Permutation Number	1	2	3	4	5	6	7	8

Decoded Real Values   0   100/7   200/7   300/7   400/7   500/7   600/7   100

FIGURE 2.3   Decoding of a variable ($l_{str} = 3$) into real values with $x_i^L = 0$ and $x_i^U = 100$ for $2^{l_{str}}$ ($= 8$) permutations.

Further, the fitness functions are modified with the addition of constraints. Constraints are taken care by adding (for minimization of F_m) or subtracting (for maximization of F_m) penalty terms to the fitness functions. Thus, the modified fitness functions for the formulation given in Eqs. (1)–(4) can be written as:

$$\min F_m = \frac{1}{1+\max I_m} + \sum_{k=1}^{r+s} w_k \left[\frac{\text{extent of } k^{th}}{\text{constraint voilation}} \right]^2 \quad OR$$

$$\max F_m = \frac{1}{1+\min I_m} - \sum_{k=1}^{r+s} w_k \left[\frac{\text{extent of } k^{th}}{\text{constraint voilation}} \right]^2 \quad (12)$$

$$w_k = \begin{cases} \text{large positive number} & \text{if constraint is violated} \\ 0 & \text{otherwise} \end{cases}$$

Equation (12) is used to evaluate the fitness functions of all N_p chromosomes one after another.

In the next step, all N_p chromosomes are ranked using the concept of non-dominance. For this, each chromosome is transferred to a new location (file) one by one. If a chromosome to be transferred is nondominated with respect to chromosomes already present in this new location, the transfer is completed. However, if this new chromosome dominates any of the existing chromosomes, the former replaces the latter whereas the latter comes back to the original location. In this way, only nondominated chromosomes are retained in the new location and these are ranked $I_R = 1$. The procedure is repeated successively for the chromosomes remained in the original location to classify all the chromosomes in successive ranks $I_R = 2, 3, 4, \ldots$.

In traditional GA, it has been found that the Pareto optimal solutions obtained are often clustered in various locations and are not uniformly distributed. Such clustering is not desirable because it gives limited options to a decision maker. This limitation is removed in NSGA-II by using a concept of crowding distance. To measure the crowding distance, the chromosomes having the same rank are arranged in increasing order of their fitness values. This helps in identifying the nearest neighbors having same rank, which encapsulate a given chromosome in fitness function space. These nearest neighbors form a rectangle for two objectives (or hypercuboid for more than two objectives). The sum of the sides (one on each axis) of a hypercuboid gives the crowding distance. An example of crowding distance measurement for a two-objective optimization problem with population size equals to 10 is shown in Figure 2.4. For an identified point 'g' in Figure 2.4, the crowding distance (I_D) is equal to $D_1 + D_2$, a sum of the sides of an enclosing rectangle on x and y axis.

In the next step, the population of chromosomes undergoes the selection operation. In this, a tournament is played between two randomly selected chromosomes in which the one with the lower I_R is declared as the winner. If I_R is same, the one with the higher I_D is declared as the winner. The winner is then copied into a mating pool. This process is repeated N_p times to copy N_p chromosomes in the mating pool. The above process copies the good chromosomes multiple times while some inferior chromosomes (better of the two worse chromosomes) also find their way into the mating pool. Selecting such inferior chromosomes is necessary in the sense that two worst chromosomes can produce a better one in successive generations.

After the selection operation, the mating pool population undergoes crossover operation. In this operation, two randomly selected chromosomes

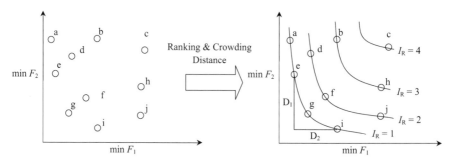

FIGURE 2.4 Ranking and crowding distance of chromosomes illustrated for a two-objective minimization problem with $N_p = 10$.

from the mating pool are checked for the possible crossover operation using the crossover probability, P_{cros}. For this, an RN is generated and if it is less than P_{cros} then the genetic contents of two selected parent chromosomes are swapped at the randomly selected crossover site among $(l_{chrom} - 1)$ internal locations as shown in Figure 2.5, otherwise, the parents are copied as it is to produce two offspring chromosomes. This process is repeated $N_p/2$ times to produce a population of N_p offspring chromosomes.

Despite the large reshuffling of genetic content offered by a crossover operation, it has a severe limitation that it can vary genetic content of a chromosome maximally to that already present in the mating pool. Suppose all chromosomes in the mating pool have 0 as their first binary bit then there is no way to get 1 in this location irrespective of how many times and for how many generations crossover is carried out. This becomes a serious problem if the optimal solution requires a 1 in the first location. This limitation is overcome by another genetic operation, known as mutation. It is a bit-wise operation in which every offspring chromosome is checked bit by bit for possible mutation with a small probability, P_{mut}. If a bit is selected for mutation operation then its associated binary is changed from its existing value of 1 to 0 or vice versa (see Figure 2.6). The mutation operation helps in maintaining genetic diversity.

In a traditional GA, the offspring population becomes the parent population for the next generation. Thus, the better parents in the current generation do not find their way into the next generation. In order to conserve these better parents, NSGA-II employs a new operation, known as elitism. In this operation, the offspring population is mixed with the original parent population. These $2N_p$ chromosomes are then re-ranked with the previously described procedure. The best N_p chromosomes from these are selected as the parents for the next generation and the rest are discarded.

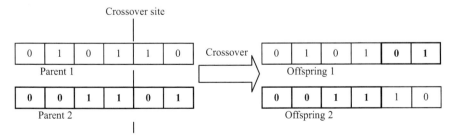

FIGURE 2.5 Crossover operation illustrated for two parent chromosomes with $n = 2$ and $l_{str} = 3$.

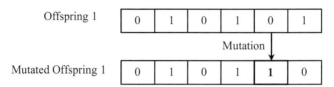

Offspring 1

| 0 | 1 | 0 | 1 | 0 | 1 |

Mutation ↓

Mutated Offspring 1

| 0 | 1 | 0 | 1 | **1** | 0 |

FIGURE 2.6 Mutation operation illustrated for offspring 1 produced in crossover operation (see Figure 2.5).

As described above, the algorithm involves following steps: (i) initialization, (ii) decoding, (iii) fitness evaluation, (iv) ranking, (v) selection, (vi) crossover, (vii) mutation, and (viii) elitism. The first generation involves all the above-mentioned steps whereas only (ii)–(viii) are repeated in each subsequent generation till the convergence (condition where results start showing little improvement over the generations) or the user defined maximum number of generations is achieved. Also, the performance of the algorithm depends on various parameters associated with above steps such as N_p, l_{str}, P_{cros}, P_{mut}. Similar to other meta-heuristic algorithms, values of these parameters are problem-specific. However, the good starting values of these, especially for a wide range of applications in chemical engineering are reported [29–31] as $N_p = 100$, $l_{str} = 20$, $P_{cros} = 0.9$, $P_{mut} = 1/l_{chrom}$. If the optimization problem has both continuous and integer variables, all continuous variables are generated as described above whereas the integer variables are often used as a nearest integer of the decoded real values. In some applications, binary variables are present [32, 33] and are generated directly.

As described above, GA was originally developed in binary-coded form. However, it shows certain limitations in handling real-life continuous search space problems. For example, it is very difficult to alter a binary-string [0, 1, 1, 1, 1, 1, 1, 1] to obtain another binary-string [1, 0, 0, 0, 0, 0, 0, 0] using genetic operators of crossover and mutation. However, the real values associated with these binary-strings are 127 and 128, respectively, which are consecutive integers. This situation is called Hamming cliff. It leads to poor performance of binary-coded GA. Other limitations include fixed mapping of problem variables and lack of precision in the solutions. To overcome these limitations, real-coded NSGA-II (RNSGA-II) is developed. In this, initialization, crossover, and mutation steps differ from NSGA-II and decoding step is not required. In initialization step, real-variable strings constitute chromosomes and these real-variables are generated as follows:

$$x_i = x_i^L + RN \times \left(x_i^U - x_i^L \right) \tag{13}$$

Several schemes of crossover and mutation for real-coded GAs are reported in literature [8]. Among these, most commonly used are simulated binary crossover [34] (SBX) and polynomial mutation (PM) [35] operators. In SBX operation, two parent chromosomes P_1 and P_2 are selected randomly to produce offspring O_1 and O_2 in the following manner:

$$x_{O_1,J} = \frac{1}{2}\left[(1-\beta_J)x_{P_1,J} + (1+\beta_J)x_{P_2,J}\right] \tag{14}$$

$$x_{O_2,J} = \frac{1}{2}\left[(1+\beta_J)x_{P_1,J} + (1-\beta_J)x_{P_2,J}\right] \tag{15}$$

$$\beta = \begin{cases} (2RN)^{\frac{1}{\eta+1}} ; RN < 0.5 \\ \dfrac{1}{\left[2(1-RN)\right]^{\frac{1}{\eta+1}}} ; \text{otherwise} \end{cases} \tag{16}$$

Here, η is crossover index, which is used to generate the distribution function β [see, Eq. (16)]. In general, the value of η is selected in order to keep value of β close to 1. Thus, the perturbation in Eqs. (14)–(15) produces offspring with variable values close to that of parent chromosomes. Similar to SBX, PM operation is also applied variable-wise. The operation is described as follows:

$$x_I = x_I^L + \delta_I \times \left(x_I^U - x_I^L\right) \tag{17}$$

$$\delta = \begin{cases} \left[(2RN)^{\frac{1}{\eta_{mut}+1}}\right] - 1; RN < 0.5 \\ 1 - \left[2(1-RN)\right]^{\frac{1}{\eta_{mut}+1}} ; \text{otherwise} \end{cases} \tag{18}$$

Here, η_{mut} is mutation index, which is used to generate the distribution function δ [Eq. (17)]. In general, the value of η_{mut} is selected in order to keep value of δ close to 0. This maintains the perturbation of variable at local level. Also, it is to be noted that PM is applied selectively on those parent chromosomes, which are unsuccessful in SBX unlike NSGA-II. The other steps in RNSGA-II are same as NSGA-II. The parameters used in RNSGA-II are also same as that of NSGA-II except η, P_{mut}, η_{mut}. The good starting values of these are reported [8] as 20, $1/n$, 20, respectively.

A simple chemical engineering application of NSGA-II is demonstrated using the example of parallel reactions in MFR (see Section 2.1). For this, the objective functions given in Eqs. (5) and (6) are first converted to the fitness functions as follows:

$$\max F_1(C_R) = \frac{0.2C_R}{0.025 + 0.2C_R + 0.4C_R^2} - w_1\left[1 - \theta(T/R)\right]^2 \quad (19)$$

$$\max F_2(C_R) = \frac{1}{1+V} - w_1\left[1 - \theta(T/R)\right]^2 \quad (20)$$

$$w_1 = \begin{cases} \text{large positive number} & \text{if } \theta(T/R) \geq 1 \\ 0 & \text{otherwise} \end{cases} \quad (21)$$

For the same values of parameters (mentioned above), the results are obtained (see Figure 2.1). The results show that the increase in instantaneous fractional yield $\theta(T/R)$ requires larger reactor volume V, which matches well with the physical reality. The decision maker can select the appropriate operating point from the Pareto optimal front shown in Figure 2.1 to design the given MFR system.

Unlike the simple illustrative problem solved above, the real-life chemical engineering problems are highly complex in nature. Solving such problems, till convergence, requires rigorous computational efforts and therefore these are often attempted for the restricted number of generations. In such cases, mostly the prematurely converged solutions are obtained and the quality of such solutions depends on the convergence speed of the algorithm. In order to improve the convergence speed of GA, several bio-mimetic adaptations are developed over the period. These are described in detail in the next section.

2.3 BIO-MIMETIC ADAPTATIONS OF GA

GA mimics the most elementary evolutionary operations such as selection, crossover and mutation in its framework. Over the decades, several discoveries in the field of evolutionary biology have revealed that besides these elementary operations, other altruistic operations are also seen in nature. It is obvious to expect that adapting these operations in GA framework can enhance its performance. In recent times, several researchers have worked on this line and developed various GA adaptations based on interesting biological concepts. These adaptations claim to provide better convergence than basic GA and generally termed as bio-mimetic adaptations of GA. In this section, such GA adaptations are discussed in detail.

2.3.1 JUMPING GENES ADAPTATIONS OF GA

In natural genetics, jumping genes (JG) or transposon is a small segment of DNA that can jump in and out of chromosomes. Its existence was first predicted by McKlintock [36] and widely studied thereafter. These are responsible for developing the resistance in bacteria against anti-bodies and drugs and also help in maintaining genetic diversity in organisms. The adaptation of this concept in the framework of NSGA-II (NSGA-II-JG [37]) has shown the significant improvement in its convergence speed.

In NSGA-II-JG, JG operation is applied to offspring population obtained after crossover and mutation. For this, each offspring chromosome is checked for possible jumping gene operation with probability P_{JG}. Once a chromosome is selected, two JG sites are randomly identified in a chromosomal string using integer random numbers. Finally, all binaries between these two sites are replaced by the same number of randomly generated binaries (see Figure 2.7). This process is repeated for all selected offspring chromosomes. It is to be noted that selection of JG sites varies from chromosome to chromosome (and so the number of binaries to be replaced) due to which it is also called variable length JG operation. It involves a macro-macro-mutation leading to higher genetic diversity. This basic JG operation is modified in several versions of NSGA-II-JG over the years. These include NSGA-II-aJG [38, 39] (fixed length JG operation), NSGA-II-mJG [40] (for optimization problems in which the global optimal of some decision variables may

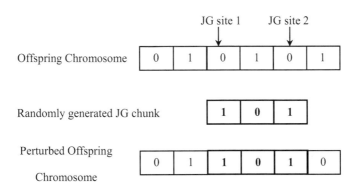

FIGURE 2.7   JG operation in NSGA-II-JG.

lie exactly at their bounds), NSGA-II-sJG [32, 41] (replacement of a single randomly selected variable) and RNSGA-II-SBJG [42] (simulating the effect of binary-coded JG operation in real-coded NSGA-II).

Another JG adaptation of NSGA-II is JGGA [43, 44] (Jumping Gene Genetic Algorithm). In this adaptation, JG operation differs from that adapted in NSGA-II-JG. Here, JG is performed just after the selection process. In this, a chromosome is selected for possible JG operation based on a jumping gene rate (probability). Two or more adjacent binaries of this chromosome are identified randomly and either copied and pasted to a new location or switched with the binaries in the new location within the chromosome or with another randomly selected chromosome. In another notable adaptation, Ripon et al. [45] developed RJGGA (Real JGGA) for real-coded NSGA-II by modifying the occurrence procedure of existing SBX and PM operators. RJGGA is further modified by Furtuna et al. [46] to develop NSGA-II-RJG. In this adaptation crossover and mutation schemes differ from those used in RJGGA.

The detailed analysis of above-mentioned JG adaptations of NSGA-II and their performance comparison is reported in some excellent literature reviews [47, 48].

2.3.2   ALTRUISTIC GA

This bio-mimetic adaptation [49] of GA is based on the altruistic nature of honeybees. There are three types of bees in the bee hives: a queen bee, a large number of female worker bees (daughters) and few male drones (sons). Honeybees show a strange behavior against the theory of natural selection that is female worker bees raises their siblings (queen's daughters) rather than producing their own offspring. The reason behind this altruistic behavior is the haplo-diploid character of honeybees. In bees, males have n chromosomes (haploid) whereas females have $2n$ chromosomes (diploid), which differ from the diploid character of humans (both males and females have $2n$ chromosomes). In a bee colony, sons are produced from unfertilized eggs of queen bee and thus have n chromosomes randomly selected from $2n$ chromosomes of the queen. However, daughters are produced by crossing of father-drone and queen bee and thus have $2n$ chromosomes (all n chromosomes of father and n chromosomes randomly selected from $2n$ chromosomes of the queen). Since, chromosomal relation of sons and daughters with queen bee is 100 and 50%, respectively, the average chromosomal relation among the sibling worker bees (female) is 75% [= 0.5 × (100 + 50)]. However, if these worker bees produce their own daughters by mating with drones, their chromosomal relation with their daughters will be 50%. Hence, worker bees sacrifice their motherhood and prefer to raise their sisters (altruistic behavior). This increases the inclusive fitness of the entire bee colony.

In altruistic NSGA-II-aJG, altruism is adapted by performing the crossover preferentially between queen chromosomes and rest of the population. The queen chromosomes are basically the best chromosomes present in the parent population. Generally, best 1/10 of the parent chromosomes are denoted as queens in each generation based on their ranking and crowding distance. Thus, the queen chromosomes are updated in every generation. If a chromosome is good, it survives in the process of elitism and is consistently selected as a queen over the generations. In crossover operation, one of the queens selected randomly swaps the genetic content with the chromosome selected randomly from the entire population to produce two offspring. The rest of the procedure is same as that of the original algorithm.

2.3.3   GA BASED ON BIOGENETIC LAW OF EMBRYOLOGY

Biogenetic law of embryology [50] states that 'ontogeny recapitulates phylogeny.' It means that the developmental phase of an embryo (ontogeny) shows all the steps of evolution (phylogeny) of the respective species. For example, the developmental phase of nine months in humans shows all the steps of evolution starting from protozoa to mammals. This indicates that the information about the entire evolution process is embedded in an embryo. This observation can be used intelligently in GA by representing a starting population in an embryonic form rather than generating it randomly. Such adaptations can be very useful for cases where an optimization problem is required to be solved repeatedly for the modified conditions on the same system. In industrial systems, MOPs are repeatedly modified because objectives, decision variables and constraints associated keep changing due to economic and environmental factors. For such cases, biogenetic NSGA-II-aJG (B-NSGA-II-aJG [51]) is developed in which N_p chromosomes are picked randomly from the solution history of a MOP to constitute the embryo. This embryo is then used as a starting population for solving all subsequent modifications.

2.3.4   GAs BASED ON BIOLOGICAL IMMUNE SYSTEM

In animals, the biological immune system provides protection against harmful pathogens. This system can detect infinite types of pathogens by producing infinitely diverse antigen receptors that exist on the surface of immune cells. It also preserves the memory of pathogens to speed up their resistance in case they attack the body again. Thus, we can say that mechanisms based on diversity and memory play an important role in the survival of organisms in ever changing environment. In GA, diversity is maintained through various schemes of crossover and mutation while memory can be added by mimicking the concept of immunity. Several immunity based GA adaptations [52–58] are reported in literature, which have been proved helpful in solving those MOPs in which objectives, decision variables and constraints keep changing due to various factors (economy, environment, system-breakdown, change in consumer demand, etc.). One such popular adaptation, IGA, is proposed by Jiao and Wang [52]. In this adaptation a new operator, known as immune operator is added involving two steps: (i) vaccination, and (ii) immune selection after crossover and mutation. In vaccination, one or more vaccines (genes) with higher fitness are generated by detailed analysis of problem at hand and 'injected' in the population to increase fitness of the population. In immune selection, if an offspring has lower fitness than its parent, the parent is included in the next generation. However, the offspring move to the next generation with an annealing probability. This algorithm was used to solve traveling salesman problem (TSP) with 75 and 442 cities and found to be converging faster than GA. It is to be noted that all scheduling problems including those associated with chemical engineering systems resemble to TSP. All immunity based GAs consist some type of memory mechanism that uses problem specific information to find global optima.

2.3.5   LAMARCKIAN GAs

According to Lamarck's theory of evolution, fitness as well as genetic structure of an individual can change through learning from local environment. Following this approach, Lamarckian GAs involves a local search strategy that improves the genetic structure of an individual. This improved individual is then inserted in the population again. This can speed up the search process of GAs [59], but at the same time exploration skill is compromised due to change in genetic content based on local optimization. This can cause premature convergence. GAs in which chromosomes are repaired to satisfy the constraints fall in the category of Lamarckian GAs and found to be particularly useful for solving combinatorial optimization problems [60].

2.3.6   GAs BASED ON BALDWIN EFFECT

Biologist Baldwin [61] proposed a hypothesis about the effect of individual learning on evolution. According to him the individuals with learning ability does not rely completely on their genetic code for the traits necessary for survival. Instead, they can learn various traits according to the local environment and improve their fitness without a change in genetic content. Also, these individuals can maintain a more diverse gene pool as the non-optimal or 'absent' traits in their genetic code can be learnt. In this way individual learning indirectly accelerates the evolutionary adaptation of species by increasing genetic diversity and survivability. This phenomenon is called Baldwin effect. Firstly, Hinton and Nolan [62] simulated the Baldwin effect in GA. These authors used a GA and a simple random learning procedure to develop a simple neural network with the best connection among the 20 possibilities. They found that Baldwin effect can decrease the difficulty level of the fitness landscape of a complex optimization problem as the solutions found by learning direct the genetic search. Besides these benefits, there is a drawback that the computational cost of learning sometimes negates the benefits gained in search process [62]. However, several hybrid GAs based on Baldwin learning [62–66] are reported in literature due to its efficiency in finding global optimum.

2.3.7   GAs BASED ON DNA AND RNA COMPUTING

Tao and Wang [67] integrated the concepts of DNA [68] and RNA [69] computing in GA to develop RNA-GA. In this algorithm, the coding pattern is 0123 instead of binaries 0 and 1 where 0, 1, 2, and 3 represent the four-nucleotide bases of the RNA sequence, Cytosine (c), Uracil (U), Adenine (A) and Guanine (G). The crossover and mutation operations are also modified according to the genetic operations of RNA sequence. The crossover includes translocation, transformation and permutation operators whereas mutation includes reversal, transition and exchange operators. In biology, the genetic operations change the length of RNA sequence, which is neglected in the above adaptation in order to maintain the fixed length of the chromosomes. In another adaptation, namely, DNA-GA [70], the coding pattern similar to RNA-GA is followed with the inclusion of DNA sequence methods from biology. In this algorithm, a two-point crossover scheme is used and three mutation operators are applied on each individual, namely, inverse-anticodon (IA) operator, maximum-minimum (MM) operator and normal-mutation (NM) operator, respectively. Several modifications of both of the above-mentioned algorithms are reported in literature [71–77] with their applications to various optimization problems of industrial importance.

2.3.8   COCKROACH GA (CGA)

In this GA adaptation [78], the competitive behavior of cockroaches during scarcity of food is mimicked to develop advanced genetic operators. It is a real-coded algorithm and involves a population of N_p artificial cockroach individuals. These cockroaches are divided in different groups using golden section search method with each group having a separate leader (individual with best fitness). During the scarcity of food (i.e., during non-optimal generations) the leader from each group finds a chance to snatch the food. The strongest (i.e., the cockroach which occupies the place with most food sources where the place with most food sources indicates the global optimum) of them rushes to the food in a derivative-based way (steepest descent method) while other leaders crawl to the food in a derivative-free way. This leads the strongest leader to a better position if the exploitation succeeds otherwise it stays at the same place. In the next step selfish hungry leaders drive out the weak individuals (weak individuals are moved away from leader) and if the latter reach food-rich places by luck the former replace them. The current best individuals are further improved by neighborhood exploitation in the winner walkabout step. The individuals obtained after these local search steps undergo advanced genetic operations. First of these is replacement in which some invaders (the strongest cockroaches) replace the victims (weaker cockroaches). In the next step, a two-point crossover is performed with mixed inbreeding (crossover between two offspring to produce two more offspring) and backcrossing (crossover between one parent and one child to produce two more offspring). In mutation operation, the competition follows the mutation. It means that the mutated individuals compete with the original ones and winners or losers lying within a prescribed limit of deviation are accepted. Finally, a periodic step (occurring after a cycle of iterations) namely, habitat migration is also included in CGA in which except the strongest cockroach, all cockroaches are made to move away from it. However, the strongest cockroach can determine if it stays at its original place or take tumbling movements (random search) to reach a food-rich place.

Though, the concept of CGA seems similar to PSO and ACO, the main difference is that these algorithms are based on cooperative behavior (each individual accelerates toward local-best or global-best positions) whereas CGA is based on competitive behavior and use of genetic operations in it provides a balance of exploitation and exploration. CGA has been used to optimize the S-type biological systems (expressing protein regulation, gene interaction, and metabolic reactions as power functions) and found to converge faster than various modern versions [79–81] of GA.

2.3.9   GAs WITH HETEROSEXUAL CROSSOVER

Although, gender based sexual reproduction plays a crucial role in natural evolution, it is neglected in traditional GAs. However, several researchers have used this concept to improve the performance of GA. Lis and Eiben [82] developed a multi-sexual GA for multiobjective optimization in which there were as many genders as the objectives and each individual had an extra gender feature. The multi-parent crossover was performed between the parents belonging to different genders. Several other similar adaptations [83–86] are reported in literature where individuals are divided as males and females to perform crossover between individuals of opposite genders. A recent adaptation [87] incorporates Baldwin effect in GA with heterosexual crossover. In this algorithm, mutation rate for males is kept higher than that of females based on a research of genetic biology [88]. Male and female subgroups have different genetic parameters to improve the exploration ability of male subgroup and local search ability of female subgroup. Also, fitness information is transmitted from parents to offspring and Baldwin learning is used to guide the acquired fitness of an individual. This algorithm has been used to solve ten benchmark functions and results are found to be converged in fewer generations as compared to simple genetic algorithm (SGA) and adaptive genetic algorithm (AGA) [89].

2.4   APPLICATIONS OF BIO-MIMETIC ADAPTATIONS OF GA IN CHEMICAL ENGINEERING

The optimization problems of chemical engineering can be characterized by multiple objectives, complex model equations, a large number of decision variables and strict constraints imposed by economic and environmental factors. These challenges associated with chemical engineering MOPs have drawn the attention of researchers and engineers in this field towards the use of meta-heuristic algorithms and particularly GA. Since the last two decades, GA and its various modifications (e.g., NSGA, NSGA-II, NSGA-II-JG, etc.) have been applied to a variety of chemical engineering applications such as, process design and operation, separation processes, operations of petroleum and petrochemical industry, polymerization processes, pharmaceuticals and biotechnology, conventional and renewable energy production, etc. These have been reviewed extensively in literature [29, 30, 90]. However, the present article focuses on applications of bio-mimetic adaptations for chemical engineering systems. From these, the multiobjective optimization of fluid catalytic cracking unit (FCCU) [37, 91] is discussed in detail and thereafter the other applications are discussed briefly.

Catalytic reactors play an important role in the chemical industry, especially for petroleum refining, petrochemical and polymer production. Interestingly, the first industrial MOP that has been solved with NSGA-II-JG is multiobjective optimization of a FCCU (see Figure 2.8), which is an integral part of a petroleum refinery. In this unit, hydrocarbons of high molecular weight such as heavy gas oils are catalytically converted to hydrocarbons of low molecular weight mainly gasoline with significant amounts of middle distillates. Kasat et al. [91] developed a mathematical model based on five lump kinetic model proposed by Dave [92]. On the basis of this mathematical model they formulated a two objective optimization problem [Eqs. (22) and (23)] and solved it with NSGA-II-JG. The objectives were maximization of gasoline which is the main desired product and minimization of the percentage of coke which is responsible for catalyst deactivation.

Objective:

$$\max I_1 \left(T_{feed}, T_{air}, F_{cat}, F_{air} \right) = \text{gasoline yield} \tag{22}$$

$$\min I_2 \left(T_{feed}, T_{air}, F_{cat}, F_{air} \right) = \% \text{coke} \tag{23}$$

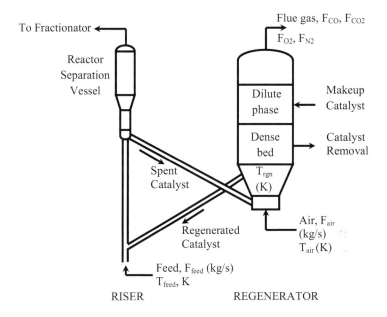

FIGURE 2.8   Schematic representation [91] of a typical FCCU.

Constraints:

$$700 \le T_{rgn} \le 950\ K \qquad (24)$$

$$\text{Model equations [91]} \qquad (25)$$

Bounds:

$$575 \le T_{feed} \le 670\ K \qquad (26)$$

$$450 \le T_{air} \le 525\ K \qquad (27)$$

$$115 \le F_{cat} \le 290\ \text{kg/s} \qquad (28)$$

$$11 \le F_{air} \le 46\ \text{kg/s} \qquad (29)$$

Here, T_{rgn}, T_{feed}, T_{air}, F_{cat} and F_{air} are the temperature of dense bed, temperature of feed to riser, temperature of air fed to the regenerator, catalyst flow rate and airflow rate to the regenerator, respectively.

Kasat and Gupta [37] solved this problem using NSGA-II-JG. The results of this problem are compared with that of NSGA-II as shown in Figure 2.9. The results show that the former gives better spread and convergence than the latter in earlier generations (=10) even with different random seeds and both the algorithms converge to near global optimal results as the number of generations are increased to 50. This clearly illustrates the usefulness of JG adaptation for the complex industrial MOPs for which the number of generations is often restricted.

In another application, the multiobjective optimization of phthalic anhydride (PA) catalytic reactor [93] has been carried out using

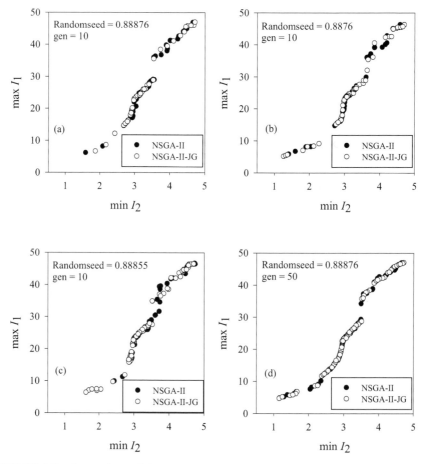

FIGURE 2.9 Comparison of the Pareto optimal fronts for the MOP of FCCU, obtained by NSGA-II-JG and NSGA-II with random-seed numbers (a) 0.88876, (b) 0.8876 and (c) 0.88855 for 10 generations and with random-seed number 0.88876 for (d) 50 generations.

NSGA-II-aJG. PA is the principal raw material for the manufacture of polyester resins, insect repellent, dyes, etc. It is produced by gas phase catalytic oxidation of o-xylene in a multi-tubular reactor with catalyst packed in several layers in each tube. Bhat and Gupta [93] developed a mathematical model for such a reactor that consists of several identical tubes with each tube having nine layers of catalyst and eight intermediate inert cooling zones. They formulated a two-objective optimization problem based on this model where the objectives were the maximization of PA yield and the minimization of the cumulative length of catalyst bed (sum of lengths of all nine layers). The same MOP [49] has been solved with Alt-NSGA-II-aJG and the results obtained were better than that obtained with NSGA-II-aJG. Furthermore, B-NSGA-II-aJG has also been used to solve this MOP [51] with the embryonic population formulated from the solution history of a simpler problem having only seven catalyst layers. This algorithm gave optimal results in less than half of generations taken by NSGA-II-aJG.

The other notable catalytic reactor applications such as many-objective (upto four objectives) optimization of an industrial low-density polyethylene tubular reactor [94, 95] using NSGA-II-aJG, multiobjective optimization of an industrial LPG thermal cracker using NSGA-II-aJG [96, 97], multiobjective optimization of a nylon-6 semi-batch reactor [98] using NSGA-II-aJG, multiobjective optimization of a fixed bed maleic anhydride reactor [99] using Alt-NSGA-II-aJG, single-objective optimization of a combustion side reaction of p-Xylene oxidation using MSIGA [56], etc., are also reported in literature. The details can be found in the respective references.

Besides, catalytic reactors, other interesting areas of chemical engineering where MOPs are commonly seen are separation processes, networking of equipments and utilities, kinetic parameter estimation, planning and scheduling, etc. One of such interesting applications is multiobjective optimization of a froth flotation circuit [40]. Froth flotation is a separation process used in the mining industry to separate valuable constituents from finely ground ore. This process is carried out in a network of flotation cells. Guria et al. [40] studied a simple two-cell circuit and optimized it for the two objectives (1) maximization of recovery (flow rate of solids in the concentrated stream/flow rate of solids in the feed stream) and (2) maximization of grade (fraction of valuable minerals in the concentrated stream). Optimal solutions for this problem have been generated using NSGA-II-mJG and found to be better than that previously reported in literature.

Multiobjective optimization of a heat exchanger network [32] has also been reported using two JG adaptations of NSGA-II, namely, NSGA-II-sJG and NSGA-II-saJG. The objectives chosen were the minimization of the total annual cost and the minimization of total hot (steam) and cold (water) utility. This problem involved variable length of the chromosomes, which had been successfully handled using above two adaptations. The complete details of problem and results can be found in Agarwal and Gupta [32].

The estimation of optimal values of kinetic and model parameters is often required in chemical engineering. In such problems, the objective function is formulated as a sum of the square error of actual outputs (obtained from the experiments) and that estimated from the simulation. GA selects the values of the parameters in order to minimize this objective function. Tao and Wang [67] solved one such SOP to estimate the parameters for a three-lump model of heavy oil thermo-cracking using RNA-GA. Total eight parameters were estimated by randomly selecting 20 groups of data from literature [100]. These parameters were verified using 56 samples of published data [100]. Also, the deviation from actual values was found to be lesser than that obtained with SGA. Several other problems of similar nature such as parameter estimation of FCCU main fractionator [67], parameter estimation of hydrogenation reaction [70], parameter estimation of dynamic systems [71, 74], parameter estimation in hydrocracking of heavy oil [73], parameter estimation of the 2-chlorophenol oxidation in supercritical water [75], modeling of proton exchange membrane fuel cells [76], etc., are solved over the years using different variants of RNA- and DNA-GA.

An SOP is also reported for short-time scheduling of gasoline blending [72] using DNA-HGA. Gasoline blending is an important process in a petroleum refinery. In this process, various gasoline products are produced by mixing different feedstock with small amounts of additives. There are some quality specifications for these products along with restrictions on the operation including bounds on the availability of blending constituents and storage facilities. These factors give rise to a non-linear optimization problem with several linear and non-linear constraints. In the above-mentioned study, two such problems were formulated with the objective of finding an optimized recipe for blending process with maximum profit for one-day and three-days, respectively. The results show that the profits obtained with DNA-HGA were better than that obtained with GA and PSO.

All the above applications along with the techniques used for problem solving are listed in Table 2.2.

TABLE 2.2 Chemical Engineering Applications of Bio-Mimetic Adaptations of GA

S. No.	Year	Chemical Engineering Application	Bio-mimetic Adaptation of GA used	Reference
1	2003	Multiobjective optimization of industrial FCCU	NSGA-II-JG	Kasat and Gupta [37]
2	2005	Multiobjective optimization of a froth flotation circuit	NSGA-II-mJG	Guria et al. [40]
3	2006	Multiobjective optimization (up to four objectives) of industrial low-density polyethylene tubular reactor	NSGA-II-aJG	Agrawal et al. [94]
4	2007	Multiobjective optimization of industrial low-density polyethylene tubular reactor	NSGA-II-aJG with constrained-dominance principle []	Agrawal et al. [95]
5	2007	Parameter estimation of a three-lump model of heavy oil thermo-cracking and FCCU main fractionator	RNA-GA	Tao and Wang [67]
6	2008	Multiobjective optimization of industrial PA reactor	NSGA-II-aJG	Bhat and Gupta [93]
7	2008	Multiobjective optimization of a nylon-6 semi-batch reactor	NSGA-II-aJG	Ramteke and Gupta [98]
8	2008	Multiobjective optimization of a heat exchanger network	NSGA-II-sJG and NSGA-II-saJG	Agarwal and Gupta [32]
9	2009	Multiobjective optimization of industrial PA reactor	Alt-NSGA-II-aJG	Ramteke and Gupta [49]
10	2009	Multiobjective optimization of industrial PA reactor	B-NSGA-II-aJG	Ramteke and Gupta [51]
11	2009	Multiobjective optimization of industrial LPG thermal cracker	NSGA-II-aJG	Nabavi et al. [96]
12	2009	Parameter estimation of hydrogenation reaction	DNA-GA	Chen and Wang [70]
13	2010	Parameter Estimation of Dynamic Systems	Novel RNA-GA (NRNA-GA)	Wang and Wang [71]
14	2010	Short-time scheduling of gasoline blending	DNA based hybrid genetic algorithm (DNA-HGA)	Chen and Wang [72]

S. No.	Year	Chemical Engineering Application	Bio-mimetic Adaptation of GA used	Reference
15	2011	Multiobjective optimization (up to six objectives) of industrial LPG thermal cracker	NSGA-II-aJG	Nabavi et al. [97]
16	2011	Single-objective optimization of a combustion side reaction of p-Xylene oxidation	Modified self-adaptive immune GA (MSIGA)	Qian et al. [56]
17	2011	Parameter estimation in hydrocracking of heavy oil	Protein inspired RNA-GA (PIRGA)	Wang and Wang [73]
18	2012	Multiobjective optimization of a fixed bed maleic anhydride reactor	Alt-NSGA-II-aJG	Chaudhari and Gupta [99]
19	2012	Parameter Estimation of Dynamic Systems	Hybrid DNA-GA (H DNA-GA)	Dai and Wang [74]
20	2013	Parameter estimation of the 2-chlorophenol oxidation in supercritical water	Modified DNA-GA (M DNA-GA)	Zhang and Wang [75]
21	2013	Parameter estimation for modeling of proton exchange membrane fuel cells	Adaptive RNA-GA (ARNA-GA)	Zhang and Wang [76]

2.5 SUMMARY

GA is undoubtedly one of the most popular stochastic optimization techniques primarily due to its ability to handle multiple objectives in a derivative-free environment. Very often, the number of generations attempted in GA to solve the MOPs of chemical engineering systems is severely restricted owing to highly complex model equations. Thus, one has to work with prematurely converged solutions. Several bio-mimetic adaptations are developed to improve the quality of these solutions by incorporating different facets of biological systems. These adapted versions have been successfully used for solving computationally intensive problems of chemical engineering. Also, several challenges such as handling a large number of objectives, combinatorial nature and inter-disciplinary applications are continually being posed which foster the development of more robust algorithms.

KEYWORDS

- **bio-mimetic adaptations**
- **chemical engineering applications**
- **genetic algorithm**
- **multiobjective optimization**

REFERENCES

1. Trambouze, P. J., & Piret, E. L. Continuous Stirred tank Reactors: Design for maximum conversions of raw material to desired product *AIChE J.* 1959, 5, 384–390.
2. Levenspiel, O. *Chemical Reaction Engineering*, 3rd ed., Wiley: New York, 1999.
3. Pareto, V. *Cours d'economie Politique*; F. Rouge: Lausanne, Switzerland, 1896.
4. Charnes, A., & Cooper, W. *Management Models and Industrial Applications of Linear Programming*; Wiley: New York, 1961.
5. Beveridge, G. S. G., & Schechter, R. S. *Optimization: Theory and Practice*; McGraw Hill: New York, 1970.
6. Ray, W. H., & Szekely, J. *Process Optimization with Applications in Metallurgy and Chemical Engineering*; Wiley: New York, 1973.
7. Chankong, V., & Haimes, Y. V. *Multiobjective Decision Making-Theory and Methodology*; Elsevier: New York, 1983.
8. Deb, K. *Multiobjective Optimization Using Evolutionary Algorithms*; Wiley: Chichester, UK, 2001.
9. Coello Coello, C. A., Veldhuizen, D. A. V., & Lamont, G. B. *Evolutionary Algorithms for Solving Multiobjective Problems*, 3rd ed., Springer: New York,. 2007.
10. Holland, J. H. *Adaptation in Natural and Artificial Systems*; University of Michigan Press: Ann Arbor, MI, USA, 1975.
11. Fogel, L. J., Owens, A. J., & Walsh, M. J. *Artificial Intelligence through Simulated Evolution*, Wiley: New York, 1966.
12. Storn, R., & Price, K. Differential evolution—a simple and efficient heuristic for global optimization over continuous spaces *J. Global Optim.* 1997, 11, 341–359.
13. Kennedy, J., & Eberhart, R. In *Particle Swarm Optimization*, Proceedings of IEEE International Conference on Neural Networks, Perth, Australia, November 27-December 01, 1995; IEEE: New Jersey, 1995.
14. Dorigo, M. Optimization, Learning and Natural Algorithms. PhD dissertation, Politecnico di Milano, Italy, 1992.
15. Karaboga, D., & Basturk, B. A powerful and efficient algorithm for numerical function optimization: Artificial Bee Colony (ABC) algorithm *J. Global Optim..* 2007, 39, 459–471.
16. Schaffer, J. D. In *Multiple Objective Optimization with Vector Evaluated Genetic Algorithm*, Proceedings of the 1st International Conference on Genetic Algorithm and their Applications, Pittsburgh, PA, July 24–26, 1985; Grenfenstett, J. J., Ed., Lawrence Erlbaum Associates: New Jersey, 1988.

17. Goldberg, D. E. *Genetic Algorithms in Search, Optimization and Machine Learning*; Addison-Wesley: Reading, MA, 1989.
18. Hajela, P., & Lin, C. Genetic Search Strategies in Multicriterion Optimal Design *Structural Optimization* 1992, 4, 99–107.
19. Fonseca, C. M., & Fleming, P. J. In *Genetic Algorithm for Multiobjective Optimization: Formulation, Discussion And Generalization*, Proceedings of the 5th International Conference on Genetic Algorithm, San Mateo, California, July 17–21, 1993; Forrest, S., Ed., Morgan Kauffman: Massachusetts, 1993.
20. Horn, J. N., Nafpliotis, N., & Goldberg, D. In: *A Niched Pareto Genetic Algorithm for Multiobjective Optimization*, Proceeding of the 1st IEEE Conference on Evolutionary Computation, Orlando, Florida, June 27–29, 1994; IEEE: New Jersey, 1994.
21. Srinivas, N., & Deb, K. Multiobjective Function Optimization using Nondominated Sorting Genetic Algorithm *Evol. Comput.* 1994, 2, 221–248.
22. Zitzler, E., & Thiele, L. Multiobjective Evolutionary Algorithms: A Comparative Case Study and the Strength Pareto Approach *IEEE Trans. Evol. Comput.* 1999, 3, 257–271.
23. Knowles, J. D., & Corne, D. W. Approximating the Nondominated Front using the Pareto Archived Evolution Strategy *Evol. Comput.* 2000, 8, 149–172.
24. Corne, D. W., Knowles, J. D., & Oates, M. J. In: *The Pareto Envelope-based Selection Algorithm for Multiobjective Optimization*, Proceedings of the Parallel Problem Solving from Nature, 6th Conference, Paris, France, September 18–20, 2000; Schoenauer, M., Deb, K., Rudolph, G., Yao, X., Lutton, E., Merelo, J. J., Schwefel, H. P., Eds., Springer: Paris, 2000.
25. Corne, D. W., Jerram, N. R., Knowles, J. D., & Oates, M. J. In *PESA-II: Region-based Selection in Evolutionary Multiobjective Optimization*, Proceedings of the Genetic and Evolutionary Computation Conference (GECCO'2001), San Francisco, California, July 7–11, 2001; Morgan Kauffman: Massachusetts, 2001.
26. Deb, K., Agrawal, S., Pratap, A., & Meyarivan, T. A Fast and Elitist Multiobjective Genetic Algorithm: NSGA-II *IEEE Trans. Evol. Comput.* 2002, 6, 182–197.
27. Sulfllow, A., Drechsler, N., & Drechsler, R. In: *Robust Multiobjective Optimization in High Dimen-Sional Spaces*, Proceeding of the Evolutionary Multi-criterion Optimization, 4th International Conference. Lecture Notes in Computer Science, Matsushima, Japan, March 5–8,. 2007; Obayashi, S., Deb, K., Poloni, C., Hiroyasu, T., Murata, T., Eds., Springer: Heidelberg,. 2007.
28. Jain, H., & Deb, K. In *An Improved Adaptive Approach for Elitist Nondominated Sorting Genetic Algorithm for Many-Objective Optimization*, Proceeding of the Evolutionary Multi-criterion Optimization, 7th International Conference. Lecture Notes in Computer Science, Sheffield, UK, March 19–22, 2013; Purshouse, R., Fleming, P., Fonseca, C., Greco, S., Shaw, J., Eds., Springer-Verlag: Berlin, 2013.
29. Masuduzzaman, Rangaiah, G. P. Multiobjective Optimization Applications in Chemical Engineering. In *Multiobjective Optimization: Techniques and Applications in Chemical Engineering*; Rangaiah, G. P., Ed., World Scientific: Singapore, 2009; p 27.
30. Sharma, S., & Rangaiah, G. P. Multiobjective Optimization Applications in Chemical Engineering. In *Multiobjective Optimization in Chemical Engineering: Developments and Applications*; Rangaiah, G. P., Petriciolet, A. B., Eds., Wiley: Oxford, 2013; 1st ed., p 35.
31. Gupta, S. K., & Ramteke, M. Application of Genetic Algorithm in Chemical Engineering II: Case Studies. In *Applications of Metaheuristics in Process Engineering*; Valadi, J., Siarry, P., Eds., Springer: Switzerland, 2014; p 60.

32. Agarwal, A., & Gupta, S. K. Jumping Gene Adaptations of NSGA-II and their use in the Multiobjective Optimal Design of Shell and Tube Heat Exchangers *Chem. Eng. Res. Des.* 2008, 6, 123–139.

33. Ramteke, M., & Srinivasan, R. Novel Genetic Algorithm for Short-Term Scheduling of Sequence Dependent Changeovers in Multiproduct Polymer Plants. *Comput. Chem. Eng.* 2011, *35*, 2945–2959.

34. Deb, K., & Agrawal, R. B. Simulated Binary Crossover for Continuous Search Space *Complex. Syst.* 1995, 9, 115–148.

35. Deb, K., & Agrawal, S. In *A Niched-penalty Approach for Constraint handling in Genetic Algorithms*, Proceedings of the International Conference Artificial Neural Networks and Genetic Algorithms, ICANNGA-99, Portoroz, Slovenia, 1999; Springer: Vienna, 1999.

36. McClintock, B. *The Collected Papers of Barbara McClintock*; Garland: New York, 1987.

37. Kasat, R. B., & Gupta, S. K. Multiobjective Optimization of an Industrial Fluidized-Bed Catalytic Cracking Unit (FCCU) using Genetic Algorithm (GA) with the Jumping Genes Operator *Comp. Chem. Eng.* 2003, 27, 1785–1800.

38. Bhat, S. A., Gupta, S., Saraf, D. N., & Gupta, S. K. On-Line Optimizing Control of Bulk Free Radical Polymerization Reactors under Temporary Loss of Temperature Regulation: An Experimental Study on a 1-Liter Batch Reactor *Ind. Eng. Chem. Res.* 2006, 45, 7530–7539.

39. Bhat, S. A. On-Line Optimizing Control of Bulk Free Radical Polymerization of Methyl Methacrylate in a Batch Reactor using Virtual Instrumentation. Ph. D. Dissertation, Indian Institute of Technology, Kanpur,. 2007.

40. Guria, C., Verma, M., Mehrotra, S. P., & Gupta, S. K. Multiobjective Optimal Synthesis and Design of Froth Flotation Circuits for Mineral Processing using the Jumping Gene Adaptation of Genetic Algorithm *Ind. Eng. Chem. Res.* 2005, 44, 2621–2633.

41. Agarwal, A., & Gupta, S. K. Multiobjective Optimal Design of Heat Exchanger Networks using New Adaptations of the Elitist Nondominated Sorting Genetic Algorithm, NSGA-II *Ind. Eng. Chem. Res.* 2008, 47, 3489–3501.

42. Ramteke, M., Ghune, N., & Trivedi, V. Simulated Binary Jumping Gene: A Step towards Enhancing the Performance of Real-Coded Genetic Algorithm, submitted.

43. Chan, T. M., Man, K. F., Tang, K. S., & Kwong, S. A Jumping Gene Algorithm for Multiobjective Resource Management in Wideband CDMA Systems *Comput. J.* 2005, 48, 749–768.

44. Chan, T. M., Man, K. F., Tang, K. S., & Kwong, S. In: *Multiobjective Optimization of Radio-to-Fiber Repeater Placement using A Jumping Gene Algorithm*, Proceedings of IEEE International Conference on Industrial Technology (ICIT 2005), Hong Kong, December 14–17, 2005.

45. Ripon, K. S. N., Kwong, S., & Man, K. F. A Real-Coding Jumping Gene Genetic Algorithm (RJGGA) for Multiobjective Optimization *Inf. Sci.* 2007, 177, 632–654.

46. Furtuna, R., Curteanu, S., & Racles, C. NSGA-II-RJG applied to Multiobjective Optimization of Polymeric Nanoparticles Synthesis with Silicone Surfactants *Cent. Eur. J. Chem.* 2011, 9, 1080–1095.

47. Sharma, S., Nabavi, S. R., & Rangaiah, G. P. Performance Comparison of Jumping Gene Adaptations of Elitist Nondominated Sorting Genetic Algorithm. In *Multiobjective Optimization in Chemical Engineering: Developments and Applications*; Rangaiah, G. P., Petriciolet, A. B., Eds., Wiley: Oxford, 2013; 1st ed.

48. Sharma, S., Nabavi, S. R., & Rangaiah G. P. Jumping Gene Adaptations of NSGA-II with Altruism Approach: Performance Comparison and Application, In *Recent Advances in Applications of Metaheuristics in Process Engineering*; Valadi, J., Siarry, P., Eds., Springer: Switzerland, 2014; p. 395.

49. Ramteke, M., & Gupta, S. K. Biomimicking Altruistic Behavior of Honey Bees in Multiobjective Genetic Algorithm *Ind. Eng. Chem. Res.* 2009, 48, 9671–9685.

50. Goodnight, C. J., Goodnight, M. L., & Gray, P. *General Zoology*; Reinhold: New York, 1964.

51. Ramteke, M., & Gupta, S. K. Bio-mimetic Adaptation of the Evolutionary Algorithm, NSGA-II-aJG, using the Biogenetic Law of Embryology for Intelligent Optimization *Ind. Eng. Chem. Res.* 2009, 48, 8054–8067.

52. Jiao, L., & Wang, L. A Novel Genetic Algorithm Based on Immunity *IEEE Trans. Syst., Man, Cybern. A, Syst.,Humans.* 2000, 30, 552–561.

53. Simões, A., & Costa, E. In *An Immune System-based Genetic Algorithm to Deal with Dynamic Environments: Diversity and Memory*, Proceedings of the 6th International Conference on Neural Networks and Genetic Algorithms, Roanne, France, 2003; Pearson, D. W., Steele, N. C., Albrecht, R. F. Eds., Springer Vienna: Vienna, 2003.

54. Yongshou, D., Yuanyuan, L., Lei, W., Junling, W., & Deling, Z. Adaptive Immune-Genetic Algorithm for Global Optimization to Multivariable Function *J. Syst. Eng. Electron..* 2007, 18, 655–660.

55. Ru, N., Jian-Hua, Y., Shuai-Qi, D., & Yang-Guang, L. In: *Wave Impedance Inversion in Coalfield based on Immune Genetic Algorithm*, Proceedings of the 6th International Conference on Mining Science and Technology (ICMST 2009), Xuzhou, PR China, October 18–20, 2009; Ge, S., Liu, J., Guo, C. Eds., Elsevier: Amsterdam, 2009.

56. Qian, F., Tao, L., Sun, W., & Du, W. Development of a Free Radical Kinetic Model for Industrial Oxidation of P-Xylene based on Artificial Network and Adaptive Immune Genetic Algorithm. *Ind. Eng. Chem. Res.* 2011, 51, 3229–3237.

57. Lili, T., Xiangdong, K., Weimin, Z., & Feng, Q. Modified Self-adaptive Immune Genetic Algorithm for Optimization of Combustion Side Reaction of p-Xylene Oxidation Chin. *J. Chem. Eng.* 2012, 20, 1047–1052.

58. Liu, W., & Huang, X. A Multi-population Immune Genetic Algorithm for Solving Multi objective TSP Problem *J. Chem. Pharm. Res.* 2014, 6, 566–569.

59. Whitley, D., Gordon, S., & Mathias, K. In *Lamarckian Evolution, the Baldwin Effect and Function Optimization*, Proceedings of the Parallel Problem Solving from Nature—PPSN III. Lecture Notes in Computer Science, Jerusalem, Israel, October 9–14, 1994; Davidor, Y., Schwefel, H. P., Manner, R. Eds., Springer-Verlag: Jerusalem, 1994.

60. Julstrom, B. In: *Comparing Darwinian, Baldwinian, and Lamarckian Search in a Genetic Algorithm for the 4-Cycle Problem*, Proceedings of the 1999 Genetic and Evolutionary Computation Conference, Late Breaking Papers, Orlando, USA, July 13–17, 1999; Brave, S., Wu, A. S. Eds., Morgan Kauffman: Massachusetts, 1999.

61. Baldwin, J. M. A New Factor in Evolution *Am. Nat.* 1896, 30, 441–451.

62. Hinton, G. E., & Nowlan, S. J. How Learning can guide Evolution *Complex Sys.* 1987, 1, 495–502.

63. Hart, W. E., Kammeyer, T. E., & Belew, R. K. The Role of Development in Genetic Algorithms. In: *Foundations of Genetic Algorithms*; Whitley, L. D., Vose, M. D., Eds; Morgan Kaufmann: San Mateo, California, 1995; Vol. 3; p. 315.

64. Gruau, F., Whitley, D. Adding learning to the Cellular Development of Neural Networks: Evolution and the Baldwin Effect *Evol. Comput.* 1993, 1, 213–233.

65. Sun, Y., & Deng, F. In *Baldwin Effect-Based Self-Adaptive Generalized Genetic Algorithm*, Proceedings of the 8th International Conference on Control, Automation, Robotics and Vision, Kunming, China, December 6–9, 2004.

66. Yuan, Q., Qian, F., & Du, W. A Hybrid Genetic Algorithm with the Baldwin Effect *Inf. Sci.* 2010, 180, 640–652.

67. Tao, J. L., & Wang, N. DNA Computing based RNA Genetic Algorithm with Applications in Parameter Estimation of Chemical Engineering Processes *Comput. Chem. Eng.*. 2007, 31, 1602–1618.

68. Adleman, L. M. Molecular Computation of Solutions to Combinatorial Problems *Science* 1994, 266, 1021–1023.

69. Cukras, A. R., Faulhammer, D., Lipton, R. J., & Landweber, L. F. Chess Games: A Model for RNA based Computation *Biosystems* 1999, 52, 35–45.

70. Chen, X., & Wang, N. A DNA based Genetic Algorithm for Parameter Estimation in the Hydrogenation Reaction *Chem. Eng. J.* 2009, 150, 527–535.

71. Wang, K. T., & Wang, N. A Novel RNA Genetic Algorithm for Parameter Estimation of Dynamic Systems *Chem. Eng. Res. Des.* 2010, 88, 1485–1493.

72. Chen, X., & Wang, N. Optimization of Short-time Gasoline Blending Scheduling Problem with a DNA based Hybrid Genetic Algorithm *Chem. Eng. Process.* 2010, 49, 1076–1083.

73. Wang, K., & Wang, N. A Protein Inspired RNA Genetic Algorithm for Parameter Estimation in Hydrocracking of Heavy Oil *Chem. Eng. J.* 2011, 167, 228–239.

74. Dai, K., & Wang, N. A Hybrid DNA based Genetic Algorithm for Parameter Estimation of Dynamic Systems *Chem. Eng. Res. Des.* 2012, 90, 2235–2246.

75. Zhang, L., & Wang, N. A Modified DNA Genetic Algorithm for Parameter Estimation of the 2-Chlorophenol Oxidation in Supercritical Water *Appl. Math. Model.* 2013, 37, 1137–1146.

76. Zhang, L., & Wang, N. An Adaptive RNA Genetic Algorithm for Modeling of Proton Exchange Membrane Fuel Cells *Int. J. Hydrogen Energy* 2013, 38, 219–228.

77. Sun, Z., Wang, N., & Bi, Y. Type-1/Type-2 Fuzzy Logic Systems Optimization with RNA Genetic Algorithm for Double Inverted Pendulum *Appl. Math. Model.* 2014, http://dx.doi.org/10.1016/j.apm.2014.04.035.

78. Wu, S., & Wu, C. Computational Optimization for S-Type Biological Systems: Cockroach Genetic Algorithm *Math. Biosci.* 2013, 245, 299–313.

79. Kikuchi, S., Tominaga, D., Arita, M., Takahashi, K., & Tomita, M. Dynamic Modeling of Genetic Networks using Genetic Algorithm and S-System *Bioinformatics* 2003, 19, 643–650.

80. Ho, S. Y., Hsieh, C. H., Yu, F. C., & Huang, H. L. An Intelligent Two-stage Evolutionary Algorithm for Dynamic Pathway Identification from Gene Expression Profiles *IEEE/ACM Trans. Comput. Biol. Bioinf.*. 2007, 4, 648–660.

81. Caponio, A., Cascella, G. L., Neri, F., Salvatore, N., & Sumner, M. A Fast Adaptive Memetic Algorithm for Online and Offline Control Design of PMSM Drives *IEEE Trans. Syst. Man Cybern.-Part B: Cybern.*. 2007, 37, 28–41.

82. Lis, J., & Eiben, A. E. In *A Multi-sexual Genetic Algorithm for Multiobjective Optimization*, Proceedings of the 1997 IEEE International Conference on Evolutionary Computing, Indianapolis, USA, April 13–16, 1997; IEEE: New Jersey, 1997.

83. Rejeb, J., & AbuElhaija, M. In *New Gender Genetic Algorithm for Solving Graph Partitioning Problems*, Proceedings of the 43rd IEEE Midwest Symposium on Circuits and Systems, Lansing, MI, USA, August 8–11, 2000; IEEE: New Jersey, 2000.

84. Vrajitoru, D. In *Simulating Gender Separation with Genetic Algorithms*, Proceedings of the 2002 Genetic and Evolutionary Computation Conference, New York, NY, USA, July, 9–13, 2002.

85. Sanchez-Velazco, J., & Bullinaria, J. A. In *Sexual Selection with Competitive/Co-operative Operators for Genetic Algorithms*, Proceedings of the IASTED International Conference on Neural Networks and Computational Intelligence, Cancun, Mexico, May 19–21, 2003.

86. Raghuwanshi, M. M., & Kakde, O. G. In *Genetic Algorithm with Species and Sexual Selection*, Proceedings of the 2006 IEEE Conference on Cybernetics and Intelligent System, Bangkok, Thailand, June 7–9, 2006; IEEE: New Jersey, 2006.

87. Zhang, M. A Novel Sexual Adaptive Genetic Algorithm based on Baldwin Effect for Global Optimization *Int. J. Intell. Comput. Cybern.* 2011, 4, 2011.

88. Dennis, C., & Gallagher, R. *The Human Genome*; Nature Publishing Group: London, 2001.

89. Srinivas, M., & Patnaik, L. M. Adaptive Probabilities of Crossover and Mutation in Genetic Algorithm *IEEE Trans. Syst. Man Cybern.* 1994, 24, 656–67.

90. Bhaskar, V., Gupta, S. K., & Ray, A. K. Applications of Multiobjective Optimization in Chemical Engineering *Reviews Chem. Eng.* 2000, 16, 1–54.

91. Kasat, R. B., Kunzru, D., Saraf, D. N., & Gupta, S. K. Multiobjective Optimization of Industrial FCC Units using Elitist Nondominated Sorting Genetic Algorithm *Ind. Eng. Chem. Res.* 2002, 41, 4765–4776.

92. Dave, D. Modeling of a Fluidized Bed Catalytic Cracker Unit. MTech. Dissertation, Indian Institute of Technology, Kanpur, India, 2001.

93. Bhat, G. R., & Gupta, S. K. MO Optimization of Phthalic Anhydride Industrial Catalytic Reactors using Guided GA with the Adapted Jumping Gene Operator. *Chem. Eng. Res. Des.* 2008, 86, 959–976.

94. Agrawal, N., Rangaiah, G. P., Ray, A. K., & Gupta, S. K. Multiobjective Optimization of the Operation of an Industrial Low-Density Polyethylene Tubular Reactor Using Genetic Algorithm and its Jumping Gene Adaptations. *Ind. Eng. Chem. Res.* 2006, 45, 3182–3199.

95. Agrawal, N., Rangaiah, G. P., Ray, A. K., & Gupta, S. K. Design Stage Optimization of an Industrial Low-Density Polyethylene Tubular Reactor for Multiple Objectives using NSGA-II and its Jumping Gene Adaptations *Chem. Eng. Sci.*. 2007, 62, 2346–2365.

96. Nabavi, S. R., Rangaiah, G. P., Niaei, A., & Salari, D. Multiobjective Optimization of an Industrial LPG Thermal Cracker using a First Principles Model. *Ind. Eng. Chem. Res.* 2009, 48, 9523–9533.

97. Nabavi, S. R., Rangaiah, G. P., Niaei, A., & Salari, D. Design Optimization of an LPG Thermal Cracker for Multiple Objectives. *Int. J. Chem. React. Eng.* 2011, 9.

98. Ramteke, M., & Gupta, S. K. Multiobjective Optimization of an Industrial Nylon-6 Semi Batch Reactor using the Jumping Gene Adaptations of Genetic Algorithm and Simulated Annealing. *Polym. Eng. Sci.* 2008, 48, 2198–2215.

99. Chaudhari, P., & Gupta, S. K. Multiobjective Optimization of a Fixed Bed Maleic Anhydride Reactor using an Improved Bio-mimetic Adaptation of NSGA-II. *Ind. Eng. Chem. Res.* 2012, 51, 3279–3294.

100. Song, X.-F., Chen, D.-Z., Hu, S.-X., Xiao, J.-Z, & Liu F.-Z. Eugenic Evolution Strategy Genetic Algorithms for Estimating Parameters of Heavy Oil Thermal Cracking Model. *J. Chem. Eng. Chin. Univ.* 2003, 17, 411–417.

CHAPTER 3

SURROGATE-ASSISTED EVOLUTIONARY COMPUTING METHODS

SAKET KANSARA,[1] SUMEET PARASHAR,[2] and ABDUS SAMAD[3]

[1]*ESTECO Software India Pvt. Ltd., Pune, Maharashtra, India*

[2]*ESTECO North America, Novi, MI 48375, USA*

[3]*Department of Ocean Engineering, Indian Institute of Technology Madras, Chennai–600036, India*

CONTENTS

3.1 OVERVIEW

This chapter covers an overview of the use of surrogate-assisted evolutionary algorithms in engineering optimization. More specifically, it describes various aspects of engineering optimization including single as well as multiobjective optimization, evolutionary algorithms, how surrogate modeling can help during optimization. It also discusses different surrogate modeling techniques. Finally, it briefly describes how some existing approaches combine surrogate modeling and evolutionary optimization techniques to speed up engineering optimization procedures.

3.2 INTRODUCTION TO OPTIMIZATION

3.2.1 SINGLE OBJECTIVE OPTIMIZATION FORMULATION

Optimization can be used to effectively search for a system or design using automatic and intelligent algorithms. In simplest terms, optimization refers to maximization or minimization of some performance metrics. Although we do face many optimization problems in our daily life, the process we use to arrive at solutions is usually based on intuition, past experience or some preliminary calculations. Optimization of real world complex engineering problems is more involved, even though it uses the same idea of minimizing/maximizing performance metric/s. A formal engineering design optimization approach can make use of mathematically advanced techniques rather than relying on 'intuition' and 'experience.' To understand the concept of optimization problem formulation, a simple example of cantilever beam design is used throughout this chapter. The design problem is formulated as a single-objective as well as a multiobjective, constrained as well as unconstrained optimization problems.

Figure 3.1 shows a cantilever beam of rectangular cross-section made of steel with a constant force of 1000N applied at the end of the beam. The beam has a fixed length of 20 m, but its breadth (b) and height (H) are free parameters to be designed. Free parameters are also called as design variables or design parameters, which can be independently controlled within their upper and lower bounds. The bounds are also sometimes referred to as side constraints. The design variable values directly affect the system performance defined by the design goals/objectives and constraints. The objective function/s is/are design goals or performance targets that one aims to achieve. They are typically expressed as functions or performance measures to be either maximized or minimized. Apart from the design variables and the objectives, it is also common to have certain restrictions posed as constraints. The constraints are performance measures, which have to meet certain limiting values of the optimal solution to be valid.

A formal problem statement can be written as an optimization problem in the form of Eq. (1) as follows:

$$\left\{ \begin{array}{c} minf(B,H) \quad Deflection(y) \quad \dfrac{FL^3}{3EI} \\[2mm] Subject\ to \\[1mm] Cost = BHLr\ Pr_s < \$4000 \\[1mm] 2 \le B \le 5 \\[1mm] 1 \le H \le 3 \end{array} \right. \tag{1}$$

Now, from intuition, it can be quickly inferred that minimizing deflection means the height and the width of the beam have to be increased.

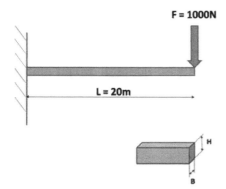

F = 1000N

L = 20m

Moment of inertia of the beam,
I= BH³/12
Force applied (F) :1000 N
Length of the beam (L) : 20 m
Young's Modulus (E): 200 GPa
Cost of Steel (Pr$_S$): $ 0.05/kg
Density: 8000 kg/m³

FIGURE 3.1   Cantilever beam optimization.

If height and width are increased, it does make the beam stronger, however, it also automatically increases volume of the beam thus increasing material required to manufacture the beam. This would clearly lead to an increase in the cost of the beam. Hence, it is necessary to choose the 'optimum' values of height and width so as to not increase cost beyond its acceptable limit.

3.2.2 MULTIOBJECTIVE OPTIMIZATION FORMULATION

In real life, most problems have multiple objectives, which were traditionally modeled as single objective problems with constraints. However, with recent advances in multiobjective optimization algorithms, it is quite common to see optimization problems being formulated and effectively solved with more than one objective [1].

For better understanding, the concept of multiobjective optimization, the above-mentioned beam optimization problem is formulated as a multiobjective problem. Along with minimizing the deflection, the second objective is to minimize cost of the beam. Minimizing deflection would lead to adding more cost and minimizing cost would result in adding more deflection to the beam. It is apparent that the two objectives of minimizing deflection and minimizing cost are conflicting with each other. Figure 3.2 explains this scenario in a better way. It shows two graphs, each showing breadth (b) and height (H) on the horizontal axes. Figure 3.2(a) shows cost on the vertical axis while Figure 3.2(b) shows deflection. It is evident from the figures that the two objectives are conflicting with each other.

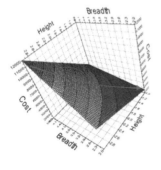

(a) Cost optimization   (b) Deflection optimization

FIGURE 3.2   Multiobjective optimization.

The mathematical formulation of the above problem can be represented as:

$$
\left\{
\begin{array}{c}
\min f_1\left(B,H\right) = y = \dfrac{FL^3}{3EI} \\[2mm]
\min f_2\left(X\right) = Cost = BHLrPr_s \\[1mm]
Subject\ to \\[1mm]
Cost = BHLrPr_s < \$4000 \\[1mm]
2 \leq B \leq 5 \\[1mm]
1 \leq H \leq 3
\end{array}
\right\}
\tag{2}
$$

3.2.3 GENERAL OPTIMIZATION PROCEDURE

A number of systematic mathematical optimization processes can be applied to solve the above-mentioned problem both in single as well as in multiobjective formulations. A general optimization procedure is shown in Figure 3.3. Once an optimization problem is formulated, it is generally fed to an optimization algorithm for generating a solution/s. Every optimization algorithm needs to be initialized, meaning, it needs at least one starting point to start optimization (initialization is not shown in the figure). The optimization algorithm then automatically takes these input values, feeds it to the solvers and evaluates output values. Once it has calculated outputs, it also computes

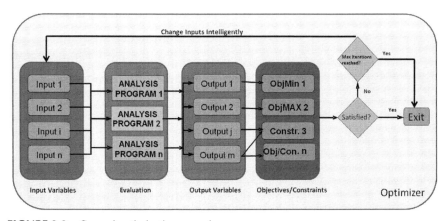

FIGURE 3.3   General optimization procedure.

values for objective functions and constraints. It then analyzes these values to check if the results have converged to a satisfactory solution. If yes, then it stops. If it has not found a satisfactory solution, it automatically and intelligently changes values of input variables and repeats the procedure until it finds a good solution. It can also be asked to exit after reaching maximum number of iterations to avoid it going into an infinite loop.

Many algorithms have been developed to solve such optimization problems. Derivate-based algorithms and evolutionary/heuristic algorithms are two of the most widely used categories of optimization algorithms. The derivative based algorithms include gradient descent [2], sequential quadratic programming [3], etc. The heuristic algorithms consist of genetic algorithm [4–6], particle swarm optimization [7], simulated annealing [8], etc. Details of these algorithms are not covered in this chapter. All the algorithms use different techniques to come up with optimum designs; however, the common theme behind all is that they change the inputs *intelligently* based on previous results. The idea of using these algorithms is that, once we provide the algorithm the information in Eq. (2), it keeps changing the values of input variables intelligently so that we reach our objectives while also satisfying the constraints. At the end of the optimization, the algorithm gives optimum values of input variables and functions. One can write his/her own optimization algorithm routine or use many available commercial software such as modeFRONTIER [9], MAT LAB [10], etc., for solving his/her own optimization problem.

3.2.4 OPTIMAL DESIGNS AND PARETO OPTIMAL FRONTIER

Different optimization formulations of the same problem can greatly affect the quality of results and the speed at which the algorithm arrives at them. Each sub-figure in Figure 3.4 represents a scatter chart of deflection (on *x*-axis) vs. cost (on *y*-axis). Each sub-figure shows the solutions obtained for the above-mentioned beam optimization problem but using different formulation. The same problem is formulated in four different ways as shown in Figure 3.4(a–d). A multiobjective genetic algorithm (MOGA-II) [11] available in modeFRONTIER [9] has been applied to solve each formulation. For each formulation, the algorithmic initialized with 20 random design points, meaning 20 random combinations of B and H are generated. MOGA-II then takes these inputs, evaluates and analyzes the outputs before intelligently creating new combinations of B and H. For each of these formulations, MOGA-II was allowed to run for 1000 generations resulting in

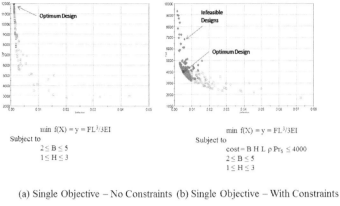

$$\min f(X) = y = FL^3/3EI$$
Subject to
$$2 \leq B \leq 5$$
$$1 \leq H \leq 3$$

$$\min f(X) = y = FL^3/3EI$$
Subject to
$$cost = B \ H \ L \ \rho \ Pr_S \leq 4000$$
$$2 \leq B \leq 5$$
$$1 \leq H \leq 3$$

(a) Single Objective – No Constraints (b) Single Objective – With Constraints

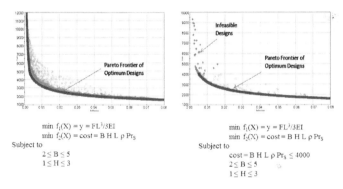

$$\min f_1(X) = y = FL^3/3EI$$
$$\min f_2(X) = cost = B \ H \ L \ \rho \ Pr_S$$
Subject to
$$2 \leq B \leq 5$$
$$1 \leq H \leq 3$$

$$\min f_1(X) = y = FL^3/3EI$$
$$\min f_2(X) = cost = B \ H \ L \ \rho \ Pr_S$$
Subject to
$$cost = B \ H \ L \ \rho \ Pr_S \leq 4000$$
$$2 \leq B \leq 5$$
$$1 \leq H \leq 3$$

(c) Multi-Objective – No Constraints (d) Multi-Objective – With Constraints

FIGURE 3.4 Results from different optimization formulations for the beam optimization problem.

a total of 20,000 design points. Optimum designs are marked in blue bubbles while infeasible designs, if any, are marked with orange diamonds. The gray squares represent designs that the genetic algorithm evaluated while trying to reach optimum designs.

The first formulation represented by Figure 3.4(a), is a single objective optimization without any constraints. This leads to a single optimum design, which has the least deflection. This can be seen in Figure 3.4(a) where the optimum design is shown in the top left corner. It is also worth noting that cost for this optimum design is very high as the only objective is to minimize deflection. Figure 3.4(b) depicts a formulation containing a single objective to minimize deflection and a constraint on cost. Here it can be seen that because of a constraint on cost, all designs which have cost >$4000 are marked as infeasible designs. The optimum design is found when cost is equal to $4000 (maximum allowable cost).

Notice that there are not many designs where cost >$4000. This is because the MOGA-II is intelligent enough to recognize that violating the constraint leads to infeasible designs. Hence, the algorithm intelligently generates values of B and H such that cost would be kept below the acceptable limit. Please note that although Figures 3.4(a) and 3.4(b) both result only in a single optimum solution, the entire history of the optimization runs is shown in both charts.

Figure 3.4(c) shows results for a pure multiobjective optimization formulation without any constraints. Here the two objectives are: minimize deflection and minimize cost. From the optimization problem statement, an ideal design should be a beam, which has minimum deflection and minimum cost, both at the same time. However, due to the nature of the problem, it is simply not possible to obtain such a design. Hence, as the two objectives are conflicting in nature, the optimization results in not one single optimum solution but a set of optimum nondominated solutions called the 'Pareto optimal solutions [12]. The set of all 'Pareto optimal' designs is also known as 'Pareto optimal frontier.' All solutions belonging to Pareto frontier are nondominated solutions. A solution is known as nondominated if there is no other solution, which is clearly better in all objectives than the current solution. Once the Pareto frontier of solutions is achieved, it becomes a subjective choice of the designer/ decision maker to choose the best design for his/her application. For example, if one wishes to give more importance to minimizing deflection, one can choose a design, which is near the top left corner of Pareto frontier in Figure 3.4(c). On the other hand, if one intends to provide more importance to minimizing weight, he/she should choose a design near the bottom right corner of the Pareto frontier in Figure 3.4(c). Figure 3.4(d) shows results from Multiobjective optimization with constraints. Notice that similar to the single objective constrained case, the optimization algorithm was intelligent to recognize that violating constraints (cost > $4000) is not useful as it leads to infeasible designs. Hence, very few designs with cost > $4000 are seen in Figure 3.4(d). Also notice that a Pareto frontier is still formed in this case, but it is trimmed at the top to avoid infeasible region. The Pareto optimal fronts seen in Figures 3.4(c) and 3.4(d) are called convex Pareto frontiers. Pareto frontier can also be non-convex, discontinuous (also known as 'broken Pareto frontier'), they can also look like inverted 'L' or of different types depending on whether the objectives are to be maximized or minimized.

3.3   EVOLUTIONARY ALGORITHMS

This section briefly discusses the algorithms that are commonly used to solve optimization problems. In particular, it focuses on a very popular and widely used class of evolutionary algorithm called genetic algorithms (GA). This section discusses the core idea behind the algorithm without delving deep into the mathematical details.

Evolutionary algorithms or GAs are a part of the family of heuristic algorithms and are based on the theory of evolution. The GAs, for example, attempts to model the principal of 'survival of the fittest' and the assumption that healthy population members will most likely lead to healthy offspring. Figure 3.5 renders the idea behind workings of a GA. As discussed before, GAs need multiple points to start with. These starting points are also called as 'starting population.' In Figure 3.5, a starting population of five designs is provided to the algorithm, it evaluates (function evaluation) each one to acquire output and objective values. This process is followed by a step called 'selection.' Selection picks pairs of designs to treat them as parents. It is a probabilistic approach, which favors the selection of "fit" or "healthy" designs, which are good in terms of their objective performance while the "weaker" or "unfit" members/designs are rarely selected. This is typically modeled as a tournament selection or roulette wheel selection. Next, the genetic information of the pairs of designs selected is exchanged to produce

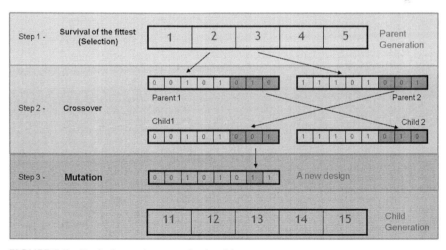

FIGURE 3.5   Typical steps in a genetic algorithm.

new designs. This is known as 'crossover' operation in GA. We can draw similarities between the selection and crossover operations in a GA and in human evolution. The 'selection' operation is similar to selecting the best "parents" out of the current population. The crossover operation is similar to combining the genetic features of parents in order to generate a new child population. A design is considered 'stronger' if it is closer to meeting objectives than previous designs. As shown in step 2 in Figure 3.5, the crossover operation breaks the designs in their binary formats and combines the two binary strings to produce new designs in binary format. The newly created designs are then treated as 'child designs' or designs belonging to a new generation. Along with selection and crossover, one more important parameter of a GA is 'mutation.' Mutation, in terms of human evolution theory, can be viewed as a random change to DNA. GAs use a tactic similar to mutation, in that they randomly change some characteristic of the child population. If mutation is not performed, the child population might end up looking very similar to the parent population. This might lead the algorithm to a local minimum and thus make it less robust. Hence, once a child population is obtained, the GA randomly makes a small change in some properties of the children. Mutation ensures that the design space is searched extensively before converging to optimum solutions. Step 3 in Figure 3.5 explains the mutation step by randomly flipping a '0' to '1' at a random location in a new design. The idea is, if the steps of selection and crossover and mutation process are repeated over a number of generations, the latest generation of designs will be stronger than the starting population and hence it will be closer to satisfying the objectives. GAs repeat this selection-crossover-mutation process until a termination criterion is reached. The termination criterion can be set to maximum number of iterations or to a specific value of objective functions. Figure 3.6 shows a flowchart explaining steps in a GA and complete schematics of a GA. Please note that the above description

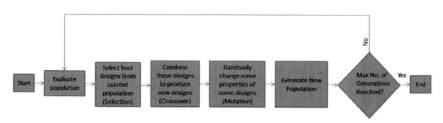

FIGURE 3.6 Schematic of a typical implementation of genetic algorithm.

of GA is a very simplistic one and actual implementation of GAs is usually much more complex. There are many articles/books such as [6–8] that describe the mathematical and implementation details of a GA.

There are many reasons why GAs are very popular in today's optimization community.

To start with, GAs are considered very "robust" in finding global optimum solutions. As opposed to this, derivative based algorithms have a high chance of getting stuck in a local optimum. While one can never be sure if the solution obtained is a local or a global optimum, a GA has fewer chances to get stuck in local optimum solutions. However, derivative based strategies are considered very fast as compared to GAs, while GAs are known to require large number of function evaluation.

Another big advantage of a GA is that it can be easily adapted to handle multiobjective problems while it is little more difficult to adapt a derivative based algorithm to handle multiple objectives.

Derivative based algorithms also need a lot of parameter tuning during optimization. For example, one has to accurately adjust the step size to calculate finite differences. If this is not set correctly, optimization might take a long time to converge or worse, it can even diverge from the optimum solution. For GAs, there are a lot of possible combinations of different settings for crossover, mutation, etc. However, once a GA is implemented, it is not mandatory to tune these parameters. The same algorithm can be used for many different problems. Hence, GAs do not require much parameter tuning.

GAs do have some disadvantages as well. Because of their slightly random nature, GAs take more time to converge to optimum solutions as compared to a derivative based algorithm. GAs are very good at exploring the space and finding the region where an optimum design can be found. Once they find the region, they are however very slow in converging to the actual optimum solution. Hence, GAs require a large number of design evaluations. This is a major drawback in using a GA for practical applications. To overcome this limitation, surrogate-modeling techniques are often used. The next sections discuss surrogate modeling in more details.

3.4   INTRODUCTION TO SURROGATE MODELING

Real world engineering problems are much more complex than the mathematical optimization problems. To start with, they are often multiobjective in nature and they also have many constraints. Also, it is very difficult to represent a real world problem in simple mathematical equations such as the beam example described earlier. In order to solve the real world problems, it is often the case that we need to use computationally intensive methods such as finite element methods (FEM), computational fluid dynamics (CFD), etc. There are many commercial simulation tools available to model and simulate such analysis. They can usually be far more computationally expensive than a set of simple mathematical equations.

Although such FEM/CFD solutions are very useful and widely used, the primary difficulty in using these tools in optimization process is that they are computationally very expensive. For example, a typical CFD analysis can easily take anywhere between a few hours to even a few days for solving a single simulation. This is even when running on high performance computing cluster of a large number of CPUs. To optimize such problems is an extremely resource intensive task because optimization involves iterations and often requires evaluation of hundreds if not thousands of designs depending on the size of the problem (which is often measured in terms of number of design variables). If one design evaluation takes, e.g., 24 h, it is impractical to run hundreds or thousands of such evaluations during optimization.

Surrogate modeling techniques are often used to tackle such problems. Automatic algorithm-based optimization removes the trial and error approach by automating the process of finding the best solution in much less time through the use of intelligent search. Even though automatic direct optimization is effective in terms of number of function evaluations, it is also possible to reduce the total clock time and computational expense required, especially when dealing with computationally expensive simulation processes. In this case, virtual or surrogate model based optimization is a valid and good alternative strategy. It can use a surrogate model in place of expensive simulation software, thus allowing to fast-track the optimization process and get similar outcome.

A surrogate model is also known as response surface models (RSM) or metamodel or transfer function or in very simple terms an approximation model. They are mathematical models that approximate the input/output behavior of the system under investigation. Starting from a dataset of a limited number of designs, an RSM algorithm creates a mathematical approximation of the actual model and using this, it can predict the output value of any unknown input combination.

In simple terms, creating a surrogate model can be viewed as fitting a curve to an existing dataset of discrete points. Consider an example of an output (y) dependent on an input (x). The mathematical relation between x and y, to express it in terms of $y=f(x)$, is not known. Actual physical experiments are performed to get observations of y based on set values of x. Five such experiments have been performed with input variable (x) set to different values from x_1 to x_5. The respective values of y ranging from y_1 to y_5 have been obtained. The five experiments are called as 'data points.' These data points are plotted as Blue bubbles in Figure 3.7. The aim of generating a response surface is to fit a mathematical function to model the curve passing through the data points such as the one shown in Figure 3.7. Once a mathematical function is obtained, it is possible to evaluate y when, e.g., $x = x_6$, without the need to perform the actual experiment again. One can simply use the curve and interpolate the values of y for any given x. This saves tremendous amount of time and efforts while evaluating new designs. One can imagine the computationally heavy and high fidelity FEA or CFD models/simulations to be similar to physical experiments, and one can benefit by creating a simple mathematical approximation model based on what can be called as computer experiments (simulations) of few design points. This mathematical function can be used as a substitute for the actual heavy simulation (e.g., FEA/CFD) during optimization studies. Notice that here we have used the mathematical function as a surrogate (substitute) for the actual simulation. Hence, the name 'surrogate modeling' is being used. While CFD and FEM codes are models of reality, a surrogate model can be seen as model of a model within the bounds of experiments. When using surrogate models, the process is to predict the value of outputs based on current

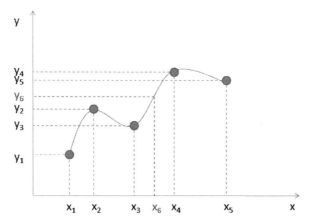

FIGURE 3.7   Curve fitting (RSM/Surrogate model generation).

inputs without running the high fidelity model. Hence, there will always be some error in values calculated by surrogate models and values obtained after running high fidelity CFD/FEM simulations. As long as the error is small or within acceptable limits, it is feasible to use a surrogate model. If the errors are too high, it is advisable to first improve accuracy of the surrogate models before using them for any further studies. There are a number of ways of increasing accuracy of the surrogate models as described in Ref. [13], although this section does not cover them in detail.

RSM accuracy depends on several factors: the complexity of the output variation, the number of points in the original training set and the choice of the type of RSM. As an example, Figure 3.8 shows an RSM using the same data used in Figure 3.7, but with different algorithms to fit a curve to the dataset. Figure 3.8(a) shows that if a very simple approximation such as linear regression is used, it might not capture the information correctly resulting in 'underfitting.' On the other hand, if a very high order polynomial is used, it might result in 'overfitting' problem. In case of 'overfitting' Figure 3.8(b), the algorithm results in a fit with increased non linearity than the original model. It can be seen as a curve with unnecessary peaks and valley which might result in incorrect predictions of the output variables. Please note that the example described using Figures 3.7 and 3.8 contains one output and one input. However, it is possible to create a surrogate model for an output based on any number of inputs. For example, Figure 3.9, which looks very similar to Figures 3.7 and 3.8, contains two plots of surrogate models for cost and deflection respectively from the beam optimization example described earlier in the chapter. Both these models are based on two variables, height and width. The black dots seen in each chart are data points, which are used to create this approximation surface. A total of 500 data points are used to

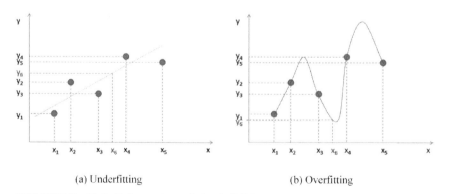

(a) Underfitting   (b) Overfitting

FIGURE 3.8   Underfitting and overfitting in RSMs.

create this surrogate model. A "3D surface" was modeled/fit to these data points. The continuous surfaces, which can be seen in both figures, are in principal, similar to the curve fit in Figure 3.7. Note that it is also possible to create a surrogate model based on more than 2 input variables. In that case, however, it becomes very difficult to visualize the response surface.

3.5 DIFFERENT SURROGATE MODELING TECHNIQUES

As is the case with optimization algorithms, there are many algorithms to create surrogate models. They range from simple regression models such as linear or quadratic equations to more sophisticated ones such as Artificial Neural Networks (ANN) [14–16], Kriging [17, 18], radial basis functions (RBF) [19–21] are some more examples of different surrogate modeling techniques. Many commercial tools such as MATLAB [10], modeFRONTIER [9] offer at least a few of these algorithms. It is advisable to create multiple RSMs for the same output variable using different algorithms and validate them one against another in order to select the best one, i.e., the RSM which best approximates the system analysis model. Figure 3.10 shows a response surface created for deflection variable in the beam optimization problem using a linear first order polynomial algorithm. The black dots represent real design points similar to Figure 3.9. It can be clearly seen that the polynomial response surface does not represent a very accurate approximation to the outputs.

It must be noted that, it is a very important and essential step to always check quality of surrogate model before making any further use of it. This is

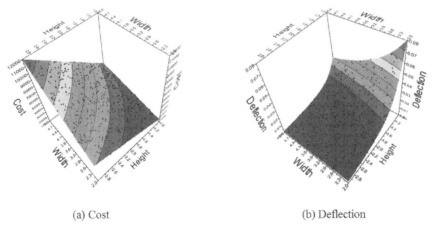

(a) Cost   (b) Deflection

FIGURE 3.9 Cost and deflection surrogate models and data points used to create the models.

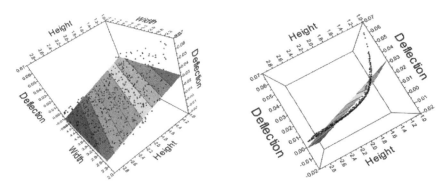

FIGURE 3.10   Example of a bad surrogate model.

because if the quality of the model is not good, there are high chances that the predicted values will be way off than actual values for the same set of inputs and it is likely to produce not so useful results during optimization.

If the quality of obtained RSM is unsatisfactory, there are many methods that can be adopted to improve its accuracy. Generating a well-distributed set of data points in the first place is a commonly used technique to obtain an accurate surrogate model. Generating more number of data points is one method, choosing a different type of algorithm is another one. There are also special algorithms called adaptive RSM algorithms where data points are iteratively added where the RSM error is high [22].

A few surrogate-modeling techniques are described briefly below.

3.5.1   POLYNOMIAL REGRESSION

As described in [23], a second order polynomial regression metamodel can be represented as:

$$\hat{y} = \beta_o + \sum_{i=1}^{k} \beta_i x_i + \sum_{i=1}^{k} \beta_{ii} x_i^2 + \sum_i \sum_j \beta_{ij} x_i x_j \tag{3}$$

In optimization, the smoothing capability of polynomial regression allows quick convergence of noisy functions [24]. There are many drawbacks when applying Polynomial Regression to model highly nonlinear behaviors. Using higher-order polynomials for problems with a large number of design variables may lead to instabilities [25]. Also, it may be too difficult to take sufficient sample data to estimate all of the coefficients in the polynomial equation in case of large dimensional problems.

3.5.2 KRIGING

Kriging is a Bayesian methodology named after professor Daniel Krige, which is the main tool for making previsions employed in geostatistics, e.g., for soil permeability, oil and other minerals extraction etc. The Kriging behavior is controlled by a covariance function, called a variogram, which rules how the correlation varies between the values of the function at different points. Kriging is particularly suitable for highly nonlinear responses and for virtual optimization. The Kriging is a computationally intensive method, so it could be slow or fail to converge on large datasets (>1000). For details on Kriging algorithm, the reader is referred to Refs. [17, 18].

3.5.3 RADIAL BASIS FUNCTIONS

Radial basis functions are a powerful tool for multivariate scattered data interpolation. Scattered data means that the training points do not need to be sampled on a regular grid. A radial basis function (RBF) is a real-valued function whose value depends only on the distance from the origin, so that

$$\phi(X) = \phi(\| X \|) \tag{4}$$

or alternatively on the distance from some other point c, called a *center*, so that

$$\phi(x, c) = \phi(\| x - c \|) \tag{5}$$

Radial basis functions are typically used to build up function approximations of the form

$$y(\chi) = \sum_{\iota=1}^{N} w_\iota \phi(\| \chi - \chi_\iota \|) \tag{6}$$

TABLE 3.1 Radial Basis Functions With Expressions

Function Name	Expression
Gaussian	$\phi(r) = e^{-(\varepsilon r)^2}$
Hardy's Multiquadrics	$\phi(r) = \sqrt{1 + (\varepsilon r)^2}$
Inverse Multiquadrics	$\phi(r) = \dfrac{1}{\sqrt{1 + (\varepsilon r)^2}}$

where the approximating function $y(x)$ is represented as a sum of N radial basis functions, each associated with a different center x_i, and weighted by an appropriate coefficient w_i. A set of commonly used Radial Functions are writing $r = (\| x - x_i \|)$.

3.5.4   EVONN ALGORITHMS

Evolutionary algorithms can also be used to train surrogate models more accurately. Evolutionary neural networks (EvoNN) are good examples of this. Although ANNs are very powerful and are used for solving a variety of problems, they still have some limitations. One of the most common limitations in ANN is associated with training the network. Usually a back-propagation algorithm is used for training the network. However, the back-propagation learning algorithm might converge to a local optimum. In real world applications, the back-propagation algorithm might converge to a set of sub-optimal weights from which it cannot escape. Hence, the ANN is often unable to find an accurate fit to the data. Another difficulty in ANNs is related to the selection of proper architecture of the network. The number of hidden layers in a neural network and also the number of neurons in each hidden layer is usually decided by intuition, experience or some previously set rules. In EvoNN algorithms, GAs are used to train the neural network instead of the standard back propagation algorithms. This helps avoid neural network getting stuck in a local optimum and hence increase its accuracy.

3.6   SURROGATE-ASSISTED EVOLUTIONARY ALGORITHMS

All steps described in previous sections can be put together into a surrogate-assisted optimization procedure. Figure 3.11 describes this optimization procedure. The first step is to generate a starting population or design samples using a suitable DoE technique. The DoE generation step involves creating different combination of input variables, which are well spread across the design space. Once the design samples are generated, the next step is to use these designs and run the analysis codes to generate output values for each DoE sample. After completing the DoE runs, the next step is to create response surfaces for each output using these data points. It is always better to create response surfaces using many different algorithms and select the best among them. Checking accuracy is a good way to select the best RSM.

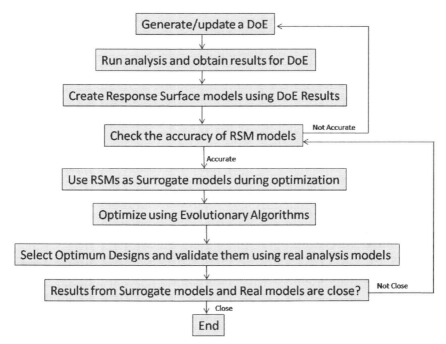

FIGURE 3.11 General optimization procedure using surrogate models.

If no RSM is accurate enough, the DoE data can be modified and more design points can be added/removed to improve accuracy. Good quality RSMs can be used as surrogate models during optimization. Many optimization algorithms can now be used to perform optimization. Once the optimization is finished, a few designs are selected from the best designs (Pareto optimal designs). Once these designs are selected, the real analysis can be run and the results can be validated. If the difference between outputs from real analysis and from surrogate model runs is small, the solutions are considered good. If the difference is large, it signals a need to check accuracy of surrogate models and rerun optimization.

Note that this is just one possible and quite commonly used procedure in surrogate-assisted optimization. Usually the surrogate model-building step has to be done manually. Recently, a lot of research has been focused on surrogate-assisted evolutionary computing methods. Loshchilov [27] highlights the growth in research publications on surrogate models and surrogate-assisted evolutionary computing methods. It is mentioned that, while the number of publications per year on surrogate-assisted evolutionary algorithms was in order of 10 s in 1990 s, it has grown to the order of

1000 s in 2012, representing the growing amount of attention being given to the field. Ong et al. [28] report various versions of surrogate-assisted evolutionary computing methods for engineering optimization problems. Ratle [29] proposes the use of Kriging interpolation for function approximation in order to speed up evolutionary optimization. El-Beltagy [30] also propose a similar framework combining Gaussian processes with evolutionary algorithms. Jin et al [23] present an approach to combine neural networks with evolutionary strategies and also compare the results using academic as well as real world problems. The RBF surrogate-assisted co-evolutionary search procedure as described in Ref. [31] is an attempt to tackle the 'curse of dimensionality,' which hampers the accuracy of surrogate modeling. Ong et al. [32] present an approach to combine local surrogate models based on radial basis functions to speed up evolutionary optimization search. Multivariate adaptive cross-validating Kriging (MACK) and Lipschitz [22] approaches available in modeFRONTIER [9] are aimed towards improving quality of surrogate models. These algorithms are capable of detecting regions of non-linearity in a response and intelligently add new design points in these regions so as to quickly improve surrogate model accuracy. A hybrid approach of combining various metamodel techniques with strength Pareto evolutionary algorithm (SPEA2) is described in Ref. [33]. The approach is also applied to various structural optimization problems to demonstrate the effectiveness. Although there has been considerable research in this field for several years, most of it focuses on small-scale problems (upto 10–15 design variables). Liu et al. [34] propose a Gaussian process surrogate model assisted evolutionary algorithm for medium-scale (i.e., 20–50 design variables) computationally expensive optimization problems. In order to tackle the curse of dimensionality, dimensionality reduction techniques are applied before building surrogate models [34]. The importance of managing surrogates is emphasized to prevent the evolutionary algorithms from being misled to a false optimum that can be introduced in a surrogate [35]. Sun et al. [36] have proposed a surrogate-assisted interactive genetic algorithm to handle uncertainties in fitness assignment by human beings. A surrogate-assisted mimetic co-evolutionary algorithm to handle expensive constrained optimization problems is presented in Ref. [37]. A cooperative co-evolutionary mechanism is adopted as the backbone of the framework to improve the efficiency of surrogate-assisted evolutionary techniques. The idea of random problem decomposition is introduced to handle interdependencies between variables, eliminating the need to determine the decomposition in an ad-hoc manner. Gambarelli and Vincenzi [38] propose an algorithm for dynamic structural identification problems. They combined robustness of differential evolution (DE) strategy with computation efficiency using a second-order surrogate approximation of objective function. Ray and Smith [39] put forward a surrogate-assisted evolutionary algorithm combining RBF metamodels with GAs.

3.6.1   ADAPTIVE RSM BASED OPTIMIZATION

Rigoni et al. [40] presented the idea of automatic adaptive RSMs used for evolutionary computing. The approach is capable of combining any of the polynomial regression, Kriging, RBF and ANN metamodeling techniques with evolutionary algorithms such as MOGA-II, NSGA-II, MOSA, etc. A brief description of the implementation of algorithm in commercial software modeFRONTIER [9] is given below. The process of generation of RSM and optimization as well as validation can be iterated in an automated way by using FAST optimization algorithms as described in the Ref. [40]. At every iteration, new designs evaluated by real optimization are used for creating new training points for RSM. This method helps in generation of accurate metamodels in an automated and adaptive way.

In an iterative loop, after evaluation of DoE, RSMs are created on the evaluated design database. The RSM is used for function evaluation for a complete optimization run. Few good solutions are selected from the optimization run and these designs are evaluated using the real function evaluation. Next step in the iterative loop is to virtually explore region around the obtained real Pareto optimal designs. Some points are also selected from the virtual exploration phase to form the new validation set. The validation set, which consists of points picked from optimization run and exploration run are appended to the previous real runs to form the training set for RSM. Hence, every iteration the RSM keeps improving not just in the optimal zone but also in the overall design space to cover global search. Thus in each iteration better designs are obtained, continuing till the algorithm reaches total number of generations. In modeFRONTIER [9], such an adaptive RSM based strategy combined with various optimization algorithm is called FAST. For example, FAST strategy combined with MOGA-II optimizer will be termed FMOGA-II (FAST-MOGA-II).

To demonstrate the effectiveness of this automatic adaptive RSM technique, a connecting rod optimization problem is taken as an example case as described

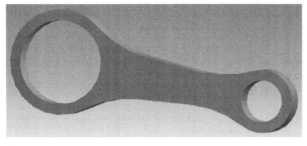

Model Parameters
1.Upper Fillet Radius [1 , 100] mm
2.Lower Fillet Radius [1 , 150] mm
3.Thickness [5 , 30] mm
4.Upper Width [3 , 10] mm
5.Lower Width [3 , 10] mm

Objectives:
1.Min Mass
2.Min Stress

FIGURE 3.12   Connecting Rod Optimization.

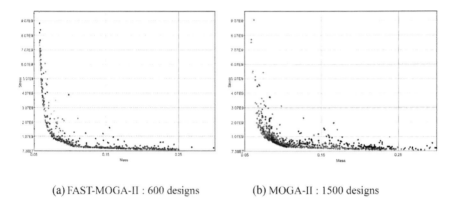

(a) FAST-MOGA-II : 600 designs   (b) MOGA-II : 1500 designs

FIGURE 3.13   Comparison of results from surrogate-assisted evolutionary algorithm and evolutionary algorithm.

in Figure 3.12. Commercial tools ANSYS [41] and modeFRONTIER [9] have been used for structural analysis and optimization respectively. Figure 3.14 shows the history of optimization run in a two dimensional scatter chart with the design number plotted on a color scale. Newer designs are depicted by red color while earlier designs are closer to blue color. As can be seen from Figure 3.13, the combination of surrogates and evolutionary algorithm FMOGA-II (FAST-MOGA-II) is able to find a much better solution in just 600 total evaluations as compared to the ones found by traditional MOGA-II in 1500 iterations.

3.7   EXAMPLE PROBLEM

Figure 3.14 shows a three bar planar truss hinged at node 1. It is constrained in y direction at node 2 and constrained at 45° angle at node 3. A constant force of 1000kN is applied at node 2 in horizontal direction. Modulus of elasticity of the material is 210GPa. Elements 1 and 2 are of length 1 m

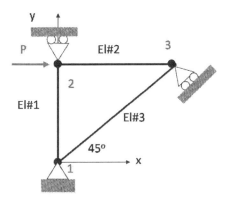

FIGURE 3.14   Planar truss optimization problem.

each while element 3 has a length of 1.41 m. The lengths of all elements are fixed. Each element has a rectangular cross-section. The width and breadth of every member of the truss can vary between 0.01 m to 0.06 m. The objective is to find the width and breadth of each element so that the mass of the truss is minimized and the displacement at node 2 is also minimized.

The CD with this book contains a solution to the above planar truss optimization problem. It contains an excel sheet which will guide you step by step through the optimization procedure described in Figure 3.11. Learners are suggested to try creating different DoEs, RSMs and optimizing the problem so that you can understand the optimization procedure.

3.8   CONCLUSION

This chapter has reviewed basics of single and multiobjective optimization as well as the concept of a Pareto Optimal Frontier. The concept behind evolutionary algorithms and Genetic algorithms in particular has been discussed. The notion of surrogate models as well as the need for surrogate modeling and different types of surrogate modeling techniques has been described. An extensive review of current applications of surrogate-assisted evolutionary algorithms is presented. Finally, an adaptive RSM based optimization algorithm is described in more details and results are compared with a traditional evolutionary algorithm.

In future, attention could be focused on reviewing the application of surrogate-assisted evolutionary algorithms for real world problems. Also, emphasis has to be given to create more accurate surrogate models with fewer samples in order to reduce the computational effort as well as to increase the reliability of the algorithms.

KEYWORDS

- **multiobjective optimization**
- **Pareto Optimal Frontier**
- **single optimization**

REFERENCES

1. Sobieszczanski-Sobieski, J., "A linear decomposition method for optimization problems – blueprint for development." NASA TM 83248, Feb. 1982.
2. Vanderplaats, G. N., "Numerical optimization techniques for engineering design: With applications." Vol. 1. New York: McGraw-Hill, 1984.
3. Powell, M. J. D., "A fast algorithm for nonlinearly constrained optimization calculations," in: G. A. Watson (Ed.), Numerical Analysis, Dundee, Springer-Verlag, Berlin, 1977, pp. 144–157.
4. Deb, K., Pratap, A., Agarwal, S., & Meyarivan, T., "A fast and elitist multiobjective genetic algorithm: NSGA-II." IEEE Transactions on Evolutionary Computation, April 2002, 6(2), 182–197.
5. Kirkpatrick, S., Gelatt, C. D., & Vecchi, M. P. "Optimization by simulated annealing." Science 1983, 220(4598), 671–680.
6. Reyes-Sierra, M., & Coello, C. A. C., "Multiobjective particle swarm optimizers: A survey of the state-of-the-art." International Journal of Computational Intelligence Research, 2006, Vol. 2, No. 3, pp. 287–308.
7. Poles, S., Rigoni, E., & Robic, T., "MOGA-II performance on noisy optimization problems." International Conference on Bioinspired Optimization Methods and their Applications, Ljubljana, Slovenia. 2004.
8. Goldberg, D. E., "Genetic Algorithms in Search, Optimization and Machine Learning," Addison-Wesley, Reading Mass, USA, 1988.
9. modeFRONTIER. http://www.esteco.com/ (accessed March 25, 2015).
10. MATLAB. http://in.mathworks.com/ (accessed March 25, 2015).
11. Rigoni, E., & Poles, S. "NBI and MOGA-II, two complementary algorithms for Multiobjective optimizations." Practical Approaches to Multiobjective Optimization, 2005 (04461).
12. Vincent, T. L., "Game theory as a design tool." Journal of Mechanical Design, 1983, 105(2), 165–170.
13. Xue, Z. L. R., Rigoni, E., Parashar, S., & Kansara, S., "RSM improvement methods for computationally expensive industrial CAE analysis." 10th World Congress on Structural and Multidisciplinary Optimization, May 19–24, 2013, Orlando, Florida, USA.

14. Hagan, M. T. A. M., "Training feed forward networks with the Marquardt algorithm," IEEE Trans. on Neural Networks, 1994, Vol. 5(6).

15. Haykin, S., "Neural Networks—A comprehensive foundation." Prentice Hall International Editions, 1999.

16. Nguyen, D. A. W. B., "Improving the learning speed of 2-layer neural networks by choosing initial values of the adaptive weights." IJCNN International Joint Conference on Neural Networks, San Diego, CA, USA, 1990.

17. Matheron, G., "Les variables régionalisées et leur estimation: une application de la théorie des fonctions aléatoires aux sciences de la nature." Masson, Paris, France, 1965.

18. Rasmussen, C. E. A. W., "Gaussian processes for machine learning," MIT Press, 2006.

19. Buhmann, M. D., "Radial basis functions: Theory and implementations." Cambridge University Press, Cambridge, UK, 2003.

20. Iske, A., "Multiresolution methods in scattered data modeling." Springer-Verlag Berlin Heidelberg, 2004.

21. Wendland, H., "Scattered data approximation," Cambridge University Press, 2004.

22. Alberto, L. E. R., "Adaptive sampling with a Lipschitz criterion for accurate metamodeling." Communications in Applied and Industrial Mathematics 2010. 1, No. 2, DOI: 10.1685/2010CAIM545.

23. Jin, R., Chen, W,, & Simpson. T. M., Comparative studies of metamodeling techniques under multiple modeling criteria. Structural and Multidisciplinary Optimization 2001, 23(1), 1–13.

24. Giunta, A. A., Dudley, J. M., Narducci, R., Grossman, B., Haftka, R. T., Mason, W. H., & Watson, L. T., "Noisy aerodynamic response and smooth approximation in HSCT design," Proceedings Analysis and Optimization (Panama City, FL), Vol. 2, AIAA, Washington, DC, 1994, pp. 1117–1128.

25. Barton, R. R., "Metamodels for simulation input-output relations," Proceedings of the 1992 Winter Simulation Conference (Swain, J. J., et al., eds.), Arlington, VA, IEEE, 1992, December 13–16, pp. 289–299.

26. Queipo, N. V., Haftka, R. T., Shyy, Wei, Goel, T, Vaidyanathan, R, Tucker, P. K., "Surrogate-Based Analysis and Optimization." Progress in Aerospace Sciences 2005, 41, 1–28.

27. Loshchilov, I. "Surrogate-sssisted evolutionary algorithms. optimization and control." Universite Paris Sud—Paris XI; Institut national de recherche en informatique et en automatique—INRIA, https://tel.archives-ouvertes.fr/tel-00823882/document, accessed on March 25, 2015.

28. Ong, Y. S., et al., "Surrogate-assisted evolutionary optimization frameworks for high-fidelity engineering design problems." Knowledge incorporation in evolutionary computation. Springer Berlin, Heidelberg, 2005, 307–331.

29. Ratle, A., Kriging as a surrogate fitness landscape in evolutionary optimization." AI EDAM January 2001, Vol. 15, Issue 01, pp. 37–49.

30. El-Beltagy, M. A., & Keane, A. J. "Metamodeling techniques for evolutionary optimization of computationally expensive problems: Promises and limitations." Proceedings of the Genetic and Evolutionary Computation Conference (GECCO-99), Orlando, USA, 13–17 Jul 1999.

31. Ong, Y., Keane A. J., & Nair P. B., 2002, "Surrogate-assisted co-evolutionary search." Neural Information Processing, 2002. ICONIP'02. Proceedings of the 9th International Conference on. Vol. 3. IEEE.

32. Yew S, O., Nair P. B., & Keane A. J., "Evolutionary optimization of computationally expensive problems via surrogate modeling." AIAA Journal 2003, 41(4) 687–696.

33. Kunakote, T., & Sujin Bureerat, "Surrogate-assisted multiobjective evolutionary algorithms for structural shape and sizing optimization." Mathematical Problems in Engineering, 2013.

34. Liu, B., Qingfu Zhang, & Georges Gielen, "A Gaussian process surrogate model assisted evolutionary algorithm for medium scale expensive optimization problems." IEEE Transactions on Evolutionary Computation, April 2014, Vol. 18, No. 2, pp. 180–192.

35. Jin, Y. "A comprehensive survey of fitness approximation in evolutionary computation." Soft Computing 9(1), 3–12.

36. Sun, X., Gong, D., & Jin Y., "A new surrogate-assisted interactive genetic algorithm with weighted semisupervised learning." IEEE Transactions on Cybernetics 2013, 43(2), 685–698.

37. Goh, C. K., Lim, D., Ma, L., Ong, Y. S., & Dutta, P. S., "A surrogate-assisted mimetic co-evolutionary algorithm for expensive constrained optimization problems." IEEE Congress on Evolutionary Computation (CEC), 5–8 June 2011, New Orleans, LA.

38. Gambarelli, P., & Vincenzi, L. "A surrogate-assisted evolutionary algorithm for dynamic structural identification." Engineering Optimization 2014, 93–98, DOI: 10.1201/b17488-18.

39. Ray, T., & Smith W., "Surrogate-assisted evolutionary algorithm for multiobjective optimization." Engineering Optimization, 2006, Vol. 38, Issue 8, pp. 997–1011.

40. Rigoni, E., Metamodels for fast multiobjective optimization: Trading of global exploration And local exploration, simulated evolution and learning," 8th International Conference, SEAL 2010, Kanpur, India, December 1–4, 2010.

41. *ANSYS*. Available from: http://www.ansys.com/ (accessed March 25, 2015).

CHAPTER 4

EVOLUTIONARY ALGORITHMS IN IRONMAKING APPLICATIONS

TAMOGHNA MITRA,[1] HENRIK SAXÉN,[1] and
NIRUPAM CHAKRABORTI[2]

[1]*Thermal and Flow Engineering Laboratory, Åbo Akademi University, Biskopsgatan 8, FI-20500 Åbo, Finland*

[2]*Department of Metallurgical and Materials Engineering, Indian Institute of Technology, Kharagpur, 712 302, West Bengal, India*

CONTENTS

4.1   BACKGROUND

Iron is probably the single most important metal for the world's industrial economy. In nature iron is mostly available in the form of oxides (mainly hematite, Fe_2O_3, and magnetite, Fe_3O_4) or hydroxides ($Fe(OH)_x$). Ironmaking refers to a number of processes, which are implemented for extracting metallic iron from these oxides, predominantly using a reductant like carbon monoxide. In order to cater to the various applications of this metal, the metallic iron is often alloyed with other elements, which can improve properties specific for an application. These alloys are known as steel and the process is called "steelmaking." Typically, steel is an iron-carbon alloy of variable composition, where other alloying elements are added to obtain various useful properties. For example, to improve the corrosion resistance, chromium is added to iron and the product is commercially available as stainless steel. World average steel use per capita has been steadily increasing. In the last decade it has increased from 150 kg in 2001 to 225 kg in 2013 [58].

Increasing global competition and a need for a relatively pollution free production now necessitates considerable optimization tasks in the ferrous industry. The application of classical optimization routines is, however, often cumbersome and sometimes impossible for a real-life ironmaking process, due to their inherent complexity. Although first principles models have been applied in such optimization [40], they have proven to have several limitations for this use, and optimization in tandem with data-driven modeling is increasingly emerging as powerful alternate [22]. In this context biologically inspired genetic and evolutionary algorithms [56] play a major role. This chapter aims to provide a state-of-the-art overview of the use of evolutionary techniques for optimization of ironmaking. In the next section the background of the ironmaking process is presented. The section contains a brief description of the history of the process and some facts about the modern process. Next, a brief description of different optimization methods applied to the ironmaking process is provided. Subsequently, the major process units in the ironmaking process, which form the focus of optimization, are described, followed by a presentation of various applications pertaining to evolutionary algorithms. Finally, conclusions are drawn regarding the studies carried out in the field and some future prospects of these methods are suggested.

4.2   IRONMAKING PROCESS

4.2.1   HISTORY

Earliest ironmaking dates back to the unrecorded part of human history, and therefore the origin has been claimed by different civilizations all over the world. The earliest use of this metal was in form of weapons or tools with superior strength compared to that of bronze. Earliest ironmaking techniques involved burning of the iron ore with wood in a covered oven. It was difficult to reach the melting point of the iron, but it separated the impurities that formed a crust on the pure iron. This crust could be hammered off to get iron. The process was gradually improved by blowing in air to increase the temperature and the production rate. These furnaces were known as bloomeries and were extensively used in early industrial age. Since then the production process has been improved steadily and the use has been extensively diversified [14].

4.2.2   MODERN IRONMAKING

Reduction of iron ore in modern industries is carried out predominantly through two alternative routes. The first option is the smelting route where the ore is reduced, then the melted, and finally stored or transported to the steelmaking units for steel production. The other option is to reduce the ore in solid form using reducing gases in lower temperatures. This is called direct reduction and the end product is known as Directly Reduced Iron (DRI).

The blast furnace route is one of the most important ironmaking smelting techniques and has existed for more than 500 years and accounts for about 70% of the iron used for crude steel production, which was 1164 million tons in 2013. This showed a 73% increase from the 669 million tons in 2003 [59, 60]. The most important reason for the success compared to alternative techniques has been fuel efficiency, productivity and scalability of the process. There are about 21 furnaces in the world, which have a volume of more than 5000 m³ and the largest of them till date is at POSCO, South Korea, with a volume of 6000 m³ and annual production capacity of 5.48 million tons of pig iron [21]. The other and more recent alternatives of producing liquid iron from ore include the COREX, FINEX and Hismelt processes [28], which are steadily finding their niche in ironmaking smelting techniques. It should be stressed that the remaining 30% of the steel production mostly uses recycled steel scrap or DRI as feed material. Midrex, HYL and Rotary kiln are the most important choices for producing DRI [14]. In 2013, world DRI production amounted to 70 million tons 2003 [60].

4.2.3 PROCESS OPTIMIZATION

The operation of an ironmaking blast furnace is extremely challenging due to complex interactions between different concurrent physical and chemical processes, but there are a limited number of variables, which may be directly influenced by the operator. With increasing demand of steel in the world, higher quality requirements and decreasing availability of resources, increasing material costs and enormous environmental challenges, optimization of different parameters involved in the process in parts and in entirety becomes extremely important for achieving an efficient production of good quality steel.

The earliest optimization methods applied to the process were seen in the late 1950 s [12] for optimizing the production sequences in an integrated steel plant. Until this time very little was known about the conditions inside a blast furnace during operation, but a series of dissections of quenched furnaces in Japan provided a deeper understanding which led to the development a large number of mathematical models [38].

In the next decades along with the advent of computers, modeling became an increasingly important tool for scientists to gain a deeper understanding of the various interrelated and extremely complicated processes involved in the production of iron and steel. Most of the early work was based upon analytical closed-form expressions, which required extensive derivation and computation and were not well suited for optimization. Interestingly, some of the problems still remain for advanced modeling techniques, such as discrete element modeling, DEM, which require computationally prohibitive efforts [2].

However, by the early 1990 s a large number of soft computing techniques, such as Artificial Neural Networks and fuzzy logic, were used to describe ironmaking processes. These models provided decent accuracy and were computationally tractable and could, furthermore, address the strong nonlinearities of the process [18].

In the later years, raw material shortages and awareness about the climate change threat caused by carbon dioxide emissions became key drivers for the industry to focus intensely on operating at high efficiency and low carbon rate. The evolutionary algorithms were already being used for various real life problems in the domains of classification, fuzzy controllers, scheduling, etc. By the turn of the century as the computational hardware grew more powerful, computationally demanding optimization techniques like those of evolutionary algorithms were being increasingly used to solve problems in industry. Evolutionary algorithms provide a distinct advantage over other optimization techniques due to their versatility and simplicity. The biggest issue with the gradient-based methods is that they, by contrast to evolutionary algorithms, do not work on non-differentiable cases, which form a big section of the real life problems and they cannot efficiently tackle problems with multiple minima. Evolutionary approaches may also be used for multiobjective optimization, where more than one objective should be optimized simultaneously, and the strategies that are available for this purpose are now very efficient and far less cumbersome than the existing gradient based methods [7, 29].

4.3   PROCESS UNITS

4.3.1   BLAST FURNACE

In an integrated steel plant, the main unit processes are a coke plant, one or several blast furnace, basic oxygen furnaces, reheating furnaces and a rolling mill. In addition to these, each blast furnace has a set of regenerative heat exchangers for heating the combustion air (termed "blast") and the steelmaking site also has a power plant where residual gases are burned to produce power (and heat). These residual gases are mainly coke oven gas, blast furnace gas and basic oxygen furnace gas, part of which is used for internal heating of hot stoves, for ladle preheating and for steel slab reheating purposes [62].

The blast furnace is a counter current heat exchanger for smelting of iron oxide using an oxygenated blast. Figure 4.1 shows the cross-section of a typical blast furnace. The raw materials, i.e., iron ore (often in preprocessed form, as sinter or pellets), coke and limestone are charged into the furnace from the top. The iron ore, which is mainly hematite (Fe_2O_3) is reduced mainly by the ascending CO produced when the charged coke is oxidized by the hot blast in the lower part of the furnace. The reduction of iron ore occurs in three steps to obtain the final hot metal, which mainly consists of liquid metallic iron.

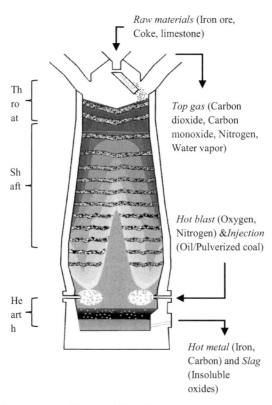

FIGURE 4.1   Cross-section of a typical blast furnace.

$$Fe_2O_3 \rightarrow Fe_3O_4 \rightarrow FeO \rightarrow Fe \qquad (1)$$

Production of a ton of hot metal requires about 500 kg of reductants, of which coke is the major part. Coking coal is the second most expensive component of the raw materials after iron ore [53]. To reduce the amount on coking coal, hydrocarbons, such as pulverized coal, plastic, oil etc., are injected into the furnace along with the hot blast.

The components of the raw materials are charged in the form of layers. These layers slowly descend into the furnace as they are "consumed" (melted, gasified, or combusted) in the lower parts of the furnace. The burden distribution plays a major role for a smooth operation of the furnace [47]. At the center of the lower parts of the furnace a "deadman" of slowly reacting and dissolving coke is formed. This coke bed, which sits in a pool of molten slag and hot metal, supports the upper burden layers. The hot metal and the slag are tapped at regular intervals. The slag is usually cooled and sold as a byproduct for cement industries, insulation etc. [13], whereas the hot metal is transported to the steel shop to be post processed further.

4.3.2   ROTARY KILN

Rotary kiln furnaces (Figure 4.2) are used for producing DRI using reducing gases, mainly carbon monoxide and hydrogen. The raw materials mainly consist of iron ore and coal, which are fed continuously from one end and the DRI is discharged from the other end of the cylindrical reactor. Air is introduced into reactor and reacts with the coal to produce the reducing gases. The cylinder rotates around its axis and mixes the content in the furnace.

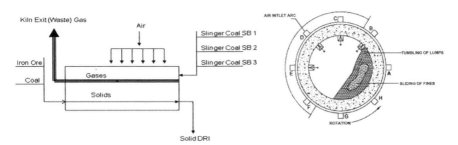

FIGURE 4.2   Left: Schematic of a rotary kiln. Right: Cross-section of the cylinder.

4.4   SOFT COMPUTING TECHNIQUES

4.4.1   ARTIFICIAL NEURAL NETWORKS

Artificial Neural Networks (ANN) are computational tools inspired by the action in biological nervous systems, where neurons are the basic decision making entities (Figure 4.3). ANN, which can be for function approximation, classification, data processing, pattern recognition, control, is a very good tool for creating data-driven computational models of complex systems.

In nature the neurons distribute electrical pulses along their connections and transfer information to their neighboring neurons, if the received impulse exceeds a certain threshold. Similarly, computational neuron (i) transmits a signal (y_i), which is a function of the accumulated incoming signals (y_i). The action of a node can be represented mathematically as

$$y_i = f(a_i) \tag{2}$$

where the input is a weighted sum of external inputs or outputs from other nodes.

$$a_i = \sum_{j=0}^{n} w_{ij} x_j \tag{3}$$

Weights of an input x_j to node i are denoted by w_{ij}. The activation function f_j may be any function, e.g., linear, piecewise linear, sigmoidal, etc. The output from a node can serve as output from the system to the environment or as an input to node in the next layer, together forming a network as illustrated in

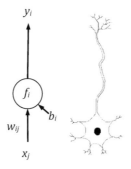

FIGURE 4.3   Left: Computational neuron i. Right: Biological neuron.

Figure 4.4. The input and the output layers correspond to the set of nodes, which receive the input signals to and generate the output signals from the system. The layers between these two layers communicating with the environment are called "hidden layers." Each such layer receives as inputs the output signals from the previous layer and transmits the outputs to the next layer. The number of nodes in the hidden layers, as well as the number of hidden layers, are internal model parameters which affect the performance of the network. The weights of all connections and the activation functions and biases of all nodes together define the ANN.

For example, the simple networks shown in Figure 4.5 realize an AND and an OR operation between two inputs, x_1 and x_2, described in Table 4.1. In this network $n = 1$, $i = 3$ and therefore $y = x_3$. The activation function for the output node (node 3) is in this case a threshold function defined by h, the Heaviside function

$$f_3 = x_3 = h(a_3) = \begin{cases} 1 \text{ if } a_3 \geq 0 \\ 0 \text{ otherwise} \end{cases} \tag{4}$$

The values of the weights and biases in more complex networks is difficult to determine intuitively, so the networks must be "trained" by an algorithm to estimate the weights which would satisfy the goal of the task. There are several algorithms available for training neural networks. The most common of is the back propagation algorithm, which was used extensively after

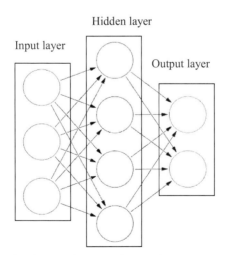

FIGURE 4.4   Schematic of an artificial neural network.

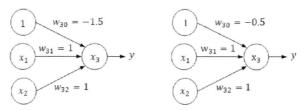

FIGURE 4.5 ANN implementation for AND (left) and OR (right) operators.

TABLE 4.1 Truth Table Showing the Working of the Networks in Figure 4.5

x_1	x_2	AND		OR	
		$w_{30} + w_{31}x_1 + w_{32}x_2$	x_3	$w_{30} + w_{31}x_1 + w_{32}x_2$	x_3
0	0	−1.5	0	−0.5	0
0	1	−0.5	0	0.5	1
1	0	−0.5	0	0.5	1
1	1	0.5	1	1.5	1

the publication of the paper by Rumelhart et al. [48], but higher order methods (such as the Levenberg-Marquardt method) are commonly used today because of their higher efficiency.

Neural networks are good tools for function approximation. According to the universal approximation theorem proved by Cybenko [9] for sigmoid activation function, a feed-forward network with a single hidden layer containing a finite but sufficient number of neurons can approximate any continuous (twice differentiable) function to any accuracy. However, in practice the "curse of dimensionality" plagues such approximators and often makes their use problematic: As the dimensions of a problem increases the number of nodes that are required to approximate the function increases exponentially. Along with this need, the possibility of over-fitting the data, incorporating errors in the measurements and noise, increases.

4.4.2 EVOLUTIONARY ALGORITHMS

The previous chapters of this book describe the general evolutionary algorithms in detail. Therefore, only some specialized evolutionary algorithms, which have been used in ironmaking processes are described briefly in the following paragraphs.

4.4.2.1 Evolutionary Neural Network Algorithm

Pettersson et al. [41] introduced the *Evolutionary Neural Network Algorithm* (EvoNN), which used multiobjective genetic algorithms to achieve optimum combination of network complexity and network accuracy so that noisy data from a physical process can be represented by a neural network without the risk of over-fitting or under-fitting. The multiobjective part of the problem used an adapted version of the Predator Prey algorithm as suggested by Li [27]. Each prey in the neighborhood matrix represents a network describing a network with a certain number of nodes and the weights for each connection. The crossover operation in this algorithm involves swapping of similarly positioned hidden nodes of two individuals with a certain probability, as schematically depicted in Figure 4.6. The mutation of any weight w_{ij}^{m} of a population member m results in

$$w_{ij}^{m\mu} = w_{ij}^{m} + \lambda\left(w_{ij}^{k} - w_{ij}^{l}\right) \tag{5}$$

where $w_{ij}^{m\mu}$ is the mutated weight and k and l are randomly chosen population members. The extent of mutation is governed by the second half of the equation.

A fitness function $(\varphi_{p,q})$ is computed for predator q evaluating a prey p to decide if the prey stays alive in the next generation.

$$\varphi_{p,q} = \chi_q E_p + \left(1 - \chi_q\right) N_p \tag{6}$$

where E_p is the training error of the ANN and N_p is the number of active connections joining the input layer to the hidden layer. In Eq. (6) χ_q is a weight value attributed randomly to the predator q such that $\chi_q = [0, 1]$.

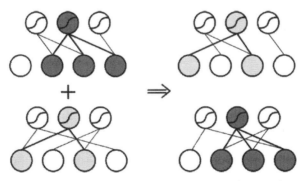

FIGURE 4.6 The crossover mechanism in Evolutionary Neural Network.

The authors used this algorithm to train a neural network using data from an actual blast furnace operation for prediction of changes in the carbon, silicon and sulfur content of hot metal. The right panel of Figure 4.7 presents the Pareto front, expressing the minimum training error as a function of the number of weights in the networks, obtained at the end of the training. The left panel of the figure depicts the networks' predictions and the measured changes in the silicon content for three different individuals at different positions of the Pareto front (labeled A, B and C). Each of the results fit the data to various extents and the best solution is often chosen on the basis of a desired tradeoff between goodness of fit and the complexity of model. A very sparse network (e.g., A) would execute fast but would fail to reproduce the data accurately, whereas a very dense network (e.g., C) would tend to reproduce many peaks and valleys in the data and therefore also fit noise. The choice of the network in such cases should depend much on the quality and quantity of training data.

4.4.2.2   Genetic Programming

Genetic programming [7, 45] is a powerful alternative to neural networks for creating a data-based model of any system. As mentioned above, neural networks are flexible approximators, but they are limited by the overall structure of the function. Genetic Programming (GP), on the other hand, provides the flexibility of selecting the pertinent mathematical operators for modeling

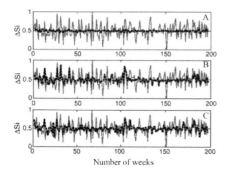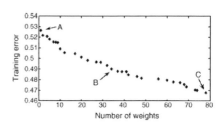

FIGURE 4.7   Network training from Pettersson et al. [41]. Right: The Pareto front at the end of the simulation. Left: Darker lines each show the simulation results predicting the weekly change in silicon levels in hot metal using the networks from the Pareto front using three individuals A, B and C indicated in the right panel. Lighter line is the industrial measurement over a period of 200 weeks.

a particular system [23]. GPs are similar to genetic algorithms in that they apply operators like crossover, mutation and selection, though they need to be reformulated as GP essentially uses a tree representation rather than binary or real encoding. Here the tree structure is a representation of a candidate function linking the variables in the optimization problem. Figure 4.8 represents one such possible candidate function linking the four variables a, b, c and d using the operators addition (+), subtraction (−), multiplication (*) and division (/). The aim of this optimization routine would be to find the most representative tree structure for a particular set of data points.

Each node in the tree (Figure 4.8) represents a member of the function set and the terminal nodes represent the input variables or constants. The standard crossover operation between two individuals of the population in practice means exchanging two sub-trees from each of them. Another form of crossover is height fair crossover, when the sub-trees at selected depths are exchanged. The standard mutation operator deletes a sub-tree and then grows it randomly. Small mutation operator replaces a terminal and perturbs a numerical constant in the terminal slightly. Sometimes, this operator also exchanges one of the operators with another randomly chosen operator of same parity. A mono-parental exchange operator for mutation exchanges two sub-trees belonging to the same tree. The selection technique is usually similar to that of a genetic algorithm, though a large population size and a large tournament size are usually employed and the calculations continue for a lesser number of generations, as compared to genetic algorithms. The fitness is usually evaluated as the root mean square of the error of the tree to represent the actual data.

Kovačič et al. [24] used genetic programming to predict the particulate matter (PM10) concentrations at a couple of residential areas close to a steel

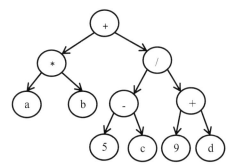

FIGURE 4.8 A typical genetic programming tree representing function $(a.b) + ((5 − c)/(9 + d))$.

plant as a function of production and meteorological conditions. These factors formed the terminal nodes for each of the solutions [weekday, month, wind speed (m/s), wind direction (°), air temperature (°C), rainfall (ml), electric arc furnace efficiency (min/hour)]. The other nodes were chosen from the function set consisting of addition, subtraction, multiplication and division. The month and weekday were considered to reflect the influence of seasons and traffic, respectively. Each individual tree from a population 500 individuals was evaluated by comparing the prediction with the measured PM10 concentration. The population was then modified using some genetic operations to generate the next generation. After about 100 generations the best model was selected for the particulate matter prediction with a mean deviation of approximately 5 $\mu g/m^3$.

Bi-Objective Genetic Programming (BioGP): This is a special variant of the GP proposed by [15, 16] which tackles different issues in the evolution in GP trees along with EvoNN. The problematic issues addressed include unmanageable growth of a sub-tree or growth without any improvement in accuracy. In BioGP, each tree representing a solution evolves in parts as distinct sub-trees, which are aggregated with some weights and biases. These weights and biases are optimized with a linear least square algorithm. A variable measures the performance of each sub-tree and if any of the sub-trees show issues while evolving, then they are pruned and regrown. The individual solutions are then evolved similar to EvoNN using predator-prey algorithm. The complexity of the network and the error in prediction are minimized using this bi-objective GA.

4.4.2.3 Differential Evolution

Differential evolution is an evolutionary algorithm for real coded genetic algorithm and is an algorithm with fast convergence and reasonable robustness [26, 54]. This algorithm is especially suited for problems with a large number of variables. Mapping of all the variables in binary format for binary coded GAs creates a Hamming Cliff problem [10]. This problem states that in situations where the binary coded chromosome is made of a large number of genes, any small change in the real space requires a significantly larger change in binary chromosome. Therefore, it takes larger time to obtain the solution and often makes it impossible to reach the optimal solution despite being at a near optimal scenario.

The algorithm starts with initialization of the population and then proceeds to mutation, crossover and selection. The last three steps are repeated until the population converges to minima in the objective function. The key element distinguishing differential evolution from genetic algorithms is its mutation technique. Each mutated individual x_i^μ is given by

$$x_i^\mu = x_i + F_m \left(x_a - x_b \right) \tag{7}$$

where F_m is a constant and x_a and x_b are randomly chosen individuals from the population. This mutation is self-adjusting in nature: If the population is very diverse, the mutation is quite large, but as the population approaches the optimum solution the mutation decreases. F_m is here a constant, which can be adjusted to expand or contract the search space.

For example, in phase equilibria research the calculation of structures of various compounds has had very far-reaching consequences. However, it is difficult to ascertain the minimum energy configuration among atoms especially for bigger molecules. The objective in such cases is to find the ground state (minimum total energy) for a system of atoms. It is a function of the sum of potentials for each pair of atoms and can therefore have a large number of variables. Thus, the optimization problem is extremely challenging but it can be readily tackled by differential evolution. Chakraborti et al. [6] used this method for evaluating the minimum energy configuration of some Si-H clusters. Figure 4.9 shows the ground state of a Si_6H cluster calculated using differential evolution.

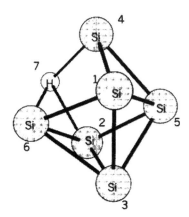

FIGURE 4.9 Final configuration with minimum energy for Si_6H molecule evaluated using differential evolution.

4.5   APPLICATIONS IN IRONMAKING

4.5.1   MODELING OF SILICON CONTENT IN HOT METAL

Silicon enters the blast furnace as an impurity in iron ore and coke, mainly in the form of silicates. Typically the silicon content is about 3.5% in sinter and 2% in pellets and in coke the value is 5–6% as the main component in coke ash is silica. The hot metal silicon content is an indicator of the thermal state of the process; high silicon content indicates a surplus of coke (high energy reserve) in the blast furnace. Therefore, optimum operation requires low hot metal silicon content, still avoiding cooling down the hearth, which can have catastrophic consequences. Modeling of the silicon content in the hot metal is extremely difficult because of the complexity of the silicon transfer processes and the flow conditions in the furnace hearth. A lack of direct measurements of the internal conditions in the lower furnace further complicates the modeling.

A number of data-driven models for prediction of hot metal silicon content in the blast furnace have been presented based on neural networks, fuzzy logic, support vector machines, etc. [49]. These models typically correlate the hot metal silicon with inputs such as blast volume, temperature and pressure, hydrocarbon injection levels, overall pressure loss and ore-to-coke ratio. Sheel and Deo (1999) created a linear fuzzy regression model on blast furnace data, applying a genetic algorithm to determine the fuzzy coefficients of the linear equations describing the silicon content. This method was found to be much better than the linear programming approach.

Time series approach: In the time series approach, models also consider the input values of previous time steps. This is especially important in silicon prediction, as the levels tend to be a manifestation of a large number of complex processes. In this case the silicon level is given by,

$$\text{Si}(t) = f\left(\mathbf{x}(t), \mathbf{x}(t-1), \ldots, \mathbf{x}(t-n)\right) \tag{8}$$

where $t-1, t-2, \ldots, t-n$ are the n preceding time steps.

Pettersson et al. [41] and Saxén et al. [51] used an evolutionary neural network algorithm for creating a model from noisy data from a blast furnace by correlating the real time input variables with the silicon level measured in the hot metal. The problem of selecting model complexity and input variables were thus tackled simultaneously, and the results demonstrated that parsimonious models could be evolved by the approaches.

4.5.2 OPTIMIZATION OF BURDEN DISTRIBUTION IN BLAST FURNACE

The efficiency of a blast furnace is largely dictated by how well the iron ore is reduced in the upper part of the furnace, which is strongly connected to the radial distribution of the gas and burden. The burden distribution also affects the size of the coke slits in the cohesive zone and the shape of the deadman lower down in the process. A primary method to control the conditions in the furnace is, therefore, to adjust the radial distribution of the burden materials, due to the large differences in the permeability and density of ore and coke. Bell and bell-less charging equipment are the two primary means of feeding the raw materials into the process. Bell-top systems are used in the older blast furnaces while newer installations are based on the more flexible bell-less approach (cf. Figure 4.1).

The flow and distribution of the burden is complex and several attempts have been made to capture the behavior of burden as it flows out of the chute [30, 63], distributes on the burden surface [61] and descends in the shaft [39]. Many recent models are based on discrete element modeling or computational fluid dynamics. These models provide very elaborate results but have limited application to control due to prohibitive computational requirements. Therefore, simple models may provide a feasible alternative for optimization. In particular, in evolutionary approaches, which require a huge number of computations of individual cases the complexity of the model is a crucial issue.

Figure 4.10 describes the setup of one simple burden distribution model [50]. The left panel shows a schematic of a bell-type charging system and the layer formation as the charged material slides along the bell to hit the movable armor, forming a layer on the burden surface. The right side of the figure shows the mathematical representation of the arising burden layer, assuming a piecewise linear profile with coefficients for layer i.

$$z_i(r) = a_{1,i}r + a_{2,i}, r_{c,i} \leq r \leq R_T$$
$$z_i(r) = a_{3,i}r + a_{4,i}, 0 \leq r \leq r_{c,i} \tag{9}$$

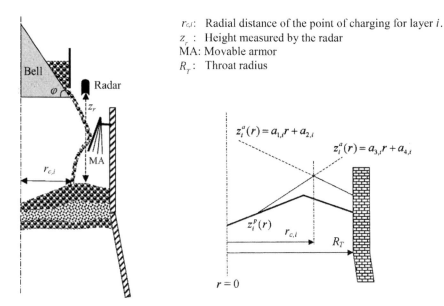

$r_{c,i}$: Radial distance of the point of charging for layer i.
z_r : Height measured by the radar
MA: Movable armor
R_T : Throat radius

FIGURE 4.10 Left: Schematic of the upper part of the blast furnace with bell charging equipment. Right: Schematic of the implementation of the layer formation.

By expressing the falling trajectories the impact point ($r_{c,i}$) can be solved, and with given repose angles the coefficients can be determined if the previous burden layer profile is known.

Pettersson et al. [44] used the above model for evolving charging program to minimize the difference between the calculated and the desired radial ore-to-coke distribution. The charging program consisted of two coke dumps followed by a pellet dump and pellet and coke mixture. The dump sizes and the armor positions were evolved using the Simple Genetic Algorithm as devised by Holland (1975).

The mathematical model was later significantly improved to work on bell-les top charging, incorporating thinner layers and ensuring robustness of the results. This gave more degrees of freedom to evolve the layers [34]. The improved model was applied (Mitra et al., 2013) with similar targets as used by Pettersson et al. [44] and it was demonstrated that the genetic algorithm could evolve feasible charging programs that fulfilled the given requirements. Some results are presented in Figure 4.11.

A main problem with the above approach is that not only the burden distribution but also the gas distribution plays an important role, and since gas temperatures are measured radially in and/or above the burden surface,

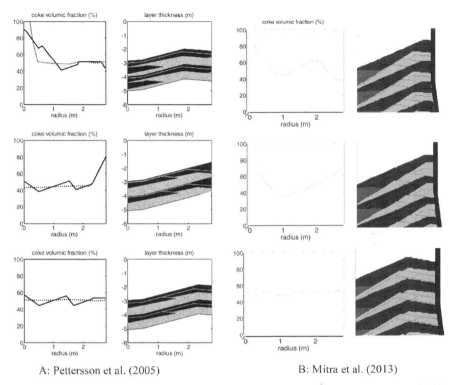

A: Pettersson et al. (2005)   B: Mitra et al. (2013)

FIGURE 4.11 A, B – Left: Target and achieved coke volume fraction represented with discontinuous and continuous lines, respectively. A, B – Right: Respective layer formation in the furnace for each case. Dark layers represent coke, light ones represent iron ore.

these measurements are used for interpretation by the operators. Mitra and Saxén [33], therefore, developed a gas distribution model based on the burden distribution model to simulate the gas distribution in the shaft of the furnace (Figure 4.12). This gave a significant advantage compared to the burden distribution model as the targets could be related to the gas distribution.

In another approach for modeling the burden layer formation, Pettersson et al. [42] used an evolutionary neural network algorithm to estimate the layer thickness of a charged layer in a furnace with a bell charging system. The inputs to the ANN included movable armor position, mass and volume of the dump, impact point of trajectory. The observed layer thickness for these variables from a real furnace was used to train the network. This eliminated the dependence on mathematical models and used data directly from the process.

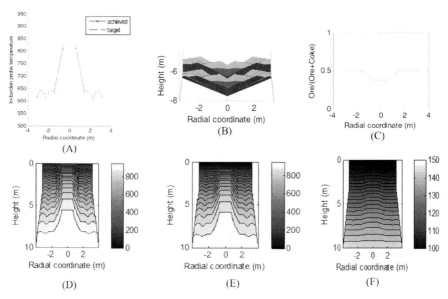

FIGURE 4.12 Some results from Mitra and Saxén [33]. A: Target temperature for the GA and the corresponding evolved temperature, B: Corresponding layer formation, C: ore-to-coke ratio, D: Gas temperature, E: Burden temperature, and F: Pressure contours for a cross-section of the furnace.

4.5.3 OPTIMIZATION OF PRODUCTION PARAMETERS

4.5.3.1 Carbon Dioxide Emissions

Carbon dioxide emissions have been a serious concern for the steel industry. Around 1.7 tons of CO_2 is emitted for every ton of hot metal produced in the blast furnace [20]. Reduced emissions can be achieved by lowering the coke rate, but the coke rate is usually minimized for reasons of process economics. Other parameters, which control the efficiency are the material properties, gas flow rate, blast oxygen levels, etc. There have been several attempts to model the CO_2 emissions from blast furnaces [37] and the steel plant as a whole [37, 57] for different operation scenarios.

CO_2 emissions depend on the carbon content of the charge and injectants per unit of production and on the efficiency of the process. Therefore, decreasing the carbon rate or increasing the efficiency can lead to the decrease in the emissions. This, on the other hand, leads to an increase in production cost as the process needs to run with better quality of raw materials or, alternatively, investments in fundamentally new (and less proven) technology, such as blast furnace top gas recycling. Therefore, this problem is inherently bi-objective.

Pettersson et al. [43] used a linear model derived from the first principle thermodynamic model [45] to get a cost function (F_1) and emission function (F_2)

$$F_1 = \frac{1}{\dot{m}_{hm}} \left(\sum_i \dot{m}_i c_i + \max \left\{ \dot{m}_{hm} F_2 - A, 0 \right\} \right) \tag{10}$$

$$F_2 = \frac{1}{\dot{m}_{hm}} \sum_i \dot{m}_i e_i \tag{11}$$

Here $\dot{m}_i$ are the mass flow rates for resource i, c_i and e_i are the corresponding cost (€/t) and emission (t_{CO2}/t) terms, $\dot{m}_{hm}$ is the hot metal production rate and A is the legal emission allowance. The authors used the Predator prey genetic algorithm to obtain the Pareto front describing the best combinations for the cost and emission for different production rates as described in Figure 4.13.

Instead of relying on thermodynamic models, Agarwal et al. (2010) used the *Evolutionary Neural Networks* (cf. Subsection 4.2.1) to create models of the carbon dioxide emissions and productivity using actual data from the blast furnace. The best network to describe the process was selected using the Bayesian, Akaike [1] and corrected Akaike criteria [4, 5]. After this, a predator prey algorithm, described by Pettersson et al. [41], was used to compute the Pareto fronts for different silicon levels in a blast furnace, as shown in Figure 4.14.

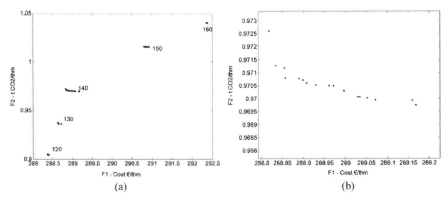

FIGURE 4.13   (a) Pareto frontier for cases with target production of $\dot{m}_{hm}$ = 120, 130, 140, 150 and 160 t_{hm}/h. (b) Pareto frontier for production capacity of $\dot{m}_{hm}$ = 140 t_{hm}/h.

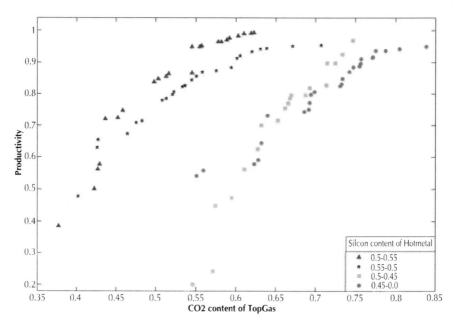

FIGURE 4.14   Pareto frontier for minimizing CO_2 emissions and maximizing productivity at different Si levels in hot metal. Variables are normalized.

In order to provide a better understanding of different methods of optimizing the emissions, Jha et al. [22] performed an extensive comparison of some of the aforementioned methods *(Bi-objective Genetic programming* and *Evolutionary Neural Networks)* in comparison with some established algorithms (Nondominated Sorting Genetic algorithm II, Deb et al., 2002) and available commercial optimization alternatives (modeFRONTIER and KIMEME) for the same set of data. All the models were trained using data from a real blast furnace covering about a year. Figure 4.15 shows the Pareto fronts obtained using modeFRONTIER and the discussed methods. For low and medium silicon levels the results are comparable, but Evolutionary Neural Network and Bi-objective Genetic Programming fares better. For high silicon case Nondominated Sorting Genetic algorithm II in modeFRONTIER fared best.

Unlike the previous models which optimized blast furnace units, Porzio et al. [46] attempted to optimize the use of residual gases in the whole integrated steel plant in order to suppress emissions under various operating conditions. The authors approached the problem by modeling the distribution of gases posing the task as an optimization problem where both costs and emissions are minimized. A non-evolutionary technique based on linear

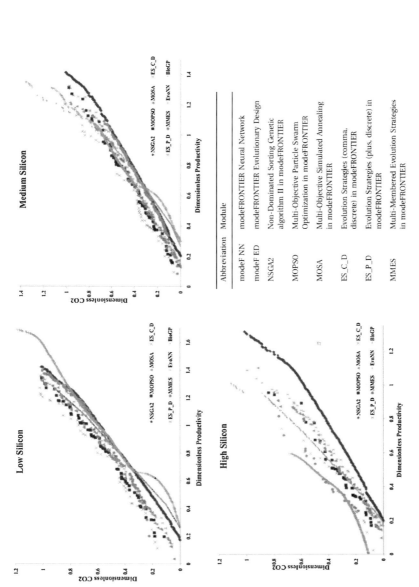

Abbreviation	Module
modeF NN	modeFRONTIER Neural Network
modeF ED	modeFRONTIER Evolutionary Design
NSGA2	Non-Dominated Sorting Genetic algorithm II in modeFRONTIER
MOPSO	Multi-Objective Particle Swarm Optimization in modeFRONTIER
MOSA	Multi-Objective Simulated Annealing in modeFRONTIER
ES_C_D	Evolution Strategies (comma, discrete) in modeFRONTIER
ES_P_D	Evolution Strategies (plus, discrete) in modeFRONTIER
MMES	Multi-Membered Evolution Strategies in modeFRONTIER

FIGURE 4.15 Pareto frontiers for minimizing CO_2 emissions and maximizing productivity at different Si levels in hot metal. Variables are normalized.

model (ε-constraint Linear Programming) was compared with a Strength Pareto Evolutionary Algorithm 2 (SEPA2, Zitzler et al., 2002) formulation. The non-evolutionary technique was found to exhibit much better performance for the task at hand since linearizing the problem made it easier to find the optimum operating points.

In order to further reduce the CO_2 emission in a blast furnace the top gases may be stripped of CO_2 and fed back in the tuyere level, termed top gas recycling. While a linear model can describe normal blast furnace operation, the top gas recycling introduces severe non-linearities into the system. Linearizing an inherently non-linear problem introduces errors especially at the end points of the objective space, where the optimum lies. Therefore, methods based on linearization may give inaccurate results as the solutions are found in the vertices. Mitra et al. [31] handled this non-linearity by using a genetic algorithm and obtained the Pareto optimal solutions for different top gas recycling conditions for the whole integrated steel plant.

Deo et al. [11] compared different non-evolutionary and evolutionary methods (Sequential Quadratic Programming, Annealed Simplex Method, Genetic Algorithm) for minimizing the cost of steel production at an integrated steel plant. The authors found that the genetic algorithm outperformed other classical methods. Hodge et al. [19] extended the work by adding another objective of minimizing constraint violation. Their work used Nash Genetic Algorithm [52] for getting Pareto optimal solutions between the two objectives and they achieved operating points with lower cost of operation.

4.5.3.2   Coal Injection

Coal injection is an important technique for reducing the coke rate and increasing the production rate in the blast furnace. Furthermore, the pulverized coal injected through the tuyeres is cheaper than coke. When coal is injected into the raceway, it combusts first and furnishes heat and carbon monoxide for the reduction reactions. Therefore, it is in general favorable to maximize the injection rate, which minimizes the coke solution loss. Giri et al. [16] tackled this problem using *Bi-objective Genetic Programming*. Figure 4.16 shows the Pareto frontier for the solutions maximizing coal injection and minimizing solution loss.

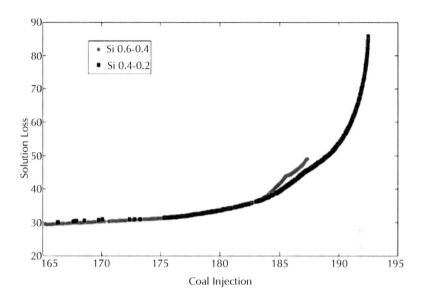

FIGURE 4.16   Pareto frontiers between solution loss and coal injection calculated by Giri et al. [16].

4.5.4   ESTIMATION OF THE INTERNAL HEARTH PROFILE

The blast furnace hearth is a hot region of the blast furnace with flows of molten iron and slag and therefore the wall and bottom lining experience wear and erosion. The hot metal temperatures can reach up to 1500°C and that of slag is usually 50–100°C higher. At these high temperatures and in this hostile environment it is impossible to continuously carry out measurements, so the internal state of the hearth is largely unknown. Mathematical models have therefore been used to shed light on the hearth conditions. For instance, the hearth wear and possible buildup formation has been estimated by solving inverse heat conduction problems in two dimensions [17, 25, 55].

Bhattacharya et al. [3] applied differential evolution in such an inverse problem formulation to evolve the position of the hot face. The fitness was evaluated as the differences between temperatures measured by thermocouples in the lining and those obtained from a numerical solution of the heat transfer equations. The top left panel of Figure 4.17 describes the computational grid, which is used to solve the heat conduction equations, the position of the thermocouples and the control points, which define the boundary of

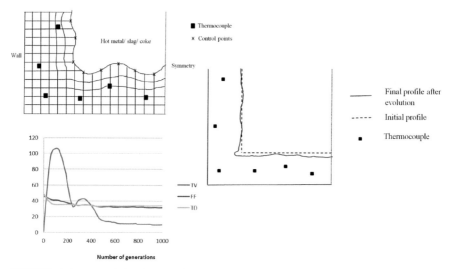

FIGURE 4.17 Top left: Computational grid representing the hearth for evaluating the heat flow equations, showing the control points for differential evolution. Bottom left: Convergence history of the profile. TD: Total Deviation; TV: Total variation; FF: Fitness function. Right: Converged solution for a set of the thermocouple reading. (adapted from Bhattacharya et al. (2013)).

the hearth region. The lower left portion of the figure shows how the solution converges after a sufficient number of generations. Different seed points provided marginally different results, as there may be multiple, yet similar profiles which describe a particular set of thermocouple readings. The profiles achieved after such computations are never very accurate as they only provide an approximate profile. From the figure it may be argued that this problem is ill-posed as the formulation attempts to solve the location of too many points on the internal surface compared to the number of residuals in the objective function.

4.5.5 OPTIMIZATION OF THE PERFORMANCE OF A ROTARY KILN

The quality of DRI produced by rotary kiln is judged in terms of its metallic iron content. This and the production rate are functions of different operating parameters like iron ore quality, particle size and feed rate, coal quality and feed rate and the kiln rotation speed. In a study by Mohanty et al. [35] an ANN was used to model the production rate and metallic iron content using data from an actual operation. Thereafter, a predator-prey algorithm [27] was used to obtain the Pareto front for the quantity and quality of the product. It was shown that the current point of operation was non-optimal and that it is possible to improve reactor performance by changing some of the operating parameters.

4.6   CONCLUSIONS AND FUTURE PROSPECTS

This chapter has reviewed papers on evolutionary algorithms used to tackle problems in the field of ironmaking. Even though the number of papers on this specific topic is still relatively limited, the results have already demonstrated the potential of the algorithms for tackling complex modeling and optimization problems. Evolutionary algorithms have a clear advantage in non-linear optimization and the methods are especially useful for solving multiobjective problems and for solving problems where the gradients are ill-defined or do not exist. Therefore, also mixed problems with both integer and real numbered variables can be tackled. The evolutionary methods have found their use in various ironmaking applications, particularly since the turn of the century. However, real-world applications of them are still very few, probably due to the complexity of the process and a lack of fast and accurate models upon which the optimization could be based.

In the future, attention should be focused on the use of evolutionary techniques in control and decision making systems for the blast furnace. Faster evolutionary algorithms should be developed which require less calculations before the optima are achieved. By such methods complex and time consuming process models that are available now could be used for improved control and optimization of the ironmaking process.

ACKNOWLEDGEMENTS

The work was supported by the SIMP research program under FIMECC. Financial support of Tekes and the participating companies: The financial support from the project, as well as from the Graduate School in Chemical Engineering of Åbo Akademi University is gratefully acknowledged.

KEYWORDS

- **genetic algorithm**
- **evolutionary algorithm**
- **ironmaking**

REFERENCES

1. Akaike, H. A new look at the statistical model identification. *IEEE T. Automat. Contr.* 1974, *19* (6), 716–723.
2. Ariyama, T., Natsui, S., Kon, T., Ueda, S., Kikuchi, S., & Nogami, H. Recent Progress on Advanced Blast Furnace Mathematical Models Based on Discrete Method. *ISIJ Int.* 2014, *54* (7), 1457–1471.
3. Bhattacharya, A. K., Aditya, D., & Sambasivam, D. Estimation of operating blast furnace reactor invisible interior surface using differential evolution. *Appl. Soft. Comput.* 2013, *13*, 2767–2789.
4. Bhattacharya, B., Dinesh Kumar, G. R., Agarwal, A., Erkoç, S., Singh, A., & Chakraborti, N. Analyzing Fe-Zn system using molecular dynamics, evolutionary neural nets and multiobjective genetic algorithms. *Comp. Mater. Sci.* 2009, *46*, 821–827.
5. Burnham, K. P., & Anderson, D. R. *Model selection and multimodel interface*; Springer, 2002.
6. Chakraborti, N., Misra, K., Bhatt, P., Barman, N., & Prasad, R. Tight-binding calculations of Si-H clusters using genetic algorithms and related techniques: studies using differential evolution. *J. Phase Eqilib.* 2001, *22* (5), 525–530.
7. Coello Coello, C. A., Van Veldhuizen, D. A., & Lamont, G. B. *Evolutionary algorithms for solving multiobjective problems*; Kluwer Academic Publishers, New York, 2002.
8. Collet, P. Genetic programming. In *Handbook of Research on Nature Inspired Computing for Economics and Management*; Rennard, J.-P. Ed., Idea, Hershey, PA,. 2007, p. 59–73.
9. Cybenko, G. Approximation by superpositions of a sigmoidal function. *Math. Control Signal* 1989, *2*, 303–314.
10. Deb, K., & Agrawal, R. M. Simulated binary crossover for continuous search space. *Complex Systems* 1995, *9*, 115–148.
11. Deo, B., Deb, K., Jha, S., Sudhakar, V., & Sridhar, N. V. Optimal operating conditions for the primary end of an integrated steel plant: Genetic adaptive search and classical techniques. *ISIJ Int.* 1998, *38* (1), 98–105.
12. Fabian, T. A Linear Programming Model of Integrated Iron and Steel Production. *Manag. Sci.* 1958, *4*, 415–449.
13. Geiseler, J. Use of steelworks slag in Europe. *Waste. Manage.* 1996, *16* (1–3), 59–63.
14. Ghosh, A., & Chatterjee, A. *Ironmaking and steelmaking: Theory and Practice*; Prentice-Hall of India, New Delhi, 2008.

15. Giri, B. K., Hakanen, J., Miettinen, K., & Chakraborti, N. Genetic programming through bi-objective genetic algorithms with a study of a simulated moving bed process involving multiple objectives. *Appl. Soft Comput.* 2013a, *13* (5), 2613–2623.

16. Giri, B. K., Pettersson, F., Saxén, H., & Chakraborti, N. Genetic programming evolved through bi-objective genetic algorithms applied to a blast furnace. *Mater. Manuf. Process.* 2013b, *28*, 776–782.

17. Gonzalez, M., & Goldschmit, M. B. Inverse heat transfer problem based on a radial basis functions geometry representation. *Int. J. Num. Met. Eng.* 2006, *65*, 1243–1262.

18. Haykin, S. S. *Neural networks and learning machines*, Prentice-Hall, 2009. Holland, J. H. *Adaptation in natural and artificial systems;* The University of Michigan Press: Ann Arbor, MI, 1975.

19. Hodge, B., & Pettersson, F., Chakraborti, N. Re-evaluation of the optimal operating conditions for the primary end of an integrated steel plant using multiobjective genetic algorithms and nash equilibrium. *Steel. Res. Int.* 2006, *77* (7), 459–461.

20. Hu, C., Han, X., Li, Z., & Zhang, C. Comparison of CO_2 emission between COREX and blast furnace iron-making system. *J. Environ. Sci.* 2009, *21* (Supplement 1), S116-S120.

21. Japan Metal Bulletin. POSCO blows in no.1 blast furnace in Gwangyang steel works. Jun 10, 2013.

22. Jha, R., Sen, P. K., & Chakraborti, N. Multiobjective genetic algorithms and genetic programming models for minimizing input carbon rates in a blast furnace compared with a conventional analytic approach. *Steel. Res. Int.* 2014, *85* (2), 219–232.

23. Kobashigawa, J., Youn, H., Iskander, M., & Yun, Z. Comparative study of genetic programming vs. neural networks for the classification of buried objects. In Antennas and Propagation Society International Symposium, Charleston, SC, 2009, p. 1–4.

24. Kovačič, M., Senčič, S., & Župerl, U. Artificial intelligence approach of modeling of PM10 emission close to a steel plant. In Quality 2013, Neum, Bosnia and Herzegovina, 2013, p. 305–310.

25. Kumar, S. Heat Transfer Analysis and Estimation of Refractory Wear in an Iron Blast Furnace Hearth Using Finite Element Method. *ISIJ Int.* 2005, *45*, 1122–1128.

26. Lampinen, J., Storn, R. Differential evolution. *Stud. Fuzz. Soft. Comp.* 2004, *141*, 123–166.

27. Li, X. A real-coded predator-prey genetic algorithm for multiobjective optimization. In *Evolutionary Multi-Criterion Optimization*; Fonseca, C. M., Fleming, P. J., Zitzler, E., Deb, K., Thiele, L., Eds., Proceedings of the Second International Conference on Evolutionary Multi-Criterion Optimization; *Lecture Notes in Computer Science* 2003, *2632*, 207–221.

28. Meijer, K., Zeilstra, C., Teerhuis, C., Ouwehand, M., & van der Stel, J. Developments in alternative ironmaking. *T. Indian I. Metals* 2013, *66* (5–6), 475–481.

29. Miettinen, K. M. *Nonlinear Multiobjective Optimization*; Kluwer Academic Publishers, Boston, Massachusetts, 1998.

30. Mio, H., Komatsuki S., Akashi, M., Shimosaka, A., Shirakawa, Y., Hidaka, J., Kadowaki, M., Matsuzaki, S., & Kunitomo, K. Validation of particle size segregation of sintered ore during flowing through laboratory-scale chute by discrete element method. *ISIJ. Int.* 2008, *48* (12), 1696–1703.

31. Mitra, T., Helle, M., Pettersson, F., Saxén, H., & Chakraborti, N. Multiobjective optimization of top gas recycling conditions in the blast furnace by genetic algorithms. *Mater. Manuf. Process.* 2011, *26* (3), 475–480.

32. Mitra, T., Mondal, D. N., Pettersson, F., & Saxén, H. Evolution of charging programs for optimal burden distribution in the blast furnace. *Computer Methods in Material Science* 2013, *13* (1), 99–106.

33. Mitra, T., & Saxén, H. Evolution of charging programs for achieving required gas temperature profile in a blast furnace. *Mater. Manuf. Process.* 2015, *30* (4), 474–487.

34. Mitra, T., & Saxén, H. Model for fast evaluation of charging programs in the blast furnace. *Metall. Mater. Trans. B.* 2014, *45* (6), 2382–2394.

35. Mohanty, D., Chandra, A., & Chakraborti, N. Genetic algorithms based multiobjective optimization of an iron making rotary kiln. *Comp. Mater. Sci.* 2009, *45*, 181–188.

36. Nogami, H., Chu, M., & Yagi, J. Multi-dimensional transient mathematical simulator of blast furnace process based on multi-fluid and kinetic theories. *Comput. Chem. Eng.* 2005, *29* (11–12), 2438–2448.

37. Nogami, H., Yagi, J., Kitamura, S., & Austin, P. R. Analysis on material and energy balances of ironmaking systems on blast furnace operations with metallic charging, top gas recycling and natural gas injection. *ISIJ. Int.* 2006, *46* (12), 1759–1766.

38. Omori, Y. *Blast furnace phenomena and modeling*; Elsevier, London, 1987.

39. Park, J-I., Baek, U-H., Jang, K-S., Oh, H-S., & Han, J-W. Development of the burden distribution and gas flow model in the blast furnace shaft. *ISIJ. Int.* 2011, *51* (10), 1617–1623.

40. Paul, S., Roy, S. K., & Sen, P. K. Approach for minimizing operating blast furnace carbon rate using Carbon-Direct Reduction (C-DRR) Diagram. *Metall. Mater. Trans. B* 2013, *44* (1), 20–27.

41. Pettersson, F., Chakraborti, N., & Saxén, H. A genetic algorithms based multiobjective neural net applied to noisy blast furnace data. *Appl. Soft. Comput.* 2007, *7*, 387–397.

42. Pettersson, F., Hinnelä, J., & Saxén, H. Evolutionary neural network modeling of blast furnace burden distribution. *Mater. Manuf. Process.* 2003, *18* (3), 385–399.

43. Pettersson, F., Saxén, H., Deb, K. Genetic algorithm-based multicriteria optimization of ironmaking in the blast furnace. *Mater. Manuf. Process.* 2009, *24* (3), 343–349.

44. Pettersson, F., Saxén, H., & Hinnelä, J. A genetic algorithm evolving charging programs in the ironmaking blast furnace. *Mater. Manuf. Process.* 2005, *20* (3), 351–361.

45. Poli, R., Langdon, W. B., & McPhee, N. F. A field guide to genetic programming. Published via http://lulu.com and freely available at http://www.gp-field-guide.org.uk (With contributions by J. R. Koza), 2008.

46. Porzio, G. F., Nastasi, G., Colla, V., Vannucci, M., & Branca, T. A. Comparison of multiobjective optimization techniques applied to off-gas management within an integrated steelwork. *Appl. Energ.* 2014, *136*, 1085–1097.

47. Poveromo, J. J. Blast furnace burden distribution fundamentals. *Iron and Steelmaker* 1995–1996, *22–23*.

48. Rumelhart, D. E., Hinton, G. E., & Williams, R. J. Learning representations by back-propagating errors. *Nature* 1986, 323, 533–536.

49. Saxén, H., Gao, C., & Gao, Z. Data-driven time discrete models for dynamic prediction of the hot metal silicon content in the blast furnace – A review. *IEEE T. Ind. Inform.* 2013, *9* (4), 2213–2225.

50. Saxén, H., & Hinnelä, J. Model for burden distribution tracking in the blast furnace. *Miner. Process. Extr. M.* 2004, *25*, 1–27.

51. Saxén, H., Pettersson, F., & Gunturu, K. Evolving nonlinear time-series models of the hot metal silicon content in the blast furnace. *Mater. Manuf. Process..* 2007, *22*, 577–584.

52. Sefrioui, M., & Periaux, J. Nash genetic algorithms: examples and applications. In proceedings of the 2000 congress on evolutionary computation, La Jolla, CA, 2000, p. 509–516.

53. Steelonthenet.com. Basic Oxygen Furnace Route Steelmaking Costs 2014. http://www.steelonthenet.com/cost-bof.html (accessed Sep 19, 2014).

54. Storn, R., & Price, K. Differential evolution – a simple and efficient heuristic for global optimization over continuous spaces. *J. Global Optim.* 1997, *11*, 341–359.

55. Torrkulla, J., & Saxén, H. Model of the state of the blast furnace hearth. *ISIJ International* 2000, *40*, 438–447.

56. Valadi, J., & Siarry, P. *Applications of metaheuristics in process engineering*; Springer, 2014.

57. Wang, C., Larsson, M., Ryman, C., Grip, C. E., Wikström, J.-O., Johnsson, A., & Engdahl, J. A model on CO_2 emission reduction in integrated steelmaking by optimization methods. *Int. J. Energy Res.* 2008, *32*, 1092–1106.

58. World Steel Association. Key facts about the world steel industry. http://www.worldsteel.org/media-center/key-facts.html (accessed Jul 17, 2014a).

59. World Steel Association. Resource efficiency. http://www.worldsteel.org/steel-by-topic/sustainable-steel/environmental/efficient-use.html (accessed Jul 17, 2014b).

60. World Steel Association. Annual iron production archive. http://www.worldsteel.org/statistics/statistics-archive/annual-iron-archive.html (accessed Jul 17, 2014c).

61. Xu, J., Wu, S., Kou, M., Zhang, L., & Yu, X. Circumferential burden distribution behaviors at bell-less top blast furnace with parallel type hoppers. *Appl. Math. Model.* 2011, *35*, 1439–1455.

62. Yang, Y., Raipala, K., & Holappa, L. Ironmaking. In *Treatise on Process Metallurgy*; Seetharaman, S., Ed., Elsivier, Oxford, 2013.

63. Yu, Y., & Saxén, H. Particle flow and behavior at bell-less charging of the blast furnace. *Steel Res. Int.* 2013, *84* (10), 1018–1033.

CHAPTER 5

HARMONY SEARCH OPTIMIZATION FOR MULTILEVEL THRESHOLDING IN DIGITAL IMAGES

DIEGO OLIVA,[1,2] ERIK CUEVAS,[2] GONZALO PAJARES,[1] DANIEL ZALDÍVAR,[2] MARCO PÉREZ-CISNEROS,[2] and VALENTÍN OSUNA-ENCISO[3]

[1]*Dpto. Ingeniería del Software e Inteligencia Artificial, Facultad Informática, Universidad Complutense de Madrid, 28040 Madrid, Spain, E-mail: doliva @ucm.es; pajares@ucm.es*

[2]*Departamento de Ciencias Computacionales, Universidad de Guadalajara, CUCEI, CU-TONALA, Av. Revolución 1500, Guadalajara, Jal, México, E-mail: diego.oliva@cucei.udg.mx, erik.cuevas@cucei.udg.mx, daniel.zaldivar@cucei.udg.mx, marco.perez@cucei.udg.mx*

[3]*Departamento de Ingenierías, CUTONALA, Universidad de Guadalajara, Sede Provisional Casa de la Cultura – Administración, Morelos #180, Tonalá, Jalisco 45400, México, E-mail: valentin.osuna@cutonala.udg.mx*

CONTENTS

5.1   INTRODUCTION

Evolutionary computation (EC) is a relatively new field of research in computer sciences, it becomes popular due to its application in several fields to solve problems where is necessary to find the best solution in a considerably reduced time. Numerous methods have been proposed and they are inspired by different natural behaviors, for instance Genetic Algorithms (GA) [1] that their operators are based on the species evolution or Electromagnetism-like Optimization (EMO) that imitates the attraction-repulsion forces in electromagnetism theory [2]. The use of these methods has been extended in recent years, they can be used and applied in areas as science, engineering, economics, and other that mathematically fulfill the requirements of the algorithms.

Image segmentation is a problem where the EC methods can be used. Moreover this is considered one of the most important tasks in image processing. Segmentation permits to identify whether a pixel intensity corresponds to a predefined class. For example, this process is used to separate the interest objects from the background in a digital image, using the information provided by the histogram (see Figure 5.1). Thresholding is the easiest method for segmentation. On its simplest explanation, it works taking a threshold (*th*) value and apply it in the image histogram. The pixels whose intensity value is higher than *th* are labeled as first class, while the rest correspond to a second class label. When the image is segmented into two classes, it is called bi-level thresholding (BT) and requires only one *th* value. On the other hand, when pixels are separated into more than two classes, the task is named as multilevel thresholding (MT) and demands more than one *th* value [3].

There exist two classical thresholding methods. Considering a *th* value, the first approach, proposed by Otsu in [4] maximizes the variance between classes while the second method, submitted by Kapur in [5], uses the maximization of the entropy to measure the homogeneity among classes. They were originally proposed for bi-level segmentation and keep balance between efficiency and accuracy. Although both Otsu and Kapur can be expanded for MT, however the computational complexity increases exponentially when a new threshold is incorporated [6]. An alternative to classical methods is the use of evolutionary computation techniques. They have demonstrated to deliver better results than those based on classical techniques in terms of accuracy, speed and robustness. Numerous evolutionary approaches have been reported in the literature. Methods as Particle Swarm Optimization (PSO) [7, 8] and Bacterial Foraging Algorithm (BFA) [9], have been employed to face the segmentation problem using both Otsu's and Kapur's objective functions.

The Harmony Search Algorithm (HSA) introduced by Geem, Kim, and Loganathan [10] is an evolutionary computation algorithm, which is based on the metaphor of the improvisation process that occurs when a musician searches for a better state of harmony. The HSA generates a new candidate solution from a feasible search space. The solution vector is analogous to the harmony in music while the local and global search schemes are analogous to musician's improvisations. HSA imposes fewer mathematical requirements as it can be easily adapted for solving several sorts of engineering optimization challenges. It has been successfully applied to solve a wide range of practical optimization problems such as discrete and continuous structural optimization [11] and design optimization of water distribution networks [12] among others.

In this chapter, a multi-thresholding segmentation algorithm is introduced. The proposed approach, called the Harmony Search Multi-thresholding Algorithm (HSMA), combines the original HSA and the Otsu´s and Kapur's methodologies. The proposed algorithm takes random samples (*th* values) from a feasible search space defined by the image histogram distribution. Such samples build each harmony (candidate solution) in the HSA context, whereas its quality is evaluated considering the objective function that is employed by the Otsu's or the Kapur's method. Guided by these objective values, the set of candidate solutions are evolved using the HSA operators until the optimal solution is found. The approach generates a multilevel segmentation algorithm, which can effectively identify the threshold values of a digital image within a reduced number of iterations. Experimental results over several complex images have validated the efficiency of the proposed technique regarding accuracy, speed and robustness. An example of the implementation of HSMA using the Otsu's method is presented in Figure 5.1, it shows the original image in gray scale, the best threshold values and the evolution of the objective function values for a predefined number of iterations.

The remainder chapter is organized as follows: In Section 5.2, the HSA is introduced and some examples of optimization are shown. Section 5.3 gives a description of the Otsu's and Kapur's methods for thresholding. Section 5.4 explains the implementation of the proposed algorithm. Section 5.5 discusses experimental results after testing the proposed method over a set benchmark images. Finally, some conclusions are discussed in Section 5.6.

5.2 HARMONY SEARCH ALGORITHM

5.2.1 *THE HARMONY SEARCH ALGORITHM (HSA)*

In the theory of the harmony search algorithm, each solution is called a "harmony" and is represented by an *n*-dimension real vector. First is necessary the generation of an initial population of harmony vectors that are

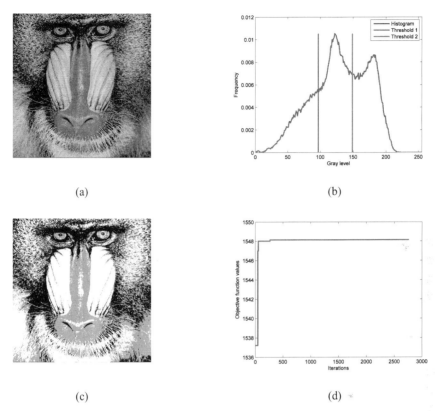

(a)   (b)

(c)   (d)

FIGURE 5.1   (a) Test image in gray scale, (b) histogram of (a) and the thresholds obtained by HSMA, (c) segmented image and (d) evolution of the objective function values in this case Otsu function.

stored within a Harmony Memory (HM). A new candidate harmony is thus generated from the elements in the HM by using a memory consideration operation either by a random re-initialization or a pitch adjustment operation. Finally, the HM is updated by comparing the new candidate harmony and the worst harmony vector in the HM. The worst harmony vector is replaced by the new candidate vector when the latter delivers a better solution in the HM. The above process is repeated until a certain termination criterion is met. The basic HS algorithm consists of three main phases: HM initialization, improvisation of new harmony vectors and updating of the HM. The following discussion addresses details about each stage and their operators.

5.2.2 INITIALIZATION OF THE PROBLEM AND ALGORITHM PARAMETERS

In general terms, a global optimization problem can be defined as follows:

$$\begin{aligned} \text{minimize} \quad & f(\mathbf{x}), \quad \mathbf{x} = (x_1, x_2, \ldots, x_n) \in \mathfrak{R}^n \\ \text{subject to:} \quad & x_j \in \left[l_j, u_j \right], \quad j = 1, 2, \ldots, n \end{aligned} \tag{1}$$

where $f(\mathbf{x})$ is the objective function that depends on the problem to be treated, $\mathbf{x} = (x_1, x_2, \ldots, x_n)$ is the set of design variables, n is the number of design variables, and j_j and u_j are the lower and upper bounds for the design variable x_j, respectively. The parameters for HSA are the harmony memory size, such value is the number of solution vectors lying on the harmony memory (HM), the harmony-memory consideration rate ($HMCR$), the pitch adjusting rate (PAR), the distance bandwidth (BW) and the number of improvisations (NI), which represents the total number of iterations. The performance of HSA is strongly influenced by values assigned to such parameters, which in turn, depend on the application domain [13].

5.2.3 HARMONY MEMORY INITIALIZATION

In this stage, the initial values of a predefined number (HMS) of vector components in the harmony memory are configured. Considering $\mathbf{x} = (x_1, x_2, \ldots, x_n)$ that represents the i-th randomly generated harmony vector, each element of x is computed as follows:

$$x_j^i = l_j + \left(u_j - l_j \right) \gamma rand(0,1), \quad \text{for} \quad \begin{array}{l} j = 1, 2, \ldots n, \\ i = 1, 2, \ldots, HMS \end{array} \tag{2}$$

where $rand(0,1)$ is a uniform random number between 0 and 1, the upper and lower bounds of the search space are defined by l_j and u_j respectively. Then, the HM forms a matrix that is filled with the HMS harmony vectors as follows:

$$HM = \begin{bmatrix} \mathbf{x}_1 \\ \mathbf{x}_2 \\ \vdots \\ \mathbf{x}_{HMS} \end{bmatrix} \tag{3}$$

5.2.4 IMPROVISATION OF NEW HARMONY VECTORS

Once the vectors of the HM are generated, the next step is create a new harmony using the improvisation operators. Thus in this phase a new harmony vector $\mathbf{x}^{new}$ is built by applying the following three operators of HSA: memory consideration, random re-initialization and pitch adjustment.

The process of generating a new harmony is known as 'improvisation' under the HSA context. In the memory consideration step, the value of the first decision variable x_1^{new} for the new vector is chosen randomly from any of the values already existing in the current HM, i.e., from the set $\{x_1^1, x_2^1, \ldots, x_{HMS}^1\}$. For this operation, a uniform random number r_1 is generated within the range $[0, 1]$. If r_1 is less than $HMCR$, the decision variable x_1^{new} is generated through memory considerations; otherwise, x_1^{new} is obtained from a random re-initialization between the search bounds $[l_1, u_1]$. Values of the other decision variables $x_2^{new}, x_3^{new}, \ldots, x_n^{new}$ are also chosen accordingly. Therefore, both operations, memory consideration and random re-initialization, can be modeled as follows:

$$x_j^{new} = \begin{cases} x_j^i \in \{x_j^1, x_j^2, \ldots, x_j^{HMS}\} & \text{with probability } HMCR \\ l_j + (u_j - l_j) \cdot rand(0,1) & \text{with probability } 1\text{-}HMCR \end{cases} \quad (4)$$

Every component obtained by memory consideration is further examined to determine whether it should be pitch-adjusted. For this operation, the Pitch-Adjusting Rate (PAR) is defined as to assign the frequency of the adjustment and the Bandwidth factor (BW) to control the local search around the selected elements of the HM. Hence, the pitch adjusting decision is calculated as follows:

$$x_j^{new} = \begin{cases} x_j^{new} = x_j^{new} \pm rand(0,1) \cdot BW & \text{with probability } PAR \\ x_j^{new} & \text{with probability } (1\text{-}PAR) \end{cases} \quad (5)$$

Pitch adjusting is responsible for generating new potential harmonies by slightly modifying original variable positions. Such operation can be considered similar to the mutation process in evolutionary algorithms. Therefore, the decision variable is either perturbed by a random number between 0 and BW or left unaltered. In order to protect the pitch adjusting operation, it is important to assure that points lying outside the feasible range $[l, u]$ must be re-assigned i.e., truncated to the maximum or minimum value of the interval.

Similar to the *HMCR* for Eq. (5), a uniform random number r_2 is generated within the range [0, 1]. If r_2 is less than *PAR* the decision new harmony is adjusted using the *BW*, otherwise it preserves its value.

5.2.5   UPDATING THE HARMONY MEMORY

After a new harmony vector $\mathbf{x}^{new}$ is generated, the harmony memory is updated by the survival of the fit competition between $\mathbf{x}^{new}$ and the worst harmony vector $\mathbf{x}^w$ in the HM. Therefore, $\mathbf{x}^{new}$ will replace $\mathbf{x}^w$ and become a new member of the HM in case the fitness value of $\mathbf{x}^{new}$ is better than the fitness value of $\mathbf{x}^w$.

5.2.6   COMPUTATIONAL PROCEDURE

The computational procedure of the basic HSA can be summarized as follows [14]:

Step 1:   Set the parameters *HMS, HMCR, PAR, BW* and *NI*

Step 2:   Initialize the HM and calculate the objective function value of each harmony vector.

Step 3:   Improvise a new harmony $\mathbf{x}^{new}$ as follows:

for ($j = 1$ to n) do

    if ($r_1 < HMCR$) then

        $x_j^{new} = x_j^a$ where $a \in (1, 2, \ldots, HMS)$

        if ($r_2 < PAR$) then

            $x_j^{new} = x_j^{new} \pm r_3 \cdot BW$ where $r_1, r_2, r_3 \in rand(0,1)$

        end if

        if $x_j^{new} < l_j$

            $x_j^{new} = l_j$

        end if

        if $x_j^{new} > u_j$

            $x_j^{new} = u_j$

        end if

    else

        $x_j^{new} = l_j + r \cdot (u_j - l_j)$, where $r \in rand(0,1)$

    end if

end for

Step 4:   Update the HM as $\mathbf{x}^w = \mathbf{x}^{new}$ if $f(\mathbf{x}^{new}) < f(\mathbf{x}^w)$

Step 5:   If NI is completed, the best harmony vector $\mathbf{x}^b$ in the HM is returned; otherwise go back to Step 3.

This procedure is implemented for minimization. If the intention is to maximize the objective function, a sign modification of Step 4 ($\mathbf{x}^w = \mathbf{x}^{new}$ if $f(\mathbf{x}^{new}) > f(\mathbf{x}^w)$) is required. In the implementation proposed the HSA is used for maximization.

5.2.7 A NUMERICAL EXAMPLE OF HSA IMPLEMENTATION

Usually the evolutionary computing methods are tested using mathematical functions with different complexity. One of the most popular benchmark functions using for testing the optimization performance is the Rosenbrock function [15], which is a 2-D function defined as follows:

$$f(x,y) = (1-x)^2 + 100(y-x^2)^2 \tag{6}$$

The optimization features of the Rosenbrock function, used for this example, are defined in Table 5.1.

Therefore, the intervals of the search space for each parameter under the optimization context are: $\mathbf{u} = [5, 5]$ and $\mathbf{I} = [-5, -5]$ that are the upper and lower limits for each dimension, respectively. The parameter configuration of HSA used for the example is suggested in [16]. Such values are presented in Table 5.2. In the literature there exist several methods for parameter calibration, some methods are automatic. However, here are used the values proposed for the original HSA.

Considering the values of Table 5.2, the harmony memory is initialized with HM random elements (harmonies). The surface of the Rosenbrock function in the 2-D space is presented in Figure 5.2 (a) and the distribution of harmonies in the search space is shown in Figure 5.2 (b). The HSA runs NI times and the values of HM are modified with the improvisation of new harmonies. Figure 5.2 (c) shows the position of the HM elements when $NI = 100$, the red point is the best harmony in the HM and the green point represents the current improvisation. Finally the positions of the HM elements when the stop criterion is achieved ($NI = 1200$) are shown in Figure 5.2 (d).

TABLE 5.1 Features of Rosenbrock Function

Higher limit	Lower limit	Global minima
$x = 5$	$x = -5$	$xbest = 1$
$y = 5$	$y = -5$	$ybest = 1$

TABLE 5.2 Parameter Configuration of HSA

HM	HMCR	PAR	BW	NI
20	0.95	0.7	0.1	1200

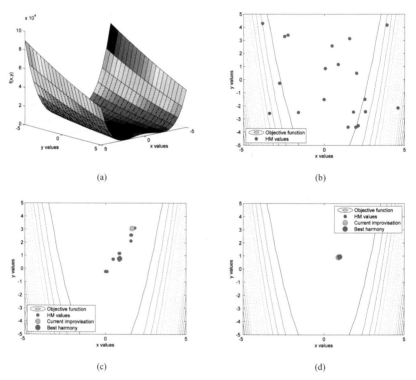

(a)   (b)

(c)   (d)

FIGURE 5.2   (a) Surface of Rosenbrock function, (b) initial positions of HM, (c) position of HM, current improvisation and best harmony after 100 iterations and (d) final positions of HM, current improvisation and best harmony.

5.3   MULTILEVEL THRESHOLDING (MT) FOR DIGITAL IMAGES

The thresholding methods can be divided in two groups parametric and nonparametric. These methods select thresholds searching the optimal value of objective functions applied to image processing. In parametric techniques

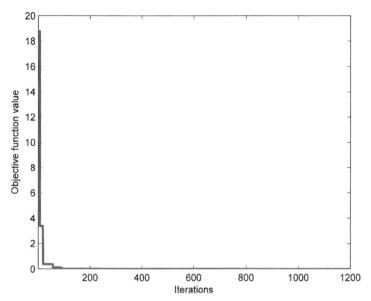

FIGURE 5.3   Evolution of the objective function (Rosenbrock) using HSA.

the distribution of the intensity levels for each class has a probability density function that for definition is considered as a Gaussian distribution. For these approaches is necessary to estimate the distribution parameters that provides a best fitting between the proposed model and the original image histogram. On the other hand, nonparametric approaches find the thresholds that separate the intensity regions of a digital image in optimal classes using a metric that determines if a value belongs to a class depending of its intensity value. Some of these metrics commonly used for thresholding are the between class variance and the cross entropy. The popular method, Otsu's method [4], selected optimal thresholds by maximizing the between class variance. Shoo et al. [17] found that the Otsu's method is one of the better threshold selection methods for real world images with regard to uniformity and shape measures. There exist several entropy measures as minimum cross entropy, maximum entropy and their variants. One of this entropy metrics is the proposed by Kapur et al. [5] for gray images thresholding. Basically this method depends of the maximum entropy value to obtain the best thresholds for the image histogram.

5.3.1   THEORY OF MULTILEVEL THRESHOLDING

Thresholding is a process in which the pixels of a gray scale image are divided in sets or classes depending on their intensity level (L). For this classification it is necessary to select a threshold value (th) and follows the simple rule of Eq. (7).

$$C_1 \leftarrow p \quad \text{if} \quad 0 \leq p < th$$
$$C_2 \leftarrow p \quad \text{if} \quad th \leq p < L-1 \tag{7}$$

where p is one of the $m \times n$ pixels of the gray scale image I_g that can be represented in L gray scale levels $L = \{0,1,2,...,L-1\}$. C_1 and C_2 are the classes in which the pixel p can be located, while th is the threshold. The rule in Eq. (7) corresponds to a bi-level thresholding and can be easily extended for multiple sets:

$$
\begin{aligned}
C_1 &\leftarrow p \quad \text{if} \quad 0 \leq p < th_1 \\
C_2 &\leftarrow p \quad \text{if} \quad th_1 \leq p < th_2 \\
C_i &\leftarrow p \quad \text{if} \quad th_i \leq p < th_{i+1} \\
C_n &\leftarrow p \quad \text{if} \quad th_n \leq p < L-1
\end{aligned}
\tag{8}
$$

where $\{th_1 \ th_2 \ ... \ th_i \ th_{i+1} \ th_k\}$ represent different thresholds. The problem for both bi-level and MT is to select the th values that correctly identify the classes. Although, Otsu's and Kapur's methods are well-known approaches for determining such values, both propose a different objective function which must be maximized in order to find optimal threshold values, just as it is discussed below.

Figure 5.4 presents an example for bi-level thresholding and for multilevel thresholding. The values for each example are: $L = 2$ and $L = 4$ for bi-level and multilevel thresholding respectively. Figure 5.4 (a) shows the original image in gray scale and its corresponding histogram is shown in Figure 5.4 (b). The segmented (binarized) image using only one threshold is presented in Figure 5.4 (c) and the histogram with the $th = \{94\}$ in Figure 5.4 (d). Finally the resultant image using the values **TH** = {56, 107, 156} are shown in Figure 5.4 (e) and Figure 5.4 (f).

5.3.2 OTSU'S METHOD (BETWEEN-CLASS VARIANCE)

This is a nonparametric technique for thresholding proposed by Otsu [4] that employs the maximum variance value of the different classes as a criterion to segment the image. Taking the L intensity levels from a gray scale image or from each component of a RGB (red, green, blue) image, the probability distribution of the intensity values is computed as follows:

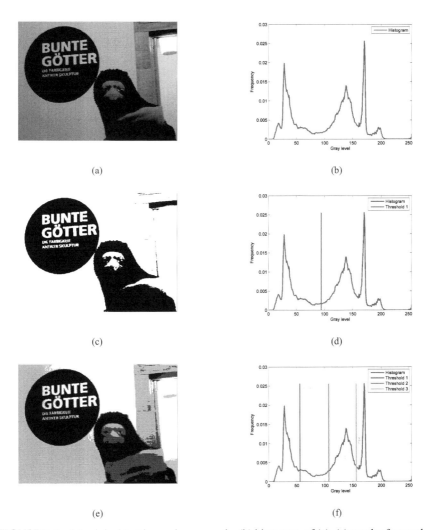

FIGURE 5.4 (a) original test image in gray scale, (b) histogram of (a), (c) result after apply the bi-level segmentation process, (d) histogram and threshold for bi-level segmentation, (e) result after apply the multilevel segmentation process, (f) histogram and thresholds for multilevel segmentation.

$$Ph_i^c = \frac{h_i^c}{NP}, \quad \sum_{i=1}^{NP} Ph_i^c = 1, \quad c = \begin{cases} 1,2,3 & \text{if} \quad \text{RGB Image} \\ 1 & \text{if} \quad \text{Gray scale Image} \end{cases} \quad (9)$$

where i is a specific intensity level $(0 \le i \le L-1)$, c is the component of the image which depends if the image is gray scale or RGB whereas NP is the total number of pixels in the image. h_i^c (histogram) is the number of pixels that corresponds to the i intensity level in c. The histogram is normalized within a probability distribution Ph_i^c. For the simplest segmentation (bi-level) two classes are defined as:

$$C_1 = \frac{Ph_1^c}{\omega_0^c(th)}, \ldots, \frac{Ph_{th}^c}{\omega_0^c(th)} \quad \text{and} \quad C_2 = \frac{Ph_{th+1}^c}{\omega_1^c(th)}, \ldots, \frac{Ph_L^c}{\omega_1^c(th)} \tag{10}$$

where $\omega_0(th)$ and $\omega_1(th)$ are probabilities distributions for C_1 and C_2, as it is shown by Eq. (11).

$$\omega_0^c(th) = \sum_{i=1}^{th} Ph_i^c, \quad \omega_1^c(th) = \sum_{i=th+1}^{L} Ph_i^c \tag{11}$$

It is necessary to compute mean levels μ_0^c and μ_1^c that define the classes using Eq. (12). Once those values are calculated, the Otsu variance between classes σ^{2^c} is calculated using Eq. (13) as follows:

$$\mu_0^c = \sum_{i=1}^{th} \frac{iPh_i^c}{\omega_0^c(th)}, \quad \mu_1^c = \sum_{i=th+1}^{L} \frac{iPh_i^c}{\omega_1^c(th)} \tag{12}$$

$$\sigma^{2^c} = \sigma_1^c + \sigma_2^c \tag{13}$$

Notice that for both equations, Eqs. (12) and (13), c depends on the type of image. In Eq. (13) the number two is part of the Otsu's variance operator and does not represent an exponent in the mathematical sense. Moreover σ_1^c and σ_2^c in Eq. (13) are the variances of C_1 and C_2 which are defined as:

$$\sigma_1^c = \omega_0^c \left(\mu_0^c + \mu_T^c \right)^2, \quad \sigma_2^c = \omega_1^c \left(\mu_1^c + \mu_T^c \right)^2 \tag{14}$$

where $\mu_T^c = \omega_0^c \mu_0^c + \omega_1^c \mu_1^c$ and $\omega_0^c + \omega_1^c = 1$. Based on the values σ_1^c and σ_2^c, Eq. (15) presents the objective function.

$$J(th) = \max(\sigma^{2^c}(th)), \quad 0 \le th \le L-1 \tag{15}$$

where $\sigma^{2^c}(th)$ is the Otsu's variance for a given th value. Therefore, the optimization problem is reduced to find the intensity level (th) that maximizes Eq. (15).

Otsu's method is applied for a single component of an image. In case of RGB images, it is necessary to apply separation into single component images. The previous description of such bi-level method can be extended for the identification of multiple thresholds. Considering k thresholds it is possible separate the original image into k classes using Eq. (8), then it is necessary to compute the k variances and their respective elements. The objective function $J(th)$ in Eq. (15) can thus be rewritten for multiple thresholds as follows:

$$J(\mathbf{TH}) = \max(\sigma^{2^c}(\mathbf{TH})), \quad 0 \le th_i \le L - 1, \quad i = 1, 2, ..., k \quad (16)$$

where $\mathbf{TH} = [th_1, th_2, ..., th_{k-1}]$, is a vector containing multiple thresholds and the variances are computed through Eq. (17) as follows.

$$\sigma^{2^c} = \sum_{i=1}^{k} \sigma_i^c = \sum_{i=1}^{k} \omega_i^c \left(\mu_i^c - \mu_T^c \right)^2 \quad (17)$$

Here, i represents and specific class, ω_i^c and μ_j^c are respectively the probability of occurrence and the mean of a class. In multilevel thresholding, such values are obtained as:

$$\omega_0^c(th) = \sum_{i=1}^{th_1} Ph_i^c$$

$$\omega_1^c(th) = \sum_{i=th_1+1}^{th_2} Ph_i^c$$

$$\vdots \qquad \vdots$$

$$\omega_{k-1}^c(th) = \sum_{i=th_k+1}^{L} Ph_i^c \quad (18)$$

And for the mean values:

$$\mu_0^c = \sum_{i=1}^{th_1} \frac{iPh_i^c}{\omega_0^c(th_1)}$$

$$\mu_1^c = \sum_{i=th_1+1}^{th_2} \frac{iPh_i^c}{\omega_0^c(th_2)}$$

$$\vdots \qquad \vdots$$

$$\mu_{k-1}^c = \sum_{i=th_k+1}^{L} \frac{iPh_i^c}{\omega_1^c(th_k)} \quad (19)$$

Similar to the bi-level case, for multilevel thresholding using the Otsu's method c corresponds to the image components, RGB $c = 1, 2, 3$ and gray scale $c = 1$.

5.3.3 KAPUR'S METHOD (ENTROPY CRITERION METHOD)

Another nonparametric method that is used to determine the optimal threshold values has been proposed by Kapur [5]. It is based on the entropy and the probability distribution of the image histogram. The method aims to find the optimal th that maximizes the overall entropy. The entropy of an image measures the compactness and separability among classes. In this sense, when the optimal th value appropriately separates the classes, the entropy has the maximum value. For the bi-level example, the objective function of the Kapur's problem can be defined as:

$$J(th) = H_1^c + H_2^c, \quad c = \begin{cases} 1,2,3 & \text{if} \quad \text{RGB Image} \\ 1 & \text{if} \quad \text{Gray scale Image} \end{cases} \tag{20}$$

where the entropies H_1 and H_2 are computed by the following model:

$$H_1^c = \sum_{i=1}^{th} \frac{Ph_i^c}{\omega_o^c} \ln\left(\frac{Ph_i^c}{\omega_o^c}\right), \qquad H_2^c = \sum_{i=th+1}^{L} \frac{Ph_i^c}{\omega_1^c} \ln\left(\frac{Ph_i^c}{\omega_1^c}\right) \tag{21}$$

Ph_i^c is the probability distribution of the intensity levels which is obtained using Eq.(9). $\omega_0(th)$ and $\omega_1(th)$ are probabilities distributions for C_1 and C_2. $\ln(\gamma)$ stands for the natural logarithm. Similar to the Otsu's method, the entropy-based approach can be extended for multiple threshold values; for such a case, it is necessary to divide the image into k classes using the similar number of thresholds. Under such conditions, the new objective function is defined as:

$$J(\mathbf{TH}) = \max\left(\sum_{i=1}^{k} H_i^c\right), \quad c = \begin{cases} 1,2,3 & \text{if} \quad \text{RGB Image} \\ 1 & \text{if} \quad \text{Gray scale Image} \end{cases} \tag{22}$$

where $\mathbf{TH} = [th_1, th_2, ..., th_{k-1}]$, is a vector that contains the multiple thresholds. Each entropy is computed separately with its respective th value, so Eq. (23) is expanded for k entropies.

$$H_1^c = \sum_{i=1}^{th_1} \frac{Ph_i^c}{\omega_0^c} \ln\left(\frac{Ph_i^c}{\omega_0^c}\right),$$

$$H_2^c = \sum_{i=th_1+1}^{th_2} \frac{Ph_i^c}{\omega_1^c} \ln\left(\frac{Ph_i^c}{\omega_1^c}\right),$$

$$\vdots \qquad\qquad \vdots$$

$$H_k^c = \sum_{i=th_k+1}^{L} \frac{Ph_i^c}{\omega_{k-1}^c} \ln\left(\frac{Ph_i^c}{\omega_{k-1}^c}\right) \qquad (23)$$

The values of the probability occurrence $(\omega_0^c, \omega_1^c, ..., \omega_{k-1}^c)$ of the k classes are obtained using Eq. (18) and the probability distribution Ph_i^c with Eq. (12). Finally, it is necessary to use Eq. (8) to separate the pixels into the corresponding classes.

5.3.4 THRESHOLDING EXAMPLE USING OTSU'S AND KAPUR'S METHODS

In order to better explain the thresholding processes using the techniques proposed by Otsu and Kapur, there is presented a graphical example. The image to be segmented is presented in Figure 5.5 (a), meanwhile it histogram is in Figure 5.5 (b). For this example are used two different number of thresholds first for bi-level thresholding where $L = 2$ and the second one for multilevel thresholding where $L = 5$.

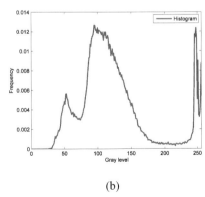

(a)   (b)

FIGURE 5.5 (a) original test image in gray scale, (b) histogram of (a).

Otsu's and Kapur's methods are applied over the histogram of Figure 5.5 (a), they use as search strategy the exhaustive search, that is time consuming and computationally expensive. The results obtained are presented in Figures 5.6 and 5.7. The value for bi-level segmentation using Otsu is $th = \{170\}$, the binarized image and the histogram are shown in

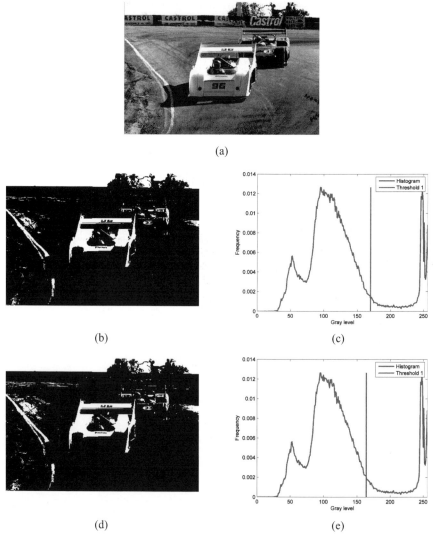

FIGURE 5.6   (a) Original image in gray scale, (b) segmented image using Otsu's method for bi-level thresholding, (b) histogram and the best threshold value, (d) segmented image using Kapur's method for bi-level thresholding, (e) histogram and the best threshold value.

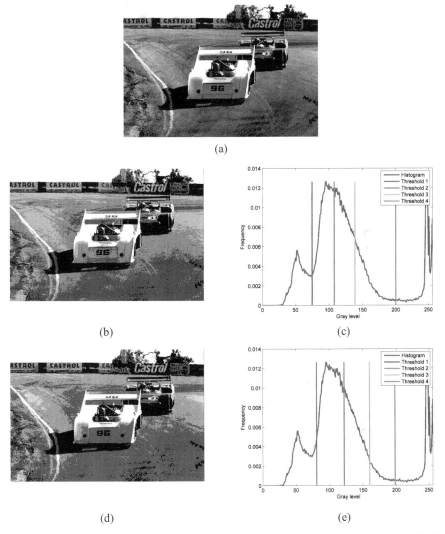

FIGURE 5.7 (a) Original image in gray scale, (b) segmented image using Otsu's method for multilevel thresholding, (c) histogram and the best threshold values, (d) segmented image using Kapur's method for multilevel thresholding, (e) histogram and the best threshold values.

Figure 5.6 (b) and 5.6 (c). For Kapur threshold value found is $th = \{164\}$ and the resultant binarized image and the histogram are exposed in Figures 5.6 (d) and 5.6 (e).

The results obtained for bi-level thresholding using Otsu's and Kapur's techniques are very similar, however the differences between both methods can be observed in multilevel thresholding. In this case the threshold values obtained using Otsu are $\mathbf{TH} = \{75,108,139,200\}$ and for Kapur $\mathbf{TH} = \{81,122,160,198\}$. In Figure 5.7 (b) and 5.7 (c) are exposed the segmented image and the histogram with the threshold values obtained by Otsu's function, meanwhile Figures 5.7 (d) and 5.7 (e) presents the result using the Kapur's function. From the histograms is possible to see how the classes are defined by the threshold values, this is one of the main differences between both methods.

5.4   MULTILEVEL THRESHOLDING USING HARMONY SEARCH ALGORITHVM (HSMA)

In this section the implementation of the harmony search algorithm as a search strategy for multilevel thresholding using as objective function one of the methods proposed by Otsu or Kapur is introduced. First is necessary to define how the harmonies are generated by the algorithm. Finally a description of each step of the proposed technique is provided.

5.4.1   HARMONY REPRESENTATION

Each harmony (candidate solution) uses k different elements as decision variables within the optimization algorithm. Such decision variables represent a different threshold point th that is used for the segmentation. Therefore, the complete population is represented as:

$$\mathbf{HM} = \left[\mathbf{x}_1^c, \mathbf{x}_2^c, ..., \mathbf{x}_{HMS}^c \right]^T, \quad \mathbf{x}_i^c = \left[th_1^c, th_2^c, ..., th_k^c \right] \tag{24}$$

where T refers to the transpose operator, HMS is the size of the harmony memory, $\mathbf{x}_i$ is the i-th element of HM and $c = 1,2,3$ is set for RGB images while $c = 1$ is chosen for gray scale images. For this problem, the boundaries of the search space are set to $l = 0$ and $u = 255$, which correspond to image intensity levels.

5.4.2   IMPLEMENTATION OF HSA FOR MULTILEVEL THRESHOLDING

The proposed segmentation algorithm has been implemented considering two different objective functions: Otsu and Kapur. Therefore, the HSA has been coupled with the Otsu and Kapur functions, producing two different segmentation algorithms. The implementation of both algorithms can be summarized into the following steps:

Step 1: Read the image I and if it is RGB separate it into I_R, I_G and I_B. If I is gray scale store it into I_{Gr}. $c = 1,2,3$ for RGB images or $c = 1$ for gray scale images.

Step 2: Obtain histograms: for RGB images h^R, h^G, h^B and for gray scale images h^{Gr}.

Step 3: Calculate the probability distribution using Eq. (9) and obtain the histograms.

Step 4: Initialize the HSA parameters: HMS, k, $HMCR$, PAR, BW, NI, and the limits l and u.

Step 5: Initialize a HM $\mathbf{x}_i^c$ of HMS random particles with k dimensions.

Step 6: Compute the values ω_i^c and μ_j^c. Evaluate each element of **HM** in the objective function $J(\mathbf{HM})$ Eq. (16) or Eq. (22) depending on the thresholding method (Otsu or Kapur, respectively).

Step 7: Improvise a new harmony $\mathbf{x}_{new}^c$ as follows:

    for $(j = 1$ to $n)$ do

        if $(r_1 < HCMR)$ then

$$x_j^{new^c} = x_j^{a^c} \text{ where } a \in (1,2,\dots,HMS)$$

           if $(r_2 < PAR)$ then

$$x_j^{new^c} = x_j^{a^c} \pm r_3 \gamma BW \text{ where } r_1, r_2, r_3 \in rand(0,1)$$

           end if

           if $x_j^{new^c} < l_j$

$$x_j^{new^c} = l_j$$

           end if

           if $x_j^{new^c} > u_j$

$$x_j^{new^c} = u_j$$

           end if

        else

$$x_j^{new^c} = l_j + r\, \gamma \,(u_j - l_j) \text{ whew } r \in rand\,(0,1)$$

        end if

    end for

Step 8: Update the *HM* as $\mathbf{x}^c_{worst} = \mathbf{x}^c_{new}$ if $f\left(\mathbf{x}^c_{new}\right) > f\left(\mathbf{x}^c_{worst}\right)$

Step 9: If *NI* is completed or the stop criteria is satisfied, then jump to Step 10; otherwise go back to Step 6.

Step 10: Select the harmony that has the best $\mathbf{x}^c_{best}$ objective function value.

Step 11: Apply the thresholds values contained in $\mathbf{x}^c_{best}$ to the image *I* Eq. (8).

5.5 EXPERIMENTAL RESULTS

The HSMA has been tested under a set of 5 benchmark images. Some of these images are widely used in the image processing literature to test different methods (Lena, Cameraman, Hunter, Baboon, etc.) [18, 19]. All the images have the same size (512 × 512 pixels) and they are in JPEG format. In Figure 5.8 the five images and their histograms are presented, and the numerical outcomes are analyzed later considering the complete set of benchmark images.

(a)

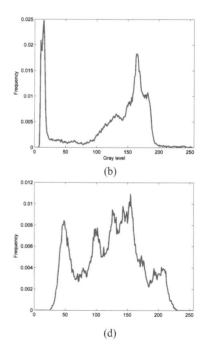

(b)

(c)

(d)

FIGURE 5.8 Continued

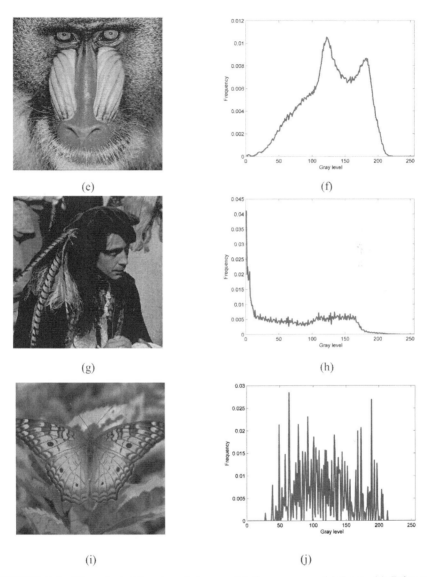

FIGURE 5.8 The selected benchmark images (a) Camera man, (c) Lena, (e) Baboon, (g) Hunter and (j) Butterfly. (b), (d), (f), (h), (j) histograms of the images.

Since HSMA is stochastic, it is necessary to employ an appropriate statistical metrics to measure its efficiency. Hence, the results have been reported executing the algorithm 35 times for each image. In order to maintain compatibility with similar works reported in the literature [8, 9, 20, 21], the number of thresholds points used in the test are $th = 2, 3, 4, 5$. In the experiments, the stop criterion is the number of times in which the best fitness values remains with no change. Therefore, if the fitness value for the best harmony remains unspoiled in 10% of the total number of iterations (NI), then the HSA is stopped. To evaluate the stability and consistency, it has been computed the standard deviation (STD) from the results obtained in the 35 executions. Since the STD represents a measure about how the data are dispersed, the algorithm becomes more instable as the STD value increases [9]. Equation (25) shows the model used to calculate the STD value.

$$STD = \sqrt{\sum_{i=1}^{NI} \frac{(bf_i - av)}{Ru}} \tag{25}$$

where bf_i is the best fitness of the i-th iteration, av is the average value of bf and Ru is the number of total executions ($Ru = 35$).

On the other hand, as an index of quality of the segmentation, the peak-to-signal ratio (PSNR) is used to assess the similarity of the segmented image against a reference image (original image) based on the produced mean square error (MSE) [8, 22]. Both PSNR and MSE are defined as:

$$PSNR = 20\log_{10}\left(\frac{255}{RMSE}\right), \quad (\text{dB})$$

$$RMSE = \sqrt{\frac{\sum_{i=1}^{ro}\sum_{j=1}^{co}\left(\mathbf{I}_o^c(i,j) - \mathbf{I}_{th}^c(i,j)\right)}{ro \times co}} \tag{26}$$

where $\mathbf{I}_o^c$ is the original image, $\mathbf{I}_{th}^c$ is the segmented image, $c = 1$ for gray scale and $c = 3$ for RGB images whereas ro, co are the total number of rows and columns of the image, respectively.

The parameters of HSA are configured experimentally guided by the values proposed in the related literature [16]. Table 5.3 presents the values used for all the test of the proposed HSMA in all the benchmark images and for Otsu's and Kapur's methods.

TABLE 5.3 Parameter Configuration of HSMA

HM	HMCR	PAR	BW	NI
100	0.75	0.5	0.5	300

5.5.1 RESULTS USING OTSU'S METHOD

This section analyzes the results of HSMA after considering the variance among classes (Eq. 16) as the objective function, just as it has been proposed by Otsu [4]. The approach is applied over the complete set of benchmark images whereas the results are registered in Table 5.4. Such results present the best threshold values after testing the proposed method with four different threshold points th = 2,3,4,5. The table also features the *PSNR* and the *STD* values. It is evident that the *PSNR* and *STD* values increase their magnitude as the number of threshold points also increases.

The processing results for the original images are presented in Tables 5.5–5.9. Such results show the segmented images considering four different threshold points th = 2,3,4,5. The tables also show the evolution of the objective function during one execution.

TABLE 5.4 Results After Apply the HSMA Using Otsu's Function to the Set of Benchmark Images

Image	k	Thresholds	PSNR	STD
Camera man	2	70, 144	17.247	2.30 E-12
	3	59, 119, 156	20.211	1.55 E-02
	4	42, 95, 140, 170	21.533	2.76 E-12
	5	36, 82, 122, 149, 173	23.282	5.30 E-03
Lena	2	91, 150	15.401	9.22 E-13
	3	79, 125, 170	17.427	2.99 E-02
	4	73, 112, 144, 179	18.763	2.77 E-01
	5	71, 107, 134, 158, 186	19.443	3.04 E-01
Baboon	2	97, 149	15.422	6.92 E-13
	3	85, 125, 161	17.709	1.92 E-02
	4	71, 105, 136, 167	20.289	5.82 E-02
	5	66, 97, 123, 147, 173	21.713	4.40 E-01
Hunter	2	51, 116	17.875	2.30 E-12
	3	36, 86, 135	20.350	2.30 E-12
	4	27, 65, 104, 143	22.203	1.22 E-02
	5	22, 53, 88, 112, 152	23.703	1.84 E-12
Butterfly	2	99, 151	13.934	7.30 E-02
	3	82, 119, 160	16.932	6.17 E-01
	4	71, 102, 130, 163	19.259	3.07 E+00
	5	62, 77, 109, 137, 167	21.450	3.87 E+00

TABLE 5.5   Results After Apply the HSMA Using Otsu's Over the Camera Man Image

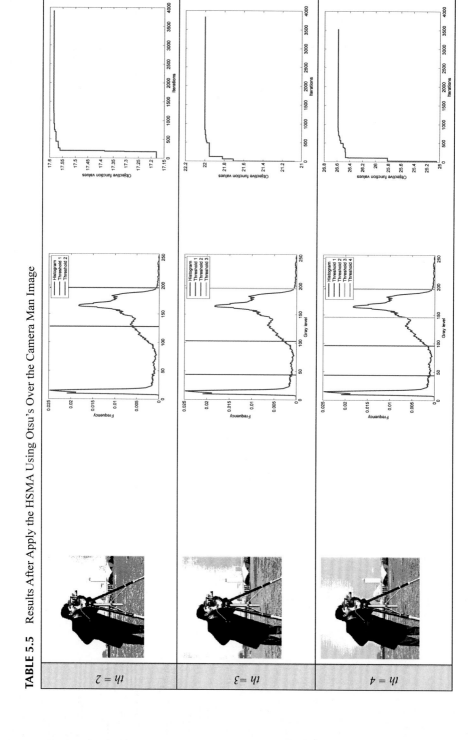

TABLE 5.5 Continued

$th = 5$

TABLE 5.6 Results After Apply the HSMA Using Otsu's Over Lena Image

$th = 2$

TABLE 5.6   Continued

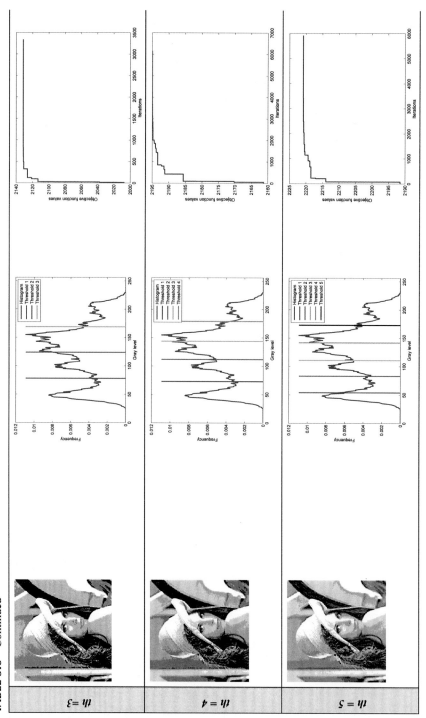

TABLE 5.7   Results After Apply the HSMA Using Otsu's Over Baboon Image

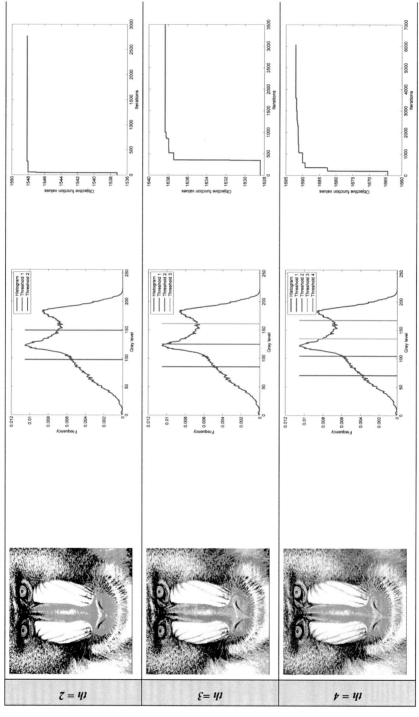

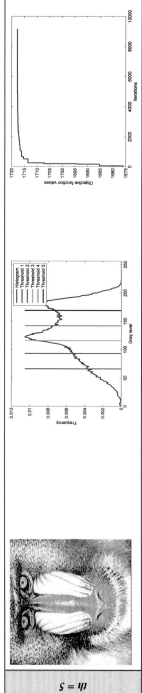

th = 5

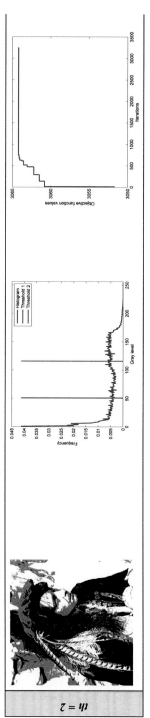

TABLE 5.8   Results After Apply the HSMA Using Otsu's Over Hunter Image

th = 2

TABLE 5.8 Continued

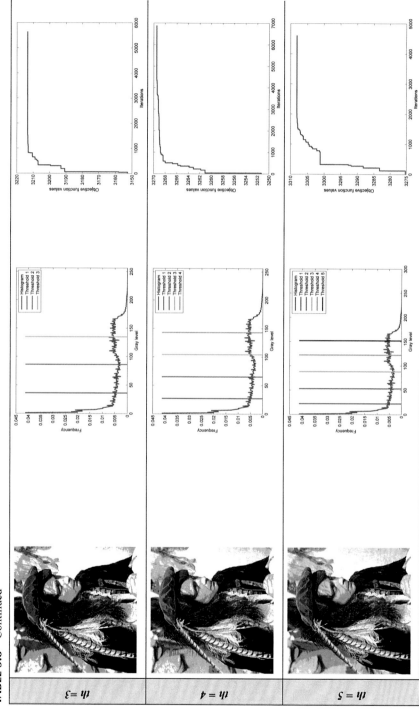

TABLE 5.9　Results After Apply the HSMA Using Otsu's Over Butterfly Image

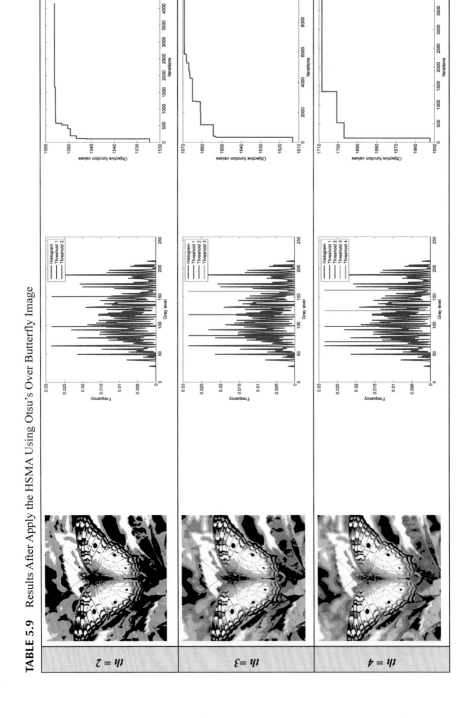

TABLE 5.9 Continued

th = 5

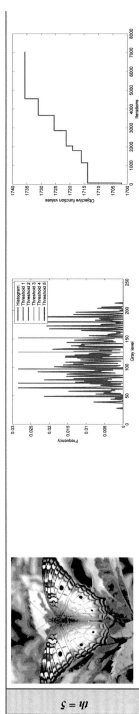

5.5.2 RESULTS USING KAPUR'S METHOD

This section analyzes the performance of HSMA after considering the entropy function (Eq. 22) as objective function, as it has been proposed by Kapur in [5]. Table 5.10 presents the experimental results after the application of HSMA over the entire set of benchmark images. The values listed are: *PSNR*, *STD* and the best threshold values of the last population ($\mathbf{x}_t^B$). The same test procedure that was previously applied to the Otsu's method is used with the Kapur's method, also considering the same stop criterion and a similar HSA parameter configuration.

The results after apply the HSMA to the selected benchmark images are presented in Tables 5.11–5.15. Four different threshold points have been employed: *th* = 2,3,4,5. All tables exhibit the segmented image,

TABLE 5.10 Results After Apply the HSMA Using Kapur's to the Set of Benchmark Images

Image		Thresholds	*PSNR*	*STD*
Camera man	2	128, 196	13.626	3.60 E-15
	3	44, 103, 196	14.460	1.40 E-03
	4	44, 96, 146, 196	20.153	1.20 E-03
	5	24, 60, 98, 146, 196	20.661	2.75 E-02
Lena	2	96, 163	14.638	3.60 E-15
	3	23, 96, 163	16.218	7.66 E-02
	4	23, 80, 125, 173	19.287	1.44 E-14
	5	23, 71, 109, 144, 180	21.047	1.22 E-02
Baboon	2	79, 143	16.016	1.08 E-14
	3	79, 143, 231	16.016	7.19 E-02
	4	44, 98, 152, 231	18.485	8.47 E-02
	5	33, 74, 114, 159, 231	20.507	1.08 E-14
Hunter	2	92, 179	15.206	1.44 E-14
	3	59, 117, 179	18.500	4.16 E-04
	4	44, 89, 133, 179	21.065	4.31 E-04
	5	44, 89, 133, 179, 222	21.086	3.43 E-02
Butterfly	2	27, 213	8.1930	2.25 E-02
	3	27, 120, 213	13.415	8.60 E-04
	4	27, 96, 144, 213	16.725	3.80 E-03
	5	27, 83, 118, 152, 213	19.413	3.90 E-03

TABLE 5.11 Results After Apply the HSMA using Kapur's Over the Camera Man Image

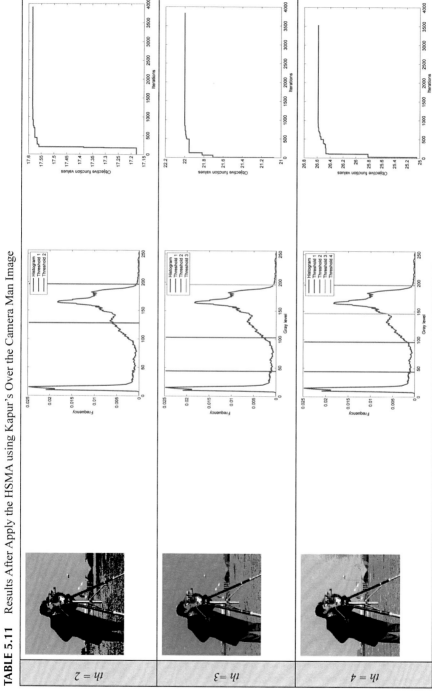

TABLE 5.11   Continued

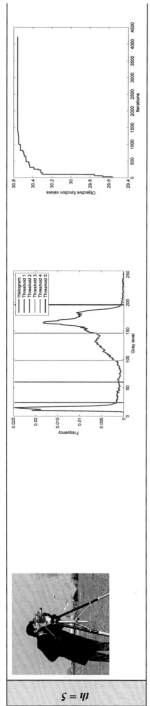

th = 5

TABLE 5.12   Results After Apply the HSMA Using Kapur's Over the Lena Image

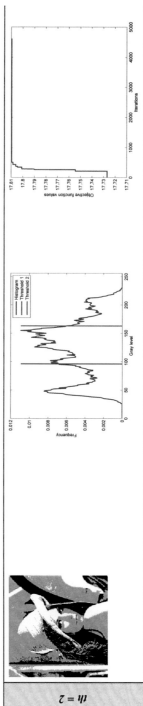

th = 2

TABLE 5.12 Continued

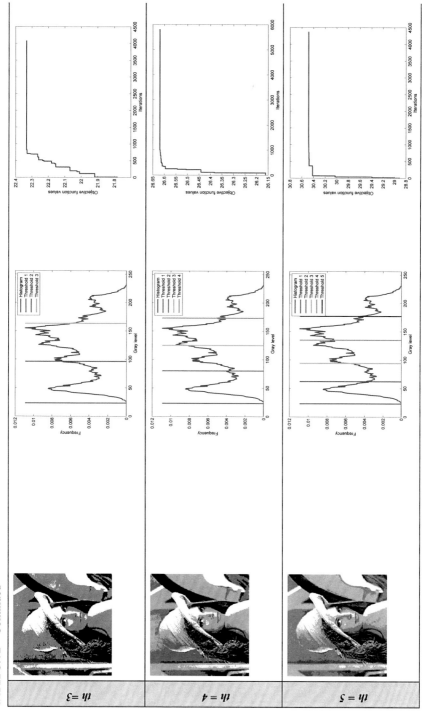

TABLE 5.13   Results After Apply the HSMA Using Kapur's Over the Lena Image

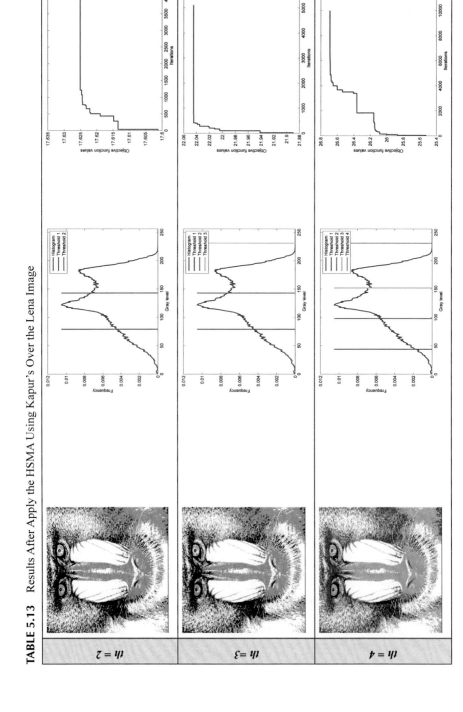

TABLE 5.13 Continued

th = 5

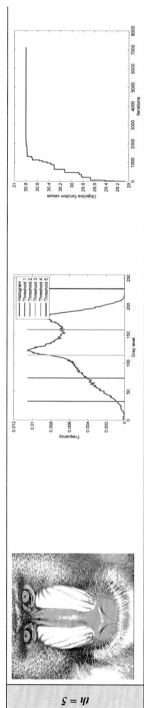

TABLE 5.14 Results After Apply the HSMA Using Kapur's Over the Hunter Image

th = 2

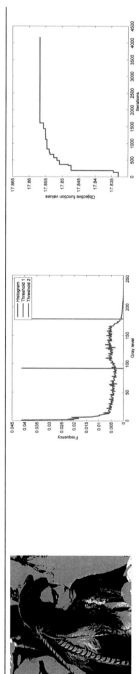

TABLE 5.14 Continued

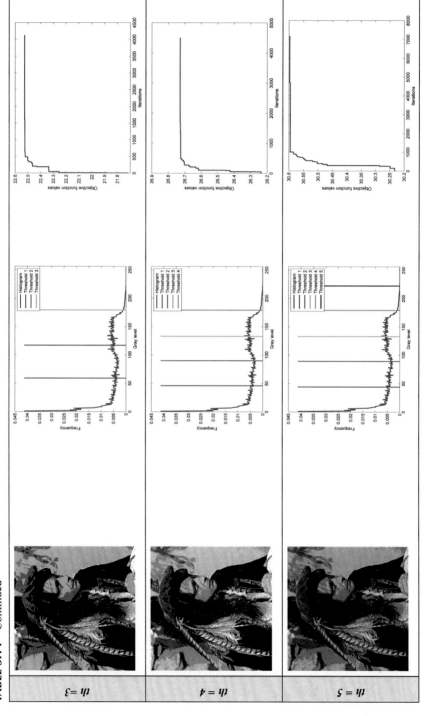

TABLE 5.15 Results After Apply the HSMA Using Kapur's over the Butterfly Image

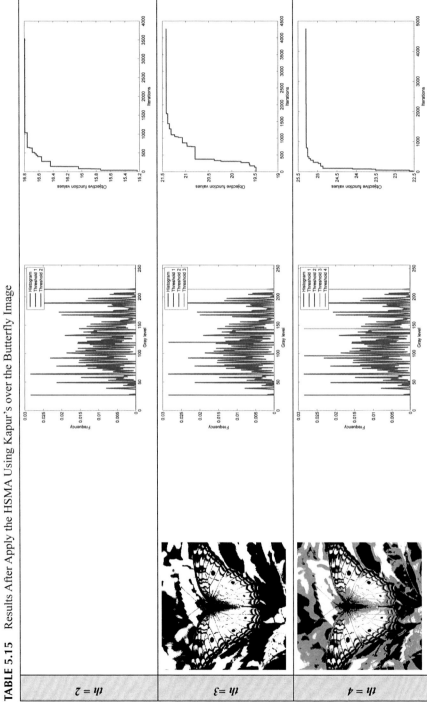

TABLE 5.15   Continued

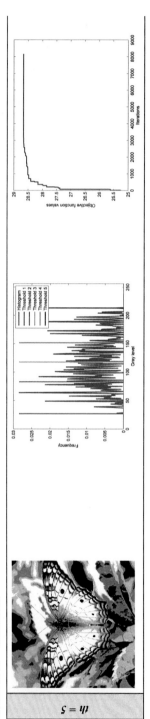

the approximated histogram and the evolution of the fitness value during the execution of the HSA method.

From the results of both Otsu's and Kapur's methods, it is possible to appreciate that the HSMA converges (stabilizes) after a determined number of iterations depending on the value. For experimental purposes HSMA continues running still further, even though the stop criterion is achieved. In this way, the graphics show that convergence is often reached in the first iterations of the optimization process. The segmented images provide evidence that the outcome is better with $th = 4$ and $th = 5$; however, if the segmentation task does not require to be extremely accurate then it is possible to select $th = 3$.

5.5.3 COMPARISONS OF THE USE OF OTSU AND KAPUR IN HSMA

In order to statistically compare the results from Tables 5.3 and 5.9, a nonparametric significance proof known as the Wilcoxon's rank test [23, 24] for 35 independent samples has been conducted. Such proof allows assessing result differences among two related methods. The analysis is performed considering a 5% significance level over the peak-to-signal ratio (*PSNR*) data corresponding to the five threshold points. Table 5.16 reports the *p*-values

TABLE 5.16 *p*-Values Produced by Wilcoxon's Test Comparing Otsu vs. Kapur Over the Averaged PSNR from Tables 5.4 and 5.10

Image	k	p-Value Otsu vs. Kapur
Camera man	2	1.0425 E-16
	3	2.1435 E-15
	4	2.6067 E-16
	5	6.2260 E-16
Lena	2	1.0425 E-16
	3	9.4577 E-15
	4	9.7127 E-15
	5	1.2356 E-12
Baboon	2	1.0425 E-16
	3	1.7500 E-02
	4	5.3417 E-14
	5	1.4013 E-14

TABLE 5.16   Continued

Image	k	p-Value Otsu vs. Kapur
Hunter	2	1.0425 E-16
	3	2.6067 E-16
	4	6.6386 E-14
	5	6.4677 E-15
Butterfly	2	1.1615 E-14
	3	2.5697 E-14
	4	3.7190 E-13
	5	1.7941 E-06

produced by Wilcoxon's test for a pair-wise comparison of the *PSNR* values between the Otsu and Kapur objective functions. As a null hypothesis, it is assumed that there is no difference between the values of the two objective functions. The alternative hypothesis considers an existent difference between the values of both approaches. All p-values reported in the Table 5.16 are less than 0.05 (5% significance level), which is a strong evidence against the null hypothesis, indicating that the Otsu *PSNR* mean values for the performance are statistically better and it has not occurred by chance.

5.5.4   COMPARISONS AMONG HSMA AND OTHER SIMILAR APPROACHES

The results produced by HSMA have been compared with the generated by state-of-the-art thresholding methods such Genetic Algorithms (GA) [29], Particle Swarm Optimization (PSO) [8] and Bacterial Foraging (BF) [9]. All the algorithms run 35 times over benchmark image. For each image, the *PSNR*, the *STD* and the mean of the objective function values are calculated. Moreover, the entire test is performed using both Otsu's and Kapur's objective functions.

Table 5.17 presents the computed values for benchmark images. It is clear that the HSMA delivers better performance than the others. Such values are computed using the Otsu's method as the objective function. On the other hand, the same experiment has been performed using the Kapur's method. The results of this experiment are presented in Table 5.18 and show that the proposed HSMA algorithm is better in comparison with the GA, PSO and BF.

TABLE 5.17 Comparisons Between HSMA, GA, PSO and BF, Applied Over the Test Images Using Otsu's Method

Image	k	HSMA			GA			PSO			BF		
		PSNR	STD	Mean	PSNR	STD	Mean	PSNR	STD	Mean	STD	PSNR	Mean
Camera man	2	17.247	2.30 E-12	3651.9	17.048	0.0232	3604.5	17.033	0.0341	3598.3	0.0345	17.058	3590.9
	3	20.211	1.55 E-02	3727.4	17.573	0.1455	3678.3	19.219	0.2345	3662.7	0.2459	20.035	3657.5
	4	21.533	2.76 E-12	3782.4	20.523	0.2232	3781.5	21.254	0.3142	3777.4	0.4560	21.209	3761.4
	5	23.282	5.30 E-03	3813.7	21.369	0.4589	3766.4	22.095	0.5089	3741.6	0.5089	22.237	3789.8
Lena	2	15.401	9.22 E-13	1964.4	15.040	0.0049	1960.9	15.077	0.0033	1961.4	2.99 E-04	15.031	1961.5
	3	17.427	2.99 E-02	2131.4	17.304	0.1100	2126.4	17.276	0.0390	2127.7	0.0061	17.401	2128.0
	4	18.763	2.77 E-01	2194.9	17.920	0.2594	2173.7	18.305	0.1810	2180.6	0.0081	18.507	2189.0
	5	19.443	3.04 E-01	2218.7	18.402	0.3048	2196.2	18.770	0.2181	2212.5	0.0502	19.001	2215.6
Baboon	2	15.422	6.92 E-13	1548.1	15.304	0.0031	1547.6	15.088	0.0077	1547.9	8.88 E-04	15.353	1548.0
	3	17.709	1.92 E-02	1638.3	17.505	0.1750	1633.5	17.603	0.0816	1635.3	0.0287	17.074	1637.0
	4	20.289	5.82 E-02	1692.1	18.708	0.2707	1677.7	19.233	0.0853	1684.3	0.0336	19.654	1690.7
	5	21.713	4.40 E-01	1717.5	20.203	0.3048	1712.9	20.526	0.1899	1712.9	0.1065	21.160	1716.7
Hunter	2	17.875	2.30 E-12	3054.2	17.088	0.0470	3064.1	17.932	0.2534	3064.1	0.0322	17.508	3064.1
	3	20.350	2.30 E-12	3213.4	20.045	0.1930	3212.9	19.940	0.9727	3212.4	0.9627	20.350	3213.4
	4	22.203	1.22 E-02	3269.5	20.836	0.6478	3268.4	21.128	2.2936	3266.3	2.2936	21.089	3266.3
	5	23.703	1.84 E-12	3308.1	21.284	1.6202	3305.6	22.026	4.1811	3276.3	3.6102	22.804	3291.1
Butterfly	2	13.934	7.30 E-02	1553.0	13.007	0.0426	1553.0	13.092	0.0846	1553.0	0.0643	13.890	1553.0
	3	16.932	6.17 E-01	1669.2	15.811	0.3586	1669.0	17.261	2.6268	1665.7	1.2113	17.285	1667.2
	4	19.259	3.07 E+00	1708.3	17.104	0.6253	1709.9	17.005	3.7976	1702.9	2.2120	17.128	1707.0
	5	21.450	3.87 E+00	1728.0	18.593	0.5968	1734.4	18.099	6.0747	1730.7	3.5217	18.9061	1733.0

TABLE 5.18 Comparisons Between HSMA, GA, PSO and BF, Applied Over the Test Images Using Kapur's Method

Image	k	HSMA			GA			PSO			BF		
		PSNR	STD	Mean	PSNR	STD	Mean	PSNR	STD	Mean	PSNR	STD	Mean
Cameraman	2	13.626	3.60 E-15	17.584	11.941	0.1270	15.341	12.259	0.1001	16.071	12.264	0.0041	16.768
	3	14.460	1.40 E-03	22.007	14.827	0.2136	20.600	15.211	0.1107	21.125	15.250	0.0075	21.498
	4	20.153	1.20 E-03	26.586	17.166	0.2857	24.267	18.000	0.2005	25.050	18.406	0.0081	25.093
	5	20.661	2.75 E-02	30.553	19.795	0.3528	28.326	20.963	0.2734	28.365	21.211	0.0741	30.026
Lena	2	14.638	3.60 E-15	17.809	12.334	0.0049	16.122	12.345	0.0033	16.916	12.345	2.99 E-4	16.605
	3	16.218	7.66 E-02	22.306	14.995	0.1100	20.920	15.133	0.0390	20.468	15.133	0.0061	20.812
	4	19.287	1.44 E-14	26.619	17.089	0.2594	23.569	17.838	0.1810	24.449	17.089	0.0081	26.214
	5	21.047	1.22 E-02	30.485	19.549	0.3043	27.213	20.442	0.2181	27.526	19.549	0.0502	28.046
Baboon	2	16.016	1.08 E-14	17.625	12.184	0.0567	16.425	12.213	0.0077	16.811	12.216	8.88 E-4	16.889
	3	16.016	7.19 E-02	22.117	14.745	0.1580	21.069	15.008	0.0816	21.088	15.211	0.0287	21.630
	4	18.485	8.47 E-02	26.671	16.935	0.1765	25.489	17.574	0.0853	24.375	17.999	0.0336	25.446
	5	20.507	1.08 E-14	30.800	19.662	0.2775	29.601	20.224	0.1899	30.994	20.720	0.1065	30.887
Hunter	2	15.206	1.44 E-14	17.856	12.349	0.0148	16.150	12.370	0.0068	15.580	12.373	0.0033	16.795
	3	18.500	4.16 E-04	22.525	14.838	0.1741	21.026	15.128	0.0936	20.639	15.553	0.1155	21.860
	4	21.065	4.31 E-04	26.728	17.218	0.2192	25.509	18.040	0.1560	27.085	18.381	0.0055	26.230
	5	21.086	3.43 E-02	30.612	19.563	0.3466	29.042	20.533	0.2720	29.013	21.256	0.0028	28.856
Butterfly	2	8.1930	2.25 E-02	16.791	10.470	0.0872	15.481	10.474	0.0025	14.098	10.474	0.0014	15.784
	3	13.415	8.60 E-04	21.417	11.628	0.2021	20.042	12.313	0.1880	19.340	12.754	0.0118	21.308
	4	16.725	3.80 E-03	25.292	13.314	0.2596	23.980	14.231	0.2473	25.190	14.877	0.0166	25.963
	5	19.413	3.90 E-03	28.664	15.756	0.3977	27.411	16.337	0.2821	27.004	16.828	0.0877	27.980

5.6 CONCLUSIONS

The proposed approach combines the good search capabilities of HSA algorithm and the use of some objective functions that have been suggested by the popular MT methods of Otsu and Kapur. The peak signal-to-noise ratio (PSNR) is used to assess the segmentation quality by considering the coincidences between the segmented and the original images. In this work, a simple HSA implementation without any modification is considered in order to demonstrate that it can be applied to image processing tasks. The study explores the comparison between two versions of HSMA: one employs the Otsu objective function while the other uses the Kapur criterion. Results show that the Otsu function delivers better results than the Kapur criterion. Such conclusion has been statistically proved considering the Wilcoxon test.

The proposed approach has been compared to other techniques that implement different optimization algorithms like GA, PSO and BF. The efficiency of the algorithm has been evaluated in terms of the PSNR index and the STD value. Experimental results provide evidence on the outstanding performance, accuracy and convergence of the proposed algorithm in comparison to other methods. Although the results offer evidence to demonstrate that the standard HSA method can yield good results on complicated images, the aim of this work is not to devise an MT algorithm that could beat all currently available methods, but to show that harmony search algorithms can be effectively considered as an attractive alternative for this purpose.

ACKNOWLEDGMENTS

The first author acknowledges The National Council of Science and Technology of Mexico (CONACyT) for the doctoral Grant number 215517. The Youth Institute of Jalisco (IJJ), The Ministry of Education (SEP) and the Mexican Government for partially support this research.

KEYWORDS

- **evolutionary algorithms**
- **harmony search**
- **image segmentation**

REFERENCES

1. Holland, J. H. Adaptation in Natural and Artificial Systems, University of Michigan Press, Ann Arbor, MI, 1975.
2. İlker Birbil, S., & Shu-Cherng Fang. An Electromagnetism-like Mechanism for Global Optimization. Journal of Global Optimization. 2003, Volume 25, 263–282.
3. Gonzalez, R. C., & Woods, R. E. Digital Image Processing, Addison Wesley, Reading, MA, 1992.
4. Otsu, N. A threshold selection method from gray-level histograms. IEEE Transactions on Systems, Man, Cybernetics 1979, SMC-9, 62–66.
5. Kapur, J. N., Sahoo, P. K., & Wong, A. K. C., A new method for gray-level picture thresholding using the entropy of the histogram. Computer Vision Graphics Image Processing, 1985, 2, 273–285.
6. Snyder, W., Bilbro, G., Logenthiran, A., & Rajala, S. "Optimal thresholding: A new approach," Pattern Recognit. Lett., 1990, vol. 11, pp. 803–810.
7. Kennedy, J., & Eberhart, R. Particle swarm optimization, in Proceedings of the 1995 IEEE International Conference on Neural Networks, December 1995, vol. 4, pp. 1942–1948.
8. Akay, B. A study on particle swarm optimization and artificial bee colony algorithms for multilevel thresholding, Applied Soft Computing (2012), doi: 10.1016/j.asoc.2012.03.072.
9. Sathya, P. D., & Kayalvizhi, R. Optimal multilevel thresholding using bacterial foraging algorithm, Expert Systems with Applications. 2011, Volume 38, 15549–15564.
10. Geem, Z. W., Kim, J. H., & Loganathan, G. V. A new heuristic optimization algorithm: harmony search. Simulations 2001, 76, 60–68.
11. Lee, K. S., Geem, Z. W., Lee, S. H., & Bae, K.-W. The harmony search heuristic algorithm for discrete structural optimization. Eng. Optim. 2005, 37, 663–684.
12. Geem, Z. W., Optimal cost design of water distribution networks using harmony search. Eng. Optim. 2006, 38, 259– 280.
13. Fernando G. Lobo, Cláudio F. Lima, & Zbigniew Michalewicz (Eds). Parameter Setting in Evolutionary Algorithms, Studies in Computational Intelligence,. 2007, Volume 54, Springer-Verlag Berlin Heidelberg.
14. Geem, Z. W., Kim, J. H., & Loganathan, G. V., A new heuristic optimization algorithm: harmony search. Simulations 2001, 76, 60–68.
15. Dixon, L. C. W., & Szegö, G. P., The global optimization problem: An introduction. In: Towards Global Optimization 2, pp. 1–15. North-Holland, Amsterdam (1978).
16. Yang, X.-S. Engineering Optimization: An Introduction with Metaheuristic Application, Wiley, USA, 2010.
17. Horng, M. Multilevel thresholding selection based on the artificial bee colony algorithm for image segmentation. Expert Systems with Applications, 2011, Volume 38, 13785–13791.
18. Pal, N. R., & Pal, S. K. "A review on image segmentation techniques," Pattern Recognit., 1993, vol. 26, pp. 1277–1294.
19. Hammouche, K., Diaf, M., & Siarry, P. A comparative study of various meta-heuristic techniques applied to the multilevel thresholding problem, Engineering Applications of Artificial Intelligence 2010, 23, 676–688.
20. Lai, C., & Tseng, D. A Hybrid Approach Using Gaussian Smoothing and Genetic Algorithm for Multilevel Thresholding. International Journal of Hybrid Intelligent Systems. 2004, 1, 143–152.

21. Yin, Peng-Yeng. A fast scheme for optimal thresholding using genetic algorithms. Signal Processing, 1999, 72, 85–95.
22. Pal, S. K., Bhandari, D., & Kundu, M. K. Genetic algorithms, for optimal image enhancement. Pattern Recognition Lett. 1994, Volume 15, 261–271.
23. Wilcoxon, F. Individual comparisons by ranking methods. Biometrics 1945, 1, 80–83.
24. Garcia, S., Molina, D., Lozano, M., & Herrera, F. A study on the use of non-parametric tests for analyzing the evolutionary algorithms' behavior: a case study on the CEC'2005 Special session on real parameter optimization. J. Heurist. (2008). doi: 10.1007/s10732-008-9080-4.

CHAPTER 6

SWARM INTELLIGENCE IN SOFTWARE ENGINEERING DESIGN PROBLEMS

TARUN KUMAR SHARMA[1] and MILLIE PANT[2]

[1]*Amity School of Engineering & Technology, Amity University Rajasthan, Jaipur, India*

[2]*Department of Applied Science and Engineering, Saharanpur Campus, IIT Roorkee, India*

CONTENTS

ABSTRACT

Being a social animal we have learned much by studying the behavior of Swarm of biological organisms. The fascinating aspect of such swarms is the fact that they reveal complex social co-operative behavior in spite of the simplicity of the individuals that forms Swarm. The collective interacting behavior of individuals in a Swarm focuses on the discipline called Swarm Intelligence (SI). SI is a new research paradigm that mimics the natural phenomenon of species, to solve successfully, complex real world optimization problems. Optimization problems exist in almost every sphere of human activities. Here our focus is on the problems of optimization in software engineering design process.

Quality Software is a sturdy foundation of Information Technology (IT) and developing the tactical competence among nationalities. It has been experienced that the working and life style is changed drastically with the emergence of Software. However, developing the quality software, involves several key issues like software reliability, costs, redundancy, user requirements, time, manpower, etc. and needs to be done in a very judicious manner. Software crises can be defused with the enhancement of software process. As it plays a significant role in influencing the management and software development system. A lot of research has been done and is still continuing to overcome the various issues in software development and design process. Hence software engineering design process possesses wide range of scope like software cost estimation, parameter estimation of reliability growth models, reducing redundancy level in modular software system models where optimization can be applied.

This chapter concentrates on the recent model instigated by the scavenge behavior of honey bee swarm, initiated by Karaboga in 2005 and employed to solve optimization problems in Software Engineering Design.

6.1 INTRODUCTION

Software engineering coined in late 1970's describes the collection of techniques that apply an engineering approach to the construction and support of software products. Software is a strong foundation of Information Technology to develop the strategic competence among nationalities [1, 2]. The development of software involves several issues and goals like reliability, overrun of costs, user requirements, etc. The improvement of software process has important practical significance to defuse software crisis, as it is influences the development and management of software [3].

Software engineering activities include managing, costing, planning, modeling, analyzing, specifying, designing, implementing, testing and maintaining software products. In the past decade numerous metaheuristic or swarm intelligence techniques have been applied to a wide range of software engineering topics namely requirement analysis, design (redundancy level), testing, estimation of reliability growth models, cost estimation.

This chapter focuses on the application of recently introduced Artificial Bee Colony (ABC) [4] and one of its novel variant to optimize three software engineering design problems (SEDP). These problems are (a) parameter estimation of software reliability growth models, (b) optimizing redundancy level in modular software system models, and (c) estimating the software cost parameters. The swarm intelligence based ABC and its variant is discussed in Section 6.3 followed by applications of ABC in SEDP. The brief description of the above named SEDP is given in Section 6.4. Experimental and Parameter settings are presented in Section 6.5. Simulation results description and conclusion drawn are presented in Sections 6.6 and 6.7, respectively.

6.2   APPLICATIONS OF ARTIFICIAL BEE COLONY ALGORITHM IN SOFTWARE ENGINEERING DESIGN OPTIMIZATION

In the field of software engineering Artificial Bee Colony (ABC) has been widely applied to software testing, cost estimation, and software reliability. Brief descriptions of the applications of ABC in software engineering are given below.

- Mala et al. [5, 6] applied ABC in optimization of software test suite.
- Bacanin et al. [7] proposed modified variant of basic ABC to describe an object-oriented software system for continuous optimization.
- Dahiya et al. [8] introduced automatic structural software tests generation using a novel search technique based on ABC.
- Kilic et al. [9] presented a solution for solving hard combinatorial automated software refactoring problem which is the lies in the domain of search-based software engineering. Adi Srikanth et al. [10] introduced optimal software test case generation to attain better path coverage using ABC algorithm.
- Liang and Ming [11] discussed and studied the use of ABC optimization technique with two-tier bitwise interest oriented QRP to reduce message flooding and improve recall rate for a small world peer-to-peer system.
- Suri and Kalkal [12] presented a review of software testing applications of ABC and its variants.
- Li and Ma [13] a solution method for logic reasoning using ABC was proposed.
- Bacanin et al. [14] improved ABC optimization was studied on the performance of object-oriented software system.
- Sharma et al. [15] applied modified version of ABC to parameter estimation of software reliability growth models.
- Koc et al. in [16] proposed a solution for automated maintenance of object-oriented software system designs via refactoring using ABC.
- Sharma and Pant [17] estimated the software cost parameters using halton based ABC optimization.
- Singh and Sandhu [18] presented a survey of ABC on software testing environment and its advantages over the Genetic Algorithm.
- Suri and Mangal [19] introduced a hybrid algorithm using ABC to reduce the test suite.
- Lam et al. [20] proposed automated generation of independent paths and test suite optimization using ABC algorithm.
- Sharma et al. [21] applied ABC to optimize the redundancy model in modular software system.

6.3 ARTIFICIAL BEE COLONY OPTIMIZATION ALGORITHM AND ITS VARIANT

6.3.1 ARTIFICIAL BEE COLONY

ABC is relatively a new swarm intelligence based optimizer proposed by Karaboga to solve continuous optimization problems [4]. ABC and its variants have been implemented on various real world problems arising in different disciplines [22–31] and have proved their mettle over various variants of Genetic Algorithm (GA), Particle Swarm Optimization (PSO), and Differential Evolution (DE) [32]. ABC divides the bee swarm into three groups namely Scout, Employed and Onlooker. Scout bees randomly search the space for the new food sources, once they get the food sources they become Employed bees. Then the Employed bees share the information about the food source to the Onlooker bees, which are waiting in the hive. This information sharing is done by performing a special kind of dance known as waggle dance, which describes the quality and quantity of the food source in particular direction. There are equal number of employed and onlooker bees in the colony. The number of employed bees represents the food sources. The position of food source represents solution to the optimization problem. And the quality of the nectar represents the fitness of the solution. The iterative process followed by ABC algorithm is discussed in the following steps:

a) Food sources are generated randomly i.e., population of randomly distributed solutions.
b) Nectar amount is calculated by sending employed bees to the food sources.
c) After getting information about food sources from employed bees, onlooker bees select the food sources and determine the quality of the nectar.
d) If the food source is abandoned, Scout bees come into play to search for the new food sources.

6.3.2 FORMULATION OF ARTIFICIAL BEE COLONY ALGORITHM

6.3.2.1 Initialization Process

FS denotes the randomly generated D-dimensional vector (food Sources), let the *ith* food source in population be represented by $X_i = (x_{i1}, x_{i2, ...,} x_{iD})$ and then each food source is generated by:

$$x_{ij} = lb_j + rand(0,1) \times (ub_j - lb_j) \qquad (1)$$

for j = 1, 2, ..., D and i = 1, 2, ..., FS, lb_j and ub_j are the lower and upper bounds respectively for the dimension j.

6.3.2.2 Employed Bee Process

Then the candidate (new) food source v_{ij} is generated by each employed bee x_i in the neighborhood of its present position by using:

$$v_{ij} = x_{ij} + \phi_{ij}(x_{ij} - x_{kj}) \qquad (2)$$

where $k \in \{1,2,..., FS\}$ such that k is not equals to i and $j \in \{1, 2, ..., D\}$ are randomly chosen indexes. Ø is a uniformly distributed random number in the range of $[-1, 1]$.

Then greedy selection is done by comparing v_i with the position of x_i, the better one become the member of the food source (population).

6.3.2.3 Onlooker Bee Process

At this stage an Onlooker bee selects the food source x_i based on its probability value p_i, which is calculated by using:

$$p_i = \frac{fitness_i}{\sum\limits_{k=1}^{FS} fitness_k} \qquad (3)$$

where $fitness_i$ is the fitness value of the food source x_i.

6.3.2.4 Scout Bee Process

If the food sources x_i gets abandoned or not improved through predetermined number of trials (*limit*) then corresponding employed bee becomes Scout bee and discovers the new food sources using Eq. (1).

For solving constrained optimization problems, Karaboga and Basturk [32] proposed the variant of ABC that employs the given Deb's rule for constraint handling [33]:

a) If we have two feasible food sources, we select the one giving the best objective function value;

b) If one food source is feasible and the other one is infeasible, we select the feasible one;

c) If both food sources turn out to be infeasible, the food source giving the minimum constraint violation is selected.

The new food sources are generated using Eq. (4):

$$v_{ij} = \begin{cases} x_{ij} + \varphi_{ij} \times (x_{ij} - x_{kj}) & if \ R_j \leq MR \\ x_{ij} & otherwise \end{cases} \tag{4}$$

where MR is a modification rate, a control parameter that controls whether the parameter x_{ij} will be modified or not.

The general algorithmic structure of the ABC optimization approach is given as follows:

STRUCTURE OF ABC ALGORITHM

Initialization of the Food Sources

Evaluation of the Food Sources

Repeat

       Produce new Food Sources for the employed bees

       Apply the *greedy* selection process

       Calculate the probability values for Onlookers

       Produce the new Food Sources for the onlookers

       Apply the *greedy* selection process

       Send randomly scout bees

       Memorize the best solution achieved so far.

Until termination criteria are met.

ALGORITHM 1.1   PSEUDOCODE OF CONSTRAINED ABC

Step 1: Initialize the D-dimensional food sources using equation (1)

Step 2: Evaluate the food sources.

Step 3: *cycle* = 1

Repeat

Step 4: Produce new solutions (food source positions) using Eq. (4).

Step 5: Apply the selection process between $x_{ij,}$ and v_{ij} based on Deb's method.

Step 6: Calculate the probability values p_i for the solutions x_{ij}, using fitness of the solutions (food sources) and constraint violations (CV):

$$p_i = \begin{cases} \left(0.5 + \left(\dfrac{fit_i}{\sum\limits_{i=1}^{SN} fit_i}\right) * 0.5\right) & if \ \ solution \ \ is \ \ feasible \\[2em] \left(1 - \dfrac{CV}{\sum\limits_{i=1}^{SN} CV_i} * 0.5\right) & if \ \ solution \ \ is \ \ inf \, easible \end{cases}$$

CV is defined by:

$$CV = \sum_{j=1, if \ g_j(x) > 0}^{q} g_j(x) + \sum_{j=q+1}^{m} h_j(x)$$

Step 7: Produce the new solutions (new food source positions) v_{ij}, for the onlookers from the solutions x_{ij} using Eq. (4), selected depending on p_i, and evaluate them.

Step 8: Apply the selection process between x_{ij}, and v_{ij} based on Deb's method.

Step 9: If Scout Production Period (SPP) is completed, determine the abandoned food sources (solutions) by using limit parameter, if exists, and replace it with a new randomly produced solution x_i for the scout using the Eq. (1).

Step 10: Memorize the best food source position (solution) achieved so far.

Step 11: $cycle = cycle + 1$.

Until cycle = Maximum Cycle Number (MCN).

6.3.3 *INTERMEDIATE ABC GREEDY (I – ABC GREEDY)*

The convergence speed of ABC algorithm is typically slower than those of representative population-based algorithms e.g., PSO and DE, when handling unimodal problems [34]. ABC algorithm can easily get trapped in the local optima when solving multimodal problems [35]. While, it is observed that the solution search equation of ABC algorithm which is used to generate new candidate solutions based on the information of previous solutions, is good at exploration but poor at exploitation, which results in the above two insufficiencies [35].

The first algorithm, I-ABC *greedy* suggested in this Chapter incorporates the concept of OBL [36] for generating initial food sources and search is always forced to move towards the solution vector having the best fitness value in the population to achieve above-mentioned two goals.

I-ABC *greedy* differs from the basic ABC in the initial step when the population (called food sources in case of ABC) is generated. Here, two sets of populations are considered each having *SN* food sources out of which *SN* are generated using the uniform random numbers while the remaining *SN* food sources are generated using opposition based learning (OBL).

After that, intermediate food locations are generated by taking the mean of the food sources generated by uniform random numbers and by OBL. Once the potential food sources are located, the usual ABC algorithm is applied to determine the best solution (optimal solution in this case). The term *greedy* here refers that the population is always forced to move towards the solution vector having best fitness value.

Before explaining the proposed algorithms in detail the concept of Opposition Based Learning (OBL), used in I-ABC and I-ABC *greedy* is briefly described.

6.3.3.1 Opposition Based Learning (OBL)

The main idea behind OBL is the simultaneous consideration of an estimate and its corresponding opposite estimate (i.e., guess and opposite guess). Opposition based initial population is based on the concept of opposite numbers generated as follows:

if $x \in [l, u]$ is a real number, then its opposite number x' is defined as:

$$x' = l + u - x \qquad (5)$$

where l and u indicates the lower and upper bounds of the variables. This definition can be extended for higher dimensions. Thus it can be said that if $X = (x_1, x_2, ..., x_n)$ is a point in n-dimensional space, where $x_1, x_2, ..., x_n \in R$

and $x_i \in [l_i, u_i] \forall i \in \{1, 2, ..., n\}$, then the opposite point is completely defined by its components

$$x_i' = l_i + u_i - x_i \qquad (6)$$

In case of optimization algorithms, OBL is used for generating the initial population as follows:

- Let $X' = (x_1, x_2, ..., x_n)$ be a candidate solution in an n-dimensional space with $x_i \in [l_i, u_i]$ and assume $f(x)$ is a fitness function which is used to measure candidate's optimality. According to the opposite point definition, $X' = (x_1', x_2', ..., x_n')$ is the opposite of $X = (x_1, x_2, ..., x_n)$.
- Evaluate the fitness of $f(X)$ and $f(X')$
- If $f(X') \leq f(X)$ (for minimization problem), then replace X with X'; otherwise, continue with X.

6.3.3.2   Proposed I-ABC greedy

I-ABC *greedy* differs from basic ABC in the initial step when the initial population (food sources) is generated with the help of following easy steps:

- In the beginning, generate food sources (population) of size $2SN$ consisting of uniformly distributed random vectors. Out of these, store SN vectors in a set P_1 and denote P_1 as $\{U_1, U_2, ... U_{SN}\}$.
- The remaining SN vectors are stored in the set P_2 where OBL rules are applied to generate the corresponding opposition vector and the one having better fitness value is retained in the set P_2. Suppose, P_2 is denoted as $\{O_1, O_2, ..., O_{SN}\}$ then the initial population consisting of intermediate locations (IL) of $P_1`$ and P_2 is generated as:

$$IL = \{(U_1+O_1)/2, (U_2+O_2)/2, ..., (U_{SN}+O_{SN})/2\} \qquad (7)$$

The vectors generated by X_{IL} will serve as the initial food sources for the bees. After initializing the food locations the usual procedure of the ABC algorithm described in the previous section is employed to find the optimal solution.

The initial population generated using the above scheme for sphere and Griekwank functions are shown in Figure 6.1.

To make ABC greedy a small change is made in Eq. (2), by replacing x_{ij} with $x_{best,j}$, where $x_{best,j}$ represents the solution vector having best fitness value. It will be called '*greedy bee*' since the search is always forced to move towards the best solution obtained so far. Equation (2) can now be written as:

$$v_{ij} = x_{best,j} + rand(0,1) \times (x_{best,j} - x_{kj}) \tag{8}$$

The remaining notations have the same meaning as defined in the previous section. Figure 6.2 demonstrates the flow chart of I-ABC *greedy* algorithm.

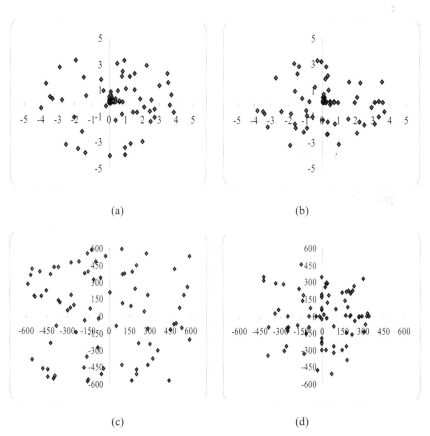

(a)   (b)

(c)   (d)

FIGURE 6.1 Food source (population) generation graphs of (a) Sphere function using random numbers (b) Sphere function using intermediate locations (c) Griekwank function using random numbers (d) Griekwank function using intermediate locations.

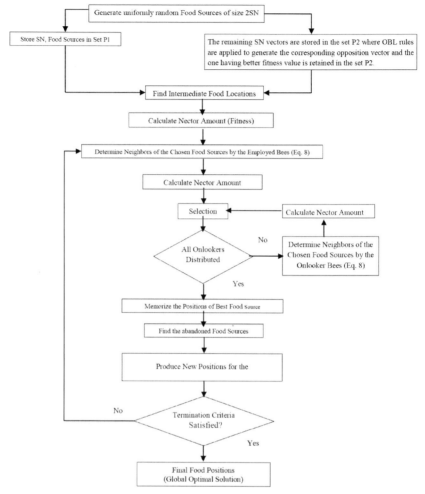

FIGURE 6.2 Flow chart of I-ABC *greedy* algorithm.

ALGORITHM 1.2   PSEUDOCODE OF BASIC I-ABC GREEDY
ILLUSTRATING THE PROCEDURE OF
INITIALIZING FOOD LOCATIONS AND
PRODUCING NEW FOOD POSITIONS

Step 1: Initialize the D-dimensional food sources using Eq. (1).

Follow Step 2 to 3 as Algorithm 1.1

Repeat

Step 4: Produce new solutions (food source positions) using Eq. (8).

Follow Step 5 to 6 as Algorithm 1.1

Step 7: Produce the new solutions (new food source positions) v_{ij} for the onlookers from the solutions x_{ij}, using Eq. (8) selected depending on p_i, and evaluate them.

Follow Step 8 to 11 as Algorithm 1.1

Until cycle = Maximum Cycle Number (MCN).

6.4 SOFTWARE ENGINEERING DESIGN PROBLEMS

In this Chapter three Software Engineering design problems are taken for experiment and to test the performance of I-ABC *greedy* algorithm. The problems are:

6.4.1 SOFTWARE RELIABILITY GROWTH MODELS

Software reliability is the probability of failure free operation of a computer program in a specified environment for a specified period of time [37]. Failure process modeling represents a challenge because of the various natures of faults discovered and the methodologies to be used in order to isolate the faults [38, 39]. Many software techniques were developed to assist in testing the software before its final release for public use. Most of these techniques are simply based on building prediction models that have the ability to predict future faults under different testing conditions [40, 41]. These models normally called 'software reliability growth models' and are defined as:

A. Power Model (POWM)

The model objective is to compute the reliability of a hardware system during testing process. It is based on the non-homogeneous Poisson process model and was provided in Ref. [42]. The equations which govern the relationship between the time t and $\mu(t; \beta)$ and $\lambda(t; \beta)$ are:

$$\mu(t;\beta)=\beta_0 t^{\beta_1}$$

$$\lambda(t;\beta)=\beta_0\beta_1 te^{\beta_1-1} \qquad (9)$$

B. Exponential Model (EXMP)

This model is known as a finite failure model was first provided in Refs. [43, 44].

The relationship among various parameters is given as:

$$\mu(t;\beta)=\beta_0(1-e^{-\beta_1 t})$$

$$\lambda(t;\beta)=\beta_0\beta_1 e^{-\beta_1 t}) \tag{10}$$

where $\mu(t;\beta)$ and $\lambda(t;\beta)$ represent the mean failure function and the failure intensity function, respectively. The parameters β_0 is the initial estimate of the total failure recovered at the end of the testing process (i.e., v_0). β_1 represents the ratio between the initial failure intensity λ_0 and total failure $v0$. Thus, $\beta_1 = \lambda_0/v_0$. It is important to realize that:

$$\lambda(t;\beta)=\frac{\partial\mu(t;\beta)}{\partial t} \tag{11}$$

C. Delayed S-Shaped Model (DSSM)

This model describes the software reliability process as a delayed S-shaped model [45]. It is also a finite` failure model. The system equation for $\mu(t;\beta)$ and $\lambda(t;\beta)$ are:

$$\mu(t;\beta)=\beta_0(1-(1+\beta_1 t)e^{-\beta_1 t})$$

$$\lambda(t;\beta)=\beta_0\beta_1^2 t^{-\beta_1 t} \tag{12}$$

6.4.2 SOFTWARE EFFORT ESTIMATION

Software effort estimation is a form of problem solving and decision making, and in mostly cases the effort estimate for a software project is too complex to be considered in a single piece. The accuracy prediction of effort estimation depends on various things like the size of the project, estimation to transform the size estimate into human effort, time duration and money.

Some of the options for achieving reliable costs and efforts estimate include:

- Estimates on similar type of software already been developed or completed.
- Decomposition techniques (software sizing, problem based estimation, lines of code (LOC) based estimation, functional points (FP) based estimation, and process based estimation) used to generate cost and effort estimation.
- More than one empirical method used to estimate the software cost and effort estimation.

A model based on experience takes the form:

$$d = f\ (v_i) \tag{13}$$

where d is one of the estimated values like – effort, cost in dollars, project duration and are selected independent parameters like – estimated LOC or FP.

Constructive Cost Model (COCOMO) is one of the most famous and effective model estimate the effort, and was developed by Boehm [46, 47]. The model helps in defining the mathematical relationship between the software development time, the effort in man-months and the maintenance effort [48]. An evolutionary model for estimating software effort using genetic algorithms was developed by Alaa F. Sheta [49].

6.4.3 OPTIMIZING REDUNDANCY LEVEL IN MODULAR SOFTWARE SYSTEM MODELS

In terms of software system, reliability can be defined as the probability that software operates without failure in a specified environment, during a specified exposure period [44]. A discrepancy between expected and actual output is called failure.

Failure is a consequence of fault, also called a defect in the program that, when executed results a failure. The reliability of the software can be improved by carefully implementing the application of redundancy, but it requires additional resources. A number of reliability models have been proposed and developed for the prediction and the assessment of the reliability of fault-tolerant software systems [50]. In the present work, the models proposed by Berman and Ashrafi [51] are considered. The following notations are used unless otherwise mentioned.

Notations:

K: is the number of functions that the software system is required to perform.

n: Number of modules within the software system.

F_k: Frequency of the use of function k, $k = 1, 2, \ldots, K$.

m_i: is the number of available versions for module i, $i = 1, \ldots, n$.

R_{ij}: Reliability estimation of version j of module i.

X_{ij}: Binary variable, i.e., 1 if version j is selected for module i, else 0.

R_i: Estimated reliability of module i.

R: Estimated reliability of the software system.

C_{ij}: Development cost for version j for module i.

B: Available budget.

A. Model 1

In this model an optimal set of modules are selected to perform one function without redundancy. The model that describes the situation of optimal selection of modules for a single program in order to optimize (maximize) reliability with respect to the development budgetary constraints is given below:

$$Maximize\ R = \prod_{i=1}^{n} R_i \tag{14}$$

with respect to

$$\sum_{j=1}^{m_i} X_{ij} = 1,$$

$$\sum_{i=1}^{n} \sum_{j=1}^{m_i} X_{ij} C_{ij} \leq B$$

$$X_{ij} = 0,1; \quad i = 1,\ldots,n; \quad j = 1,\ldots,m_i$$

where

$$R_i = \sum_{j=1}^{m_i} X_{ij} R_{ij} \tag{14a}$$

B. Model 2

Here in this model an optimal set of modules are selected to perform one function with redundancy. The key objective that arises in this situation is to determine the optimal set of modules, allowing redundancy, and to maximize the reliability of the software system with the budgetary constraint. The model is presented below:

$$Maximize\ R = \prod_{i=1}^{n} R_i \tag{15}$$

with respect to $\qquad X_{ij} = 0,1;\quad i = 1,...,n;\quad j = 1,...,m_i$

where $$\sum_{j=1}^{m_i} X_{ij} \geq 1,$$

$$\sum_{i=1}^{n} \sum_{j=1}^{m_i} X_{ij} C_{ij} \leq B$$

$$R_i = 1 - \sum_{j=1}^{m_i} (1 - R_{ij})^{X_{ij}} \tag{15a}$$

The probability that at least one of the m_i, versions is performing correctly defines the reliability of the module i (given as one minus the probability that none of the m_i, versions is performing correctly). Constraint set mentioned for the model assures that for each module i at least one version is selected.

C. Model 3

In the model 3 the set of modules without having redundancy, are selected for a system with K functionality. The objective of this model is again just like discussed above in two models i.e., to determine the optimal set of modules for the programs, without allowing redundancy, and in such a way that the reliability of the software system is maximized within the budgetary constraints. Model 3 is presented below:

Let S_k symbolize the set of modules corresponding to program k. For each module $i \in S_k$ there are m_i, versions available. Here different programs can call the same module. All the modules to be called by all programs are numbered from 1 to n. The problem can be formulated as follows:

$$Maximize\ R = \sum_{k=1}^{K} F_k \prod_{i \in S_k} R_i \tag{16}$$

with respect to $\qquad \sum_{j=1}^{m_i} X_{ij} = 1, \qquad i = 1,....,n$

$$\sum_{i=1}^{n}\sum_{j=1}^{m_i} X_{ij} C_{ij} \leq B$$

$$X_{ij} = 0,1; \quad i = 1,...,n; \quad j = 1,...,m_i$$

where R_i is referred from Model 1.

D. Model 4

This model is similar to the model 3 above. In this case models are selected with redundancy, i.e., the choice of more than one version for each one of the modules are allowed. The problem is presented as follows:

$$Maximize\ R = \sum_{k=1}^{K} F_k \prod_{i \in S_k} R_i \tag{17}$$

with respect to $\sum_{j=1}^{m_i} X_{ij} \geq 1, \qquad i = 1,....,n$

$$\sum_{i=1}^{n}\sum_{j=1}^{m_i} X_{ij} C_{ij} \leq B$$

$$X_{ij} = 0,1; \quad i = 1,...,n; \quad j = 1,...,m_i$$

where R_i is referred from Model 2.

6.5   PARAMETER AND EXPERIMENTAL SETTINGS

After conducting several experiments and referring to various literatures, in order to make the comparison with other algorithm(s), settings given in Table 6.1, have been taken for all the experiments unless otherwise mentioned.

TABLE 6.1   Parameter and Experimental Settings

Parameter	Symbol	Value	Description
Colony Size	SN	100 for traditional problems 500 for nontraditional problems	Total number of solutions (food sources), employed bees and onlooker bees.
Maximum Cycle Number	MCN	5000	Total number of cycles (iterations).

TABLE 6.1   Continued

Parameter	Symbol	Value	Description
Limit	*limit*	$MCN/(2 * SN)$	Number of cycles a non improved solution will be kept before being replaced by a new one generated by a scout bee
Maximum Number of Function Evaluation	*NFE*	10^6, for traditional problems 5×10^6 for nontraditional problems (D is Dimension of the problem)	Performance measure
Accuracy	ε	10^{-9}, for all the test problems except noisy function (f7) for which it is set as 10^{-2}	Measuring the reliability of the algorithm.
Number of trials	–	30	Number of times experiment (algorithm) executed.
PC Configuration		Processor: Intel dual core RAM: 1 GB Operating System: Windows XP	
Software and Package Used		DEV C++ SPSS 16 (Software Package for Social Science)	

Random numbers are generated using inbuilt rand () function with same seed for every algorithm

6.6   EXPERIMENTAL RESULTS

6.6.1   *SOFTWARE RELIABILITY GROWTH MODELS*

Test/Debug data for estimating the parameters of software reliability growth models

I-ABC *greedy* is used to find the best parameters to tune the exponential model, power model and Delayed S-Shaped model. A Test/Debug dataset of 111 measurements (Tables 6.2(a) and 6.2(b)) presented in Ref. [49] was used for the experiments.

Root-mean-square error (RMSE) criterion is used to measure the performance of the proposed I-ABC *greedy*. RMSE is frequently used to measure differences between values predicted by a model or estimator and the values actually observed from the thing being modeled or estimated. It is just the square root of the mean square error as shown in equation given below:

$$RMSE = \sqrt{\frac{1}{N}\sum_{i=1}^{N}(y_i - \hat{y}_i)^2}$$

(18)

where y_i represents the ith value of the effort, $\hat{y}_i$ is the estimated effort and N is the number of measurements used in parameter estimation of growth

TABLE 6.2(A) Test Data (1–52) Those Marked with * Are Interpolated Data

Days	Faults	Cumulative Faults	Number of Test Workers tester(i)	Days	Faults	Cumulative Faults	Number of Test Workers tester(i)
1	5*	5*	4'	27	3	243	5
2	5*	10*	4*	28	9	252	6
3	5*	15*	4*	29	2	254	6
4	5*	20'	4*	30	5	259	6
5	6*	26*	4*	31	4	263	6
6	8	34	5	32	1	264	6
7	2	36	5	33	4	268	6
8	7	43	5	34	3	271	6
9	4	47	5	35	6	277	6
10	2	49	5	36	13	290	6
11	31	80	5	37	19	309	8
12	4	84	5	38	15	324	8
13	24	108	5	39	7	331	8
14	49	157	5	40	15	346	8
15	14	171	5	41	21	367	8
16	12	183	5	42	8	375	8
17	8	191	5	43	6	381	8
18	9	200	5	44	20	401	8
19	4	204	5	45	10	411	8
20	7	211	5	46	3	414	8
21	6	217	5	47	3	417	8
22	9	226	5	48	8	425	4
23	4	230	5	49	5	430	4
24	4	234	5	50	1	431	4
25	2	236	5	51	2	433	4
26	4	240	5	52	2	435	4

TABLE 6.2(B) Test Data (53–111) Those Marked with * Are Interpolated Data

Days	Faults	Cumulative Faults	Number of Test Workers tester(i)	Days	Faults	Cumulative Faults	Number of Test Workers tester(i)
53	2	437	4	83	0	473	2*
54	7	444	4	84	0	473	2*
55	2	446	4	85	0	473	2*
56	0	446	4*	86	0	473	2*
57	2	448	4*	87	2	475	2*
58	3	451	4	88	0	475	2*
59	2	453	4	89	0	475	2*
60	7	460	4	90	0	475	2*
61	3	463	4	91	0	475	2*
62	0	463	4*	92	0	475	2*
63	1	464	4*	93	0	475	2*
64	0	464	4*	94	0	475	2*
65	1	465	4*	95	0	475	2*
66	0	465	3*	96	1	476	2*
67	0	465	3*	97	0	476	2*
68	1	466	3*	98	0	476	2*
69	1	467	3	99	0	476	2*
70	0	467	3*	100	1	477	2*
71	0	467	3*	101	0	477	1*
72	1	468	3*	102	0	477	1*
73	1	469	4	103	1	478	1*
74	0	469	4*	104	0	478	1*
75	0	469	4*	105	0	478	1*
76	0	469	4*	106	1	479	1*
77	1	470	4*	107	0	479	1*
78	2	472	2	108	0	479	1*
79	0	472	2*	109	1	480	1*
80	1	473	2*	110	0	480	1*
81	0	473	2*	111	1	481	1*
82	0	473	2*				

models. The convergence graph of the three growth models are shown in Figure 6.3. The computed parameters and RMSE (training & testing) of all the three software reliability growth models using ABC and proposed I-ABC *greedy* algorithms are given in the Tables 6.3 and 6.4, respectively.

It can clearly be analyzed that Delayed S-Shaped model provided the minimum RMSE in comparison of other models.

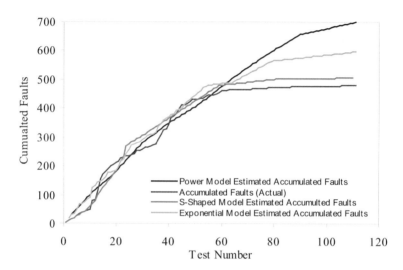

FIGURE 6.3 Actual and accumulated failures for the three growth models using test/debug data (111 Measurements).

TABLE 6.3 Estimated Reliability Growth Model Parameters Using ABC and I-ABC *greedy*

EXMP	ABC	$\mu(t;\beta)=681.096(1-e^{-0.158177})$
	I-ABC *greedy*	$\mu(t;\beta)=317.167(1-e^{-0.188819})$
POWM	ABC	$\mu(t;\beta)=28.8965t^{0.0134281}$
	I-ABC *greedy*	$\mu(t;\beta)=22.3989t^{0.234013}$
DSSM	ABC	$\mu(t;\beta)=779.724(1-(1+0.01144\ t)e^{-0.01144})$
	I-ABC *greedy*	$\mu(t;\beta)=659.012(1-(1+0.02176t)e^{-0.05176})$

TABLE 6.4 Computed RMSE for Test/Debug Data

Model	Algorithm	RMSE Training	RMSE Testing (Validation)
EXMP	ABC	27.0912	87.0934
	I-ABC *greedy*	19.7196	32.0492

TABLE 6.4   Continued

Model	Algorithm	RMSE Training	RMSE Testing (Validation)
POWM	ABC	41.0482	98.953
	I-ABC *greedy*	34.3901	60.8045
DSSM	ABC	21.0830	22.7126
	I-ABC *greedy*	18.9215	16.9342

6.6.2   SOFTWARE EFFORT ESTIMATION

In this problem, the data is taken from Bailey and Basili [52]. The dataset consist of three variables: Developed Line of Code (DLOC), the Methodology (ME) and the measured effort. DLOC is described in Kilo Line of Code (KLOC) and the effort is in man months. The dataset is given in Table 6.5.

Effort Model Based KLOC:

The Constructive Cost Model (COCOMO) was provided by Boehm [46, 47]. This model structure is classified on the type of projects to be handled. These

TABLE 6.5   NASA Data of Effort Estimation

Project No.	KLOC	Methodology	Actual Effort
1	90.2	30	115.8
2	46.2	20	96
3	46.5	19	79
4	54.5	20	90.8
5	31.1	35	39.6
6	67.5	29	98.4
7	12.8	26	18.9
8	10.5	34	10.3
9	21.5	31	28.5
10	3.1	26	7
11	4.2	19	9
12	7.8	31	7.3
13	2.1	28	5
14	5	29	8.4
15	78.6	35	98.7
16	9.7	27	15.6
17	12.5	27	23.9
18	100.8	34	138.3

include the organic, semidetached and embedded projects. This model structure is as follows:

$$E = a*(KLOC)^b \tag{19}$$

The parameters are generally fixed for these models based on the software project type. New parameters are estimated using I-ABC *greedy* for the COCOMO model parameters. Consequently, the effort developed for the NASA software projects is computed.

I-ABC *greedy* is used to develop the following model:

$$E = 3.62911*(KLOC)^{0.426643} \tag{20}$$

Table 6.6, presents the actual measured effort over the given 18 NASA projects and the effort estimated based on the GA [49], ABC and by our proposed algorithm I-ABC *greedy* model. The search space domain of 'a' is taken as 0:10, and for 'b' is 0.3:2.

Figure 6.4 illustrates the measured efforts and estimated efforts using GA and I-ABC *greedy* algorithms w.r.t project number and KLOC.

TABLE 6.6   Measured and Estimated Efforts Using GA, ABC, I-ABC and I-ABC *greedy*

Project No.	Measured Efforts	GA	ABC	I-ABC *greedy*
1	115.8	131.9154	118.577	24.7728
2	96	80.8827	75.5418	18.6212
3	79	81.2663	75.872	18.6727
4	90.8	91.2677	84.4389	19.9812
5	39.6	60.5603	57.857	15.7281
6	98.4	106.7196	97.5333	21.8907
7	18.9	31.6447	31.8078	10.7692
8	10.3	27.3785	27.8333	9.8965
9	28.5	46.2352	45.1144	13.4362
10	7	11.2212	12.2325	5.8808
11	9	14.0108	15.0104	6.6943
12	7.3	22.0305	22.7807	8.7178
13	5	8.4406	9.4087	4.9805
14	8.4	15.9157	16.8819	7.2112
15	98.7	119.285	108.071	23.3598
16	15.6	25.8372	26.3858	9.5675
17	23.9	31.1008	31.3035	10.6608
18	138.3	143.0788	127.796	25.9754

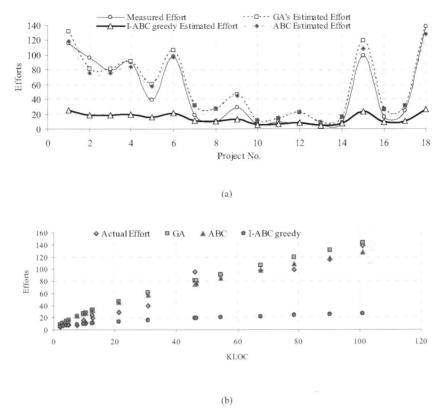

(a)

(b)

FIGURE 6.4 Measured Efforts and Estimated Efforts Using GA and I-ABC *greedy* Algorithms with respect to (a) Project Number and (b) KLOC.

6.6.3 OPTIMIZING REDUNDANCY LEVEL IN MODULAR SOFTWARE SYSTEM MODELS

To solve the above-discussed software system structure models, numerical examples have been taken.

A. Numerical example for the Model 1 is as: Let there are three modules in the software, i.e., $n = 3$ ($m_1=3$, $m_2=3$, $m_3=3$). The cost and the reliability of the modules are taken as $R_{11}=0.90$, $R_{12}=0.80$, $R_{13}=0.85$; $C_{11}=\$3$, $C_{12}=\$1$, $C_{13}=\$2$; $R_{21}=0.95$, $R_{22}=0.80$, $R_{23}=0.70$; $C_{21}=\$3$, $C_{22}=\$2$, $C_{23}=\$1$; $R_{31}=0.98$, $R_{32}=0.94$; $C_{31}=\$3$, $C_{32}=\$2$ and given budget, i.e., $B = 6$, the Model 1 can be formulated as follows:

Maximize
$$(0.9X_{11} + 0.8X_{12} + 0.85X_{13})*$$
$$(0.95X_{21} + 0.8X_{22} + 0.7X_{23})*$$
$$(0.98X_{31} + 0.94X_{32}) \tag{21}$$

with respect to
$$X_{11} + X_{12} + X_{13} = 1$$
$$X_{21} + X_{22} + X_{23} = 1$$
$$X_{31} + X_{32} = 1$$
$$3X_{11} + X_{12} + 2X_{13} + 3X_{21} + 2X_{22} + X_{23} + 3X_{31} + 2X_{32} \leq 6$$

where
$$X_{11}, X_{12}, X_{13}, X_{21}, X_{22}, X_{23}, X_{31}, X_{32} = 0, 1.$$

The optimal solution found by ABC and the proposed variant called I-ABC *greedy* is presented in the Table 6.7. The objective function value obtained using basic I-ABC *greedy* is better than that of basic ABC. Also I-ABC *greedy* took only 27913 function evaluations to solve Model 1, which is about 30% faster than ABC.

B.   The same problem discussed above is considered for solving Model 2, with a difference of budget only. Here the budget is taken as $10. The optimal solution of the model using ABC and modified variant is given in Table 6.8. In this case I-ABC *greedy* performs 34% faster than basic ABC in achieving objective function value.

C.   To solve Model 3, the numerical example considered is as follows: $K = 2$, $F_1 = 0.70$, $F_2 = 0.30$, $n = 3$, $s_1 = (1, 2\}$, $s_2 = (2, 3\}$; $m_1 = 2$, $m_2 = 2$, $m_3 = 2$; $R_{11} = 0.80$, $R_{12} = 0.85$; $R_{21} = 0.70$, $R_{22} = 0.90$; $R_{31} = 0.95$ $R_{32} = 0.90$; $C_{11} = \$2$, $C_{12} = \$3$; $C_{21} = \$1$, $C_{22} = \$3$; $C_{31} = \$4$, $C_{32} = \$3$ and the available budget is taken as $8. Mathematically, Model can be formulated as:

Maximize $(0.392X_{11}X_{12} + 0.504X_{11}X_{22} + 0.4165X_{12}X_{21} + 0.5355X_{12}X_{22} +$
$$0.1995X_{21}X_{31} + 0.189X_{21}X_{32} + 0.2565X_{22}X_{31} + 0.243X_{22}X_{32}) \quad (22)$$

with respect to
$$X_{11} + X_{12} = 1$$
$$X_{21} + X_{22} = 1$$
$$X_{31} + X_{32} = 1$$
$$2X_{11} + 3X_{12} + X_{21} + 3X_{22} + 4X_{31} + 3X_{32} \leq 8$$

where
$$X_{ij} = 0,1; i = 1, 2, 3 \; j = 1, ..., m_i.$$

In this model both the algorithms achieved the same optimal solution and is presented in Table 6.9. But I-ABC *greedy* as above again performed better in terms of function evaluation i.e., 20% faster than basic ABC.

D.   For the Model 4 the same problem discussed above is again considered with the budget of $9 and the optimal result for this model is presented in Table 6.10. Here, once again the proposed variant performed better in achieving better objective function value as well as better convergence of 19% when compared with basic ABC.

The number of function evaluation taken to execute the considered Models 1 - 4 using ABC and I-ABC *greedy* are shown graphically in Figure 6.5.

TABLE 6.7 Optimal Solution of Model 1

Algorithm	Decision Variables	Obj. Func. Value	Cost ($)	Func. Eval. Number
ABC	X_{31}, X_{22}, X_{12}	0.6272	6	36,176
I-ABC *greedy*	X_{12}, X_{21}, X_{32}	0.714	6	27,913

TABLE 6.8 Optimal Solution of Model 2

Algorithm	Decision Variables	Obj. Func. Value	Cost ($)	Func. Eval. Number
ABC	$X_{11}, X_{12}, X_{22}, X_{23}, X_{31}$	2.499	10	36219
I-ABC *greedy*	$X_{31}, X_{21}, X_{23}, X_{12}, X_{13}$	2.6680	10	26984

TABLE 6.9 Optimal Solution of Model 3

Algorithm	Decision Variables	Obj. Func. Value	Cost ($)	Func. Eval. Number
ABC	X_{11}, X_{22}, X_{32}	0.747	8	39134
I-ABC *greedy*	X_{11}, X_{22}, X_{32}	0.747	8	32546

TABLE 6.10 Optimal Solution of Model 4

Algorithm	Decision Variables	Obj. Func. Value	Cost ($)	Func. Eval. Number
ABC	$X_{11}, X_{21}, X_{22}, X_{32}$	0.8052	9	39034
I-ABC *greedy*	$X_{11}, X_{12}, X_{21}, X_{32}$	0.9975	9	32612

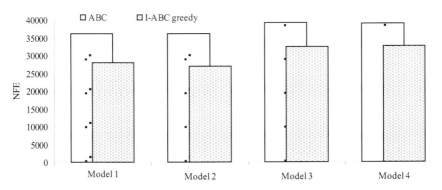

FIGURE 6.5 NFE taken to solve Model 1–Model 4 using ABC & I-ABC *greedy*.

6.7 SUMMARY

In this chapter, the performance of I-ABC *greedy* algorithm is analyzed on real world problems like parameter estimation of software reliability growth models, optimizing redundancy level in modular software system models and estimating the software cost parameters arising in the field of Software Engineering. Conclusions drawn at the end of this chapter can be summarized as follows:

- In case of software effort estimation, it can be analyzed that the proposed I-ABC *greedy* performs comparably better, specially in case of higher KLOC when compared with the measured efforts, estimated efforts by GA and ABC.
- I-ABC *greedy* can be easily modified for solving the problems having integer or/and binary restrictions imposed on it.
- I-ABC *greedy* outperformed basic ABC in terms of solution quality as well as convergence rate for the considered Software design problems.

From the above-mentioned points, a conclusion that can be drawn at this stage is that the proposed swarm intelligence I-ABC *greedy* is competent in dealing with such problems, which is evident from the solution quality and rate of convergence.

KEYWORDS

- **Artificial Bee Colony**
- **Convergence**
- **Optimization**
- **Software Engineering Design Problems**
- **Swarm Intelligence**

REFERENCES

1. Carbone, P., Buglione, L., & Mari, L. A comparison between foundations of metrology and software measurement. *IEEE T. Instrum. Meas.* 2008, 57(2), 235–241.
2. Wang, Y. X., & Patel, S. Exploring the cognitive foundations of software engineering. *Int. J. Soft. Sci. Comp. Intel.* 2009, 1(2), 1–19.

3. Hagan, P., Hanna, E., & Territt, R. Addressing the corrections crisis with software technology. *Comp.* 2010, 43(2), 90–93.

4. Karaboga, D. An Idea based on Bee Swarm for Numerical Optimization,Technical Report, TR-06, Erciyes University Engineering Faculty, Computer Engineering Department 2005.

5. Mala, D. J., Kamalapriya, M., & Shobana, R., Mohan, V. A non-pheromone based intelligent swarm optimization technique in software test suite optimization. In: *IAMA: 2009 International Conference on Intelligent Agent and Multi-Agent Systems.* 2009, 188–192.

6. Mala, D. J., Mohan, V., & Kamalapriya, M. Automated software test optimization framework—an artificial bee colony optimization-based approach. *IET Softw.* 2010, 4(5), 334–348.

7. Bacanin, N., Tuba, M., & Brajevic, I. An object-oriented software implementation of a modified artificial bee colony (abc) algorithm. In: *Recent Advances in Neural Networks, Fuzzy Systems and Evolutionary Computing.* 2010, 179–184.

8. Dahiya, S. S., Chhabra, J. K., & Kumar, S. Application of artificial bee colony algorithm to software testing. In: *2010 21st Australian Software Engineering Conference (ASWEC)*, 2010, 149–154.

9. Kilic, H., Koc, E., & Cereci, I. Search-based parallel refactoring using population-based direct approaches. In: *Search Based Software Engineering.* 2011, 271–272.

10. Adi Srikanth, Kulkarni, N. J., Naveen, K. V., Singh, P., & Srivastava, P. R. Test case optimization using artificial bee colony algorithm. In: *Advances in Computing and Communications, Communications in Computer and Information Science.* 2011, 570–579.

11. Liang, C. Y., & Ming, L. T. Using two-tier bitwise interest oriented QRP with artificial bee colony optimization to reduce message flooding and improve recall rate for a small world peer-to-peer system. In: *2011 7th International Conference on Information Technology in Asia (CITA 11)*, 2011, 1–7.

12. Suri, B., & Kalkal, S. Review of artificial bee colony algorithm to software testing. *Int J Res Rev Comput Sci.* 2011, 2(3), 706–711.

13. Li, L. F., & Ma, M. Artificial bee colony algorithm based solution method for logic reasoning. *Comput Technol Dev.* 2011. (doi:CNKI:SUN:WJFZ.0.2011–06–035).

14. Bacanin, N., Tuba, M., & Brajevic, I. Performance of object-oriented software system for improved artificial bee colony optimization. *Int J Math Comput Simul.* 2011, 5(2), 154–162.

15. Sharma, T. K., & Pant, M. Dichotomous search in ABC and its application in parameter estimation of software reliability growth models. In: *Proceedings of Third World Congress on Nature and Biologically Inspired Computing (NaBIC), Salamica, Spain.* 2011, 207–212.

16. Koc, E., Ersoy, N., Andac, A., Camlidere, Z. S., Cereci, I., & Kilic, H. An empirical study about search-based refactoring using alternative multiple and population-based search techniques. In: *Computer and information sciences II*, 2012, 59–66.

17. Sharma, T. K., & Pant, M. Halton Based Initial Distribution in Artificial Bee Colony Algorithm and its Application in Software Effort Estimation. In: *Proceedings of Bio-Inspired Computing: Theories and Applications (BIC-TA), Penang, Malaysia.* 2011, 80–84.

18. Singh, T., & Sandhu, M. K. An Approach in the Software Testing Environment using Artificial Bee Colony (ABC). *Optimization. International Journal of Computer Applications.* 2012, 58(21), 5–7.

19. Suri, B., & Mangal, I. Analyzing Test Case Selection using Proposed Hybrid Technique based on BCO and Genetic Algorithm and a Comparison with ACO. *Computer Science and Software Engineering International Journal.* 2012, 2(4), 206–211.

20. Soma, S. B. L., Raju, M. L. H. P., Uday, K. M., Swaraj, Ch., & Srivastav, P. R. Automated Generation of Independent Paths and Test Suite Optimization Using Artificial Bee Colony. In: *Proc. of International Conference on Communication Technology and System Design – 2011. Procedia Engineering* 2012, 30, 191–200.

21. Sharma, T. K., & Pant, M. Redundancy Level Optimization in Modular Software System Models using ABC. *International Journal of Intelligent Systems and Applications (IJISA).* 2014, 6(4), 40.

22. Singh, A. An artificial bee colony algorithm for the leaf constrained minimum spanning tree problem. *Applied Soft Computing Journal.* 2009, 9(2), 625–631.

23. Pan, Q. K., Tasgetiren, M. F., Suganthan, P., & Chua, T. A discrete artificial bee colony algorithm for the lot-streaming flow shop-scheduling problem. *Information Sciences.* 2011, 181(12), 2455–2468.

24. Kalayci, C. B., & Surendra, Gupta, M. Artificial bee colony algorithm for solving sequence-dependent disassembly line balancing problem. *Expert Systems with Applications.* 2013, 40(18), 7231–7241.

25. Pan, Q. K., Wang, L., Li, J. Q., & Duan, J. H. A novel discrete artificial bee colony algorithm for the hybrid flow shop-scheduling problem with make-span minimization. *Omega.* 2014, 45, 42–56.

26. Sharma, T. K., & Pant, M. Enhancing the food locations in an artificial bee colony algorithm. *Soft Computing.* 2013, 17(10), 1939–1965.

27. Sharma, T. K., Pant, M., & Singh, M. Nature-Inspired Metaheuristic Techniques as Powerful Optimizers in the Paper Industry. *Materials and Manufacturing Processes.* 2013, 28(7), 788–802.

28. Kumar, S., Kumar, P., Sharma, T. K., & Pant, M. Bi-level thresholding using PSO, Artificial Bee Colony and MRLDE embedded with Otsu method. *Memetic Computing,* 2013, 5(4), 323–334.

29. Sharma, T. K., & Pant, M. Enhancing the food locations in an Artificial Bee Colony algorithm, In: *Proceedings of IEEE Swarm Intelligence Symposium (IEEE SSCI 2011) Paris, France.* 2011, 1–5.

30. Sharma, T. K., Pant, M., & Bansal, J. C. Some Modifications to Enhance the Performance of Artificial Bee Colony. In: *Proceedings of IEEE World Congress on Computational Intelligence (CEC), Brisbane, Australia.* 2012, 3454–3461.

31. Sharma, T. K., Pant, M., & Ahn, C. W. Improved Food Sources in Artificial Bee Colony. In: *Proceedings of IEEE Swarm Intelligence Symposium, IEEE SSCI 2013, Singapore.* 2013, 95–102.

32. Karaboga, D., & Basturk, B. A Powerful and Efficient Algorithm for Numerical Function Optimization: Artificial Bee Colony (ABC) algorithm. *Journal of Global Optimization, Springer Netherlands..* 2007, 39, 459–471.

33. Deb, K. An efficient constraint handling method for genetic algorithms. *Computer Methods in Applied Mechanics and Engineering.* 2000, 186(2/4), 311–338.

34. Karaboga, D., & Basturk, B. On the performance of artificial bee colony (abc) algorithm. *Appl Soft Comput.* 2008, 8(1), 687–697.

35. Zhu, G., & Kwong, S. Gbest-Guided Artificial Bee Colony Algorithm for Numerical Function Optimization. *Appl Math Computing.* 2010, 217(7), 3166–3173.

36. Rahnamayan, S., Tizhoosh, H. R., & Salama, M. M. A. A Novel Population Initialization Method for Accelerating Evolutionary Algorithms. *Computer and Applied Mathematics with Application.*. 2007, 53, 1605–1614.

37. Musa, JD., Iannino, A., & Okumoto, K. Software Reliability: Measurement, *Prediction, Applications.* McGraw Hill, 1987. H. Pham. Software Reliability. Springer-Verlag, 2000.

38. Bishop, P. G., & Bloomfield, R. Worst case reliability prediction on a prior estimate of residual defects. In: *13th IEEE International Symposium on Software Reliability Engineering (ISSRE-2002).* 2002, 295–303.

39. Xie, M. Software reliability models – past, present and future. In: *Recent Advances in Reliability Theory: Methodology, Practice and Inference.* 2002, 323–340.

40. Yamada, S., & Osaki, S. Optimal software release policies with simultaneous cost and reliability requirements. *European J. Operational Research.* 1987, 31(1), 46–51.

41. Yamada, S., & Somaki, H. Statistical methods for software testing-progress control based on software reliability growth models (in Japanese). *Transactions Japan SIAM.* 1996, 317–327.

42. Crow, L. H. Reliability for complex repairable systems. *Reliability and Biometry, SIAM.* 1974, 379–410.

43. Moranda, P. B. Predictions of software reliability during debugging. In: *Annual Reliability and Maintainability Symposium.* 1975, 327–332.

44. Musa, J. D. A theory of software reliability and its application. *IEEE Trans. Software Engineering.* 1975, 1, 312–327.

45. Yamada, S., Ohba, M., Osaki, S. S-Shaped software reliability growth models and their applications. IEEE Trans. Reliability. 1984, R-33(4), 289–292.

46. Boehm, B. Software Engineering Economics, Englewood Cliffs, NJ. Prentice-Hall, 1981.

47. Boehm, B. Cost Models for Future Software Life Cycle Process: COCOMO2 *Annals of Software Engineering*, 1995.

48. Kemere, C. F. An empirical validation of software cost estimation models. *Communication ACM.* 1987, 30, 416–429.

49. Sheta, A. F. Estimation of the COCOMO Model Parameters Using Genetic Algorithms for NASA Software Projects. Journal of Computer Science. 2006, 2(2),118–123.

50. Belli, F., & Jedrzejowicz, P. An approach to reliability optimization of software with redundancy. *IEEE Transactions on Software Engineering.* 1991, 17(3), 310–312.

51. Berman, O., & Ashrafi, N. Optimization Models for Reliability of Modular Software Systems. IEEE Transactions on Software Engineering. 1993, 19(11), 1119–1123.

52. Bailey, J. W., & Basili, V. R. A meta model for software development resource expenditure. In: *Proceedings of ICSE '81, 5th International Conference on Software Engineering,* 1981, 107–116.

APPENDIX

Benchmark Functions

1. The Sphere function described as:

$$f_1(x) = \sum_{i=1}^{n} x_i^2$$

where the initial range of x is $[-100, 100]^n$, and n denotes the dimension of the solution space. The minimum solution of the Sphere function is $x^* = [0,0,\ldots, 0]$ and $f_1(x^*)=0$.

2. The Griewank function described as:

$$f_2(x) = \frac{1}{4000}\left(\sum_{i=1}^{n}(x_i - 100)^2\right) - \left(\prod_{i=1}^{n}\cos\left(\frac{x_i - 100}{\sqrt{i}}\right)\right) + 1$$

where the initial range of x is $[-600, 600]^n$. The minimum of the Griewank function is $x^* = [100, 100, \ldots, 100]$ and $f_6(x^*)=0$.

CHAPTER 7

GENE EXPRESSION PROGRAMMING IN NANOTECHNOLOGY APPLICATIONS

PRAVIN M. SINGRU,[1] VISHAL JAIN,[1] NIKILESH KRISHNAKUMAR,[1] A. GARG,[2] K. TAI,[2] and V. VIJAYARAGHAVAN[2]

[1]*Department of Mechanical Engineering, Birla Institute of Technology and Science, BITS Pilani, K.K. Birla Goa Campus, Zuarinagar, Goa 403726, India*

[2]*School of Mechanical and Aerospace Engineering, Nanyang Technological University, 50 Nanyang Avenue, 639798, Singapore*

CONTENTS

Research in nanomaterials has gained enormous interest in recent years due to its wide spread applications. The desirable qualities of nanomaterials have enabled them to find promising applications in emerging areas such as nano-fluidics, nano-biotechnology and nano electro mechanical systems (NEMS). In this chapter, an integrated Genetic Programming (GP) simulation approach for modeling the material properties of nanoscale materials is proposed. A sensitivity and parametric analysis will also be conducted to validate the robustness of the proposed theme and for identifying the key input factors that govern the material characteristics at nanoscale.

7.1   BACKGROUND

Investigating the properties of Boron Nitride Nanotubes (BNNT) has attracted significant interest in material science [1, 2]. The exceptional qualities of BNNT has been widely studied and investigated to explore its diverse possible applications in real world. These include applications in electric circuits such as BNNT-based integrated circuits (ICs), structural composite materials and hydrogen storage applications [3–5]. In addition, BNNT is an ideal candidate for potential applications in nano-biological and nano-level drug delivery devices [6, 7]. Numerous studies have been undertaken to predict the mechanical properties of BNNT using experimental and computational techniques. Tang et al [8] determined the mechanical properties of BNNT under tension using *in situ* Transmission Electron Microscopy (TEM) and Molecular Dynamics (MD) simulation approach. They found that the mechanical properties and deformation behaviors are correlated with the interfacial structure under atomic resolution, which clearly demonstrates a geometry strengthening effect. Liew and Yuan [9] studied the structural performance of double-walled BNNT under compression at high temperatures using computational modeling approach. They found that the tensile strength and thermal stability of BNNTs are superior to carbon nanotubes (CNT). Shokuhfar et al. [10] studied the buckling strength of BNNTs at various temperatures using MD simulation technique. They found that the buckling strength generally decreases at high temperatures. Furthermore, the buckling resistance of BNNT was also found to decrease with increasing the length of BNNT.

The effect of vacancy defects on the structural properties of BNNT were studied by Shokuhfar and Ebrahimi-Nejad [11]. It was found from their analysis that the tensile strength of BNNT decreases with increasing vacancy defect concentration in BNNT structure. The tensile strength of BNNT for hydrogen storage applications was analyzed by Ebrahimi-Nejad et al. [12]. They found that hydrogen storage decreases the room temperature buckling strength of BNNT. The above-mentioned literature studies clearly indicate that the tensile strength of BNNTs depends on various factors such as system size, chirality, temperature and defects. Hence, understanding the influence of each factor on the tensile strength of BNNTs is important for optimizing the elastic properties of BNNT. One way of optimizing system properties of nanoscale materials is to form an explicit model formulation, which can then be used to extract system input variables for desirable material performance.

Therefore, there is a need to develop an integrated GP simulation technique for modeling the material properties of BNNT. The new hybrid approach combines powerful advantages of accuracy and low cost of MD simulation with the explicit model formulation of GP approach. These methods require input training data, which can be obtained from the MD simulations, which is based on a specific geometry and temperature. Considering the input data, the GP technique can then be able to generate meaningful solutions for the complicated problems [13–16]. The parametric and sensitivity analysis is conducted in this chapter to validate the robustness of the proposed model by unveiling important hidden parameters and non-linear relationships.

7.2 PROPOSED COMPUTATIONAL METHODOLOGY

The tensile characteristics of BNNTs described in this work are modeled entirely using an integrated GP simulation approach as shown in Figure 7.1. In this approach, the MD is integrated in the paradigm of popular GP approach. The data obtained from the MD simulation is further fed into GP cluster. GP based on Darwin's theory of 'survival of the fittest,' finds the optimal solution by mimicking the process of evolution in nature [17].

The initial population of models is obtained by randomly combining the elements from the function and terminal sets. The elements in the function set can be arithmetic operators (+, −, /, ×), non-linear functions (sin, cos, tan, exp, tanh, log) or Boolean operators. The elements of the terminal set are input process variables and random constants. The present study has three input process variables, and, therefore these are chosen as elements of terminal set. The constants are chosen randomly in the range as specified by the user since these accounts for the human or experimental error.

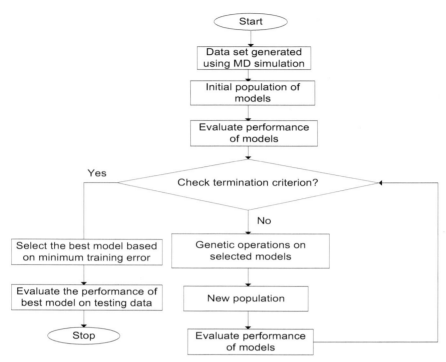

FIGURE 7.1   Mechanism of Integrated GP approach for modeling tensile strength of BNNT.

The performance of the initial population is measured using the fitness function, which compares the predicted values of the Multi-gene Genetic Programming (MGGP) model to that of the actual values. Fitness function must be minimized for obtaining better solutions. Typically used fitness function, namely, root mean square error (RMSE) is given by:

$$RMSE = \sqrt{\frac{\sum_{i=1}^{N} |G_i - A_i|^2}{N}} \tag{1}$$

where G_i is the valued predicted of ith data sample by the MGGP model, A_i is the actual value of the ith data sample and N is the number of training sample.

The performance of the initial population is evaluated and the termination criterion is checked. The termination criterion is specified by the user and is the maximum number of generations and/or the threshold error of the model. If the performance does not match the criterion, the new population is generated by performing the genetic operators on the selected individuals of the initial population. Genetic operators applied are crossover, mutation and reproduction. Tournament selection method is used to select the individuals for the genetic operations. This selection method maintains genetic diversity in the population and thus avoids local/premature convergence. Tournament sizes of 2, 4 and 7 are preferred. The models with lowest fitness value reproduce or copied in the next generation. The crossover operation, namely, subtree crossover is used. Figure 7.2 shows the functioning of subtree crossover in which the branch of the two models is chosen randomly and swapped. The mutation used is subtree mutation (Figure 7.3) in which the branch of the model is replaced with the newly randomly generated model/tree. By Koza [17], the probability rate of reproduction, crossover and mutation kept is 85%, 10% and 5%, respectively. This indicates that most of the new population came from the application of crossover operation. The iterative phenomenon of generating new population continues as long as the above-mentioned termination criterion is met. The best model is selected based on minimum RMSE and its performance is evaluated on testing data.

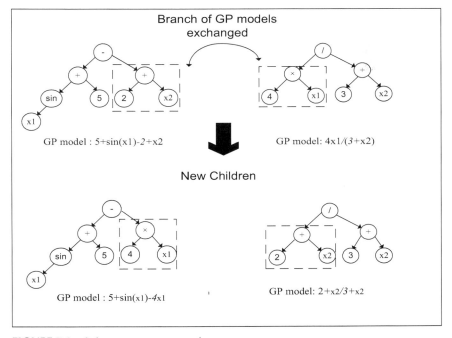

FIGURE 7.2   Subtree crossover operation.

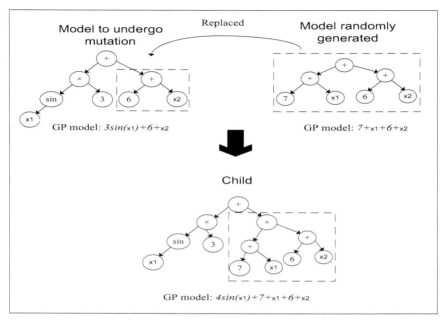

FIGURE 7.3 Subtree Mutation Operation.

7.3 DESIGN OF TENSILE LOADING IN BNNT MATERIAL

The BNNT structure is first thermally equilibrated in an NVT ensemble (Number of particles, Volume and Temperature is conserved) to release any residual stresses. The simulations are carried out by maintaining the desired system temperature. Six temperatures ranging from 0 K to 1500 K are considered in our study to gather the required data of mechanical strength. The mechanical strength in our study is defined as the maximum tensile force that BNNT structure can sustain under compression. The temperature stability of the system is attained by using the Nose-Hoover thermostat [18, 19]. Following equilibration, the single-walled carbon nanotube (SWCNT) is subjected to tensile loading as shown in Figure 7.4. It can be seen from Figure 7.4 that the end atoms enclosed inside the red rectangle is subjected to a constant inward displacement (strain rate = 0.001 ps^{-1}). The system is allowed to relax after every 1000 time steps such that the atoms attain the favorable minimum energy positions. The inward velocity and the trajectories of end atoms are calculated and the atoms are subsequently shifted to the new position. The remaining atoms are relaxed in an NVT

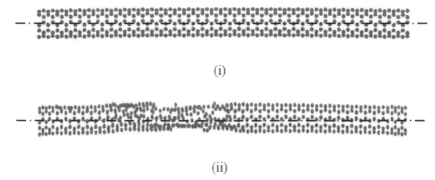

(i)

(ii)

FIGURE 7.4 Procedure of mechanical loading of BNNT under compression. The end atoms enclosed within the red colored rectangle is subjected to inward displacement to effect compression.

ensemble and the procedure is repeated until failure occurs. We used a total of 600,000 time steps (including 200,000 time steps for initial equilibration) with each time step equivalent to 1 fs. The effect of size on mechanical strength of BNNTs is studied by varying the aspect ratio (ratio of length to diameter of the BNNT). The diameter of BNNT is varied by changing the chirality of BNNT structure. The effect of temperature is studied by carrying out the mechanical loading of BNNT at six different temperatures, viz. 0 K, 300 K, 600 K, 900 K, 1200 K and 1500 K. The influence of vacancy defects on the mechanical strength of BNNT is studied by manually reconstructing vacancy defects ranging from 1 to 4 missing atoms in the perfect hexagonal lattice of BNNT.

7.3.1 DESCRIPTION OF DATA

Data obtained from the MD simulations comprise of three input process variables i.e., Aspect ratio of BNNTs (x_1), temperature (x_2), number of vacancy defects (x_3) and the one output process variable, namely, the tensile strength (y_1). 47 data points for BNNTs are obtained from the MD simulations. Nature of the data points collected is shown in Table 7.1. Selection of the training samples affect the learning phenomenon of the proposed model. In this work, 80% of the total samples are chosen randomly as training samples with the remaining used as the set of the test samples. Data is then fed into cluster of the proposed model.

In the present work, GPTIPS [20, 21] is used to perform the implementation of proposed approach for the evaluation of tensile strength of BNNTs. Approach is applied to the dataset as shown in Table 7.1. For the effective implementation of the proposed approach, the parameter settings are adjusted using the trial-and-error method (Table 7.2). Wide range of elements is chosen in the function set so as to generate the mathematical models of different sizes. Depending on the problem, the values of

TABLE 7.1 Descriptive Statistics of the Input and Output Process Variables Obtained from MD Simulations for BNNTs

Statistical Parameter	Aspect ratio (x_1)	Temperature (x_2)	Number of vacancy defects (x_2)	Tensile strength (y)
Mean	3.04	627.65	5.36	3562.38
Standard error	0.19	64.98	0.50	280.88
Median	3.31	550	6	4271.71
Standard deviation	1.36	445.48	3.46	1925.63
Variance	1.85	198457.44	12.01	3.70
Kurtosis	−1.59	−0.86	−1.02	−0.85
Skewness	0.01	0.47	−0.10	−0.26
Minimum	1.48	0	0	61.00
Maximum	4.74	1500	12	7016

TABLE 7.2 Parameter Settings for Proposed Integrated GP Approach

Parameters	Values assigned
Runs	8
Population size	100
Number of generations	100
Tournament size	3
Max depth of tree	6
Max genes	8
Functional set (F)	(multiply, plus, minus, plog, tan, tanh, sin, cos exp)
Terminal set (T)	$(x_1, x_2, x_3, [-10\ 10])$
Crossover probability rate	0.85
Reproduction probability rate	0.10
Mutation probability rate	0.05

population size and generations are set. Based on having good number of 47 data points, the value of population size and generations is kept lower (100) to avoid any over-fitting. The size and variety of forms of the model to be searched in the solution space is determined by the maximum number of genes and depth of the gene. Based on collection of good number of data samples for the BNNTs, the maximum number of genes and maximum depth of gene is chosen at 8 and 6 respectively. The performance of the best model (see Eq. (2)) selected is shown in the following section. In this model, x1, x2 and x3 are aspect ratio, temperature and number of vacancy defects for the BNNTs.

Integrated_GP_Model = 7075.9395 + (0.15325) × ((exp((4.751686))) + (exp(plog(x2)))) + (29.9508) × ((tan(cos(plog(x2)))) + (plog((cos(exp(x3))) − (sin((x3) + (x1)))))) + (−0.010533) × (((exp(x3)) × ((cos(exp(x3))) − (sin((4.798872))))) × ((cos(x3)) − (sin((4.798872))))) + (−59.0461) × (((x1) × (x1)) × (tan((4.428378)))) + (−53.8336) × (plog(plog((cos(exp(x3))) − (cos((x3) + (x1)))))) + (−85.7442) × (((x3) + ((x3) − (x1))) + (x3)) + (−242.8619) × (tanh(plog((cos(exp(x3))) − (sin((x3) + (x1)))))) + (61.3814) × ((sin((x3) + (x1))) − (sin(((x3) + (x1)) + (x1))))   (2)

7.4   STATISTICAL EVALUATION OF THE PROPOSED MODEL

The results obtained from the integrated model are illustrated in Figure 7.5 on the training and testing data. Performance of the proposed model is evaluated against the actual results [12] using the five metrics: the square of the correlation coefficient (R^2), the mean absolute percentage error (MAPE), the RMSE, relative error (%) and multiobjective error function (MO) given by:

$$R^2 = \left(\frac{\sum_{i=1}^{n}(A_i - \overline{A_i})(M_i - \overline{M_i})}{\sqrt{\sum_{i=1}^{n}(A_i - \overline{A_i})^2 \sum_{i=1}^{n}(M_i - \overline{M_i})^2}} \right)^2 \qquad (3)$$

$$MAPE(\%) = \frac{1}{n}\sum_{i} \left| \frac{A_i - M_i}{A_i} \right| \times 100 \qquad (4)$$

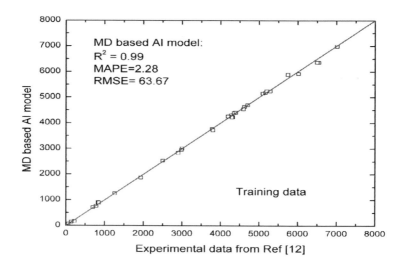

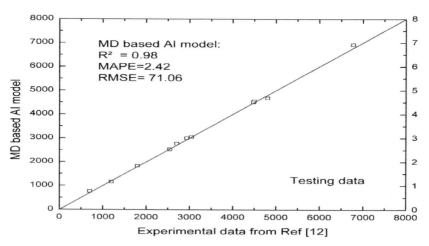

FIGURE 7.5 Performance of the Integrated GP model for the BNNTs on (a) training and (b) testing data.

$$RMSE = \sqrt{\frac{\sum_{i=1}^{N}\left|M_i - A_i\right|^2}{N}} \qquad (5)$$

$$Relative\,error\,(\%) = \frac{\left|M_i - A_i\right|}{A_i} \times 100 \qquad (6)$$

$$Multiobjective\,error = \frac{MAPE + RMSE}{R^2} \tag{7}$$

where M_i and A_i are the predicted and actual values respectively, $\overline{M_i}$ and $\overline{A_i}$ are the average values of the predicted and actual respectively and n is the number of training samples. Since, the values of R^2 do not change by changing the models values equally and the functions: MAPE, RMSE and relative error only shows the error and no correlation. Therefore, a MO error function that is a combination of these metrics is also used.

The result of the training phase shown in Figure 7.5a indicates that the proposed model have impressively learned the non-linear relationship between the input variables and tensile strength with high correlation values and relatively low error values. The result of the testing phase shown in Figure 7.5b indicates that the predictions obtained from the model are in good agreement with the actual data, with achieved values of R^2 as high as 0.98.

MO values of the proposed model are computed on the training and testing data as shown in Table 7.3. The descriptive statistics of the relative error of the proposed model are shown in Table 7.4, which illustrates error mean, standard deviation (Std dev), Standard error of mean (SE mean), lower confidence interval (LCI) of mean at 95%, upper confidence interval (UCI) of mean at 95%, median, maximum and minimum. The lower values of range (UCI-LCI) of the confidence intervals of the proposed model indicates that it is able to generalize the tensile strength values satisfactory based on the variations in aspect ratio (AR), temperature and incursion of defects.

Goodness of fit of the proposed model is evaluated based on the hypothesis tests and shown in Table 7.5. These are t-tests to determine the mean and

TABLE 7.3 Multiobjective Error of the Proposed Model

Integrated GP model	Training data	Testing data
BNNTs	54.84	59.25

TABLE 7.4 Descriptive Statistics Based on the Relative Error (%) of the Proposed Model

Integrated GP model	Count	Mean	LCI 95%	UCI 95%	Std dev	SE mean	Median	Maximum	Minimum
BNNTs	47	2.31	1.16	3.46	3.90	0.56	1.41	25.55	1.41

TABLE 7.5   *P*-Values to Evaluate Goodness of Fit of the Model

95% CI	BNNTs
Mean paired *t* test	0.98
Variance *F* test	0.97

f-tests for variance. For the *t*-tests and the *f*-tests, the *p*-value of the model is >0.05, so there is not enough evidence to conclude that the actual values and predicted values from the model differ. Therefore, the proposed model has statistically satisfactory goodness of fit from the modeling point of view.

Thus, from the statistical comparison presented, it can be concluded that the proposed model is able to capture the dynamics of the nanosystem.

7.5   MECHANICS OF BNNT MATERIAL USING SENSITIVITY AND PARAMETRIC ANALYSIS

Sensitivity and parametric analysis about the mean is conducted for validating the robustness of our proposed model. The sensitivity analysis (*SA*) percentage of the output to each input parameter is determined using the following formulas:

$$L_i = f_{max}(x_i) - f_{min}(x_i) \tag{8}$$

$$SA_i = \frac{L_i}{\sum_{j=1}^{n} L_j} \times 100 \tag{9}$$

where $f_{max}(x_i)$ and $f_{min}(x_i)$ are, respectively, the maximum and minimum of the predicted output over the *i*th input domain, where the other variables are equal to their mean values.

Figure 7.6 shows the plots of the sensitivity results of input variables in the prediction of tensile strength of BNNTs. From Figure 7.6, it is clear that the process input variable, namely the aspect ratio, has the greater impact on the tensile strength of BNNTs followed by number of defects and temperature. This reveals that by regulating the aspect ratio of BNNTs, a greatest variation in tensile strength of BNNTs can be achieved. The parametric analysis provides a measure of the relative importance among the inputs of the model and illustrates how the tensile strength of BNNTs varies in response to the variation in input variables. For this reason, on the formulated model,

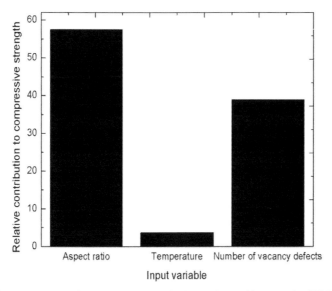

FIGURE 7.6 Amount of impact of input variables to the tensile strength of BN.

the first input is varied between its mean ± definite number of standard deviations, and the tensile strength is computed, while, the other input is fixed at its mean value. This analysis is then repeated for the other inputs. Figure 7.7 displays the plots generated for each input variable and the tensile strength of BNNTs. These plots reveal that, for example, the tensile strength decreases with an increase in all three input variables.

7.6 CONCLUSIONS

The integrated GP approach is proposed in simulating the tensile strength characteristic of BNNTs based on aspect ratio, temperature and number of defects. The results show that the predictions obtained from the proposed model are in good agreement with the actual results. Furthermore, the dominant process parameters and the hidden non-linear relationships are unveiled, which further validate the robustness of our proposed model. The higher generalization ability of the proposed model obtained is beneficial for experts in evaluation of tensile strength in uncertain input process conditions. The proposed method evolve model that represents the explicit formulation between the tensile strength and input process parameters. Thus, by using the model, the vital economic factors such as time and cost for estimating the tensile strength using the trial-and-error experimental approach can be reduced.

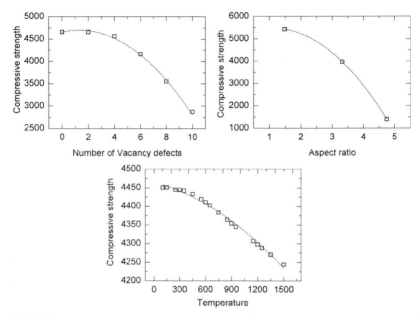

FIGURE 7.7   Parametric analysis of the Integrated GP model showing the effect of variation of tensile strength in respect to input variables for BNNT.

KEYWORDS

- **Defects**
- **Inorganic compounds**
- **Mechanical properties**
- **Nanostructures**

REFERENCES

1. Wang, J., Lee, C. H., & Yap, Y. K. *Recent Advancements in Boron Nitride Nanotubes.* Nanoscale, 2010, 2(10), p. 2028–2034.
2. Griebel, M., & Hamaekers, J. *Molecular Dynamics Simulations of Boron-Nitride Nanotubes Embedded in Amorphous Si-B-N.* Computational Materials Science,. 2007, 39(3), p. 502–517.
3. Mohajeri, A., & Omidvar, A. *Density Functional Theory Study on the Static Dipole Polarizability of Boron Nitride Nanotubes: Single Wall and Coaxial Systems.* Journal of Physical Chemistry C, 2014, 118(3), p. 1739–1745.

4. Yan, H., et al., *Enhanced Thermal-Mechanical Properties of Polymer Composites with Hybrid Boron Nitride Nanofillers.* Applied Physics A: Materials Science and Processing, 2014, 114(2), p. 331–337.

5. Lu, H., Lei, M., & Leng, J. *Significantly Improving Electro-Activated Shape Recovery Performance of Shape Memory Nanocomposite by Self-Assembled Carbon Nanofiber and Hexagonal Boron Nitride.* Journal of Applied Polymer Science, 2014.

6. Ferreira, T. H., et al., *Boron Nitride Nanotubes Coated with Organic Hydrophilic Agents: Stability and Cytocompatibility Studies.* Materials Science and Engineering C, 2013, 33(8), p. 4616–4623.

7. Del Turco, S., et al., *Cytocompatibility Evaluation of Glycol-Chitosan Coated Boron Nitride Nanotubes in Human Endothelial Cells.* Colloids and Surfaces B: Biointerfaces, 2013, 111, p. 142–149.

8. Tang, D. M., et al., *Mechanical Properties Of Bamboo-Like Boron Nitride Nanotubes by in situ TEM and MD Simulations: Strengthening effect of interlocked joint interfaces.* ACS Nano, 2011, 5(9), p. 7362–7368.

9. Liew, K. M., & Yuan, J. *High-Temperature Thermal Stability and Axial Tensile Properties of a Coaxial Carbon Nanotube Inside a Boron Nitride Nanotube.* Nanotechnology, 2011, 22(8).

10. Shokuhfar, A., et al., *The effect of temperature on the tensile buckling of boron nitride nanotubes.* Physica Status Solidi (A) Applications and Materials Science, 2012, 209(7), p. 1266–1273.

11. Ebrahimi-Nejad, S., et al., *Effects of structural defects on the tensile buckling of boron nitride nanotubes.* Physica E: Low-Dimensional Systems and Nanostructures, 2013, 48, p. 53–60.

12. Ebrahimi-Nejad, S., & Shokuhfar, A. *Tensile Buckling of Open-Ended Boron Nitride Nanotubes in Hydrogen Storage Applications.* Physica E: Low-Dimensional Systems and Nanostructures, 2013, 50, p. 29–36.

13. Vijayaraghavan, V., et al., *Estimation of Mechanical Properties of Nanomaterials Using Artificial Intelligence Methods.* Applied Physics A: Materials Science and Processing, 2013, p. 1–9.

14. Vijayaraghavan, V., et al., *Predicting the Mechanical Characteristics of Hydrogen Functionalized Graphene Sheets Using Artificial Neural Network Approach.* Journal of Nanostructure in Chemistry, 2013, 3(1), p. 83.

15. Cevik, A., et al., *Soft Computing Based Formulation for Strength Enhancement of CFRP Confined Concrete Cylinders.* Advances in Engineering Software, 2010, 41(4), p. 527–536.

16. Gandomi, A. H., & Alavi, A. H. *Multi-Stage Genetic Programming: A New Strategy to Nonlinear System Modeling.* Information Sciences, 2011, 181(23), p. 5227–5239.

17. Koza, J. R., *Genetic Programming as a Means for Programming Computers by Natural Selection.* Statistics and Computing, 1994, 4(2), p. 87–112.

18. Hoover, W. G., *Canonical Dynamics – Equilibrium Phase-Space Distributions.* Physical Review A, 1985, 31(3), p. 1695–1697.

19. Nose, S., *A Unified Formulation of the Constant Temperature Molecular-Dynamics Methods.* Journal of Chemical Physics, 1984. 81(1), p. 511–519.

20. Hinchliffe, M., et al. *Modelling Chemical Process Systems Using a Multi-Gene Genetic Programming Algorithm*, 1996.

21. Searson, D. P., Leahy, D. E., & Willis, M. J. *GPTIPS: An Open Source Genetic Programming Toolbox for Multigene Symbolic Regression.* In *Proceedings of the International MultiConference of Engineers and Computer Scientists.* 2010, Citeseer.

PART II

THEORY AND APPLICATIONS OF SINGLE OBJECTIVE OPTIMIZATION STUDIES

CHAPTER 8

AN ALTERNATE HYBRID EVOLUTIONARY METHOD FOR SOLVING MINLP PROBLEMS

MUNAWAR A. SHAIK[1] and RAVINDRA D. GUDI[2]

[1]*Associate Professor, Department of Chemical Engineering, Indian Institute of Technology Delhi, Hauz Khas, New Delhi, India, E-mail: munawar@iitd.ac.in, Tel: +91-11-26591038*

[2]*Professor, Department of Chemical Engineering, Indian Institute of Technology Bombay, Powai, Mumbai, India, E-mail: ravigudi@iitb.ac.in, Tel: +91-22-25767231*

CONTENTS

Most of the real world optimization problems in engineering involve several inherent nonlinearities in their model and often require solution of either nonlinear programming (NLP) or mixed-integer nonlinear programming (MINLP) problems [15]. The presence of discrete variables, bilinearities, and non-convexities further makes it challenging to determine global optimal solutions to these problems, which has been one of the important research topics in the literature [1, 7–11, 16, 19, 28]. Optimization techniques can be broadly classified as deterministic (or traditional) and stochastic (or non-traditional) approaches. On one hand, most of the deterministic optimization techniques such as branch-and-bound, cutting plane, and decomposition schemes either fail to obtain global optimal solutions or have difficulty in proving global optimality; while on the other hand, most of the stochastic optimization techniques have critical issues related to either slower convergence, longer computational times, and/or difficulty in handling of discrete variables.

Most of the deterministic optimization methods assume convexity and often guarantee determination of global optimum solutions. Handling of discrete variables is very cumbersome especially of the bilinear and binomial terms leading to non-convexities [5, 13, 19], and many approaches have been proposed based on relaxation, partitioning and bounding steps that result in an evolutionary refinement of the search space [9, 10, 28, 30, 31]. However, strategies for generalizing these relaxation and partition methods are yet unknown. Therefore, for non-convex MINLP problems, in general, there are no known *robust* deterministic algorithms that can guarantee global optimal solutions.

Nontraditional optimization techniques [24] such as Simulated Annealing (SA), Genetic Algorithms (GA), and Differential Evolution (DE), among many such methods, do not make any assumptions related to the nature of convexity of the problem. They have been widely used in numerous engineering applications and are known to yield global optimal solutions to complex real-life problems. Handling of integer variables is relatively easier and the solution is generally unaffected by the presence of bilinear/binomial terms involving discrete variables. For instance in DE, the algorithm works by assuming discrete variables as continuous variables during all the steps, but only for the objective function evaluation a truncation operation is used for forcing the integrality requirements. However, these methods are often slow and do not guarantee convergence.

In this chapter, we discuss both the convergence issues and alternate ways of handling discrete variables. We present application of a nonlinear transformation for representing discrete variables as continuous variables [18] and discuss an alternate method for solving MINLP problems by converting them into equivalent NLP problems. Since the resulting NLP problem is nonconvex a hybrid method involving switching between deterministic and stochastic optimization techniques is discussed [21, 23]. For the deterministic part of the solution either standard gradient based methods or commercial solvers available with optimization software (such as GAMS [12]) are used with a better starting point, which is provided by the stochastic part of the solution obtained using Differential Evolution [27]. A parameter based on rapid change in the objective function is used to aid in deciding when to switch from deterministic to stochastic solution. Selected examples from literature are used to illustrate the effectiveness of the hybrid evolutionary method.

8.1 NONLINEAR TRANSFORMATION

A binary variable $y \in \{0,1\}$ can be modeled as a continuous variable $x \in [0,1]$, using the following nonlinear binary condition from Li [18]:

$$x(1-x) = 0,\ 0 \leq x \leq 1 \tag{8.1}$$

which enforces x to take either 0 or 1. The function $x(1-x)$ is a non-convex nonlinear function as shown in Figure 8.1. For MINLP problems involving more than one binary variable the binary condition of Eq. (8.1) was generalized in Munawar and Gudi [23] as follows:

$$\sum_{i=1}^{n} x_i(1-x_i) = 0 \quad \text{OR} \quad \sum_{i=1}^{n} x_i(1-x_i) \leq 0 \quad 0 \leq x_i \leq 1\ i = 1..n \tag{8.2}$$

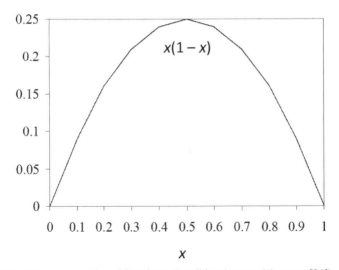

FIGURE 8.1 Non-convex Plot of the Binary Condition (source: Munawar [21]).

From Figure 8.1, which is plotted for single variable case, it can be observed that $x(1-x)$ is never negative within the specified bounds and hence the inequality in Eq. (8.2) can only be satisfied as an equality. Using this transformation any MINLP model can be converted into an equivalent NLP model. Li [18] solved the resulting NLP using a modified penalty function method, however only local optimal solutions were reported due to the presence of non-convexities. In this chapter, we demonstrate the use of differential evolution for solving the resulting NLP problem based on the hybrid evolutionary approach. The following example illustrates the use of binary condition.

Example 8.1: Consider the following simple convex MINLP problem from Floudas [8], which has one continuous variable x and one binary variable y.

$$\min F(x,\ y) = 2x - y - ln(0.5x)$$

subject to

$$G(x,y) = y - x - ln(0.5x) \leq 0$$

$$0.5 \leq x \leq 1.4; y \in \{0,\ 1\} \tag{1.3}$$

The global optimum for this problem is reported to be $(x, y; F) = (1.375, 1; 2.1247)$. The objective function $F(x, y)$ and the inequality constraint $G(x,y)$ for $y=0$ and $y=1$ are shown in Figure 8.2. The feasible region corresponds to all negative values of $G(x,y)$. For $y = 1$, F has a minimum at $x=1.375$ where $G(x,1)$ is negative. Using the binary condition on y, Example 8.1 can

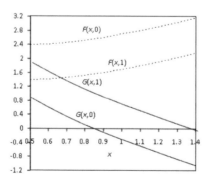

FIGURE 8.2   Objective Function and the Inequality Constraint for Example 8.1 (Source: Munawar [21]).

be transformed into an equivalent NLP problem with two continuous variables x and z as given in Eq. (8.4). The resulting nonconvex NLP problem is shown in Figure 8.3.

$$\min F(x,\,z) = 2x - z - ln(0.5x)$$

subject to

$$H(z) = z(1 - z) = 0$$

$$G(x,z) = z - x - ln(0.5x) \leq 0$$

$$0.5 \leq x \leq 1.4; \quad 0 \leq z \leq 1 \qquad (8.4)$$

8.2 APPLICATION OF DIFFERENTIAL EVOLUTION

DE is a simple and efficient evolutionary method that has been successfully used for solving numerous engineering optimization problems. It is a population-based random search technique and an improved version compared to binary coded genetic algorithms based on 'survival of the fittest' principle of nature. From the initial randomly generated populations (vectors of decision variables each analogous to a gene) newer generations are created through successive application of genetic operators such as mutation, crossover and selection. It is an enhanced version of GA as it uses addition operator for *mutation* and a non-uniform *crossover* wherein the parameter values of the child vector are inherited in unequal proportions from the parent vectors.

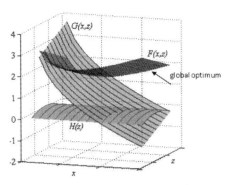

FIGURE 8.3 Objective function and constraints for the transformed problem of Example 8.1 (Source: Munawar [21]).

For *reproduction*, DE uses a tournament selection where the child vector competes against one of its parents. The overall structure of the DE algorithm resembles that of most other population based searches. The original DE algorithm [26, 27] is capable of handling only continuous variables and for solving unconstrained optimization problems. Few simple modifications were discussed in the literature [24] to extend it for optimization involving discrete variables and for handling inequality and equality constraints. Real values are converted to integer values by truncation but only for the purpose of objective function evaluation. Truncated values are not used elsewhere during the search. Discrete values can been handled by assigning the elements to another parameter and using its index, thus converting it into integer variable. Constraints are normally handled as soft constraints in DE using penalty function method where the sum of the squares of the violation of the constraints is augmented with a penalty to the objective function.

Different strategies can be adapted in DE algorithm [27] which vary based on (i) vector to be perturbed, (ii) number of difference vectors considered for perturbation, and (iii) type of crossover used. There are ten different working strategies proposed by Price and Storn [27]: (1) DE/best/1/exp, (2) DE/rand/1/exp, (3) DE/rand-to-best/1/exp, (4) DE/best/2/exp, (5) DE/rand/2/exp, (6) DE/best/1/bin, (7) DE/rand/1/bin, (8) DE/rand-to-best/1/bin, (9) DE/best/2/bin, and (10) DE/rand/2/bin.

The general convention used above is DE/*u*/*v*/*w*. DE stands for Differential Evolution, *u* represents a string denoting the vector to be perturbed, *v* is the number of difference vectors considered for perturbation of *u*, and *w* stands for the type of crossover being used (exp: exponential; bin: binomial). The perturbation can be either in the best vector of the previous generation or in any randomly chosen vector, and for perturbation either single or two vector differences can be used.

8.3 HYBRID EVOLUTIONARY METHOD

A general MINLP problem can be formulated as:

$$min\ f(X,Y)$$

subject to
$$G(X,Y) \leq 0$$

$$H(X,Y) = 0$$

$$X \in R^n\ Y \in \{0,1\}^m \tag{8.5}$$

In the above problem the scalar real-valued objective function $f(X,Y)$, subject to real constraints $G(X,Y)$ and $H(X,Y)$, is required to be minimized in the space of continuous variables X and binary variables Y. It is generally assumed here that Y can occur in a nonlinear fashion in both F and G. Using the binary condition of Eq. (8.2), the above problem can be transformed into an equivalent NLP in the space of $Z = [X,Y]$ as in Eq. (8.6)

$$min \ F(Z)$$

subject to. $\quad\quad\quad\quad\quad\quad Q(Z) \leq 0$

$$W(Z) = 0$$

$$Z \in R^{n+m} \qu\quad\quad\quad\quad\quad\quad (8.6)$$

where $F(Z)$, $Q(Z)$, $W(Z)$ are the transformed objective function and constraints set, and $W(Z)$ constitutes nonlinear binary condition in addition to $H(X,Y)$ to preserve the integrality of Y. As mentioned earlier, the NLP problem in Eq. (8.6) resulting from this transformation is necessarily non-convex and hence global optimization algorithms need to be used to solve this. If a deterministic method is used to solve this nonconvex NLP using one of the commercially available standard NLP solvers (such as CONOPT2, SNOPT, MINOS in GAMS software [12]) then mostly local optimal solutions can be expected, and global optimal solution may also be obtained depending on the right initial guess. Since a single run from an initial guess could terminate in local optima, one can use multiple potential initial guesses to search for the global optima. Such good initial guesses can be conveniently generated by stopping a population based stochastic algorithm such as DE at the right time before it begins to develop convergence issues.

The hybrid evolutionary method [21, 23] has two steps: (i) application of initial DE algorithm, and (ii) solution of the deterministic NLP. In the first step the DE algorithm is applied to the problem in Eq. (8.5) or (8.6) and the progress of the algorithm is monitored based on diversity of the population members at each generation in terms of the cost or population variance. Consider that the population at any generation i during the progress of the algorithm is denoted by G_i. The diversity of the population can be measured in terms of its cost variance σ_i from the mean μ_i. As the DE algorithm progresses to solve for the global optimum, the diversity of the population can be expected to decrease, with the rate of decrease being fairly rapid initially and then slowly as the global optimum is reached. During the early phase of rapid decrease in diversity, the DE algorithm can be expected to generate potential candidates that may be optimal in the discrete domain, which are further improved in the continuous domain in the subsequent generations towards global optimality. To overcome the slower convergence issue in the latter phase, the DE algorithm is terminated whenever there is a dramatic change in the cost/population variance (as indicated by a 'knee' in the cost variance versus generation plot), and best available population members are stored. In the second step the solution obtained from the first step is used as an initial guess for solving the deterministic NLP problem to global optimality.

Quantitatively, the *knee* can be described as an abrupt change in the cost variance with orders of magnitude difference in generation i to $i+1$ ($\sigma_i \gg \sigma_{i+1}$). Considering that there could be several abrupt changes or knees, and to avoid the omission of global solutions, the quantitative *termination* criteria for the initial DE algorithm is defined to be either of the following two cases: (i) until there is no further improvement in the objective function of the deterministic NLP solution for two successive knees, or (ii) $\sigma_i \leq 1$, whichever occurs earlier. The second termination condition is more rigorous in the sense that for the cost variance to reach less than or equal to 1, in the worst case, the complete classical DE algorithm may sometimes need to be enumerated. But in most cases the first termination criteria is general, that the initial DE would be terminated much before the completion of the classical DE algorithm, thus improving the efficiency of the overall solution.

The direct application of DE algorithm to an MINLP problem in Eq. (8.5), with integer rounding off during objective function evaluation, is termed as *Integer-DE*. The application of DE algorithm to the transformed NLP problem in Eq. (8.6) with the binary condition is termed as *NLP-DE*. The application of hybrid evolutionary method in which either Integer-DE or NLP-DE is used in the first step to generate initial solutions followed by a deterministic NLP solver to ensure faster rate of convergence, is termed as *Hybrid-DE*. Generally, Integer-DE is recommended for generation of the initial solution, because in the first few generations of the DE algorithm, NLP-DE is sometimes slower because of the satisfaction of the binary condition being expressed as penalty function in the objective function. When closer to the optimum, Integer-DE is generally slower may be because of the disconnectedness between the continuous treatment of binary variables in the algorithm and the rounding off of the integer requirements in the objective function evaluation.

8.4   SELECTED CASE STUDIES

In this section, we present selected case studies from literature to illustrate the hybrid evolutionary method based on Integer-DE, NLP-DE and Hybrid-DE approaches. For each of the example problems, the different DE versions are tried with different random seeds, strategies and different key parameters (NP, F, and CR), and the best solutions are reported after trial-and-error.

Example 8.2: The following problem with one binary and one continuous variable was proposed by Kocis and Grossmann [16]; and was solved by many authors [2, 4, 6, 9, 21, 23, 28].

$$\min f(x, y) = 2x + y$$

subject to

$$1.25 - x^2 - y \leq 0$$
$$x + y \leq 1.6$$
$$0 \leq x \leq 1.6$$
$$y \in \{0, 1\} \tag{8.7}$$

The reported global optimum is $(x, y; f) = (0.5, 1; 2)$. The first nonlinear inequality constraint has a non-convex term in continuous variable x. This MINLP problem is solved in GAMS software using SBB solver. The binary condition on y, Eq. (8.1), is applied and the problem is converted into an equivalent NLP by replacing y by a continuous variable $z \in [0, 1]$ as follows:

$$\min f(x, z) = 2x + z$$

subject to

$$1.25 - x^2 - z \leq 0$$
$$x + z \leq 1.6$$
$$z(1-z) = 0$$
$$0 \leq x \leq 1.6$$
$$0 \leq z \leq 1 \tag{8.8}$$

This resulting NLP problem is solved in GAMS using available solvers SNOPT (using SQP) and CONOPT2 (using GRG). Since the problem in Eq. (8.8) is still non-convex, depending on the initial guess as shown in Table 8.1, local solutions and sometimes global solutions are obtained. As this is a small case study, all possible initial guesses are tried. From Table 8.1 it can be seen that SNOPT yields several local minima. For $z = 0.5$,

TABLE 8.1 Results for Example 1.2 for Different Initial Guesses Using GAMS Solvers (Source: Munawar [21])

(a) NLP solver: CONOPT2 (GRG based)

Initial guess		Optimal solution		
z	x	z	x	f
any z	0	----infeasible-----		
0	≥ 0.1	0	1.118	2.236
0.25	≥ 0.1	0	1.118	2.236
0.4	≥ 0.1	0	1.118	2.236
0.5	≥ 0.1	----infeasible-----		
0.75	≥ 0.1	1	0.5	**2**	(global)
1	≥ 0.1	1	0.5	**2**	(global)

(b) NLP solver: SNOPT (SQP based)

Initial guess		Optimal solution		
z	x	z	x	f
any z	0	----infeasible-----		
0	0.1–0.4	0	1.19	2.38
0	≥ 0.5	0	1.118	2.236
0.25	0.1–0.25	1	0.507	2.015
0.25	0.3–0.4	0	1.19	2.38
0.25	≥ 0.5	0	1.118	2.236
0.4	0.1–0.25	1	0.507	2.015
0.4	0.3–0.4	1	0.5	**2**	(global)
0.4	≥ 0.5	0	1.118	2.236
0.5	0.1–0.25	1	0.507	2.015
0.5	0.3–0.9	1	0.5	**2**	(global)
0.5	≥ 1	0	1.118	2.236
0.75	0.1–0.25	1	0.507	2.015
0.75	≥ 0.3	1	0.5	**2**	(global)
1	0.1–0.25	1	0.507	2.015
1	≥ 0.3	1	0.5	**2**	(global)

CONOPT2 fails for all x, while SNOPT solves and yields global optima as well. It is also seen that the algorithms are also sensitive to the initial guesses in z. For example, for an initial guess of $z = 0.25$ and 0.4, depending on initial guess of x, it is seen that z takes either 0 or 1 in the optimal solution. These results further justify that the strategy related to rounding off to the nearest integer does not always yield satisfactory solutions.

RESULTS USING INTEGER-DE: If we use DE for directly solving the problem (8.7) as MINLP using conventional truncation operator with strategy 10 and random seed of 99 (used for generating population members randomly), the problem in Eq. (8.7) is solved to the same global optimality of a specified accuracy in about 35 generations. For some representative generations the cost variance (C_{var}) and the best solution found (C_{min}) until a given generation (G) are shown in Table 8.2.

Here it can be observed that DE has slower convergence closer to the optimum (w.r.t the results discussed later in Table 8.4); the reason for this could be attributed to the disconnectedness introduced due to rounding off the integer values only during the objective function evaluation. A plot of the cost variance (C_{var}) at each generation is shown in Figure 8.4. The key parameter values of DE used here are: NP = 20, F = 0.9, CR = 0.8.

TABLE 8.2 Performance of *Integer-DE* for Example 8.2 (Source: Munawar [21])

G	C_{var}	C_{min}
1	898.27	2.5096
7	472.76	2.0038
8	0.1424	2.0038
34	0.0044	2.00255
35	0.0029	**2.00083**
56	4.27×10^{-7}	2.000023

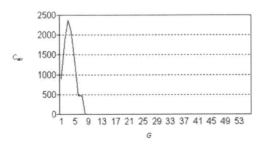

FIGURE 8.4 Cost Variance for Example 8.2 using *Integer-DE* (Source: Munawar [21]).

RESULTS USING *NLP-DE*: For the NLP problem in Eq.(8.8) if we apply DE using the binary condition, the problem is solved to the same global optimality in about 14 generations with strategy 1 and random seed of 10. For some representative generations the cost variance (C_{var}) and the best solution found (C_{min}) until a given generation (G) is shown in Table 8.3.

Since the cost function has same penalty for violations of the integrality constraints, in the initial generations the cost variance is zero here, perhaps because none of the population members satisfy the integrality constraints initially. A plot of the cost variance (C_{var}) at each generation is shown in Figure 8.5. The key parameter values of DE used are: NP = 20, F = 0.9, CR = 0.8.

RESULTS USING *HYBRID-DE*: Since the traditional DE algorithm has slower convergence as shown in Table 8.2, instead of continuing DE up to 56 generations we can prematurely stop the DE algorithm at a point where C_{var} changes dramatically (at generation 8 for this problem) and switch to the local deterministic NLP solution of (8.8) using SNOPT/CONOPT2 in GAMS. Hence the initial guess here for use in deterministic NLP algorithm $x^1 = 0.50192$, $z^1 = 1$ where $f^1 = 2.00383$. It is found that the NLP problem is easily solved to global optimality using SNOPT in GAMS in just two additional iterations of the deterministic NLP solver.

TABLE 8.3   Performance of *NLP-DE* for Example 8.2 (Source: Munawar [21])

G	C_{var}	C_{min}
1	0	100
13	908.95	2.00075
14	0.00175	**2.00075**
24	3.96×10^{-7}	2.0000099

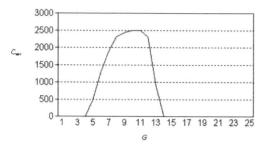

FIGURE 8.5   Cost Variance for Example 8.2 using *NLP-DE* (Source: Munawar [21]).

Example 8.3 (Cyclic Scheduling): This is a simpler instance of the cyclic scheduling problem discussed in Munawar et al. [22] and Munawar and Gudi [23]. The model is an extension of Pinto and Grossmann [25] for incorporating the slopping losses and for accounting inverted triangular inventory profiles. Consider the plant topology as shown in Figure 8.6 with two sequential stages (Stages 1 and 2) for production of three product grades (A, B and C). Finite inventory storage is considered only for the intermediate grades, while for the feed and product grades unlimited storage capacity is assumed. The nomenclature and the mathematical formulation for this problem are given in Appendix-A.

This cyclic scheduling problem has 207 variables including 24 binary variables and 208 constraints. The demand rates for the products A, B and C are 100, 150 and 250 m³/hr; and the sale price for these products is 15, 40, and 65 $/m³, respectively. The yield is considered as 1 for all grades in both the stages for simplicity. The inventory costs are assumed to be $15/m³ and the upper bound on maximum breakpoint for each intermediate grade is assumed to be 30 m³. The average processing rates, the sequence and stage dependent transition times and costs are given in Table 8.4.

Results using GAMS: The MINLP model has several bilinearities and nonlinearities. With maximization of profit as the objective function, when this

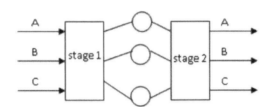

FIGURE 8.6 Problem Topology for Example 8.3 (Source: Munawar [21]).

TABLE 8.4 Problem Data for Example 8.3 (Source: Munawar [21])

Processing rates (m³/hr)			Transition time (hr)						Transition cost ($/hr)					
Grade	Stage 1	Stage 2	Stage 1			Stage 2			Stage 1			Stage 2		
			A	B	C	A	B	C	A	B	C	A	B	C
A	0.8	0.9	0	3	8	0	3	4	0	15	15	0	15	15
B	1.2	0.6	10	0	3	7	0	0	40	0	40	40	0	40
C	1.0	1.1	3	6	0	3	10	0	65	65	0	65	65	0

model is directly solved on GAMS using the available standard solvers, it was found that the solutions differed based on the MINLP solver that was chosen. Table 8.5 represents the nature of the solutions obtained with each of these solvers.

The model was initially solved with SNOPT or CONOPT2 as the solver at the root node for the RMINLP (relaxed MINLP problem) to generate a feasible solution for use in MINLP. Then SBB was used as the MINLP solver, again with SNOPT or CONOPT2 as the solvers at the root node. For the NLP sub-problems at each subsequent node, if one solver fails to obtain a feasible solution then another solver is tried as per the sequence given in Table 8.5. For different solver permutations four different solutions (local and global) were obtained after convergence. For some of the other solver combinations GAMS reports the problem to be infeasible. It can be seen that depending on the solvers used, different local and global solutions can be realized. The Gantt chart and the inventory profiles for the global solution corresponding to the objective function of \$48,800.69 are shown in Figures 8.6 and 8.7, respectively. The dark bands in the Gantt chart represent the transition times.

Results using DE: Both *Integer-DE* and *NLP-DE* took more than 20,000 generations for solving this problem to the same global optimal solution of \$48,800. Hence, the *Hybrid-DE* is used for quickly generating some good initial guesses using DE, and then the NLP problem is solved in GAMS

TABLE 8.5 MINLP Solution of Example 8.3 Using Different GAMS Solvers (Source: Munawar [21])

RMINLP	MINLP	Objective	
Solver	Root node solver	NLP sub-solver sequence	(\$)
CONOPT2	SNOPT	CONOPT2 SNOPT	44428.64
CONOPT2	CONOPT2	CONOPT2 SNOPT	45620.61
SNOPT	CONOPT2	SNOPT CONOPT2	48396.87
CONOPT2	CONOPT2	SNOPT CONOPT2	48800.69 (global)

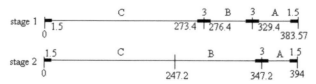

FIGURE 8.7 Gantt Chart for Example 8.3.

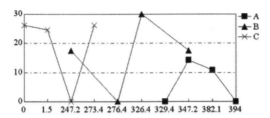

FIGURE 8.8 Inventory Profiles for Example 8.3.

from these initial guesses. For instance, the problem was solved using *Integer-DE* in about 15 generations using strategy 6 for a random seed of 9, with NP = 270, F = 0.9, CR = 0.8 to obtain an initial guess. Using this initial guess, the deterministic NLP solution in GAMS solved much faster giving the same global optimal solution. However, there is a word of caution that the choice of the DE strategies and the key parameters also has similar effect as that of the choice of the NLP solver used in SBB, and that *Hybrid-DE* is also not an exception and may get entrapped in local optima. In general, the hybrid method is found to outperform and improve the convergence rate of both versions of the DE algorithm, viz. *Integer-DE* and *NLP-DE* (Munawar and Gudi [21]).

Example 8.4 (Optimal Component Lumping): Characterization of complex fluid mixtures is an important problem in modeling, simulation, and design of chemical processes, especially in oil and gas sector. Due to the presence of large number of components in these complex mixtures it is practically not feasible to consider detailed composition based models. Instead, the overall mixture is represented using hypothetical components, in which several components with similar properties are lumped into a single pseudo component with average mixture property. There are several lumping schemes proposed in the literature [3, 20]. Lumping obviously leads to loss of information and there is a trade-off between accuracy that can be achieved and dimensionality that can be handled. Hence, it has been referred to as the *optimal component lumping* problem.

The optimal component lumping problem may be defined as follows [20]: Given a system of n components with mole fractions $\{x_i\}$, $i = 1, 2, \ldots, n$ and the corresponding ordered property of $\{p_i\}$, $p_{i-1} \leq p_i \leq p_{i+1}$, $i = 1, 2, \ldots, n$, the objective is to optimally lump the n components into m pseudo components, $\{S_k\}$, $k = 1, 2, \ldots, m$ by minimizing loss of information through a specified objective function. The basic assumption of this formulation is that one component belongs to one and only one lump. A binary variable $\delta_{i,k}$ is defined such that $\delta_{i,k} = 1$ if component i is in lump S_k, otherwise, $\delta_{i,k} = 0$. The mixture property is calculated assuming a linear weighting rule using mole fractions as the weighting factors. The average property of each lump, P_k, is thus calculated in Eq. (8.9) as a weighted sum of the component properties, p_i:

$$P_k = \frac{\sum_{i=1}^{n} x_i p_i \delta_{i,k}}{\sum_{i=1}^{n} x_i \delta_{i,k}} \quad \forall\, k \tag{8.9}$$

The objective function is defined in Eq. (8.10) through minimization of square of the difference between the property of each component of a lump (p_i, $\delta_{i,k} = 1$) and the average component property value, P_k.

$$\min\ \mathrm{obj} = \sum_{k=1}^{m} \sum_{i=1}^{n} x_i (p_i - P_k)^2 \delta_{i,k} \tag{8.10}$$

The constraints are as follow:

$$\sum_{i=1}^{n} \delta_{i,k} \geq 1 \quad \forall\, k \tag{8.11}$$

$$\sum_{k=1}^{m} \delta_{i,k} = 1 \quad \forall\, i \tag{8.12}$$

$$\sum_{k=1}^{m} \delta_{i,k} \times k \leq \sum_{k=1}^{m} \delta_{i+1,k} \times k \quad \forall\, i \tag{8.13}$$

Constraint (8.11) states that there is at least one component in each lump. Constraint (8.12) enforces each component to belong to exactly one lump. If component i is in lump S_k, component $i + 1$ must be in either in S_k or in S_{k+1} as restricted in constraint (8.13). The resulting MINLP problem needs to be solved using number of lumps (m) as a parameter. The resulting model will have very large number of variables and constraints for a typical multi-component mixture comprising large number of components. However, considering the nature of the objective function it is more important to find good feasible solutions rather than looking for a global optimal solution. Hence, the application of stochastic optimization techniques might be a good choice, since they are effective at finding good solutions quickly.

A representative case study of a crude oil comprising 42 components has been considered [3]. The distilation properties, volume fractions and True Boiling Point (TBP) data is given in Table 8.6. The objective is to reduce the problem size by combining these components into a specified number of lumps and minimize the loss of information. This problem is solved for four different cases: from 3-lumps scheme to 6-lumps scheme.

TABLE 8.6 Properties of Crude-oil for Example 8.4

Component	Volume Fraction	True Boiling Point (°C)
1	0.04	19
2	0.07	60
3	0.032	70
4	0.023	80
5	0.035	90
6	0.063	100
7	0.04	110
8	0.057	120
9	0.049	130
10	0.033	140
11	0.051	150
12	0.035	160
13	0.018	165
14	0.022	170
15	0.034	180
16	0.034	190
17	0.037	200
18	0.036	210
19	0.036	220
20	0.012	230
21	0.012	240
22	0.027	250
23	0.021	260
24	0.015	270
25	0.016	280
26	0.018	290
27	0.019	300
28	0.01	310

TABLE 8.6 Continued

Component	Volume Fraction	True Boiling Point (°C)
29	0.009	320
30	0.01	330
31	0.012	340
32	0.009	350
33	0.009	360
34	0.009	390
35	0.009	400
36	0.006	410
37	0.009	430
38	0.007	450
39	0.006	480
40	0.003	500
41	0.003	530
42	0.004	550

Tables 8.7–8.9 show the results obtained for this case study using different optimization methods: Hybrid-DE, SA, GA, and GAMS software. In Table 8.9, the original MINLP problem was directly solved in GAMS software using BARON solver. The results indicate the value of the objective function, lumping scheme, and average temperature of each lump. The objective function gives the value of the sum of squares of the deviations of TBP of all components from their respective average TBP of lumped pseudo components. Lumping scheme indicates the boundary components. For instance, in case of 4-lumps scheme solved using hybrid-DE in Table 8.7, lumping scheme obtained is (9, 21, 33), which means that components 1–9 are grouped as lump-1, 10–21 belong to lump-2, 22–33 belong to lump-3 and 34–42 to lump-4. T(lump) gives the average temperature of the lump in °C.

From these tables one can observe that as number of lumps increases the deviation (objective function) naturally decreases. The SA and GA methods yield identical solutions. If we compare Tables 8.7–8.9, we can see that all the different optimization methods yield similar results with minor variations in the lumping schemes and average temperatures. However, when we compared the CPU times it was found that all the stochastic optimization techniques had a similar behavior and solve relatively faster compared to the deterministic method used in GAMS software. For instance in Table 8.10, we compare

TABLE 8.7   Solution of Example 8.4 using Hybrid-DE

No. of lumps	3	4	5	6
Objective function	1840.4	1062	686.65	496.94
Lumping scheme	11, 25	9, 21, 33	6, 14, 21, 33	4, 9, 16, 23, 33
T (lump 1) °C	98.8	88.3	70.3	54.8
T (lump 2) °C	212.8	179.8	138.4	110.9
T (lump 3) °C	374.7	290.1	209.9	163.7
T (lump 4) °C	-	442.2	301.8	225.0
T (lump 5) °C	-	-	442.3	308.7
T (lump 6) °C	-	-	-	442.1

TABLE 8.8   Solution of Example 8.4 using SA & GA

No. of lumps	3	4	5	6
Objective function	1840.2	1060.5	686.595	496.94
Lumping scheme	11, 26	9, 20, 33	6, 14, 21, 33	4, 9, 16, 23, 33
T (lump 1) °C	98.1	88.3	70.3	54.8
T (lump 2) °C	212.8	180.5	138.3	110.9
T (lump 3) °C	374.7	290.3	209.9	163.7
T (lump 4) °C	-	442.1	301.8	225.0
T (lump 5) °C	-	-	442.1	308.7
T (lump 6) °C	-	-	-	442.1

TABLE 8.9   Solution of Example 8.4 using GAMS (BARON)

No. of lumps	3	4	5	6
Objective function	1840.2	1066.92	687.93	496.94
Lumping scheme	11, 26	9, 20, 32	6, 13, 22, 33	4, 9, 16, 23, 33
T (lump 1) °C	98.1	88.3	60.95	54.8
T (lump 2) °C	212.8	180.5	129.3	110.9
T (lump 3) °C	374.7	286.8	206.4	163.7
T (lump 4) °C	-	430.8	301.8	225.0
T (lump 5) °C	-	-	442.1	308.7
T (lump 6) °C	-	-	-	442.1

TABLE 8.10 Comparison of Hybrid-DE & GAMS Solutions for Example 8.4

No. of lumps	Hybrid-DE			GAMS		
	Objective function	Lumping scheme	CPU time (min:sec)	Objective function	Lumping scheme	CPU time (min:sec)
3	1840.4	11, 25	0:1	1840.2	11, 26	0:31
4	1062	9, 21, 33	0:6	1066.92	9, 20, 32	0:53
5	686.65	6, 14, 21, 33	0:15	687.93	6, 13, 22, 33	11:31
6	496.94	4, 9, 16, 23, 33	1:0	496.94	4, 9, 16, 23, 33	49:38

the solutions obtained using hybrid-DE and GAMS software. The CPU times given in Table 8.10 are for completeness only, since it may not be the right indicator for evaluating the performance of stochastic optimization techniques. From this table we can observe that solution time increases as the number of lumps increases. Hybrid-DE solves relatively faster compared to GAMS software especially when large number of lumps are considered.

ACKNOWLEDGEMENTS

The authors would like to thank Mr. Rahul Chauhan (B.Tech student) at IIT Delhi for help in implementing the optimal component lumping problem given in Example 8.4.

KEYWORDS

- **Binary condition**
- **Differential evolution**
- **Hybrid evolutionary method**
- **MINLP**
- **Nonlinear transformation**
- **Optimization**

REFERENCES

1. Adjiman, C. S., Androulakis, I. P., & Floudas, C. A. Global Optimization of Mixed Integer Nonlinear Problems. *AIChE. J.* 2000, 46(9), 1769–1797.

2. Angira, R., & Babu, B. V. Optimization of Process Synthesis and Design Problems: A Modified Differential Evolution Approach. *Chem. Eng. Sci.* 2006, 61, 4707–4721.

3. Behrenbruch, P., & Dedigama. T. Classification and Characterization of Crude Oils Based on Distillation Properties. *J. Petrol. Sci. Eng.*. 2007, 57, 166–180.

4. Cardoso, M. F., Salcedo, R. L., Feyo de Azevedo, S., & Barbosa, D. A Simulated Annealing Approach to the Solution of MINLP Problems. *Comp. Chem. Eng.* 1997, 21(12), 1349–1364.

5. Chang, C. T., & Chang, C. C. A Linearization Method for Mixed 0–1 Polynomial Programs. *Comp. Oper. Res.* 2000, 27, 1005–1016.

6. Costa, L., & Olivera, P. Evolutionary Algorithms Approach to the Solution of Mixed Integer Nonlinear Programming Problems. *Comp. Chem. Eng.* 2001, 25, 257–266.

7. Duran, M. A., & Grossmann, I. E., An Outer Approximation Algorithm for a Class of MINLPs. *Math. Prog.*1986, 36, 307–339.

8. Floudas, C. A. *Nonlinear and Mixed Integer Optimization: Theory, Methods and Applications*, Oxford Univ. Press, New York, 1995.

9. Floudas, C. A., Aggarwal, A., & Ciric, A. R. Global Optimum Search for Nonconvex NLP and MINLP Problems. *Comp. Chem. Eng.* 1989, 13(10), 1117–1132.

10. Floudas, C. A., & Gounaris, C. E. A Review of Recent Advances in Global Optimization, J. Glob. Optim. 2009, 45(1), 3–38.

11. Floudas, C. A., Pardalos, P. M., Adjiman, C. S., Esposito, W. R., Gumus, Z. H., Harding, S. T., Klepis, J. L., Meyer, C. A., & Schweiger, C. A. *Handbook of Test Problems in Local and Global Optimization*, Kluwer Academic Publishers, Dordrecht, Netherlands, 1999.

12. *GAMS – A User's Guide*; GAMS Development Corporation: Washington, D. C., 2012.

13. Glover, F. Improved Linear Integer Programming Formulations of Nonlinear Integer Problems. *Mgmt Sci.* 1975, 22(4), 455–460.

14. Grossmann, I. E., & Sargent, R. W. H. Optimal Design of Multipurpose Chemical Plants. *Ind. Eng. Chem. Proc. Des. Dev.* 1979, 18, 343–348.

15. Grossmann, I. E., Mixed Integer Nonlinear Programming Techniques for the Synthesis of Engineering Systems, *Res. Eng. Des.* 1990, 1, 205–228.

16. Kocis, G. R., & Grossmann, I. E. Global Optimization of Nonconvex Mixed-Integer Nonlinear Programming (MINLP) Problems in Process Synthesis. *Ind. Eng. Chem. Res.* 1988, 27, 1407–1421.

17. Lampinen, J., & Zelinka, I. Mixed Integer-Discrete-Continuous Optimization by Differential Evolution, Part 1: The Optimization Method. In P. Osmera, Eds., Proceedings of MENDEL 1999 – 5th International Mendel Conference on Soft Computing, Brno (Czech Republic). 9–12 June 1999, 71–76.

18. Li, H.-L. An Approximate Method for Local Optima for Nonlinear Mixed Integer Programming Problems. *Comp. Oper. Res.* 1992, 19 (5), 435–444.

19. Li, H.-L., & Chang, C. T. An Approximate Approach of Global Optimization for Polynomial Programming Problems. *Euro. J. Oper. Res.* 1998, 107, 625–632.

20. Lin, B., Leibovici, C. F., & Jorgensen, S. B. Optimal Component Lumping: Problem Formulation and Solution Techniques. *Comp. Chem. Eng.* 2008, 32, 1167–1172.

21. Munawar, S. A. Multilevel Decomposition based Approaches to Integrated Planning and Scheduling, PhD Dissertation, Indian Institute of Technology, Bombay, 2005.

22. Munawar, S. A., Bhushan, M., Gudi, R. D., & Belliappa, A. M. Cyclic Scheduling of Continuous Multi-product Plants in a Hybrid Flowshop Facility. *Ind. Eng. Chem. Res.* 2003, 42, 5861–5882.

23. Munawar, S. A., & Gudi, R. D. A Nonlinear Transformation Based Hybrid Evolutionary Method for MINLP Solution. *Chem. Eng. Res. Des.* 2005, 83 (A10), 1218–1236.

24. Onwubolu, G. C., & Babu, B. V. *New Optimization Techniques in Engineering*, Springer, Germany, 2004.
25. Pinto, J. M., & Grossmann, I. E., Optimal Cyclic Scheduling of Multistage Continuous Multiproduct Plants. *Comp. Chem. Eng.* 1994, 18, 797–816.
26. Price, K., & Storn, R. Differential Evolution – A Simple Evolution Strategy for Fast Optimization. *Dr. Dobb's J.* 1997, 22 (4), 18–24.
27. Price, K., & Storn, R. *Website of Differential Evolution,* Jul 2014. URL: http://www1. icsi.berkeley.edu/~storn/code.html
28. Ryoo, H. S., Sahinidis, B. P. Global Optimization of Nonconvex NLPs and MINLPs with Application in Process Design. *Comp. Chem. Eng.* 1995, 19, 551–566.
29. Salcedo, R. L. Solving Nonconvex Nonlinear Programming Problems with Adaptive Random Search. *Ind. Eng. Chem. Res.* 1992, 31, 262–273.
30. Sherali, H. D., & Wang, H. Global Optimization of Nonconvex Factorable Programming Problems. *Math. Prog.* 2001, 89, 459–478.
31. Smith, E. M. B., & Pantelides, C. C. A Symbolic Reformulation/Spatial Branch-and-Bound Algorithm for Global Optimization of Nonconvex MINLPs. *Comp. Chem. Eng.* 1999, 23, 457–478.

APPENDIX A: MATHEMATICAL MODEL FOR EXAMPLE 8.3

The following is the nomenclature and model used for cyclic scheduling problem (Munawar et al. [23]) given in Example 8.3.

Indices:

i,j	grades $\in I = \{A, B, C\}$
k	slots (1,2 and 3)
m	Stages (1 and 2)

Variables:

$I0_i, I1_i, I2_i, I3_i$	inventory breakpoints of grade i, m³
$Imax_i$	maximum of the inventory breakpoints of grade i, m³
Nc	Number of cycles
qs_i	binary variable to denote which stage (Stage 1 or Stage 2) starts processing grade i first
qe_i	binary variable to denote which stage (Stage 1 or Stage 2) ends processing grade i first
Tc	overall cycle time, maximum over the cycle times of stages 1 and 2, hr
Tep_{ikm}	end time of processing of grade i in slot k in stage m, hr
Tpp_{ikm}	processing time of grade i in slot k in stage m, hr

Tsp_{ikm}	start time of processing of grade i in slot k in stage m, hr
$Trep_{ikm}$	end time of slopping of grade i in the end of slot k in stage m, hr
$Trpp_{ikm}$	total slot time (processing + transition) of grade i in slot k in stage m, hr
$Trsp_{ikm}$	start time of slopping of grade i in the beginning of slot k in stage m, hr
Wp_i	production rate per cycle of final products i from stage 2, m³/hr
y_{ikm}	binary variable denoting allotment of grade i to slot k in stage m
z_{ijkm}	transition from grade j to i in slot k in stage m

Parameters:

$Cinv_i$	Inventory holding cost of product i, \$/m³
Ctr_{ijm}	transition cost from grade j to i in stage m, \$
Qd_i	specified bulk demand of product i, m³
Td	specified time horizon, hr
D_i	specified demand rate of product i, m³/hr ($D_i = Qd_i/Td$)
P_i	price of final products i, \$/m³
U_i^I	upper limit on inventory for grade i, m³
U^T	upper limit on time, hr
Rp_{im}	processing rate of grade i in stage m, m³/hr
α_{im}	yield or conversion of grade i in stage m
τ_{ijm}	transition time from grade j to i in stage m, hr

The mathematical formulation is as follows:

$$\sum_k y_{ikm} = 1 \quad \forall\, i \quad \forall\, m \tag{A.1a}$$

$$\sum_i y_{ikm} = 1 \quad \forall\, k \quad \forall\, m \tag{A.1b}$$

$$\sum_i z_{ijkm} = y_{j(k-1)m} \quad \forall\, j \quad \forall\, k \quad \forall\, m \tag{A.2a}$$

$$\sum_j z_{ijkm} = y_{ikm} \quad \forall\, i \quad \forall\, k \quad \forall\, m \tag{A.2b}$$

$$\left.\begin{array}{l} Tsp_{ikm} \leq U^T y_{ikm} \\ Tep_{ikm} \leq U^T y_{ikm} \\ Tpp_{ikm} \leq U^T y_{ikm} \\ Tpp_{ikm} = Tep_{ikm} - Tsp_{ikm} \end{array}\right\} \quad \forall i \quad \forall k \quad \forall m \qquad (A.3)$$

$$\sum_i Tsp_{i1m} = (1/2)\sum_i \sum_j \tau_{ijm} z_{ij1m} \quad \forall m \qquad (A.4a)$$

$$\sum_i Tsp_{i(k+1)m} = \sum_i Tep_{ikm} + \sum_i \sum_j \tau_{ijm} z_{ij(k+1)m} \quad \forall k=1,2 \quad \forall m \qquad (A.4b)$$

$$\left.\begin{array}{l} Trsp_{ikm} \leq U^T y_{ikm} \\ Trep_{ikm} \leq U^T y_{ikm} \\ Trpp_{ikm} \leq U^T y_{ikm} \\ Trpp_{ikm} = Trep_{ikm} - Trsp_{ikm} \end{array}\right\} \quad \forall i \quad \forall k \quad \forall m \qquad (A.5)$$

$$\sum_i Trsp_{ikm} = \sum_i Tsp_{ikm} - (1/2)\sum_i \sum_j \tau_{ijm} z_{ijkm} \quad \forall k \quad \forall m \qquad (A.6a)$$

$$\sum_i Trep_{ikm} = \sum_i Tep_{ikm} + (1/2)\sum_i \sum_j \tau_{ijm} z_{ij(k+1)m} \quad \forall k \quad \forall m \qquad (A.6b)$$

$$T_c \geq \sum_k \sum_i Trpp_{ikm} \quad \forall m \qquad (A.7)$$

$$\alpha_{im} Rp_{im} \sum_k Tpp_{ikm} = Rp_{i(m+1)} \sum_k Trpp_{ik(m+1)} \quad \forall i \ m=\text{stage 1} \qquad (A.8)$$

$$Wp_i = \frac{\alpha_{iM} Rp_{iM} \sum_k Tpp_{ikM}}{T_c} \quad \forall i, M = \text{stage2} \qquad (A.9a)$$

$$Wp_i \geq D_i \quad \forall i \qquad (A.9b)$$

$$(2qs_i - 1)\left[\left(\sum_k Tsp_{ikm}\right) - \left(\sum_k Trsp_{ik(m+1)}\right)\right] \leq 0 \quad \forall i \quad m=\text{stage 1} \qquad (A.10a)$$

$$(2qe_i - 1)\left[\left(\sum_k Tep_{ikm}\right) - \left(\sum_k Trep_{ik(m+1)}\right)\right] \le 0 \quad \forall i \quad m=\text{stage } 1 \quad \text{(A.10b)}$$

$$I1_i = I0_i + \left[\alpha_{im} Rp_{im} \quad \min\left\{\sum_k Trsp_{ik(m+1)} - \sum_k Tsp_{ikm}, \sum_k Tpp_{ikm}\right\}\right] qs_i$$
$$-\left[Rp_{i(m+1)} \quad \min\left\{\sum_k Tsp_{ikm} - \sum_k Trsp_{ikm+1}, \sum_k Trpp_{ikm+1}\right\}\right](1-qs_i)$$
$$\forall i \forall \quad m=\text{stage } 1 \quad \text{(A.11a)}$$

$$I2_i = I1_i + \left[\left(\alpha_{im} Rp_{im} - Rp_{i(m+1)}\right) \quad \max\left\{0, \sum_k Tep_{ikm} - \sum_k Trsp_{ik(m+1)}\right\}\right] qs_i \ qe_i$$
$$+\left[\left(\alpha_{im} Rp_{im} - Rp_{i(m+1)}\right) \sum_k Trpp_{ik(m+1)}\right] qs_i \ (1-qe_i)$$
$$+\left[\left(\alpha_{im} Rp_{im} - Rp_{i(m+1)}\right) \quad \max\left\{0, \sum_k Trep_{ik(m+1)} - \sum_k Tsp_{ikm}\right\}\right] (1-qs_i) \ (1- qe_i)$$
$$+\left[\left(\alpha_{im} Rp_{im} - Rp_{i(m+1)}\right) \sum_k Tpp_{ikm}\right] (1-qs_i) \ qe_i$$
$$\forall i \forall m=\text{stage } 1 \quad \text{(A.11b)}$$

$$I3_i = I2_i - \left[Rp_{i(m+1)} \min\left\{\sum_k Trep_{ik(m+1)} - \sum_k Tep_{ikm}, \sum_k Trpp_{ik(m+1)}\right\}\right] qe_i$$
$$+\left[\alpha_{im} \ Rp_{im} \min\left\{\sum_k Tep_{ikm} - \sum_k Trep_{ik(m+1)}, \sum_k Tpp_{ikm}\right\}\right] (1-qe_i)$$
$$\forall i \forall m=\text{stage } 1 \quad \text{(A.11c)}$$

$$\text{Im} ax_i = \max(I0_i, I1_i, I2_i, I3_i) \quad \forall i \quad \text{(A.11d)}$$

$$\text{Im} ax_i \ " \ U_i^I \quad \forall i \quad \text{(A.11e)}$$

$$profit = \sum_i P_i Wp_i Td - \sum_i \sum_j \sum_k \sum_m Ctr_{ijm} z_{ijkm} Nc - \sum_i Cinv_i \, \text{Im} ax_i Nc \quad \text{(A.12)}$$

Constraints (A.1) enforce unique allotment of a grade to a slot with no grade repetitions within a cycle. Similarly, the transition variable z_{ijkm} is uniquely defined by constraints (A.2) and is a continuous variable between 0 and 1. Whenever there is transition from grade j, being produced in slot $k-1$, to a grade i, to be produced in the current slot k, then $z_{ijkm} = 1$, else it is zero. The non-negativity inequalities of (A.3) ensure that when a product is not assigned to a slot the corresponding start, end and processing times are all zero. Constraint (A.4a) defines the start time of processing of the first slot in any cycle, while for all other slots the start time of processing constraint is defined by (A.4b). The Eqs. (A.3) and (A.4) of Tsp_{ikm} and Tep_{ikm} are written for the slopping variables $Trsp_{ikm}$ and $Trep_{ikm}$ as given in Eqs. (A.5) and (A.6). The cycle time is defined in constraint (A.7) and the material balance between Stage 1 and Stage 2 is given by constraint (A.8). The production rate per cycle is defined by constraint (A.9a) and over production of a product is allowed as given in (A.9b). The binary variable qs_i defined by Eq. (A.10a) is used to find out which of Stage 1 or Stage 2 first starts the processing of grade i. If $qs_i = 1$ then Stage 1 starts first else, if $qs_i = 0$, then Stage 2 starts processing grade i first. Similarly, the binary variable qe_i is defined by the Eq. (A.10b). The inventory breakpoints are as given in constraints (A.11) and the objective function is maximization of profit subject to penalties for grade transition costs and inventory costs as defined by Eq. (A.12), where the number of cycles $Nc = Td/Tc$.

CHAPTER 9

DIFFERENTIAL EVOLUTION FOR OPTIMAL DESIGN OF SHELL-AND-TUBE HEAT EXCHANGERS

MUNAWAR A. SHAIK[1] and B. V. BABU[2]

[1]*Associate Professor, Department of Chemical Engineering, Indian Institute of Technology Delhi, Hauz Khas, New Delhi, India, E-mail: munawar@iitd.ac.in, Tel: +91-11-26591038*

[2]*Vice Chancellor, Galgotias University, Greater Noida, Uttar Pradesh, India, E-mail: profbvbabu@gmail.com, Tel: +91-12-04806849*

CONTENTS

9.1   INTRODUCTION

Heat exchangers are used extensively in the process and allied industries and thus are very important during design and operation. The most commonly used type of heat exchanger is the shell-and-tube heat exchanger, the optimal design of which is the main objective of this chapter. The secondary objective is performance evaluation of different non-traditional optimization techniques such as Genetic Algorithms (GA) and Differential Evolution (DE) for the optimal design of shell-and-tube heat exchangers.

The design of a shell-and-tube heat exchanger involves finding an optimal heat exchanger configuration to match the desired heat transfer area required for a specified heat duty. Different design variables such as tube outer diameter, tube pitch, tube length, number of tube passes, different shell head types, baffle spacing, and baffle cut, are used in deciding a particular heat exchanger configuration. Traditionally, the design of a shell-and-tube heat exchanger involves trial and error procedure where for a certain combination of the design variables the heat transfer area is calculated and then another combination is tried to check if there is any possibility of reducing the heat transfer area. The numerous combinations of different design variables lead to a combinatorial explosion of different heat exchanger configurations with different overall heat transfer area available for heat exchange. Thus, the optimal design of a heat exchanger can be posed as a large scale, discrete, combinatorial optimization problem [8]. The presence of discrete decision variables, and the absence of a definite mathematical structure for directly determining the overall heat transfer area as a function of these design variables renders the heat exchanger design problem a challenging task. Therefore, application of non-traditional optimization techniques [19] is a natural choice for solving the optimal design problem for shell-and-tube heat exchangers. In this chapter, we illustrate the application of differential evolution for solving the optimal heat exchanger design problem and present comparison with genetic algorithms.

Chaudhuri et al. [8] used Simulated Annealing (SA) for the optimal design of heat exchangers and developed a procedure to run the HTRI design program coupled to the annealing algorithm, iteratively. They compared the results of SA with a base case design and concluded that significant savings in heat transfer area and hence the heat exchanger cost can be obtained using SA. Manish et al. [17] used GA to solve this optimal problem and compared the performance of SA and GA. They presented GA strategies to improve the performance of the optimization framework. They concluded that these algorithms result in considerable savings in computational time compared to an exhaustive search, and have an advantage over other methods in obtaining multiple solutions of the same quality, thus providing more flexibility to the designer. Munawar [18] and Babu and Munawar [5, 6] demonstrated the first successful application of DE to the optimal heat exchanger design problem and studied the effect of DE key parameters and different strategies of DE on the optimality along with presenting a comparison of GA and DE. The DE algorithm and its variants have been successfully applied in diverse applications [1, 3, 4, 13, 14].

Serna and Jimenez [26] presented a compact formulation of Bell-Delaware method for heat exchanger design and optimization. There are several applications of other techniques for solving this problem such as genetic algorithms [7, 17, 20, 23], tube count table search [9], harmony search algorithm [10], particle swarm optimization [21], artificial bee colony algorithm [25], constructal theory [2], multiobjective optimization [11], imperialist competitive algorithm [15], and biogeography-based algorithm [16].

In the next section, the general procedure of shell-and-tube heat exchanger design is discussed followed by the optimal problem formulation.

9.2   OPTIMAL DESIGN OF SHELL-AND-TUBE HEAT EXCHANGERS

The *design of a process heat exchanger* usually proceeds through the following steps [22] involving trial and error:

- Process conditions (stream compositions, flow rates, temperatures, pressures) must be specified.
- Required physical properties over the temperature and pressure ranges of interest must be obtained.
- The type of heat exchanger to be employed is chosen.
- A preliminary estimate of the size of the exchanger is made, using a heat transfer coefficient appropriate to the fluids, the process, and the equipment.
- A first design is chosen, complete in all details necessary to carry out the design calculations.
- The design chosen is now evaluated or rated, as to its ability to meet the process specifications with respect to both heat duty and pressure drop.

- Based on this result a new configuration is chosen if necessary and the above step is repeated. If the first design was inadequate to meet the required heat load, it is usually necessary to increase the size of the exchanger, while still remaining within specified or feasible limits of pressure drop, tube length, shell diameter, etc. This will sometimes mean going to multiple exchanger configurations. If the first design more than meets heat load requirements or does not use all the allowable pressure drop, a less expensive exchanger can usually be designed to fulfill the process requirements.
- The final design should meet process requirements (within the allowable error limits) at lowest cost. The lowest cost should include operation and maintenance costs and credit for ability to meet long-term process changes as well as installed (capital) cost. Exchangers should not be selected entirely on a lowest first cost basis, which frequently results in future penalties.

The corresponding flow chart given in Figure 9.1 [27] gives the sequence of steps and the loops involved in the optimal design of a shell-and-tube heat exchanger. In the present study, Bell's method of heat exchanger design is used to find the heat transfer area for a given design configuration. Bell's method gives good estimates of the shell-side heat transfer coefficient and pressure drop compared to Kern's method, as it takes into account the factors for leakage, bypassing, flow in window zone etc.

9.2.1 BELL's METHOD

The following are the details of the steps involved in Bell's method of heat exchanger design as given in Figure 9.1 [27]:

1. The first step in any heat exchanger design is to calculate the heat duty (Q) and the unspecified outlet temperatures or flow rates.
2. The next step is to collect together the fluid physical properties required: density, viscosity, thermal conductivity, and specific heat.
3. From the literature a trial value for the overall heat transfer coefficient is assumed ($U_{o'ass}$).
4. The mean temperature difference, $\Delta T_m = (\Delta T_{LMTD} \, F_t)$ is evaluated, where F_t is the temperature correction factor, which is a function of the shell and tube fluid temperatures and the number of tube and shell passes. It is normally correlated as a function of two dimensionless temperature ratios:

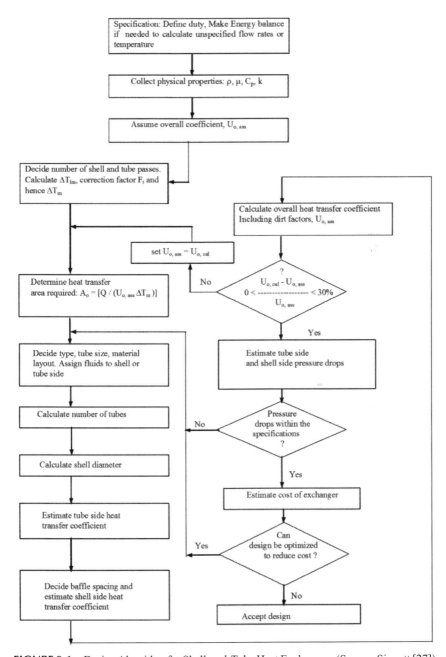

FIGURE 9.1   Design Algorithm for Shell-and-Tube Heat Exchangers (Source: Sinnott [27]).

$$R = \frac{(T_1 - T_2)}{(t_2 - t_1)}; \quad S = \frac{(t_2 - t_1)}{(T_1 - t_1)} \qquad (1)$$

5. The total heat transfer area thus required can be calculated from $A_o = Q/(U_{o'ass} \Delta T_m)$.

6. Since the shell and tube geometry and layouts are user supplied the next step is to calculate the number of tubes required to achieve the area A_o. Area of one tube is $\pi D_o L$. So, the number of tubes N_t, is A_o/A_t, rounded off to the next higher integer.

7. Once the number of tubes are calculated the bundle diameter can be calculated from $D_b = D_o (N_t/k_1)^{1/n}$ where k_1, n_1 are constants dependent on tube pitch and number of tube passes.

8. The shell diameter (D_s) must be selected now to give as close a fit to the tube bundle as is practical; to reduce bypassing round the outside of the bundle. The clearance required between the outermost tubes in the bundle and the shell inside diameter will depend on the type of exchanger and the manufacturing tolerances.

9. *Tube-side heat transfer coefficient (h_t):* The correlation used here is

$$\frac{h_t d_i}{k_f} = j_h \ \text{Re} \ \text{Pr}^{1/3} \left(\frac{\mu}{\mu_w} \right)^{0.14} \qquad (2)$$

where j_h is the tube-side heat transfer factor which is a function of Reynolds number.

Viscosity correction factor: This factor will normally only be significant for viscous fluids. To apply the correction an estimate of the wall temperature is needed. This is done by trial and error. First, h_t is calculated without the viscosity correction. Then, the tube wall temperature is estimated from the relation $h_t(t_w - t) = U(T - t)$. Now, the viscosity is evaluated at the wall temperature and h_t is calculated using the viscosity correction factor, iteratively.

10. *Shell-side heat transfer coefficient (h_s):* The main difference between Kern's method and Bell's method lies in the evaluation of shell-side heat transfer coefficient and pressure drop. In Bell's method the heat transfer coefficient (h.t.c.) and pressure drop are estimated from the correlations for flow over ideal tube banks, and the effects of leakage, bypassing and flow in the window zone are allowed for by applying

the correction factors. This approach will give more satisfactory predictions of the h.t.c. and pressure drop than Kern's method. The shell-side h.t.c. is given by

$$h_s = h_{oc} F_n F_w F_b F_L \tag{3}$$

where h_{oc} – h.t.c. calculated for cross-flow over an ideal tube bank, no leakage or bypassing; F_n – corrector factor to allow for the effect of the number of vertical tube rows; F_w – window effect correction factor; F_b – bypass stream correction factor; F_L – leakage correction factor. The total correction will vary from 0.6 for a poorly designed exchanger with large clearances, to 0.9 for a well-designed exchanger.

h_{oc}, *ideal cross-flow coefficient:* The correlation used here is

$$\frac{h_{oc}}{k_f} = j_h \, Re \, Pr^{1/3} \left(\frac{\mu}{\mu_w} \right)^{0.14} \tag{4}$$

where j_h, the factor for cross-flow tube banks is a function of Reynolds number.

Fn, Tube row correction factor: The mean h.t.c. will depend on the number of tubes crossed. For turbulent flow the correction factor F_n is close to 1.0. In laminar flow the h.t.c. may decrease with increasing rows of tubes crossed, due to the build up of the temperature boundary layer. The factor given below can be used for various flow regimes. Here, N_w = number of constrictions crossed i.e., number of tube rows between the baffle tips.

1. $Re > 2000$, turbulent, take F_n from nomograph.
2. $Re > 100$ to 2000, transition region, take $F_n = 1.0$;
3. $Re < 100$, laminar region, $F_n \propto (N'_c)^{-0.18}$.

F_w, *window correction factor:* This factor corrects for the effect of flow through the baffle window, and is a function of the heat transfer area in the window zone and the total heat transfer area. The correction factor is plotted versus. R_w, the ratio of the number of tubes in the window zone to the total number in the bundle, $R_w = 2$ Ra where Ra is the ratio of the bundle cross sectional area in the window zone to the total bundle cross sectional area. Ra can be obtained from nomograph for the appropriate bundle cut, B_b. ($B_b = H_b/D_b$).

$$F_b = \exp\left(-\alpha\, A_b/A_s\right) \tag{5}$$

where $\alpha = 1.5$ for laminar flow, Re < 100.
$= 1.35$ for transitional and turbulent flow, Re > 100.
A_b = clearance area between the bundle and the shell,
$A_b = l_B\,(D_s - D_b)$.
A_s – maximum area for cross flow.
F_L, *leakage correction factor:* This factor corrects for the leakage through the tube to baffle clearance and the baffle-to-shell clearance.

$$F_L = 1 - \left(\beta_L\,[(A_{tb} + 2A_{sb})/A_L]\right) \tag{6}$$

where β_L – a factor obtained from Figure A.8; A_{tb} – the tube-to-baffle clearance area, per baffle; A_{sb} – shell-to-baffle clearance area, per baffle; A_L – total leakage area = $(A_{tb} + A_{sb})$.

11. *Overall heat transfer coefficient:* Once the shell-side and tube-side h.t.c. are calculated the overall h.t.c. can be determined from the following relation:

$$\frac{1}{U_{o,cal}} = \frac{1}{h_s} + \frac{d_o \ln(d_o/d_i)}{2k_w} + \frac{d_o}{d_i}\frac{1}{h_t} + \frac{d_o}{d_i}\frac{1}{h_{id}} + \frac{1}{h_{od}} \tag{7}$$

where h_{id} = inside dirt coefficient (fouling factor); h_{od} = outside dirt coefficient; k_w = thermal conductivity of the tube wall material.
The calculated overall h.t.c. is compared with the assumed value and is iterated until it converges, as shown in the flow chart.

12. Now, the tube side and shell side pressure drops are to be calculated to check whether they are within the specifications required. Then, if the pressure drops are huge, accordingly the shell or the tube geometries have to be changed. But as we are considering all possible design combinations of the variables, in the Differential Evolution function itself, here we just assign a high value for heat exchanger cost or we can keep generating new design configurations in DE until the pressure drop constraints are matched. Otherwise, the design is accepted and the heat exchanger cost is calculated for the area A_o, obtained by taking the converged value for overall h.t.c. The details of shell side and tube side pressure drops are given in Appendix-A.

9.2.2   OPTIMIZATION PROBLEM

The objective function and the optimal problem of shell-and-tube heat exchanger design can be represented as shown below [17, 18].

min $C(\mathbf{X})$ or $A(\mathbf{X})$

$\quad\quad \mathbf{X} \in \{x_1, x_2, x_3, x_4, x_5, x_6, x_7\}$

where

$\quad\quad x_1 = \{1,2,...,12\}$
$\quad\quad x_2 = \{1,2\}$
$\quad\quad x_3 = \{1,2,3,4\}$
$\quad\quad x_4 = \{1,2,...,5\}$
$\quad\quad x_5 = \{1,2,...,8\}$
$\quad\quad x_6 = \{1,2,...,6\}$
$\quad\quad x_7 = \{1,2,...,7\}$

subject to

$\quad\quad$ feasibility constraints [pressure-drop] $\hspace{3cm}$ (8)

The objective function can be minimization of heat exchanger cost $C(\mathbf{X})$ or heat transfer area $A(\mathbf{X})$ and $\mathbf{X}$ is a solution string representing a design configuration. The design variable x_1 takes 12 values for tube outer diameter in the range of 0.25" to 2.5" (0.25", 0.375", 0.5", 0.625", 0.75", 0.875", 1.0", 1.25", 1.5", 1.75", 2", 2.5"). x_2 represents the tube pitch – either square or triangular – taking two values represented by 1 and 2. x_3 takes the shell head types: floating head, fixed tube sheet, U tube, and pull through floating head represented by the numbers 1, 2, 3 and 4, respectively. x_4 takes number of tube passes 1–1, 1–2, 1–4, 1–6, 1–8 represented by numbers from 1 to 5. The variable x_5 takes eight values of the various tube lengths in the range 6' to 24' (6', 8', 10', 12', 16', 20', 22', 24') represented by numbers 1 to 8. x_6 takes six values for the variable baffle spacing, in the range 0.2 to 0.45 times the shell diameter (0.2, 0.25, 0.3, 0.35, 0.4, 0.45). x_7 takes seven values for the baffle cut in the range 15 to 45 percent (0.15, 0.2, 0.25, 0.3, 0.35, 0.4, 0.45).

The pressure drop on the fluids exchanging heat is considered as a feasibility constraint. Generally a pressure drop of more than 1 bar is not desirable for the flow of fluid through a heat exchanger. For a given design configuration, whenever the pressure drop exceeds the specified limit, a high value for the heat transfer area is returned so that as an infeasible configuration it will be eliminated in the next iteration of the optimization routine. The total number of design combinations with these variables are

$12 \times 2 \times 4 \times 5 \times 8 \times 6 \times 7 = 1,61,280$. This means that if an exhaustive search is to be performed it will take at the maximum 1,61,280 function evaluations before arriving at the global minimum heat exchanger cost. So the strategy, which takes few function evaluations, is the best one. Considering minimization of heat transfer area as the objective function, differential evolution technique is applied to find the optimum design configuration as discussed in next section.

9.3   APPLICATION OF DIFFERENTIAL EVOLUTION

The overall structure of the DE algorithm resembles that of most other population based searches. The parallel version of DE maintains two arrays, each of which holds a population of NP, D-dimensional, real valued vectors. The primary array holds the current vector population, while the secondary array accumulates vectors that are selected for the next generation. In each generation, NP competitions are held to determine the composition of the next generation. Every pair of vectors $(\mathbf{X}_a, \mathbf{X}_{b)}$ defines a vector differential: $\mathbf{X}_a - \mathbf{X}_b$. When $\mathbf{X}_a$ and $\mathbf{X}_b$ are chosen randomly, their weighted differential is used to perturb another randomly chosen vector $\mathbf{X}_c$. This process can be mathematically written as $\mathbf{X}'_c = \mathbf{X}_c + F(\mathbf{X}_a - \mathbf{X}_b)$. The scaling factor F is a user supplied constant in the range $(0 < F \leq 1.2)$. The optimal value of F for most of the functions lies in the range of 0.4 to 1.0 [24]. Then in every generation, each primary array vector, $\mathbf{X}_i$ is targeted for crossover with a vector like $\mathbf{X}'_c$ to produce a trial vector $\mathbf{X}_t$. Thus, the trial vector is the child of two parents, a noisy random vector and the target vector against which it must compete. The non-uniform crossover is used with a crossover constant CR, in the range $0 \leq CR \leq 1$. CR actually represents the probability that the child vector inherits the parameter values from the noisy random vector. When CR = 1, for example, every trial vector parameter is certain to come from $\mathbf{X}'_c$. If, on the other hand, CR = 0, all but one trial vector parameter comes from the target vector. To ensure that $\mathbf{X}_t$ differs from $\mathbf{X}_i$ by at least one parameter, the final trial vector parameter always comes from the noisy random vector, even when CR = 0. Then the cost of the trial vector is compared with that of the target vector, and the vector that has the lowest cost of the two would survive for the next generation. In all, just three factors control evolution under DE, the population size, NP; the weight applied to the random differential, F; and the crossover constant, CR.

The algorithm of Differential Evolution as given by Price and Storn [24], is in general applicable for continuous function optimization. The upper and lower bounds of the design variables are initially specified. Then, after mutation because of the addition of the weighted random differential the parameter values may even go beyond the specified boundary limits. So, irrespective of the boundary limits initially specified, DE finds the global optimum by exploring beyond the limits. Hence, when applied to discrete function optimization the parameter values have to be limited to the specified bounds. In the present problem, since each design variable has a different upper bound when represented by means of integers, the same DE code given by Price and Storn [24] cannot be used. Munawar [18] used normalized values for all the design variables and randomly initialized all the design variables between 0 and 1. Whenever it is required to find the heat transfer area using Bell's method for a given design configuration, these normalized values are converted back to their corresponding boundary limits.

The pseudo code of the DE algorithm (adapted from Ref. [24]) for the optimal heat exchanger design problem as used in Refs. [6, 18] is given below:

- Choose a strategy and a seed for the random number generator.
- Initialize the values of D, NP, CR, F and MAXGEN.
- Initialize all vectors of the population randomly. Since the upper bounds are all different for each variable in this problem, the variables are all normalized. Hence generate a random number between 0 and 1 for all the design variables for initialization.

  for $i = 1$ to NP

  {for $j = 1$ to D

  $x_{i,j}$ = random number}

- Evaluate the cost of each vector. Cost here is the area of the shell-and-tube heat exchanger for the given design configuration, calculated by a separate function cal_area() using Bell' method.

  for $i = 1$ to NP

  C_i = cal_area()

- Find out the vector with the lowest cost i.e. the best vector so far.

  $C_{min} = C_1$ and best $= 1$

  for $i = 2$ to NP

  {if $(C_i < C_{min})$

  then $C_{min} = C_i$ and best $= i$}

- Perform mutation, crossover, selection and evaluation of the objective function for a specified number of generations.

    While (gen < MAXGEN)

        {for i = 1 to NP}

- For each vector X_i (target vector), select three distinct vectors X_a, X_b and X_c (select five, if two vector differences are to be used) randomly from the current population (primary array) other than the vector X_i

do

    {

    r_1 = random number * NP

    r_2 = random number * NP

    r_3 = random number * NP

    } while $(r_1 = i)$ OR $(r_2 = i)$ OR $(r_3 = i)$ OR $(r_1 = r_2)$ OR $(r_2 = r_3)$ OR $(r_1 = r_3)$

- Perform crossover for each target vector X_i with its noisy vector $X_{n,i}$ and create a trial vector, $X_{t,i}$. The noisy vector is created by performing mutation. If CR = 0 inherit all the parameters from the target vector X_i, except one which should be from $X_{n,i}$.

for exponential crossover

    {   p = random number * 1

            r = random number * D

            n = 0

do

    $\{X_{n,i} = X_{a,i} + F(X_{b,i} - X_{c,i})$ /* add two weighted vector differences for r = (r+1) % D two vector perturbation. For best / random increment r by 1 vector perturbation the weighted vector } while ((p<CR) and (r<D)) difference is added to the best / random

                vector of the current population. */

    }

    for binomial crossover

    { p = random number * 1

r = random number * D

for n = 1 to D

    { if((p<CR) or (p = D-1)) /* change at least one parameter if CR=0 */

        $X_{n,i} = X_{a,i} + F(X_{b,i} - X_{c,i})$

        r = (r+1)%D }

}

if ($X_{n,i}$ > 1) $X_{n,i}$ = 1 /* for discrete function optimization check the

if ($X_{n,i}$ < 0) $X_{n,i}$ = 0 values and restrict within limits */

/* 1 – normalized upper bound;
    0 – normalized lower bound */
- Perform selection for each target vector, X_i by comparing its cost with that of the trial vector, $X_{t,i}$; whichever has the lowest cost will survive for the next generation.

$C_{t,i}$ = cal_area()
if ($C_{t,i} < C_i$) new $X_i = X_{t,i}$
    else new $X_i = X_i$ } /* for i=1 to NP */
}

The entire scheme of optimization of shell-and-tube heat exchanger design is performed by the DE algorithm, while intermittently it is required to evaluate the heat transfer area for a given design configuration. This task is accomplished through the separate function cal_area() which employs Bell's method of heat exchanger design. Bell's method gives accurate estimates of the shell-side heat transfer coefficient and pressure drop compared to Kern's method, as it takes into account the factors for leakage, bypassing, flow in window zone etc. The various correction factors in Bell's method include: temperature correction factor, tube-side heat transfer and friction factor, shell-side heat transfer and friction factor, tube row correction factor, window correction factor for heat transfer and pressure drop, bypass correction factor for heat transfer and pressure drop, friction factor for cross-flow tube banks, baffle geometrical factors etc. These correction factors are reported in the literature in the form of nomographs [22, 27]. The data on these correction factors from the nomographs were fitted into polynomial equations and incorporated in the computer program [18].

9.4   RESULTS AND DISCUSSION

As a *case study* the following problem for the design of a shell-and-tube heat exchanger [27] is considered as presented in Munawar [18] and Babu and Munawar [6]: 20,000 kg/hr of kerosene leaves the base of a side-stripping column at 200°C and is to be cooled to 90°C with 70,000 kg/hr light crude oil coming from storage at 40°C. The kerosene enters the exchanger at a pressure of 5 bar and the crude oil at 6.5 bar. A pressure drop of 0.8 bar is permissible on both the streams. Allowance should be made for fouling by including fouling factor of 0.00035 (W/m² °C)⁻¹ on the crude stream and 0.0002 (W/m² °C)⁻¹ on the kerosene side.

By performing enthalpy balance, the heat duty for this case study is found to be 1509.4 kW and the outlet temperature of crude oil to be 78.6°C. The crude is dirtier than the kerosene and so is assigned through the tube-side and kerosene to the shell-side. Using a proprietary program (HTFS, STEP5) the lowest cost design meeting the above specifications is reported to be a heat transfer area of 55 m² based on outside diameter [27]. The result of the above program is considered as the base case design and DE is applied for the same problem with all ten different strategies (listed in Section 9.2). As a heuristic, the pressure drop in a heat exchanger normally should not exceed 1 bar. Hence, the DE strategies are applied for this case study separately with both 0.8 bar and 1 bar as the constraints. In both cases, same global minimum heat exchanger area was obtained using DE. In the subsequent analysis, the results for 1 bar as the constraint are referred here.

A seed value for the pseudo random number generator must be selected by trial and error. In principle any positive integer can be taken. Different integer values were tried (say 3, 5, 7, 10, 15 and 20) with all the strategies for a NP value of 70 (10 times D). The F values were varied from 0.1 to 1.1 in steps of 0.1 and CR values from 0 to 1 in steps of 0.1, leading to 121 combinations of F and CR for each seed. When DE algorithm was executed for all the above combinations, the global minimum heat exchanger area for the above heat duty was found to be 34.44 m² as against 55 m² for the base case design. For each seed, out of the 121 combinations of F and CR considered, the percentage of the combinations converging to this global minimum (C_{DE}) in less than 30 generations is listed for each strategy in Table 9.1 [6, 18]. The average C_{DE} for each seed as well as for each strategy are also listed in the same table. This average was considered to be a measure of the 'likeliness' in achieving the global minimum [18].

The average C_{DE} for each seed ranges from 40.7 to 64.8 and the average C_{DE} for different strategies varies from 44.2 to 59.4. If we consider a benchmark C_{DE} as 40, then seeds 5, 7 and 10 stand good from the rest, and excepting strategy numbers 2, 7 and 8 all other strategies can be considered as good for this problem.

Considering 'speed' as the other criteria, to further consolidate the effect of strategies on each seed and vice versa, the best combinations of F and CR – taking the minimum number of generations to converge to the global minimum (termed as G_{min}) – are listed in Table 9.2 [6, 18].

The criteria for choosing a good seed from 'speed' point view could be: (1) it should yield the global minimum in less number of generations, and

TABLE 9.1 Effect of Seed on DE Strategies Based on C_{DE}

S. No.	Strategy	Seed = 3 C_{DE}	Seed = 5 C_{DE}	Seed = 7 C_{DE}	Seed = 10 C_{DE}	Seed = 15 C_{DE}	Seed = 20 C_{DE}	Average of C_{DE} w.r.t. seed
1	DE/best/1/exp	41.3	62.0	52.1	48.0	57.0	45.5	51.0
2	DE/rand/1/exp	36.4	72.0	68.6	51.2	61.1	39.7	54.8
3	DE/rand-to-best/1/exp	44.6	53.0	46.3	43.0	48.8	43.0	46.5
4	DE/best/2/exp	44.6	72.7	72.0	66.1	53.7	47.1	59.4
5	DE/rand/2/exp	28.1	72.0	53.7	48.8	46.3	43.0	48.7
6	DE/best/1/bin	37.2	57.0	44.6	68.6	47.1	47.9	50.4
7	DE/rand/1/bin	44.6	81.8	59.5	37.2	60.0	38.8	53.7
8	DE/rand-to-best/1/bin	37.2	48.0	42.1	63.6	38.8	35.5	44.2
9	DE/best/2/bin	48.8	61.2	57.0	64.5	52.1	58.7	57.1
10	DE/rand/2/bin	43.8	68.6	48.0	55.4	39.7	68.6	54.0
Average of CDE w.r.t. strategies		40.7	64.8	54.4	54.6	50.5	46.8	-

TABLE 9.2   Effect of Seed on DE Strategies Based on F, CR and G_{min}

S. No.	Strategy	Seed = 3 F CR G_{min}	Seed = 5 F CR G_{min}	Seed = 7 F CR G_{min}	Seed = 10 F CR G_{min}	Seed = 15 F CR G_{min}	Seed = 20 F CR G_{min}
1	DE/best/1/ exp	0.7 0.9 3	0.9 1.0 3 1.0 1.0 3 1.1 1.0 3	0.4 0.4 5 0.5 0.4 5 0.6 0.4 5 0.7 0.4 5 0.8 0.4 5 0.9 0.4 5 0.9 0.7 5 1.0 0.4 5 1.1 0.4 5	0.8 0.8 2 0.9 0.8 2 0.9 1.0 2 1.0 1.0 2 1.1 1.0 2	0.6 0.3 4 0.7 0.3 4	0.8 0.4 3 0.9 0.4 3 1.0 0.4 3
2	DE/ rand/1/ exp	1.1 0.1 3	1.0 0.6 6 1.1 0.6 6	0.9 0.9 6 1.0 0.9 6	0.9 0.9 7	0.9 0.4 7 1.0 0.4 7 1.1 0.4 7	1.0 0.5 6 1.1 0.5 6
3	DE/rand-to-best/1/ exp	1.0 0.9 4	1.0 0.2 5 1.1 0.2 5	0.9 1.0 2 1.0 1.0 2 1.1 1.0 2	0.6 0.8 4 0.8 0.9 4 0.9 1.0 4 1.0 0.7 4	0.7 0.9 3	1.0 0.7 4 1.0 0.9 4
4	DE/best/2/ exp	0.4 0.9 4 0.7 0.9 4 0.9 1.0 4 1.1 0.7 4	1.1 1.0 2	0.8 0.7 2	1.0 0.6 4 1.1 0.6 4	0.8 0.8 4	0.7 1.0 2 0.8 1.0 2
5	DE/ rand/2/ exp	0.6 0.9 12 1.1 0.7 12	1.0 0.5 6 1.1 0.5 6	0.8 0.9 2	1.0 1.0 3	0.6 0.9 1 1.0 0.3 9	0.2 1.0 6
6	DE/best/1/ bin	0.4 0.6 7 1.0 0.9 7	0.9 1.0 3 1.0 1.0 3 1.1 1.0 3	1.0 0.8 3	0.5 0.7 2 0.6 0.4 2 0.6 0.5 2 0.6 0.9 2 0.7 0.4 2 0.9 1.0 2 1.0 0.9 2 1.0 1.0 2 1.1 0.8 2 1.1 0.9 2 1.1 1.0 2	0.9 0.8 5	0.4 0.7 5 1.0 0.9 5

TABLE 9.2 Continued

S. No.	Strategy	Seed = 3 F CR G$_{min}$	Seed = 5 F CR G$_{min}$	Seed = 7 F CR G$_{min}$	Seed = 10 F CR G$_{min}$	Seed = 15 F CR G$_{min}$	Seed = 20 F CR G$_{min}$
7	DE/ rand/1/bin	1.0 0.5 9	0.1 0.3 7 0.7 0.5 7 0.8 0.8 7	0.9 0.9 5	0.4 0.5 7	0.7 0.9 3	0.6 0.7 10
8	DE/rand- to-best/1/ bin	0.7 0.1 6 1.1 0.9 6	0.7 0.4 5 0.8 0.5 5 1.1 0.4 5	0.9 0.6 2 0.9 0.7 2 0.9 0.8 2 0.9 1.0 2 1.0 1.0 2 1.1 1.0 2	0.9 0.7 3 0.9 0.8 3	0.7 0.8 5 0.7 1.0 5	1.0 0.5 3
9	DE/best/2/ bin	0.9 1.0 4	0.6 0.8 2 0.7 0.8 2 1.1 1.0 2	1.1 0.8 3	0.5 0.5 2 0.6 0.5 2	0.7 0.4 5	0.7 1.0 2 0.8 1.0 2
10	DE/ rand/2/bin	0.6 0.8 6	0.4 0.8 5 1.1 0.5 5	0.9 1.0 6	1.0 1.0 3	1.1 0.4 9	0.2 0.9 4 0.8 0.9 4

(2) it should yield the same over a wide range of F and CR for most of the strategies. Based on these criteria, seed 10 can be considered to be good as it gives the global minimum in two generations and for more combinations of F and CR compared to other seed values. From the 'speed' point of view it can be observed from Table 9.2 that the strategy numbers 2, 5, 7, and 10 are good. Hence, from both 'more likeliness' and 'speed' point of view, for different seeds, strategy numbers 1, 3, 4, 6, and 9 are good.

To further explore the effect of the key parameters in detail, the NP values are varied along with F & CR. The maximum number of generations, MAXGEN, is taken as 15 because as can be seen from Table 9.2, a maximum of 12 generations are required to converge to the global minimum with NP=70. With MAXGEN = 15, and for the selected seed value of 10, the percentage of combinations converging to the global minimum (C_{DE}) are listed for each strategy in Table 9.3 [6, 18] for different NP values. From this table, it is seen that the individual C_{DE} values cover a wide range from 0.8 to 66.1. Considering C_{DE} of 10 as a benchmark it can be observed that NP value of 70 and above stand good from the rest. From 'more likeliness' point of view, with a benchmark of 20 for C_{DE} values, again NP values of 70 and above are good. But for NP values of 100 and above it is observed that the likeliness

TABLE 9.3　Effect of NP on DE Strategies Based on C_{DE}

S. No.	Strategy	NP = 40 C_{DE}	NP = 50 C_{DE}	NP = 60 C_{DE}	NP = 70 C_{DE}	NP = 80 C_{DE}	NP = 90 C_{DE}	NP = 100 C_{DE}	Average of C_{DE} w.r.t. NP
1	DE/best/1/exp	8.3	9.9	9.9	20.7	30.6	33.1	41.3	22
2	DE/rand/1/exp	2.5	2.5	6.6	6.6	13.2	8.3	5.0	6.4
3	DE/rand-to-best/1/exp	9.1	7.4	7.4	25.6	18.2	21.5	20.7	15.7
4	DE/best/2/exp	9.1	10.7	17.4	40.5	28.1	43.0	46.3	27.9
5	DE/rand/2/exp	0.8	5.0	7.4	0.8	10.7	9.9	7.4	6.0
6	DE/best/1/bin	9.1	7.4	18.2	13.2	47.9	53.7	33.1	26.1
7	DE/rand/1/bin	1.7	2.5	2.5	0.8	7.4	15.7	17.4	6.9
8	DE/rand-to-best/1/bin	8.3	11.6	9.1	7.4	27.3	39.7	14.3	16.8
9	DE/best/2/bin	12.4	12.4	11.6	3.3	66.1	66.1	52.9	32.1
10	DE/rand/2/bin	1.7	1.7	6.6	0.8	14.9	21.5	24.0	10.2
	Average CDE w.r.t. strategy	6.3	7.1	9.7	12.0	26.4	31.3	26.2	-

decreases. Considering 'speed' as the other criteria, the best combinations of F and CR, taking the least number of generations to converge to the global minimum (G_{min}), are listed in Table 9.4 [6, 18] for various strategies for different NP values.

From the 'speed' point of view it is evident that the NP values of 70 and above are good – indicating that at least a population size of 10 times D is essential to have more likeliness in achieving the global minimum. Combining the results of variations in seed and NP, from 'more likeliness' as well as 'speed' point of view, it can be concluded that DE/best/...(strategy numbers 1, 4, 6, and 9) are good. Hence, for the optimal heat exchanger design problem, the best vector perturbations either with a single or two vector differences are the best with either exponential or binomial crossovers. The relationship between number of function evaluations (NFE) and population size (NP) is NFE = NP * (G_{min} + 1) (plus one corresponds to the function evaluations in the initial population). It can be inferred from Table 9.4 (using NP and G_{min} values to compute NFE) that NFE varies from 100 to 1300, out of the 1,61,280 possible function evaluations considered. Hence, the best combination corresponding to the least function evaluations from Table 9.4 is for NP = 50 and DE/best/1/exp strategy, with 100 function evaluations as it converges in one generation itself. Regarding the effect of parameters F and CR, it is observed that the DE strategies are more sensitive to the values of CR than to F. Extreme values of CR are worth to be tried first. Selection of a good seed is indeed the first hurdle before investigating the right combination of the key parameters and the choice of the strategy.

9.4.1   COMPARISON OF GA AND DE

For comparison, Genetic Algorithms with binary coding for the design variables are also applied for the same case study with Roulette-wheel selection, single-point crossover, and bit-wise mutation as the operators for creating the new population. The GA algorithm is executed for various values of N – the population size, p_c – the crossover probability and p_m – the mutation probability. With a seed value of 10 for the pseudo random number generator, N is varied from 32 to 100 in steps of 4; p_c from 0.5 to 0.95 in steps of 0.05; and p_m from 0.05 to 0.3 in steps of 0.05, leading to a total of 1080 combinations. N/2 has to be an even number for single-point crossover and hence the starting value of 32 and the step size of 4 are taken. The step size is smaller for GA compared to DE, as it can be seen later that GA has less likeliness so more search space is required. For each population size, 60 combinations of p_c and p_m are possible in this range. For the case study considered, the same global minimum heat transfer area is obtained (34.44 m²) by using GA also. The minimum number of generations required by GA to converge to the global minimum (G_{min}), in the above range of the key parameters is listed in Table 9.5 [6, 18] along with the Number of Function Evaluations (NFE).

   For each combination of N and p_c listed in this table, GA is converging to the global minimum heat transfer area of 34.44 m² for all the six values of p_m from 0.05 to 0.3 in steps of 0.05. While executing the GA program, it

TABLE 9.4 Effect of NP on DE Strategies Based on G_{min}

S. No.	Strategy	NP = 40			NP = 50			NP = 60			NP = 70			NP = 80			NP = 90			NP = 100		
		F	CR	G_{min}	F	CR	G_{min}	F	CR	G_{min}	F	CR	G_{min}	F	CR	G_{min}	F	CR	G_{min}	F	CR	G_{min}
1	DE/best/1/exp	1.1	1.0	5	0.8	0.7	1	0.9	1.0	6	0.8	0.8	2	0.6	0.8	1	0.6	1.0	2	0.7	0.8	2
					0.9	0.7	1				0.9	0.8	2	0.7	0.8	1	0.9	1.0	2	0.8	0.9	2
					1.0	0.7	1				0.9	1.0	2							0.9	0.9	2
					1.1	0.7	1				1.0	1.0	2							1.0	0.9	2
											1.1	1.0	2									
2	DE/rand/1/exp	0.8	0.5	9	0.8	0.9	11	1.1	0.7	11	0.9	0.9	7	1.1	0.9	4	0.5	1.0	7	0.5	0.7	12
		0.9	1.0	9													0.7	1.0	7	0.7	1.0	12
3	DE/rand-to-best/1/exp	1.0	0.9	3	1.0	0.9	6	0.7	0.8	5	0.6	0.8	4	1.1	0.9	2	0.6	0.8	4	1.0	1.0	2
								0.9	1.0	5	0.8	0.9	4				1.1	1.0	4			
											0.9	1.0	4									
											1.0	0.7	4									
4	DE/best/2/exp	0.5	0.7	6	1.0	1.0	3	0.9	1.0	5	1.0	0.6	4	0.3	1.0	2	0.7	0.9	2	0.5	0.9	2
								1.0	1.0	5	1.1	0.6	4	0.8	0.9	2				0.6	0.9	2
														0.9	0.9	2						
														1.0	0.9	2						
5	DE/rand/2/exp	1.1	0.9	10	0.8	0.7	11	0.9	1.0	8	1.0	1.0	3	0.7	0.9	6	0.7	0.6	4	0.9	1.0	4
																	0.8	0.6	4			
																	0.9	0.6	4			
																	1.0	0.6	4			
																	1.1	0.6	4			

TABLE 9.4 Continued

S. No.	Strategy	NP = 40 F CR G_min	NP = 50 F CR G_min	NP = 60 F CR G_min	NP = 70 F CR G_min	NP = 80 F CR G_min	NP = 90 F CR G_min	NP = 100 F CR G_min
6	DE/best/1/bin	1.1 1.0 5	0.3 0.4 3	0.3 0.4 3	0.5 0.7 2 0.6 0.4 2 0.6 0.5 2 0.6 0.9 2 0.7 0.4 2 0.9 1.0 2 1.0 0.9 2 1.0 1.0 2 1.1 0.8 2 1.1 0.9 2 1.1 1.0 2	0.6 0.6 1 0.6 0.8 1 0.6 0.9 1	0.8 0.6 1 0.8 0.7 1 0.9 0.6 1 0.9 0.7 1	0.4 0.8 2 0.5 0.7 2
7	DE/rand/1/bin	0.9 1.0 9	0.7 0.8 8	0.5 0.6 7	0.4 0.5 7	1.1 0.5 4	0.6 0.6 5	1.0 0.7 4 1.1 0.4 4
8	DE/rand-to-best/1/bin	0.7 1.0 5	0.9 0.4 5	0.9 1.0 5 1.1 0.6 5	0.9 0.7 3 0.9 0.8 3	0.8 0.8 2 0.8 0.9 2	0.6 0.9 5 0.9 0.7 4 1.0 0.7 4 1.1 1.0 4	1.0 1.0 2 1.1 0.4 2
9	DE/best/2/bin	0.5 0.8 6	0.9 0.9 3 1.0 1.0 3	0.9 1.0 5 1.0 1.0 5	0.5 0.5 2 0.6 0.5 2	0.6 0.6 1 0.6 0.7 1 0.6 0.8 1 0.6 0.9 1	1.1 0.8 2	0.3 0.5 1 0.3 0.6 1 0.3 0.7 1 0.3 0.8 1
10	DE/rand/2/bin	1.1 0.9 9	0.8 0.1 8 1.0 0.1 8	0.5 0.9 8 0.9 1.0 8 1.0 0.9 8	1.0 1.0 3	0.5 0.8 12	0.9 0.2 9	0.3 0.5 1 0.3 0.6 1 0.3 0.7 1 0.3 0.8 1

TABLE 9.5   GA Parameters Converging to the Global Minimum

S. No.	N	p_c	p_m	C_{GA}	G_{min}	NFE
1	44	0.55	0.05–0.30	20	53	2376
		0.60			63	2816
2	48	0.75	0.05–0.30	10	5	288
3	52	0.75	0.05–0.30	20	67	3536
		0.90			19	1040
4	60	0.65	0.05–0.30	20	71	4320
		0.70			14	900
5	64	0.50	0.05–0.30	30	59	3840
		0.60			43	2816
		0.80			25	1664
6	68	0.65	0.05–0.30	20	64	4420
		0.85			6	476
7	72	0.50	0.05–0.30	10	47	3456
8	76	0.75	0.05–0.30	10	90	6916
9	80	0.60	0.05–0.30	50	23	1920
		0.75			18	1520
		0.80			39	3200
		0.85			35	2880
		0.95			46	3760
10	84	0.75	0.05–0.30	20	97	8232
		0.90	0.05–0.30		70	5964
11	100	0.60		20	13	1400
		0.75			4	500

is observed that more number of generations are taken by GA to converge and hence, the maximum number of generations (MAXGEN) is specified as 100. For a given N, the percentage of the combinations converging to the global minimum (C_{GA}) in less than 100 generations, out of the 60 possible combinations of p_c and p_m considered, is also listed in Table 9.5. As can be seen, C_{GA} ranges from 10 to 50 compared to C_{DE} (in Table 9.3), which varies from 0.8 to 66.1. From Table 9.5, the average C_{GA} is calculated to be 20.9, whereas the average C_{DE} from Table 9.3 is 22 for four out of the ten strategies (strategy numbers 1, 4, 6 and 9). But the C_{DE} values are reported in Table 9.3 for MAXGEN of only 15. It is interesting to note that, had the basis of MAXGEN been same (i.e., 100) for both GA and DE then, it is quite obvious that the C_{DE} values would have been very high (may be close to 100), indicating that DE has 'more likeliness' of achieving the global optimum compared to GA, as it has a wide range of the individual C_{DE} values. Also DE has more strategies to choose from, which is an advantage over GA. As a measure of 'likeliness' another criteria was identified by Babu and Munawar [6] and defined as the percentage of the key parameter combinations converging to the global minimum, out of the total number of combinations considered (C_{tot}). In Table 9.5, out of the 1080 combinations of the key parameters considered with GA, only 138 combinations (i.e., C_{tot} = 12.8) are converging to the global minimum in less than 100 generations. Whereas in DE, out of the total of 9680 possible combination of key parameters considered 1395 combinations (i.e., C_{tot} = 14.4) are converging to global minimum in less than 15 generations itself. It is also evident from Table 9.5 that the best combination corresponding to the least function evaluations is for N = 48, p_c = 0.75, and p_m 0.05 to 0.3 (entire range of p_m), with 288 function evaluations as it converges in 5 generations itself.

The relation between NFE and the number of generations in GA also remains the same as in DE, NFE = N * (G_{min} + 1). Using GA, with a seed value of 10, NFE varies from 288 to 8148 as against a small range of 100 to 1300 for DE (from Table 9.4), which is an indication of the tremendous 'speed' of the DE algorithm. The above two observations clearly demonstrate that for the case study taken up, the DE algorithm is significantly faster and has more likeliness in achieving the global optimum and so is efficient compared to GA. The summary of results for the selected seed value of 10 is listed in Table 9.6 [6, 18]. The performance of DE and GA is compared for the present problem in Table 9.7 [6, 18], with respect to the 'best' parameters – parameter values converging to the global minimum out of the entire range considered. For NP=50, with DE/best/1/exp strategy, CR=0.7 and F = 0.8 to 1.1 (any value in steps of 0.1), DE took one generation, and 100 function evaluations. But with GA, for N = 48, p_c = 0.75, and p_m = 0.05 to 0.3 (any value in steps of 0.05) it took 5 generations, and 288 function evaluations. The best design variables for the given case study are listed in Table 9.8 [6, 18] along with the best key parameters of the DE algorithm used for this optimization. However, practical difficulties, if any, in implementation of the resulting heat exchanger can only be verified through suitable experimentation after fabrication. However, if it is known a priori that some combinations of design variables may lead to difficulties in practical implementation, then such solutions can be eliminated during design by assigning a very high cost.

TABLE 9.6 Comparison of DE and GA Based on Various Criteria

S. No.	Seed = 10			
	Criteria		DE	GA
1	MAXGEN		15	100
2	Global minimum heat exchanger area (m²)		34.44	34.44
3	Parameter values converging to the global minimum	(a) Key parameters	NP: 50–100	N: 44–100
			F: 0.3–1.1	P_m: 0.05–0.30
			CR: 0.1–1.0	P_c: 0.50–0.95
		(b) Strategy	DE/best/1/exp	—
			DE/best/1/bin	
			DE/best/2/exp	
			DE/best/2/bin	
4	Measure of 'likeliness' in achieving the global minimum		C_{DE}: 0.6–66.1	C_{GA}: 10–50
	Avg. C_{DE}: 6.0–32.1		Avg. C_{GA}: 20.9	
	C_{tot}: 12.8		C_{tot}: 14.4	
5	Measure of 'speed' in achieving the global minimum	(a) G_{min}	1–12	4–97
		(b) NFE	100–1300	288–8148
		(c) CPU time (s)	0.1–2.23	0.46–15.29

TABLE 9.7 Comparison of DE and GA for the Best Parameters

S. No.	Seed = 10			
	Criteria		DE	GA
1	Global minimum heat transfer area (m²)		34.44	34.44
2	Best parameter values converging to the global minimum	(a) Key para-meters	NP : 50	N: 48
			F: 0.8–1.1	p_m : 0.05–0.30
			CR: 0.7	p_c : 0.75
		(b) Strategy	DE/best/1/exp	—
3	G_{min}		1	5
4	NFE		100	288

TABLE 9.8   Summary of the Final Design for the Case Study

S. No.	Parameters		
1	Heat duty (kW)		1509.4
2	Best design variables	(a) Tube outer diameter (inch)	½
		(b) Tube pitch	5/8" triangular
		(c) Shell head type	Fixed tube sheet
		(d) Tube passes	Single
		(e) Tube length(ft)	24
		(f) Baffle spacing	20%
		(g) Baffles cut	15%
3	Heat exchanger area (m²)		34.44
4	Pressure drop (bar)	(a) Tube-side	0.67
		(b) Shell-side	0.31
5	DE parameters	(a) Strategy	DE/best/1/exp
		(b) Seed	10
		(c) NP	50
		(d) F	0.8–1.1
		(e) CR	0.7

KEYWORDS

- **Bell's method**
- **differential evolution**
- **heat exchanger design**
- **optimization**
- **shell-and-tube heat exchangers**

REFERENCES

1. Angira, R., & Babu, B. V. Optimization of Process Synthesis and Design Problems: A Modified Differential Evolution Approach, *Chem. Eng. Sci.* 2006, 61, 4707–4721.
2. Azad, A. V., & Amidpour, M. Economic Optimization of Shell and Tube Heat Exchanger based on Constructal Theory, *Energy* 2011, 36, 1087–1096.
3. Babu, B. V., & Angira, R. Modified Differential Evolution (MDE) for Optimization of Nonlinear Chemical Processes, *Comp. Chem. Eng.* 2006, 30, 989–1002.

4. Babu, B. V., Chakole, P. G., & Mubeen, J. H. S. Multiobjective Differential Evolution (MODE) for Optimization of Adiabatic Styrene Reactor, *Chem. Eng. Sci.* 2005, 60, 48224837.

5. Babu, B. V., & Munawar, S. A. Differential Evolution for the Optimal Design of Heat Exchangers, *Proceedings of All-India Seminar on Chemical Engineering Progress on Resource Development: A Vision 2010 and Beyond*, IE (I), Bhubaneswar, India, March 11, 2000.

6. Babu, B. V., & Munawar, S. A. Differential Evolution Strategies for Optimal Design of Shell-and-Tube Heat Exchangers, *Chem. Eng. Sci..* 2007, 62, 3720–3739.

7. Caputo, A. C., Pelagagge, P. M., & Salini, P. Heat Exchanger Design based on Economic Optimization, *App. Therm. Eng.* 2008, 28, 1151–1159.

8. Chaudhuri, P. D., Urmila, M. D., & Jefery, S. L. An Automated Approach for the Optimal Design of Heat Exchangers, *Ind. Eng. Chem. Res.* 1997, 36, 3685–3693.

9. Costa, A. L. H., & Queiroz, E. M. Design Optimization of Shell-and-Tube Heat Exchangers, *App. Therm Eng.* 2008, 28, 1798–1805.

10. Fesanghary, M., Damangir, E., & Soleimani, I. Design Optimization of Shell and Tube Heat Exchangers using Global Sensitivity Analysis and Harmony Search Algorithm, *App. Therm. Eng.* 2009, 29, 1026–1031.

11. Fettaka, S., Thibault, J., & Gupta, Y. Design of Shell-and-Tube Heat Exchangers Using Multiobjective Optimization, *Int. J. Heat Mass Tran.* 2013, 60, 343–354.

12. Goldberg, D. E. *Genetic Algorithms in Search, Optimization, and Machine Learning*, Reading, MA: Addison-Wesley, 1989.

13. Gujarathi, A. M., & Babu, B. V. Optimization of Adiabatic Styrene Reactor: A Hybrid Multiobjective Differential Evolution (H-MODE), *Ind. Eng. Chem. Res.* 2009, 48, 11115–11132.

14. Gujarathi, A. M., & Babu, B. V. Multiobjective Optimization of Industrial Styrene Reactor: Adiabatic and Pseudo-Isothermal Operation, *Chem. Eng. Sci.* 2010, 65, 2009–2026.

15. Hadidi, A., Hadidi, M., & Nazari, A. A New Design Approach for Shell-and-Tube Heat Exchangers using Imperialist Competitive Algorithm (ICA) from Economic Point of View, *Ener. Conv. Mgmt.* 2013, 67, 66–74.

16. Hadidi, A., & Nazari, A. Design and Economic Optimization of Shell-and-Tube Heat Exchangers using Biogeography-based (BBO) Algorithm, *App. Therm. Eng.* 2013, 51, 1263–1272.

17. Manish, C. T., Yan, F., & Urmila, M. D. Optimal Design of Heat Exchangers: A Genetic Algorithm Framework, *Ind. Eng. Chem. Res.* 1999, 38, 456–467.

18. Munawar, S. A. Expert Systems for the Optimal Design of Heat Exchangers, M. E. Dissertation, Birla Institute of Technology and Science, Pilani, 2000.

19. Onwubolu, G. C., & Babu, B. V. *New Optimization Techniques in Engineering*, Springer, Germany, 2004.

20. Özçelik, Y. Exergetic Optimization of Shell and Tube Heat Exchangers using a Genetic based Algorithm, *App. Therm. Eng..* 2007, 1849–1856.

21. Patel, V. K., & Rao, R. V. Design Optimization of Shell-and-Tube Heat Exchanger using Particle Swarm Optimization Technique, *App. Therm. Eng.* 2010, 30, 1417–1425.

22. Perry, R. H., & Green, D. *Perry's Chemical Engineers' Handbook*, 6th ed., New York: McGHI Editions, Chem. Eng. Series, 1993.

23. Ponce-Ortega, J. M., Serna-González, M., & Jiménez-Gutiérrez, A. Use of Genetic Algorithms for the Optimal Design of Shell-and-Tube Heat Exchangers, *App. Therm. Eng.* 2009, 29, 203–209.

24. Price, K., & Storn, R. Differential Evolution – A Simple Evolution Strategy for Fast Optimization. *Dr. Dobb's J.* 1997, 22 (4), 18–24.
25. Şahin, A. S., Kiliç, B., & Kiliç, U. Design and Economic Optimization of Shell and Tube Heat Exchangers using Artificial Bee Colony (ABC) Algorithm, *Ener. Conv. Mgmt.* 2011, 52, 3356–3362.
26. Serna, M., & Jimenez, A. A Compact Formulation of the Bell–Delaware Method for Heat Exchanger Design and Optimization, *Chem. Eng. Res. Des.* 2005, 83 (A5), 539–550.
27. Sinnot, R. K. *Coulson and Richardson's Chemical Engineering (Design)*, vol. 6, 2nd ed., New York: Pergamon, 1993.

APPENDIX A: PRESSURE DROP CALCULATIONS

The pressure drop calculations in Bell's method of heat exchanger design are given below for shell-side and tube-side separately.

A.1. TUBE-SIDE PRESSURE DROP

There are two major sources of pressure loss on the tube-side of a shell-and-tube heat exchanger: the friction loss in the tubes, the losses due to the sudden contraction and expansion and flow reversals that the fluid experiences in flow through the tube arrangement. The following correlation is widely used to estimate the tube side pressure drop [27].

$$\Delta P_t = N_p \left[8 j_f \left(\frac{L}{D_i} \right) \left(\frac{\mu}{\mu_w} \right)^{-0.14} + 2.5 \right] \frac{\rho u_t^2}{2} \qquad (A.1)$$

where j_f is the friction factor obtained from nomograph plotted as a function of Reynolds number.

A.2. SHELL-SIDE PRESSURE DROP

Bell's method suggests that pressure drops in cross flow and window zones be determined separately and summed up with that in end zones to give the total shell side pressure drop.

Cross flow Zones: The pressure drop in the cross flow zones between the baffle tips is calculated from the correlations for ideal tube banks, and corrected for leakage and by passing.

$$\Delta P_c = \Delta P_i \, F'_b \, F'_L \tag{A.2}$$

where ΔP_c – the pressure drop in cross flow zone between the baffle tips, corrected for by passing and leakage; ΔP_i – the pressure drop calculated for an equivalent ideal tube bank; F'_b – bypass correction factor; F'_L – leakage correction factor.

ΔP_p *ideal tube bank pressure drop:* The correlation used here is:

$$\Delta P_i = 8 \, j_f \, N_{cv} \frac{\rho u_s^2}{2} \left(\frac{\mu}{\mu_w} \right)^{-0.14} \tag{A.3}$$

where the friction factor j_f for cross-flow tube banks can be obtained from nomograph at the appropriate Reynolds number.

F'_b, *bypass correction factor for pressure drop:* Bypassing will affect the pressure drop only in the cross flow zones. If no sealing strips are used:

$$F'_b = \exp\left(-\alpha \, A_b / A_s\right) \tag{A.4}$$

where $\alpha = 5.0$ for Laminar region, $Re < 100$
        $= 4.0$ for transition and turbulent region, $Re > 100$.

F'_L, *Leakage correction factor for pressure drop:* Leakages will effect the pressure drop in both the cross-flow and window zones.

$$F'_L = 1 - \beta'_L \left[(A_{tb} + 2A_{sb}) / A_L \right] \tag{A.5}$$

where β'_L is obtained from nomograph.

Window zone pressure drop: The correlation used here is:

$$\Delta P_w = F'_L \, (2 + 0.6 \, N_{wv}) \frac{\rho u_z^2}{2} \tag{A.6}$$

where u_z – the geometric mean velocity.

$$u_z = \sqrt{u_w \, u_s}$$

where u_w – the velocity in the window zone, based on the window area less the area occupied by the tube, A_w, ($u_w = W_s / A_w \rho$); W_s – shell-side mass flow, kg/s.

The window area is:

$$A_w = \left(\frac{\pi D_s^2}{4} Ra \right) - \left(N_w \frac{\pi D_b^2}{4} \right) \tag{A.7}$$

where R_a is obtained from nomograph for the appropriate baffle cut B_c.

End Zone Pressure Drop: There will be no leakage paths in an end zone (the zone between the tube sheet and baffle).

$$\Delta P_e = \Delta P_i \, [(N_{wv} + N_{cv})/N_{cv}] \, F'_b \tag{A.8}$$

Total shell Side Pressure Drop: Summing the pressure drops over all the zones in series from inlet to outlet gives:

$$\Delta P_s = 2 \text{ end zones} + (N_b - 1) \text{ cross flow zones} + N_b \text{ window zones}$$

$$\Delta P_s = 2 \, \Delta P_e + \Delta P_c \, (N_b - 1) + \Delta P_w \, N_b \tag{A.9}$$

where N_b is the number of baffles $= \left(\dfrac{L}{l_b} - 1 \right)$. An estimate of the pressure loss incurred in the shell inlet and outlet nozzles must be added to that calculated by Eq. (A.9).

CHAPTER 10

EVOLUTIONARY COMPUTATION BASED QoS-AWARE MULTICAST ROUTING

MANAS RANJAN KABAT, SATYA PRAKASH SAHOO, and
MANOJ KUMAR PATEL

*Department of Computer Science and Engineering, Veer Surendra Sai
University of Technology, Burla, India,
E-mail: sahoo.satyaprakash@gmail.com, kabatmanas@gmail.com,
patel.mkp@gmail.com*

CONTENTS

ABSTRACT

The real-time multimedia applications on today's Internet require the data transmission from one or more senders to a group of the intended receivers with certain Quality of Service (QoS) requirements. These applications need the underlying network to create a distribution tree structure, which spans the source and the group of receivers. This QoS multicast tree construction problem is a non-linear combinatorial optimization problem which has been proved to be NP-complete. In the event of involving more number of QoS parameters, deterministic heuristic algorithms for QoS multicast routing are usually very slow. The evolutionary computation techniques have attracted many researchers over the years to find the near optimal solution of many combinatorial optimization problems. Hence methods based on evolutionary computation such as Genetic Algorithms (GA), Ant Colony Optimization (ACO), Particle Swarm Optimization (PSO) technique, etc. are found to be more suitable to find the multi constrained QoS routing problems. In this chapter, we present the review of various evolutionary algorithms to solve the QoS-aware multicast routing problem.

10.1 INTRODUCTION

The proliferation of the Internet has led to the increase in demand of real time multimedia applications such as video/audio conferencing, video on-demand, news distribution and on-line games, etc. These applications require send information from a source to multiple destinations through a communication network. Multicasting is the ability of the network to send the same message to multiple intended receivers in a computer network. The multicast employs a tree structure to deliver the same message efficiently to a group of receivers. The membership of a host in a group is dynamic. The host may be a member of more than one group at a time, while the multicast sources need not be the members of the group. There are the following three general categories of multicast applications [20]. Those are (i) one-to-many (ii) many-to-many (iii) many-to-one. The one-to-many applications have a single sender, and multiple simultaneous receivers.

The scheduled audio/video distribution, push media, file distribution and caching, announcements, monitoring stock prices, sensor equipments are some examples of one-to-many applications. The many-to-many applications are characterized by two-way multicast communications where two or more of the receivers may act as senders. The multimedia conferencing, shared/distributed databases, distributed parallel processing, shared document editing, distance learning, distributed interactive simulations, multi-player games are real life examples pertaining many-to-many applications. The many-to-one application has multiple senders and one (or a few) receiver(s). The many to one application can either be one-way or two-way request/response types, where either senders or receiver(s) may generate the request. The data collection, auctions, polling, jukebox, accounting describes some important many-to-one applications.

The QoS is considered as an important aspect of multicasting. The various QoS parameters are delay, delay jitter, loss rate, bandwidth, etc. The cost may be either the monetary cost or any other measure of resource utilization, which must be optimized. The delay of each link is defined as the sum of switching, queuing, and transmission and propagation delays. The delay jitter of the multicast tree is the delay variation among the delays along the individual source-destination paths. The bandwidth of the link can be defined as the residual bandwidth of the physical or logical link. The real time messages involving one-to-many applications must be transmitted from the source node to their destinations within a predefined end-to-end delay for smooth delivery to the audiences. These applications also require a bound on the variation among the delays along the individual source destination paths to avoid causing inconsistence or unfairness problem among the users. The main dimension of QoS-aware multicast routing problem is the need to construct trees that will satisfy the QoS requirements of networked multimedia applications (delay, delay jitter, loss, etc.). The cost optimal solution of the QoS multicast routing problem is defined in Molnar et al. [14]. The problem is to find the minimum cost multicast tree where the end-to-end paths from source to destinations satisfying the QoS constraints.

There are several researchers who have developed several heuristics and evolutionary algorithms (EA) for QoS multicast tree generation. The heuristic algorithms first compute the least delay spanning trees. Then the least delay paths are replaced by cost sensitive paths so that the delay can be relaxed with respect to the delay bound and the cost of the multicast tree can be minimized. Though these algorithms are fast in generating multicast trees, they consider only one QoS criterion that is the delay constraint. The evolution of EAs such as genetic algorithm (GA), harmony search (HS), particle swarm optimization (PSO), ant colony optimization (ACO) has attracted the researchers to find the cost optimal solution of the multicast tree under multiple constraints. The tree generation in EA shares a common conceptual base of simulating the evolution of individual structures via processes of selection, mutation, and reproduction. More precisely, EAs maintain a population of structures that evolve according to rules of selection and other search operators, such as recombination and mutation. The EAs proposed for QoS multicast routing are either path based or tree based. In path based techniques, a set of least delay or least cost paths are computed from source to all the destinations of the multicast group. Then the candidate multicast tree structures are evolved as the random combination of the paths from the source to the destinations of the multicast group. The evolutionary algorithms generate the new multicast trees by using mutation and crossover operations. This tree evolution process continues till the optimal multicast tree is generated. These methods are slow and complex. In tree-based techniques, the candidate multicast tree structures are generated randomly or heuristically. Each individual in the population receives a measure of its fitness in the environment. Reproduction focuses attention on high fitness individuals, thus exploiting the available fitness information. Recombination and mutation perturb those individuals, providing general heuristics for exploration. Although it is simple from the viewpoint of a biologist, these algorithms are the backbone of QoS multicast routing because these algorithms not only consider multiple QoS constraints but also helps in obtaining more cost-efficient multicast trees. The next section focuses on the problem definition of QoS multicast routing. The evolutionary computation based techniques for QoS multicast routing is presented in Section 10.3. The Section 10.4 presents the simulation results to illustrate the performance of the path based and tree-based algorithms. The summary of conclusion is presented in Section 10.5.

10.2 PROBLEM STATEMENT

Let $G = (V, E)$ be an undirected weighted graph representing the communication network, where V is the set of nodes that represents routers or switches and E is the set of edges that represent the physical or logical connection between nodes. Let $s \in V$ be the source node, $M \subseteq V$ be the set of destinations such that $M = \{d_j \in V, d_j \neq s, j=1,2,...m\}$ and m be the number of destinations. Each edge is associated with the set of non-negative real numbers representing the QoS metrics such as bandwidth, delay, jitter and loss rate etc. Let R^+ is the set of all real numbers. The QoS metric for each link $e \in E$ is defined as $bandwidth(e)$, $delay(e)$, $jitter(e)$ and $lossrate(e)$ where $e \in E$ and $E \rightarrow R^+$.

The QoS can be roughly classified as additive (delay), multiplicative (lossrate) and bottleneck metrics (bandwidth). We can easily deal with the bottleneck metric, i.e., available bandwidth by pruning the links from the group that do not satisfy the QoS constraint. The end-to-end delay of the path $P_T(s,d_j)$ from source s to destination d_j of the multicast group is the sum of the delay of the links in the path.

$$Delay\left(P_T\left(s,d_j\right)\right) = \sum_{e \in \left(P_T\left(s,d_j\right)\right)} D(e) \tag{1}$$

The delay, jitter and cost of the tree, lossrate of the path $P_T(s, d)$ are defined as follows :

$$P_T\left(s,d_j\right) = 1 - \left\{\prod_{e \in \left(P_T\left(s,d_j\right)\right)}\left(1 - lossrate\left(e\right)\right)\right\} d_j \in M \tag{2}$$

$$Cost(T,(s,M)) = \sum_{e \in T(s \in M)} cost(e) \tag{3}$$

$$jitter(T,(s,M)) = \sum_{\substack{e \in T(s,M) \\ d_j \in M}} \sqrt{\left(delay\left(s,d_j\right) - delay\_avg\right)^2} \tag{4}$$

where delay_avg denotes the average value of delay of the paths from the source to the destination.

The objective of QoS multicast tree problem is to construct the least cost multicast tree under multiple QoS constraints [14]. The QoS constraints are roughly defined as link constraint ($bandwidth$), path constraints ($delay, lossrate$) and tree constraints (delay, jitter). The least cost multiconstrained multicast tree problem is defined as

Minimize $Cost(T(s,M))$, subject to :

$$delay\left(P_T\left(s,d_j\right)\right)\le DC\,\forall\,d_j\in M \tag{5}$$

$$lossrate\left(P_T\left(s,d_j\right)\right)\le LC\,\forall\,d_j\in M \tag{6}$$

$$jitter\left(T\left(s,M\right)\right)\le DJC \tag{7}$$

$$bandwidth\left(e\right)\ge BC\,\forall\,d_j\in M \tag{8}$$

where DC, LC, DJC and BC represent delay constraint, loss constraint, delay jitter constraint and bandwidth constraints, respectively.

10.3   EVOLUTIONARY COMPUTATION BASED QoS-AWARE MULTICAST ROUTING

The evolutionary algorithm is inspired by biological evolution, such as reproduction, mutation, recombination, selection of the fittest ones. This technique has been used earlier to find near-optimal solution to many combinatorial optimization problems. In this section, we present the evolutionary computation based algorithms such as GA, HS, ACO, PSO and the hybrid algorithms for QoS-aware multicast routing. The set of candidate solutions is considered as the population. The fitness function is used to find the live solutions and the evolution of the population is decided by repeatedly using this fitness function. The recombination and mutation process is used to create new solutions in the population. The selection process forces to find the best quality solution to the problem.

10.3.1   GA BASED QoS MULTICAST ROUTING

In this section we present various algorithms based on GA to solve QoS multicast routing. The GA (Hwang, 2010) has been used by many researchers to solve the optimization problems because of efficient search in complex spaces. The candidate solutions at each iteration are represented as a population. Then the GA uses three basic operations namely reproduction, cross over and mutation to generate new offspring. This process is repeated to find the fittest chromosome as the desired solution.

The solution of QoS multicast routing problem using genetic algorithm was proposed by Hwang et al. [8]. This algorithm first computes all the possible routes between source and destination pairs using k-shortest path algorithm [32]. Let there are k paths to each destination $(d_j \in M, 1 \le j \le m)$ from the source. The chromosome of the population is represented by a string of integers with length m, where the gene of the chromosome is an integer from 1 to k. The paths are sorted either with respect to the hop count or delay so that a better path will get a smaller number. Each chromosome represents the combination of paths from source to all destinations of the multicast group, which is not necessarily being a multicast tree. Though a coding system proposed in Palmer et al. [15] can be used to represent chromosome as a multicast tree. It requires complex transformation to obtain the links from the tree. The algorithm generates C different chromosomes randomly in the first generation where C is the size of the population. The Figure 10.1(a) shows the example network where v_0 is the source and v_4 is one of destinations of the multicast group. The Figure 10.1(b) shows relationship of the chromosome, gene and the routing table for the source-destination pair (v_0, v_4).

The fitness value of each chromosome h_i, $1 \le i \le |C|$ in the initial population is computed as $Fitness(h_i) = 1 - \dfrac{cost(h_i)}{cost(L)}$ where cost (h_i) is the sum of the cost of the links of the graph represented by the chromosome h_i and $cost(L)$ is the sum of the costs of all links in the network. The chromosomes are sorted according to their fitness values. Then the duplicate chromosomes are removed from the population so that the search ability of the algorithm can be improved. Then some of the chromosomes are selected on the basis of the fitness function to generate more offspring through crossover and mutation

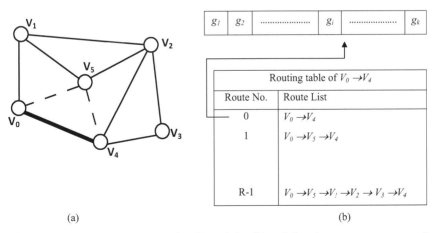

(a)   (b)

FIGURE 10.1(A)   Example Network; (b) Relationship of the chromosomes, gene and routing table.

procedure. This process helps to find the chromosomes with high fitness by killing the chromosomes with low fitness. Then a certain number of chromosomes with the best fitness values are selected from the current generation for reproduction. Another set of chromosomes with the best fitness values is selected to reproduce offspring through the crossover operation. However, the total number of chromosomes in the population remains C. The crossover operation considers two chromosome strings with larger fitness values of the population. Then the start point and length of the substring to be exchanged are randomly selected. The two new offspring is created and placed in the population as shown in Figure 10.2.

Wang et al. used a tree-based approach to construct the set of the multicast tree as the candidate solution. This method starts with the source and randomly selects the unvisited nodes by using a random depth first search. This process continues till all the destination nodes are included in the tree.

Another GA-based heuristic algorithm for bandwidth-delay-constrained least-cost multicast routing is proposed by Haghighat et al. [7]. This algorithm uses a randomized depth first search DFS to construct the multicast trees. First, a path is constructed from the source node s to one of the destination nodes. Then, one of the unvisited destinations is selected and the path to a node in the sub-tree constructed in the previous step is set up by selecting the next unvisited node. This process continues till all the destinations are included in the tree. This process is called C times to create the total population.

The fitness function considered in this algorithm is an improved version of the scheme proposed in Wang et al. [30]. The fitness functions for each individual tree $T(s, M)$ using the penalty technique is defined as follows:

$$Fitness\left(T(s,M)\right)$$

$$= \frac{\infty}{\sum_{e \in T(s,M)} cost(e)} \prod_{d \in M} \varnothing\left(Delay\left(P_T\left(s,d\right)\right) - DC\right)$$
$$\times \prod_{d \in M} \varnothing\left(bandwidth\left(P_T\left(s,d\right)\right) - BC\right) \qquad (9)$$

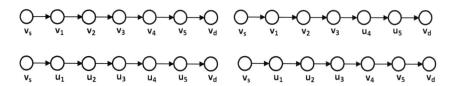

FIGURE 10.2   Crossover operation.

$$where\left(\varnothing\left(z\right)\right)=\begin{cases}1 \, if \; z\le0\\\gamma \, if \; z<0\end{cases}$$

$$bandwidth \, P_T\left(s,d\right)= \min_{e\in P_T\left(s,d\right)} \{bandwidth(e)\} \qquad (10)$$

The α is a positive real coefficient, $\varnothing(z)$ is the penalty function and γ is the degree of penalty. The optimal solution depends on the degree of penalty. Then, they have used the approach for crossover of Steiner trees [21] with some modifications. In this scheme, two multicast trees, $T_1(s, M)$ and $T_2(s, M)$ are selected as parents and the crossover operation produces an offspring $T_3(s, M)$ by identifying the links that are common to both parents. The operator selects the same links of two parents for quicker convergence of the genetic algorithm. The genetic algorithms [7, 8] for QoS multicast routing are based on mutation and crossover operator. However, both of the algorithms have some drawbacks such as lack of local search ability, premature convergence and slow convergence speed (Zhang et al., 2009).

10.3.2 GSA BASED QoS MULTICAST ROUTING

The method proposed in GSA [36] is a tree-based approach that combines GA and simulated annealing (SA) adequately to avoid the premature convergence of the GA. The SA has the capability of escaping from local optima [12]. The GA has the character of parallel processing and high convergence speed. In GSA, the multicast trees spanning the source and destinations are represented as chromosomes. The tree initialization is done in two steps: trunk creating and limb appending. The trunk-creating step starts from the source and goes on selecting links randomly till the one of the destination nodes is reached. The limb-appending step starts from an unvisited destination and goes on selecting the links till it is added to the previous tree. This process is repeated till all the destinations are added into the multicast tree. The selection operation uses the last chromosome to record the best individual in all generations to make sure that the algorithm can converge to the global optimal solution. The crossover operation is performed with an adaptive probability to improve the evolutionary efficiency.

Subsequently, Peng et al. [17] presented an adaptive genetic simulated annealing algorithm (AGSAA) to solve the QoS multicast routing problem. This is a path0based approach in which the chromosomes are created as discussed in Section 3.1 and shown in Figure 10.1. This algorithm adopts a roulette wheel selection method and best individual preservation strategy. The probability of selection of an individual is the ratio of the fitness function of that individual and the sum of the fitness values of all the individuals in the population. The improved adaptive crossover and mutation probabilities are adopted in this algorithm, which is varied depending on the fitness values of the solutions. This not only improves the convergence rate, but also prevents the GA to lock at local optima.

The SA algorithm is used to set the initial temperature at a high value so that all neighboring solutions can be accepted with high probability and a very good convergence can be ensured. The initial population of the SA algorithm is the population generated by mutation operation of the GA in the current generation. Then the neighboring solution is created by randomly replacing one path (gene) by another path (gene) from the path set to the corresponding destination.

10.3.3   ACO BASED ALGORITHMS FOR QoS MULTICAST ROUTING

The basic ACO and its formulation were presented in Dorigo et al. [12]. Gong et al. [6] presented an efficient QoS multicast routing algorithm based on ACA considering multiple QoS metrics. This is a path-based approach which finds the shortest paths from the source to each destination separately by the ant algorithm and then merging the resulting paths to form a multicast tree. This algorithm was framed up into four steps. Those are data structures, state transition rule, pheromone updating rule and the merge operation. In the data structure phase, the data structure of ant, link and node were described. In the second phase, they modified the state transition rule for the ant and defined the probability for the ant to move from one node to another node with the pheromone guide. Thirdly, after all the ants have found their path as per the probability, the pheromone amount of edges is updated and the cost of the path was calculated and finally the pheromone updating rule of an ant algorithm was applied to ensure the feasibility of a solution, and devise a flexible approach to merge the multiple paths by ants.

A modified ACO was devised by Wang et al. [6] for finding the least cost multicast tree by adding an orientation factor in it. This path based multicast tree generation algorithm is based on the orientation factor. The orientation factor is introduced while computing the probability of selecting the next hop. The orientation factor is calculated after repetitive simulation experiments by using the equation $\eta_{ij} = a + \cos\theta$, where $a = 2$, and θ is the polar coordinates of the two orientation vectors. The first one is directed from the current node to the candidate node *and* the second one is from the current node to the destination node. The orientation factor η_{ij} is computed by using the following formula.

$$p_{ij}^k(t) = \begin{cases} \dfrac{\left[\tau_{ij}(t)\right]^\alpha \left[\eta_{ij}(t)\right]^\beta}{\sum_{s\in candidates_k}\left[\tau_{ij}(t)\right]^\alpha \left[\eta_{ij}(t)\right]^\beta} & j \in candidates_k \\ 0 & Otherwise \end{cases} \qquad (11)$$

where *candidates*$_k$ denote the node set allowed to select by ant k, $\eta_{ij}(t)$ denotes the value of the orientation factor of the path from the current node i to next node j in the tth search cycle. α is the adjustment factor of pheromone $\tau_{ij}(t)$, β is the adjustable factor of the orientation factor $\tau_{ij}(t)$. The orientation factor affects the computation of the probability of the next hop. Besides the addition of the orientation factor, the ideas of Tabu search and Simulated Annealing are also integrated to speed up convergence and avoid the search result converging in locally optimal solution.

A path based ant algorithm for QoS multicast routing is proposed by Younes [34]. each part of the path set to a destination is assigned an initial value 0. In each iteration, a number of ants move on the paths in set P_i. The pheromone value left by the ants in a path p_k of P_i. Then, the local pheromone value is updated. The corresponding probability function f_k for each p_k is computed and global pheromone value is updated. This is repeated for a fixed number of iterations. After the fixed number of iterations, the values of the pheromone of the paths are computed and compared to get the best path for the destinations d_i.

A delay-constrained multicast routing algorithm based on the ant colony algorithm is proposed by Shi et.al. [23]. This is a path-based approach which first computes a set of alternate paths to the destinations. Then the candidate multicast trees are generated by integrating randomly the paths to the destinations. The fitness function is used to choose the *jth* path in an alternative path set Ω_i to a destination d_i.

$$fitness(P_{ij}) = \frac{phe_{P_j(s,d_i)}}{\sum_{P_j\in\Omega_i}phe_{P_j(s,d_i)}} \qquad (12)$$

where $phe_{pj}(s,d_i)$ is the pheromone of the jth path indicating that a higher pheromone of the path assures higher probability of being selected. The pheromone of the routes in which the ant passes is updated by using the equation

$$phe_{P_j(s,d_i)} = phe_{P_j(s,d_i)} + \frac{a}{cost\left(P_j\left(s,d_j\right)\right)}, \text{ where } a \text{ is the constant parameter.}$$

When all ants finish one path, the volatile secretions of all paths are adjusted by the following equation.

$$\begin{cases} phe_{P_j(s,d_i)} = (1-\rho) phe_{P_j(s,d_i)} & phe_{P_j(s,d_i)} > 0.5 \\ phe_{P_j(s,d_i)} = 0 \, otherwise \end{cases} \tag{13}$$

where, ρ is the volatility, and Δ is the initial information strength on each path. When ants found the entire destination nodes to form a multicast tree, and then pheromone is adjusted by using the equation $phe_{P_j(s,d_i)} = phe_{P_j(s,d_i)} + \dfrac{B}{cost(T)}$,

where B is a constant parameter, $cost(T)$ is the cost of the multicast tree.

The tree growth based ant colony algorithm for the QoS multicast routing problem is proposed by Wang et al. [29]. This is a tree-based approach, which directly constructs the multicast trees as the candidate solutions. The multicast tree starts with the source and record the link set that satisfies the bandwidth constraint, end-to-end delay and loss rate constraint from the source node to the current node. Then, the ant selects one link and adds it to the current tree according to the probability in order to make the tree grow constantly. When the tree covers all the multicast members it stops growing. The tree obtained is then pruned and rendezvous links are removed to get the real multicast tree. This algorithm forms a positive feedback mechanism by updating the pheromone of the links regularly. The ants are sent for a number of generations. The best tree generated after the generation (local best) and the global best trees are updated when updating the pheromone. The pheromone of the link $l(i, j)$ is updated by using the formula $\tau_{ij}=(1-\rho)\,\tau_{ij}+ \rho\Delta\tau_{ij}$, where ρ is used to control the evaporating speed of pheromone and $\Delta\tau_{ij}$ is the increase in pheromone on the link (i, j) obtained by local optimization and global optimal tree in that generation. This process is repeated until the algorithm converges. This algorithm has three basic operations-tree growth, tree pruning, and pheromone updating. The tree growth is critical among these three operations. Considering the effect of parameter selection of ant colony algorithm on algorithm performance, one can adopt orthogonal experiments to optimize ant colony algorithm parameters in order to achieve better performance.

Yin et al. [33] proposed a Niched ant colony optimization (NACO) algorithm which uses the colony guide algorithm (NACOg) to solve the QoS multicast routing problem. The NACO algorithm is basically a hybridization of tree-based and a path based approach. This first performs a constrained tree traversal (CTT) strategy to generate a set of feasible trees with respect to the QoS constraints. These feasible trees are constructed by considering the delay constraint of each path. The NACO algorithm for multicast routing manages multiple pheromone matrices for better diversity among individual solutions. Next, the use of the colony guides enhances the intensification search capability of each niche-colony. The NACOg algorithm used in NACO algorithm divides the colony of ants into n niche-colonies. Each candidate multicast tree is represented by a niche-colony. Each niche-colony has its own pheromone matrix and evolves to optimize an individual multicast tree. The number of ants in each niche-colony is equal to the number of destination nodes and each constrained routing path from the source node to one of the destination nodes is represented as an ant. The CTT strategy employed by every ant of a niche-colony ensures the feasibility of the obtained multi-constrained multicast tree. The cost of the multicast tree can be minimized through the evolutionary optimization process by use of two different factors. The first one is the pheromone matrix τ, which estimates the long-term experience about the desirability for using a particular transmission link. The other one is the visibility matrix η, which estimates immediate reward for traversing the link.

Let us illustrate the optimization of path-routing mechanism employed by an ant. Let the current position of the ant is at node i, and Π denotes the set of all candidate nodes for serving as the next node in the routing with reference to the CTT strategy. A selection probability is assigned to every node in Π and the probability of choosing node j as the next node for extending the routing path is calculated as follows:

$$f_{ij} = \frac{\alpha\tau_{ij} + (1-\alpha)\eta_{ij}}{\sum_{\forall k \in}(\alpha\tau_{ik} + (1-\alpha)\eta_{ik})} \forall j \in \Pi \tag{14}$$

where τ_{ij} is the pheromone amount that is sensed on edge e_{ij}, η_{ij} is the visibility on edge e_{ij}, and $\alpha \in (0,1)$ is the weighting value for tuning the trade-off significance for pheromone and visibility. The denominator is a normalization term for ensuring that the sum of all the probability values equals one. The selection of the next node in the routing process is following a learning scheme that takes into account both of the long-term experience (pheromone) and the immediate reward (visibility). The pheromone matrix is updated according to the fitness of the produced multicast tree as will be noted. The visibility value reflects the immediate reward that encourages the selection for the edges that are in the completed routing paths and for those edges with little transmission cost. The visibility value of selecting edge e_{ij} is defined as follows.

$$\tau_{ij} = R_{ij} + \frac{Q_1}{cost(e_{ij})} \tag{15}$$

where R_{ij} is the merit for the reuse of edge e_{ij} and Q_1 is the parameter for tuning the relative importance between increasing the edge-reuse rate and decreasing the incurred cost. The value of R_{ij} is inversely proportional to the level of e_{ij} in the multicast tree because the reuse of higher-level edges can avoid a dramatic change in the topology of the multicast tree after the inclusion of the current routing-path. In order to update the long-term evolutionary experience in the individual pheromone matrices, two forms of feedback have been collected. The path-level feedback estimates the fitness of the routing path constructed by each ant and the tree-level feedback appraises the value of the two colony guides (the best multicast tree obtained by each niche-colony and the overall best multicast tree obtained by the entire colony). The pheromone-updating rule of the proposed NACOg algorithm uses the following equation to update the pheromone matrix of each niche-colony.

$$\tau_{ij} = (1-\rho)\tau_{ij} + \sum_{k=1}^{m} \Delta_{ij}^{k} + k_{ij}^{ng} + k_{ij}^{cg} \tag{16}$$

where ρ is the evaporation rate for pheromone trails, Δ_{ij}^{k} is the pheromone increment by reference to the routing path constructed by the kth ant, and k_{ij}^{ng} and k_{ij}^{cg} are the additive pheromone quantities contributed by the niche-colony guide and the entire- colony guide, respectively. The pheromone entry τ_{ij} receives positive values for the three long-term rewards (Δ_{ij}^{k}, k_{ij}^{ng} and k_{ij}^{cg}) only when e_{ij} is contained in their respective routing representations. The values of these three rewards are derived as follows.

$$\Delta_{ij}^k = \frac{Q_2}{Cost_k} k_{ij}^{ng} = \frac{Q_2}{Cost_{ng}} k_{ij}^{cg} = \frac{Q_2}{Cost_{cg}} \tag{17}$$

where $Cost_k$, $Cost_{ng}$, and $Cost_{cg}$ are respectively the total transmission cost over all the edges contained in the path constructed by the k_{th} ant, the best multicast tree for each niche-colony, and the overall best multicast tree for the entire colony. Q_2 is the parameter for weighting the additive pheromone amounts.

10.3.4 HS BASED ALGORITHMS FOR QoS MULTICAST ROUTING

Forsati et al. [3] proposed a HS based algorithm for bandwidth-delay-constrained multicast routing problem. The Harmony search (HS) based algorithms are meta-heuristic algorithms, mimicking the improvisation process of music players. During the last several years, the HS algorithm has been rigorously applied to various optimization problems [4, 5]. The algorithm has several advantages with respect to the traditional optimization techniques [13]. The HS algorithm imposes fewer mathematical requirements and does not require initial value setting of decision variables. As the HS algorithm uses stochastic random searches, the derivative information is also not necessary. Further, the HS algorithm generates a new vector, after considering all of the existing vectors, whereas the GA considers two parent vectors. These features increase the flexibility of the HS algorithm and produce the best solution.

The HS algorithm is applied to find the bandwidth delay constrained least cost multicast routing tree. It assumes a source node and requests to establish a least-cost multicast tree with two constraints: bandwidth constraint in all the links of the multicast tree and end-to-end-delay constraint from the source node to each of the destinations. The HS algorithm is a centralized routing algorithm. That is, the complete network topology information available at the source node and the central node that is responsible for computing the entire routing table. The algorithm is designed in two phases. The first was HS-based algorithm called HSPR. The Prüfer number representation is modified, which was used for multicast tree and was used to encode the solution space. The authors proposed a novel representation called as node parent index (NPI) representation for encoding trees, in order to overcome the poor performance of the prufer in representing incomplete graphs [3].

The NPI representation possesses properties, which are necessary to a heuristic algorithm to function most effectively. Later, a new HS-based algorithm using NPI representation named as HSNPI to find the bandwidth and end-to-end delay constrained multicast tree was proposed [4].

10.3.5 PSO BASED ALGORITHMS FOR QoS MULTICAST ROUTING

The original PSO algorithm was proposed by J. Kennedy as a simulation of social behavior of bird flock [9]. In some cases the PSO does not suffer the difficulties met by GA and has been proven as an efficient approach for many continuous problems.

The tree-based PSO algorithm (PSOTREE) for QoS multicast routing is proposed by Wang et al. [27] which optimizes the multicast tree directly constructing the trees rather than combining the paths. The algorithm constructs the candidate multicast trees by calculating the fitness of the link considering cost, delay and bandwidth of the links. The fitness of the link is calculated by using the following formula.

$$Fitness(e) = \begin{cases} 0 \text{ if bandwidth}(e) < BC \\ a_1 \times e^{-cost(e)/averagecost} + a_1 \times e^{-delay(e)/averagedelay} \text{ Otherwise} \end{cases} \quad (18)$$

where, BC is the bandwidth constraint of the link, the parameters a_1, a_2 are the weight values of the cost and delay function of the link, and the variables *averagecost* and *averagedelay* represent the average cost and delay of the set of links in network topology. The multicast tree construction starts with the source node. Then the neighboring nodes of the current node that satisfy the bandwidth, delay requirements are placed into a candidate set. Then one of the neighbor nodes is selected randomly and added to the multicast tree. The delay up to that node is updated. This process is repeated till all the destination nodes are included in the multicast tree. The multicast tree generated is represented by a particle. Each particle flies in the best-known direction and interacts with other particle to generate a new particle. The new multicast tree is generated by merging, loop trimming and edge pruning of two candidate trees. The merging operation can be simply viewed as the addition of all the links of the two trees T_1 and T_2 into the new tree T_3, i.e., $T_3[i][j] = T_1[i][j] \vee T_2[i][j]$. The new tree obtained is not necessarily being a tree. This may contain circles or nested circles. In PSOTREE, the Depth First Search (DFS) algorithm is used for loop trimming and then the nodes with in-degree bigger than 1 is eliminated. The fitness of the new particle is computed and then the global best solution is updated according to the fitness. The fitness of the multicast tree is evaluated by using the formula given below.

$$Fitness(T(s,M)) = a_1 \times e^{-cost(T(s,M))/bestcostsofar}$$

$$+a_2 \times e^{-delay(T(s,M))/bestdelaysofar}$$

$$+a_3 \times e^{-jitter(T(s,M))/bestjittersofar} \qquad (19)$$

where *cost, delay and jitter* are the values of cost, delay, and delay jitter of the multicast tree. The *bestcostsofar, bestdelaysofar, bestjittersofar* stand for the cost, delay, jitter of the best solution of the current multicast tree, respectively. This process continues till the number of iterations reaches the maximum number of iterations or the best solution has not changed for some consecutive iteration.

The PSO algorithm based on jumping PSO (JPSO) algorithm was proposed by Qu et al. [19] for QoS multicast routing. This is a tree-based approach in which a set of multicast trees is randomly generated. These multicast trees are represented as a swarm of random particles. Each multicast tree is constructed starting from the source node and randomly adding the links into the on-tree node until all destination nodes are added to the tree. The multicast tree is represented by using a predecessor array with $|V|$ elements. The multicast tree representation of predecessor array is shown in Figure 10.3.

For each particle i in the swarm at iteration j, its position (solution) $x_{i,j}$ and velocity (rate of change) $v_{i,j}$ are updated on the evolution by using the following two equations:

$$v_{i,j+1} = c_0 v_{i,j} + c_1 r_1 (b_i - x_{i,j}) + c_2 r_2 (g_j - x_{i,j}) + c_3 r_3 (g_{i,j} - x_{i,j}) \qquad (20)$$

$$x_{i,j+1} = x_{i,j} + v_{i,j} + 1 \qquad (21)$$

The velocity update $v_{i,j+1}$ consists of four components:

- the first component, $c_0 v_{i,j}$, is the inertia and that enables the particle to maintain the flow of its previous movement to avoid abrupt moves and premature convergence;

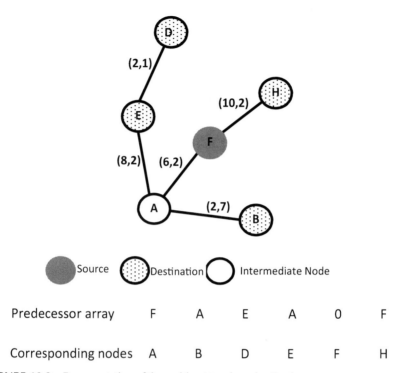

Source   Destination   Intermediate Node

Predecessor array	F	A	E	A	0	F
Corresponding nodes	A	B	D	E	F	H

FIGURE 10.3 Representation of the multicast tree by using Predecessor array.

- the second component, $c_1 r_1(b_i - x_{i,j})$, uses the particle's best achieved position so far as the reference to encourage its self learning ability
- the third component, $c_2 r_2(g_j - x_{i,j})$, is the social factor that remembers the particle's best position so far within the swarm, g_j,
- the fourth component, $c_3 r_3(g_{i,j} - x_{i,j})$, uses the particle's best position found so far in a neighborhood sub-swarm $g_{i,j}$. This is to enhance the exploration capacity of the particles and also to prevent premature convergence within the swarm.

The first part $c_0 v_{i,j}$ enables the particle to continue the exploration of its current position. The second part $c_1 r_1(b_i - x_{i,j}) + c_2 r_2(g_j - x_{i,j}) + c_3 r_3(g_{i,j} - x_{i,j})$ encourages the particle to move towards a better location with respect to other particles' positions. Therefore, the particle that performs the move *follows* the other three different *attractors*, and is thus named the *follower* in the literature.

At every iteration of the evolution the movement of each particle is selected either based on its current position or the position of the attractor that is chosen by using the weight. The local search is applied after the particle jumped to a new position. Then the best position of both the particle and the swarm are updated. This process is repeated till achieving the best position of the swarm.

In PSO base multicast routing, there are two types of moves: the first one moves towards an attractor and the second one moves around the current position without involving the attractor.

- If $r \in [0, c_0)$ then no attractor is selected, then a super path from the tree is randomly removed and the resulting two sub-tree are connected by using a random link. The super path considered in the algorithm is the longest path between two end nodes where all the internal nodes except the two end nodes of the path has a node degree of two.
- If $r \in [c_0, c_0+c_1)$, the best position achieved by the particle so far (b_i is chosen as the attractor. If $r \in [c_0+c_1, c_0+c_1+c_2)$, the attractor is the best position achieved by the whole swarm so far (g_j). If $r \in [c_0 +c_1 +c_2, 1]$, the best located particle in the neighborhood of the current particle ($g_{i,j}$) acts as the attractor.

If the attractor is selected the path replacement operator adds the least cost path from the source to the destination in the selected attractor and removes the corresponding path in the follower particle. If the added path is already in the follower particle, then a random move is applied to the follower particle. The path replacement operator is used to update a particle's position based on a chosen attractor. The Figure 10.4 shows the newly generated particle from the

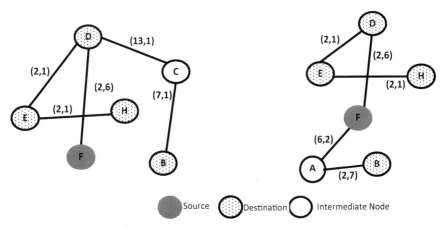

FIGURE 10.4   Path replacement operation.

follower particle (current tree) and attractor particle (attractor tree). The least cost path (F A B) in the follower tree is found by the path replacement operator and the corresponding path (F D C B) of the attractor tree is replaced.

In QPSO [26] each single particle is treated as a spin-less one in quantum space. By this modification he has attempted to explore the QPSO applicability to combinatorial optimization problems. In this method, the problem was converted to constrained integer program and then QPSO algorithm was applied to solve the problem through loop deletion operation.

The QPSO algorithm for the QoS multicast routing problem has been proposed by Sun et al. [24]. This method first converts the multicast routing problem into a constrained integer programming and then solves by the QPSO algorithm with the loop deletion operation. The main issues to be tackled are the representation of a particle's position and determination of the fitness function. This is a path-based approach in which the paths generated to a destination are numbered as an integer. Then, the multicast tree is represented by m-dimensional integral vector. The QPSO based QoS multicast routing is reduced to a m-dimensional integer programming. The fitness of the multicast tree is computed

$$Minimize\ f(T(s,M)) = f_c + \omega_1 f_b + \omega_2 f_d + \omega_3 f_{dj} + \omega_4 f_{pl} \qquad (22)$$

where ω_1, ω_2, ω_3 and ω_4 are the weights of bandwidth, delay, jitter and loss-rate respectively. The functions f_c, f_b, f_d, f_{dj} and f_{pl} are defined by

$$f_c = cost(T(s,M)) \qquad (23)$$

$$f_b = \sum_{d\in M} max\{QB - B(p(s,d)),0\} \qquad (24)$$

$$f_d = \sum_{d\in M} max\{delay(p(s,t)) - DC,0\} \qquad (25)$$

$$f_{dj} = \sum_{d\in M} max\{jitter(p(s,t)) - DJC,0\} \qquad (26)$$

$$f_{pl} = \sum_{d\in M} max\{lossrate(p(s,t)) - LC,0\} \qquad (27)$$

The personal position and the global best position are updated after the completion of fitness values of trees in each iteration. Then each component of the particle position is updated and the integer representing the path is also adjusted. In the next step, the loops existing in the tree are deleted to make the multicast tree feasible. If there exists loop between i^{th} and j^{th} path of the path serial then the costs of the routes that constitute the loop and the more expensive route is deleted.

10.3.6 HYBRID ALGORITHMS FOR QoS MULTICAST ROUTING

The hybrid discrete PSO with GA for QoS multicast routing is proposed by Abdel-Kader [1]. The hybrid PSO-GA algorithm combines the power of both PSO and GA to solve the multicast routing problem. This hybrid algorithm uses the position velocity update rules of PSO along with crossover, mutation and selection operators of GA. This is a path-based approach in which a particle represents a multicast tree. The particle (chromosome in GA) $C = (g_1, g_2,.....,g_n)$ is a string of integers where g_i is the gene of the particle. The given $g_i \in \{1, 2,...., k\}$ represents one of the k paths from the source to the *ith* destination. The particle is constructed by choosing the paths that satisfy the QoS constraints and removing the loops from the combination of paths. Then the fitness of the particle is computed by using the following equation.

$$f(T(s,M)) = Cost(T(s,M)) + \eta_1 \min\{DC - delay(T(s,M)), 0\}$$
$$+\eta_2 \min\{DJC - jitter(T(s,M)), 0\} \tag{28}$$

where η_1 and η_2 are punishment coefficients and the value of the coefficients decide the punishment extent. If $a \geq 0$, then min $(a, 0) = a$; else min $(a,0) = 0$.

This hybrid algorithm uses a two-point crossover operation to generate new particles. The two-point crossover operation is depicted in Figure 10.2. It gives single routing path (SRP) and multiple routing paths (MRP) mutation operations to generate new particles. In SRP mutation, a single path gene is selected and replaced randomly by another gene from the set of paths to the corresponding destination. In MRP mutation, multiple path genes are selected and replaced by other genes randomly. If the new particle has better fitness then it replaces the old particle. The algorithm replaces the duplicate particles by new particles generated randomly. The algorithm terminates after a fixed number of iterations if the fitness value of the particle is not improved more than a fixed threshold value.

An improved Bacteria Foraging Algorithm was proposed by Korani [10] in the year 2008, naming it BF-PSO algorithm. A unit length direction of tumble characteristic shown in bacteria is randomly generated by BFO algorithm. This generation of random direction may lead to delay in reaching the global solution. Thus, the PSO ability to exchange the social information and the ability to find a new solution by elimination and dispersal of BFO algorithm is incorporated in BF-PSO. In this algorithm, the global best position and local best position of each bacterium can decide the unit length random direction of tumble behavior of the bacteria. The tumble direction update during the process of chemotaxis loop is determined by using the following equation:

$$\varphi(j+1) = C_0 * \varphi(j) + C_1 * \gamma_1 (plbest - pcurrent)$$
$$+ C_2 * \gamma_2 (pgbest - pcurrent)$$

where, *pcurrent* is the current position of the bacteria and *plbest* is the local best position of each bacteria and *pgbest* is the global best position of each bacteria.

Pradhan et al. [18] proposed a BF-PSO algorithm for QoS multicast routing (BFQMR). In BFQMR, the QoS multicast tree is generated by setting the delay of the links to infinite which have the available bandwidth less than the required bandwidth. Then *k*-least delay paths are generated by using the *k*-Bellman-Ford algorithm. The delay constrained paths to each destination of the multicast group are selected from the *k*-least delay paths. For a set of multicast destination nodes, where $M = \{d_1, d_2, ..., d_m\}$ and a given source node *s*, a bacterium in our proposed BFO algorithm is represented as $\{x_1, x_2, ..., x_m\}$. Each bacterium represents a multicast tree and each bacterium has m number of components, i.e., $\{x_1, x_2, ..., x_m\}$, where the ith component represents the path to the ith destination in a multicast tree. Each component in bacteria selects a feasible path from *k*-number of delays constrained paths to each destination.

Each bacterium is a candidate solution of QoS multicast tree from source node to each destination. A bacterium represents a multicast tree after randomly combining the paths from the generated *k*-least cost, delay constrained paths and removing the loops, if any. After the multicast trees are randomly generated and chosen by bacteria, loop deletion and calculation of fitness function is performed.

The movement of bacteria is defined by tumbling and swimming during the process of chemotaxis. In the chemotactic process, the fitness of the ith bacterium, $1 \leq i \leq S$ in the jth chemotactic step, $1 \leq j \leq N_c$ is computed as *Fitness (P(i, j))*. The N_c is the maximum number of chemotactic steps taken during the iteration, where $P(i, j)$ is the current position of the *ith* bacterium at j^{th} chemotactic step. The bacterium at position $P(i, j)$ represents a multicast tree $T (s, M)$. The fitness function evaluated is stored in the J_{last}. The fitness value of the bacterium is computed by modifying the fitness function used in the hybrid PSO/GA algorithm for QoS multicast routing [1].

$$J(i,j) = Fitness\big(T(s,M)\big) = cost\big(T(s,M)\big)$$
$$+ \tau_1 \min\big(DC - delay\big(T(s,M),0\big)$$
$$+ \tau_2 \min\big(DJC - jitter\big(T(s,M),0\big)$$
$$+ \tau_3 \min\big(BC - bandwidth\big(T(s,M),0\big) \qquad (30)$$

where τ_1, τ_2, τ_3 are the weights predefined for the delay, delay-jitter and bandwidth respectively.

After computing the fitness value of a bacterium at current position, the local position of the bacterium with respect to the next chemotactic step is updated. During the lifetime, a bacterium must undergo one tumble behavior. The $\varphi(m, i)$ decides the direction for the tumble behavior for the bacterium. A bacterium in its chemotactic phase undergoes a maximum of N_j number of swimming steps. During swimming, if the updated local position of the bacterium is better than the previous local best position J_{last}, then J_{last} is updated by $J(i, j+1)$. The PSO parameter *pcurrent* is evaluated. The current position *pcurrent* is updated if the current position of the bacterium is better than the previous current position of the bacterium. With each movement the current position of the bacteria is updated as well as the fitness function. Thus, each bacterium moves towards the optimum solution of QoS multicast tree in the chemotactic step. The local best position *plocal* and global best position *pgbest* of each bacterium is evaluated after the loop swimming for maximum of consecutive N_s steps. If the global best position *pgbest* does not change over three chemotactic phase, then the chemotactic step is stopped and the reproduction phase starts.

In the reproduction phase, the multicast trees obtained in the chemotactic step are sorted and arranged with respect to their fitness values J_i health. The bacterium that has the best fitness value for each chemotactic step are chosen and arranged in ascending order. The bacteria with worst fitness values die and the remaining bacteria with best value remains and copies.

$$J_{health}^{i} = min_{j=1}^{n_c+1} \{Fitness(P(i,j))\} \tag{31}$$

In elimination-dispersal step the weakest or the poor bacteria are dispersed with probability.

A hybrid-swarming agent based ACO/PSO algorithm is proposed by Patel et al. (2014). This algorithm generates n multicast trees randomly where each multicast tree has m attributes for m destinations. The structure of the pattern is defined as $T_i = (a_{i1}, a_{i2}, ...a_{im})$ for $i = 1$ to n. Where T_i is the ith multicast tree pattern and a_{ij} represents the jth attribute of ith pattern. Then n numbers of fixed pattern agents are created to associate with n tree patterns. The n numbers of mobile particle agents are generated and each particle agent is randomly attached to a pattern agent. The arrangement of particle and pattern agent is shown in Figure 10.5.

The particle agents can move from one pattern to another pattern to interact with other particle agents dynamically. The particle agents are allowed to deposit pheromones and sense local attributes in each pattern. The pattern agent executes the dynamics of pheromone aggregation, dispersion and evaporation. Each pattern agent maintains the pheromones in two levels that are pattern pheromone and attribute pheromone. When a particle agent moves to another particle agent the particles attract dynamically to replace the attribute of one pattern by a better pattern. After some iteration the particle agent converges to the fittest patterns. The ACO algorithm basically works on the three ideas. First, each ant's movement is associated with the

P_1	P_2	P_3	P_4	P_5	P_6
a_5	a_9	a_{25}	a_{30}	a_2	a_8
P_7	P_8	P_9	P_{10}	P_{11}	P_{12}
a_6	a_{10}	a_{26}	a_{29}	a_{28}	a_{27}
P_{13}	P_{14}	P_{15}	P_{16}	P_{17}	P_{18}
a_7	a_{11}	a_1	a_{24}	a_{12}	a_{22}
P_{19}	P_{20}	P_{21}	P_{22}	P_{23}	P_{24}
a_4	a_{14}	a_{21}	a_{17}	a_{20}	a_{19}
P_{25}	P_{26}	P_{27}	P_{28}	P_{29}	P_{30}
a_{23}	a_{18}	a_{16}	a_{15}	a_{13}	a_3

FIGURE 10.5 Particle agents and patterns arranged in a grid.

candidate solution. Secondly, when an ant moves in a path the amount of pheromone deposited on that path is proportional to the quality of that solution. When there are two or more paths, then the ant follows the path with the best pheromone value.

The PSO algorithm represents each candidate solution as a particle and each particle's movement is the composition of an initial random velocity and two randomly weighted influences: individuality, the tendency to return to the particle's best previous position, and sociality, the tendency to move towards the neighborhood's best previous position.

The velocity and position of the particle at any iteration is updated based on the following equations:

$$v_{id}^{t+1} = c_0.v_{id}^t + c_1.r_1().(p_{id}^t - x_{id}^t) + c_2.r_2().(p_{gd}^t - x_{id}^t) \qquad (32)$$

$$x_{id}^{t+1} = x_{id}^t + v_{id}^{t+1} \qquad (33)$$

where v_{id}^t is the component in dimension d of the i-th particle velocity in iteration t, x_{id}^t is the component in dimension d of the i-th particle position in iteration t, c_1, c_2 are constant weight factors, p_i is the best position achieved by particle i, p_{gd} is the best position found by the neighbors of particle i, r_1, r_2 are random factors in the [0,1] interval, and w is the inertia weight.

10.4   SIMULATION RESULTS

We evaluate the performance of the path based and tree-based evolutionary algorithms for QoS multicast routing by implementing these algorithms in Visual C++. The experiments are performed on an Intel Core i3 @ 2.27 G.Hz. and 4 GB RAM based platform running Windows 7.0. The nodes are positioned randomly in an area of size 4000 km x 2400 km. The Euclidean metric is then used to determine the distance between each pair of nodes. The network topology used in our simulation was generated randomly using Waxman's topology [35]. The edges are introduced between the pairs of nodes u, v with a probability that depends on the distance between them. The edge probability is given by $p(u,v) = \beta \exp(-l(u,v)/\alpha L)$, where l(u,v) is the Euler distance from node u to v and L is the maximum distance between any two points in the network. The delay, loss rate, bandwidth and cost of the links are set randomly from 1 to 30, 0.0001 to 0.01, 2 to 10 Mbps and 1 to 100, respectively.

The source node is selected randomly and destination nodes are picked up uniformly from the set of nodes chosen in the network topology. The delay bound, the delay jitter bound and the loss bound are set 120 ms, 60 ms and 0.05, respectively. The bandwidth requested by a multicast application is generated randomly. We generate 30 multicast trees randomly to study and compare the performance of the tree-based algorithms. The simulation is run for 100 times for each case and the average of the multicast tree cost is taken as the output. We also generate 20 shortest paths for each destination to study and compare the performance of the path-based algorithms. We vary the network size from 20 to 140 and the number of destinations is considered as 20% of the number of nodes in the network. The performance of these algorithms is studied in terms of multicast tree cost, delay and delay-jitter.

The Figure 10.6 shows the comparison of multicast tree cost of various path based algorithms such as BFPSO [18], QPSO [24], GAPSO [1], PSO [26] with respect to varying network size .The results show that the BF-PSO performs better than QPSO, PSO and GA-PSO in terms of cost. This is because the BFPSO uses an efficient loop deletion procedure to generate a better-multicast tree while combining the paths to the destinations. Furthermore, the BFO has a powerful searching capability to select the optimal set of paths to generate the best possible tree. The convergence speed of BFO has also been improved with PSO..The Figure 10.7 shows the multicast tree cost verses the network size with 20 percent of the nodes as the group size. The multicast

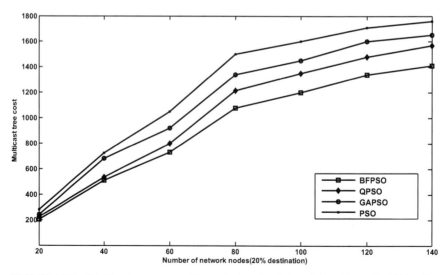

FIGURE 10.6   Multicast tree cost vs. Network size of path based algorithm with 20% nodes as destinations.

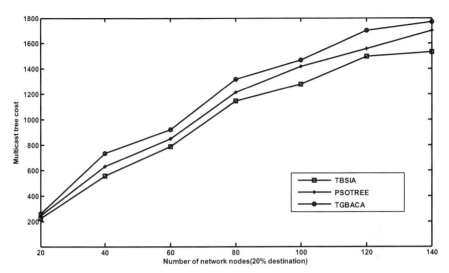

FIGURE 10.7   Multicast tree cost vs. Network size of tree-based algorithm with 20% nodes as destinations.

trees generated by PSOTREE [27], TGBACA [29], TBSIA [16] and TBCROA satisfy the delay, delay jitter, loss rate and bandwidth constraints. However, the figures clearly illustrate that the cost of the multicast tree generated by our proposed algorithm is less than the multicast trees generated by PSOTREE and TGBACA and TBSIA. The PSOTREE algorithm constructs the multicast tree by combining the multicast trees and removing directed cycles. This algorithm removes the links that are in any of the trees, but not in both and have minimum fitness. However, this approach may not generate a better tree, because the links deleted from the cycle may be better than the links not in the directed cycles. The TGBACA algorithm follows a pheromone updating strategy to construct the best-multicast tree. The algorithm updates pheromones on the links used by the global best tree and the best tree generated after each generation. Though this strategy fasts the convergence process, but the solution may fall into local optimization. The TBSIA combines two multicast tree patterns by bringing the better attributes of one pattern to another pattern. It generates a new tree pattern after each iteration, which is better than both the patterns. Since the whole path of one tree is replaced by another path from another multicast tree, some better links may be excluded from the tree. This may fail to generate an optimal tree in some cases.

The Figures 10.8 and 10.9 show the multicast tree delay of path based and tree-based algorithms versus number of network nodes with 20% nodes as destinations respectively. It is observed that both the path based and

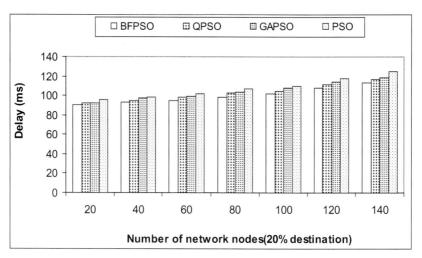

FIGURE 10.8   Multicast tree Delay vs. Network size of path based algorithms with 20% nodes as destinations.

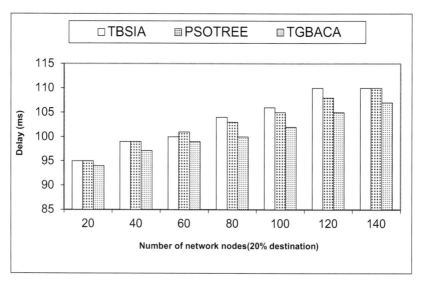

FIGURE 10.9   Multicast tree Delay vs. Network size of tree-based algorithms with 20% nodes as destinations.

tree-based algorithms satisfy the delay constraints. The BFPSO experiences less delay than QPSO, GAPSO and PSO. This is because the BFPSO computes k-least delay paths to the destinations and uses the searching power of BFO to select the best delay paths to generate the least cost delay constraint multicast tree. The TGBACA experience less delay than PSOTREE and TBSIA. The PSOTREE and TBSIA attempts to minimize the cost of the trees keeping in mind that the delay constraint should should not be violated. The TGBACA generates the candidate multicast trees considering the pheromone of the paths selected in previous iterations. The Figures 10.10 and 10.11 show the comparison of multicast tree delay jitter versus number of network nodes with 20% nodes as destinations respectively. It is observed

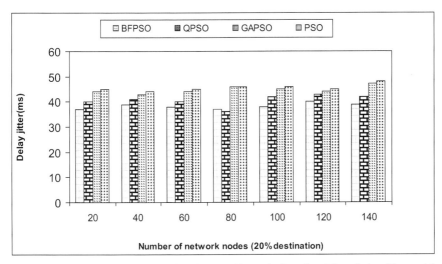

FIGURE 10.10   Multicast tree Delay Jitter vs. Network size of path based algorithms with 20% nodes as destinations.

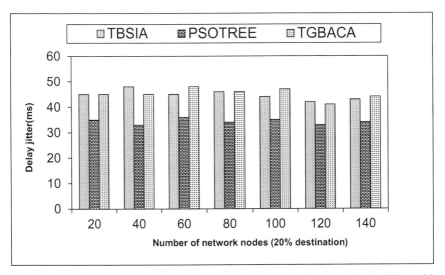

FIGURE 10.11   Multicast tree Delay Jitter vs. Network size of tree-based algorithms with 20% nodes as destinations.

that both the path based and tree-based algorithms satisfy the delay jitter constraints. The BFPSO experiences less delay jitter than QPSO, GAPSO and PSO. The PSOTREE experience less delay jitter than TGBACA and TBSIA. The PSOTREE uses a merging procedure to merge two trees and delete the loops to generate the cost optimal multicast tree. This may select longer delay paths to minimize the cost of the tree.

10.5   CONCLUSION

The cost optimal QoS-aware multicast routing problem can be reduced to constrained Steiner tree problem which have been proven to be NP-complete. Therefore, many heuristics and evolutionary algorithm have been proposed to generate cost optimal QoS multicast tree. In this chapter, we studied various evolutionary algorithms developed for the above problem. The EAs attack this problem by using either path-based techniques or tree-based techniques. The path-based techniques generate a set of constrained paths to each destination. Let there are k number of paths to each of the m destinations. Then the number of candidate trees that can be generated by combining these paths is k^m, which is very high. The EAs combined these paths either heuristically or randomly to generate a set of fixed number of candidate feasible solutions. Similarly, in a tree-based approach, the multicast trees are directly created either randomly or heuristically. The EAs and their hybridization usually combine these solutions to generate a set of new and better solution in each iteration using recombination or mutation. These algorithms run for a fixed number of iterations and after each iteration, the global best solution and local based solution are recorded. Though these algorithms run for a fixed number of iterations the termination criteria can be set very sensibly to stop the algorithm when they converge. This not only generates the best solution but also reduces the execution time.

KEYWORDS

- **Ant colony optimization**
- **Bacteria foraging**
- **Evolutionary algorithm**
- **Genetic algorithm**
- **Harmony search**

- **Multicast**
- **Particle swarm optimization**
- **QoS routing**

REFERENCES

1. Abdel-Kader, R. F., Hybrid discrete PSO with GA operators for efficient QoS-multicast routing. *Ain Shams Engineering Journal.* 2011, 2, 21–31.

2. Dorigo, M., & Caro G. D., *The ant colony optimization meta-heuristic, new ideas in optimization.* McGraw-Hill, 1999.

3. Forsati, R., Haghighat, A. T., & Mahdavi, M., Harmony search based algorithms for bandwidth-delay-constrained least-cost multicast routing. *Computer Communications.* 2008, 31, 2505–2519.

4. Geem, Z. W., Kim, J. H., & Loganathan, G. V., A new heuristic optimization algorithm: harmony search. *Simulation.* 2001, 76 (2), 60–68.

5. Geem, Z. W., Tseng, C., & Park, Y., Harmony search for generalized orienteering problem. *Lecture Notes in Computer Science.* 2005, 3412, 741–750.

6. Gong, B., Li, L., Wang, X., Multicast Routing Based on Ant Algorithm with Multiple Constraints. *IEEE International Conference on Wireless Communications, Networking and Mobile Computing,.* 2007, 1945–1948.

7. Haghighat, A. T., Faez, K., Dehghan, M., Mowlaei, A., & Ghahremani, Y., GA-based heuristic algorithms for bandwidth-delay-constrained least-cost multicast routing. *Computer Communications.* 2004, 27, 111–127.

8. Hwang, R. H., Do, W. Y., & Yang, S. C., Multicast Routing Based on Genetic Algorithms. *Journal of Information Science and Engineering.* 2000, 16, 885–901 .

9. Kennedy, J., & Eberhart, R. C., Particle Swarm Optimization. *IEEE International Conference on Neural Networks.* 1995, 1942–1948.

10. Korani, W., Bacterial Foraging Oriented by Particle Swarm Optimization Strategy for PID Tuning, *GECCO,* 2008.

11. Kosiur, D., *IP Multicasting: The Complete Guide to Interactive Corporate Networks.* Wiley, New York, 1998.

12. Kun Z., Heng W., & Liu F. Y., Distributed multicast routing for delay and delay variation-bounded Steiner tree using simulated annealing. *Computer Communications.* 2005, 28 (11), 1356–1370.

13. Lee, K. S., & Geem, Z. W., A new meta-heuristic algorithm for continues engineering optimization: harmony search theory and practice. *Computer Methods in Applied Mechanics and Engineering.* 2004, 194, 3902–3933.

14. Molnar, M., Bellabas, A., & Lahoud, S., The cost optimal solution of the multi-constrained multicast routing problem. *Computer Networks.* 2012, 56, 3136–3149.

15. Palmer, C. C., & Kershenbaum, A., Representing trees in genetic algorithms. *IEEE world Congress on Computational Intelligence.* 1993, 1, 379–384.

16. Patel, M.K; Kabat, M. R., & Tripathy, C. R., A Hybrid ACO/PSO based algorithm for QoS multicast routing problem. *Ain Shams Engineering Journal.* 2014, 5,113–120.

17. Peng, B., & Lei L., A Method for QoS Multicast Routing Based on Genetic Simulated Annealing Algorithm. *International Journal of Future Generation Communication and Networking.* 2012, 5 (1), 43–60.

18. Pradhan, R., Kabat, M. R., & Sahoo, S. P., A Bacteria Foraging-Particle Swarm Optimization Algorithm for QoS multicast Routing. *Lecture notes on Computer Science.* 2013, 8297,590–600.

19. Qu, R., Xu, Y., & Castro, J. P., Particle swarm optimization for the Steiner tree in graph and delay delay-constrained multicast routing problems. *Journal of Heuristics*, 2013, 19, 317–342.

20. Quinn, B., IP Multicast Applications: Challenges and Solutions. *IETF RFC 3170*, 2001.

21. Ravikumar, C. P., & Bajpai, R., Source-based delay-bounded multicasting in multimedia networks. *Computer Communications.* 1998, 21, 126–132.

22. Rong, Q., Ying X,; Castro, J. P., & Landa-Silva, D., Particle swarm optimization for the Steiner tree in graph and delay-constrained multicast routing problems. *Journal of Heuristics.* 2013, 19, 317–342.

23. Shi, L., Li, L., Zhao, W., & Qu, B., A Delay constrained Multicast Routing Algorithm Based on the Ant Colony Algorithm. *Lecture Notes in Electrical Engineering.* 2013, 219, 875–882.

24. Sun, J., Fang, W., Xiaojun, W; Xie, Z., & Xu, W., QoS multicast routing using a quantum-behaved particle swarm optimization algorithm. *Engineering Applications of Artificial Intelligence.* 2011, 24, 123–131,

25. Sun, J., Feng, B., & Xu, W. B., Particle swarm optimization with particles having quantum behavior, *In Proceedings of Congress on Evolutionary Computation.* 2004, 325–331.

26. Sun, J., Xu, W. B., & Feng, B., A global search strategy of quantum-behaved particle swarm optimization, *In Proceedings of IEEE Conference on Cybernetics and Intelligent Systems*, 2004, 111–116.

27. Wang, H., Meng, X., Li, S., & Xu, H., A tree-based particle swarm optimization for multicast routing. *Computer Network.* 2010, 54, 2775–86.

28. Wang, H., Shi, Z., & Li, S., Multicast routing for delay variation bound using a modified ant colony algorithm. *Journal of Network and Computer Applications.* 2009, 32, 258–272.

29. Wang, H., Xu, H., Yi, S., & Shi, Z., A tree-growth based ant colony algorithm for QoS multicast routing problem. *Expert Systems with Applications.* 2011, 38, 11787–11795.

30. Wang, Z., Shi, B., & Zhao, E., Bandwidth-delay-constrainted least-cost multicast routing based on heuristic genetic algorithm. *Computer Communications.* 2001, 24, 685–692.

31. Xi-Hong, C., Shao-Wei, L., Jiao, G., & Qiang, L., Study on QoS multicast routing based on ACO-PSO algorithm. *Proceedings of 2010 International Conference on Intelligent Computation Technology and Automation.* 2010, 534–753.

32. Yen, J. Y., Finding the *k*-shortest loop-less paths in a network, *Manage Science.* 1971, 17(11), 712–716.

33. Yin, P. Y., Chang, R. I., Chao, C. C., & Chu. Y. T., Niched ant colony optimization with colony guides for QoS multicast routing. *Journal of network and computer applications.* 2014, 40, 61–72.

34. Younes, A., An Ant Algorithm for Solving QoS Multicast Routing Problem. *International Journal of Computer Science and Security (IJCSS).* 2011, 5(1), 766–777.

35. Waxman, B. M. "Routing of multipoint connections," IEEE J Select Areas Communications 1988, 6(9), pp. 1617–1622.

36. Zhang, L., Cai, L., Li, M., & Wang, F., A method for least-cost QoS multicast routing based on genetic simulated annealing algorithm. *Computer Communications.* 2009, 32, 105–110.

CHAPTER 11

PERFORMANCE ASSESSMENT OF THE CANONICAL GENETIC ALGORITHM: A STUDY ON PARALLEL PROCESSING VIA GPU ARCHITECTURE

PAULO FAZENDEIRO[1] and PAULA PRATA[2]

Universityof Beira Interior, Department of Informatics, Portugal
Instituto de Telecomunicações (IT), Portugal,
[1]*E-mail: pprata@di.ubi.pt*
[2]*E-mail: fazendeiro@ubi.pt*

CONTENTS

ABSTRACT

Genetic Algorithms (GAs) exhibit a well-balanced operation, combining exploration with exploitation. This balance, which has a strong impact on the quality of the solutions, depends on the right choice of the genetic operators and on the size of the population. The results reported in the present work shows that the GPU architecture is an efficient alternative to implement population-based search methods. In the case of heavy workloads the speedup gains are quite impressive. The reported experiments also show that the two-dimensional granularity offered by the GPU architecture is advantageous for the operators presenting functional and data independence at the population+genotype level.

11.1   INTRODUCTION

The Genetic Algorithms (GAs) are bio-inspired population-based computational methods with application in search, (multiobjective) optimization and learning problems [1, 3, 8]. A GA requires a limited amount of knowledge about the problem being solved. Relative evaluation of the candidate solutions is enough and no derivatives of cost functions are required. This can be a definitive advantage when compared with other candidate methods of optimization, as the derivatives of the involved functional in various real world problems can be computationally demanding or have no closed form [15].

These computationally intensive population-based search methods present heavy time requirements, hence reducing their applicability especially in what concerns real time applications. The potential for acceleration of population-based, stochastic function optimizers using Graphic Processing Units (GPUs) has been already verified in a number of recent works [4, 6, 10, 13, 16–20] over a representative set of benchmark problems.

This chapter is focused on the study of the effective parallelization of the canonical GA. The chapter presents a complete characterization of the relative execution times of the atomic operators of the GA, varying the population cardinality and the genotype size. It is complemented with an analysis of the achieved speedups. The findings of the assessment of the parallelization potential at different granularity levels (population and population+genotype) altogether with the analysis of data parallelism are also reported in the experimental part.

The remaining of this chapter is organized as follows. For the sake of self-containment, Section 11.2 presents a brief introduction to the design of a parallel canonical GA. Next in Section 11.3 is introduced the programming model of the OpenCL framework and described the architecture of the used GPU. In Section 11.4, the details of the experimental setup are described and the obtained results are shown. Finally in Section 11.5, the main results and some directions for future work are discussed.

11.2 GA PARALLEL DESIGN

The basic principles of GAs were established rigorously by Holland in 1975 [9]. They operate on a population of individuals, each representing a possible solution to a given problem. Each individual receives a reward value according to how worthwhile is the solution that it represents for the problem in question. This figure of merit (or fitness) determines which the most capable individuals are; these are given more opportunities to reproduce by crossing its genetic material with other individuals in the population. This results in a new generation of descendants inheriting characteristics of each one of their parents. The least-fit individuals, with little chance of being selected to mate, die (almost invariably) without descendants. A new population of possible solutions appears as a result of selection of the best individuals of the present generation and their subsequent mating, giving birth to a new set of individuals. This new population contains a greater proportion of the characteristics of the good elements of the preceding generation. Thus, from generation to generation, these good characteristics are spread by the population and are mixed and recombined with each other. The prioritization of the superior individuals and their pairing leads the exploration of the search space for its most promising areas. Consequently the population will tend to converge to a *near-optimal solution*.

A properly implemented GA should be able to evolve the population generation after generation so that both the merits of the best individual and the average merit of the population "move" towards the global optimum.

Optimizing the population and not a single individual contributes to the robustness of these algorithms: even if inadvertently a desirable characteristic is lost in an individual it may have been retained in other elements of the population and often reappear in later generations.

Informally a GA can be conceptualized as an iterative algorithm where the successive epochs mimic in silico, in a search space context, the necessary processes in order to a successfully reproduction of the biological organisms. The evolution of potential solutions over successive generations comprises different phases that succeed each other in continuum until a stopping criterion is met. The result of each generation is a new offspring population, which replaces (or sometimes competes with) the previous population.

Most GAs use fixed population size so it is necessary to decide which offspring are inserted in the next generation. In this work, after some preliminary tests, we have followed a deterministic elitist approach combining a fitness-based replacement (the two parents of higher rank are kept) with and age-based one (the remaining parents are replaced by the offspring). Figure 11.1 presents an overview of the different modules applied in our implementation.

The parallel design of a GA implementation must consider the different granularity levels subsumed in these different phases. The fitness evaluation

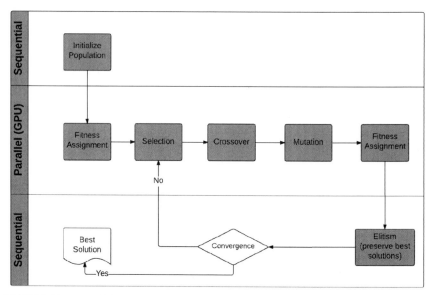

FIGURE 11.1 Modules of a Genetic Algorithm.

is eminently a one-dimensional operation (in the sense that usually each chromosome, taken as a whole, can be evaluated in parallel) whereas the crossover and mutation operators are, in the general case, 2D operators allowing the parallel treatment of not only each chromosome but also each gene inside it. Regarding the selection operator (tournament with two chromosomes) it is also a one-dimensional operation including the tournament and posterior inclusion (copy) of the winner into the population.

For each problem it is necessary to establish beforehand an adequate representation or codification of solutions and a fitness function to grade them. During the execution of the algorithm the parents are selected for reproduction and from their recombination are produced new offsprings.

It is assumed that a potential solution to a problem can be represented by a set of parameters called genes. These are grouped in a structure called chromosome (an individual).

In the biogenetic terminology each of these parameters is a feature codified by a gene whose value is referred to as allele. The set of parameters represented by a chromosome is called genotype. It contains information needed to produce an organism designated the phenotype. The merit of an individual therefore depends on the performance of the phenotype (assessed by the fitness function) inferred from its genotype (decoding it).

Traditionally the applications of GAs use a binary encoding to represent the chromosomes. This is consistent with the point of view put forward by Holland [9] and reiterated by Goldberg [8] suggesting that the ideal number of elements of the alphabet is 2.

However higher cardinality alphabets (where typically each symbol is an integer or a real number) have been used with satisfactory results in some research studies, cf. [1]. When the parameters of the problem are numerical in nature, which often happens, the direct representation as numbers (instead of strings of bits) can facilitate the definition of the crossover and mutation operators with a clearer meaning to a specific problem.

11.2.1 FITNESS ASSIGNMENT

The merit function depends on the nature of the problem to be solved. Given a chromosome, the fitness function returns a numeric value of fitness or merit that is supposed proportional to its usefulness as an individual. In optimization of a known function the obvious choice to the fitness function is the function itself or one of its simple transformations.

Since the GAs are designed to maximize the merit function, minimization problems must be transformed into problems of maximization. This can be done, for example, through the following transformation:

$$f = 1/(1 + J) \qquad (1)$$

where J is the objective function to minimize. For a successful reproduction it is necessary to have a fitness function that correctly distinguishes the best from the worst elements. The population is initialized with a random distribution of values of each gene by the set of chromosomes, so there is great variability in the individual values of merit. With the evolution of the population there are alleles in certain positions that begin to exert their dominance.

With the convergence of the population the variability of the fitness decreases, giving rise to two possible scaling problems:

- *Premature Convergence:* the genes of a group of individuals with high degree of fitness (but non-optimal) quickly dominate the population causing it to converge to a local maximum.
- *Slow Finishing:* the population evolves in general, but with a uniform degree of fitness, there are no individuals who "drag" the population towards the maximum. The possibility of the best individuals being selected is substantially the same as the one of the other individuals in the population, leading to a decrease in the rate of convergence [7].

The technique to overcome these problems consists in changing the codomain of the fitness function: a compression to avoid premature convergence and its expansion to avoid a slow finish. For a description of some of the available techniques for scaling, the interested reader is referred to Ref. [1].

The reproduction phase of the GA begins by the selection of the individuals that are going to be in the mating pool. From their recombination, typically applying crossover and mutation mechanisms will result a new generation of offsprings.

11.2.2 SELECTION

The parents are selected randomly according to a scheme that favors the fittest. This scheme can present various "nuances." Some of them are:

- Selection by tournament: there are randomly chosen pairs of individuals and the best individual of each pair passes to the next generation.

- Viability selection: selection proportional to the fitness where the proportionality constant is given by the ratio between the fitness of the individual and the total fitness of the population.
- Roulette wheel selection: selection proportional to the merit but probabilistic; the elements of greatest merit are assigned larger areas of the wheel (proportional to their relative merit) and the wheel spins randomly.
- Selection by sorting: the individuals are ordered by merit and the best ones are chosen.

In the scheme of selection by roulette wheel – one of the most popular – the best element may not be selected. The *elitist strategy* can solve this problem copying always the best element (or a percentage of the best elements) of each generation into the next. In the determination of the number of individuals who should be removed from the old generation there are two extreme views based on the observation of nature: the generational replacement, by which all members of the population are replaced by the descendants and stationary replacement that specifies a very small number of elements to be replaced in order to allow the parents to "educate" the children by competing with them.

11.2.3 CROSSOVER

The crossover operates on two individuals exchanging and recombining their genetic material in order to produce two offspring. Usually crossover is not applied to all individuals selected to mate: instead it is performed a random selection of individuals who will crossover (usually with a probability of applying crossover between 0.6 and 1. When crossover is not applied the descendant is created by simple duplication of the parent. This procedure gives selected individuals opportunity to transmit to future generations their genetic material without any tampering. Among the crossover techniques commonly used are the following:

- Crossover with a single cutoff point: given two individuals it is randomly chosen a position that breaks the chains of genes in two shares, *viz.* beginning and end. Two new individuals are created by binding the beginning of one to the end of the other and vice-versa.
- Two-point crossover: Similar to above but more generic, in that the two points define a section that does not have to coincide with the beginning or the end of the chromosomal string. Yet it is possible to define crossover operators with a greater number of cutoff points.

- Uniform crossover: a mask is generated composed by n binary digits, where n is the number of genes in the chromosomes. The first offspring is created as follows: if the i-th position of the mask has the value 1 the i-th gene is copied from the first parent, otherwise it is copied from the second parent. The second strain is created similarly, changing the order of the parents.

In non-binary representations this operator can be defined, for instance, as the average or as the geometric mean of the parents. The blend crossover operator, BLX-α, is another option specifically designed for real-valued chromosomes [5]. The resulting offsprings are distributed across a hyper-box encompassing the two parents. The parameter α extends the bounds of the hyper-box, hence to the children is given the possibility to explore new search space inside of an extended range given by their parents, see Figure 11.2.

Besides these many other crossover techniques have been suggested, however there is no consensus about the superiority of one over the other, cf. [1]. This aspect is dependent on the population characteristics and on the encoding used.

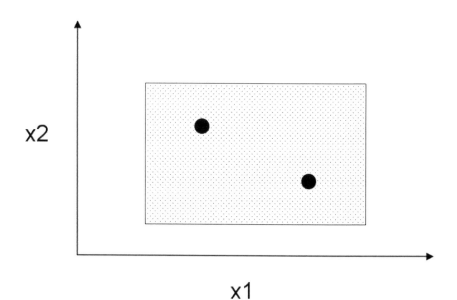

FIGURE 11.2   The BLX-α crossover operator. The big dots represent the parents, while the small dots in the rectangle indicate possible children.

11.2.4   MUTATION

Mutation reintroduces diversity in the form of new genetic material and can restore genes previously lost in the selection process, allowing them to be tested in a new context. This operator is applied to each gene after the crossover, usually with a very low probability (typically between 0.01 and 0001) because a too high mutation rate can turn the genetic search into a random search. In the binary encoding, the value of a mutated gene can be calculated as $genenew = |1 - geneold|$.

Due to its typical low probability one might be tempted to underestimate this operator. However, this operator helps to ensure that there is no point in the search space with null probability of being evaluated by the GA. Moreover Davis [2] refers that as the population converges mutation becomes a more fruitful operation as opposed to crossover that sees its importance diminished. An interesting strategy linked to this aspect is to use dynamic probabilities for each of these operators or adjust these probabilities according to the environment in order to improve the performance of the algorithm.

11.3   THE ENVIRONMENT OF THE STUDY

The need to take advantage of the parallel processing power of GPU has been promoting the development of several high level programming interfaces for the GPU architecture. Some of the most proeminent examples include CUDA (Compute Unified Device Architecture) from NVIDIA [14], Brook+ from AMD/ATI and OpenCL [11]. OpenCL has the advantage of being platform-independent, providing a cross-platform (CPU/GPU) programming language.

An OpenCL application allows the parallel execution of a set of threads launched by a host program. The host program is executed in the CPU (in the host machine) while the kernels are executed in one or more OpenCL devices. An OpenCL device is divided into one or more compute units, which are further divided into one or more processing elements, or cores. Each thread, viz. each kernel instance, is called a work-item and is identified by its point in an index space. The same code can be executed over different data items following a Single Instruction Multiple Data (SIMD) model, thus implementing a data parallel programming model. Additionally, work-items can be organized into work-groups executed in a Single Program Multiple Data (SPMD) model.

According to the OpenCL terminology, a streaming multiprocessor is a compute unit, and a processor core is a processing element. When a kernel is launched, the work-groups, and corresponding work items are numbered and automatically distributed by the compute units with capacity to execute them. Work groups are assigned to compute units, and the work items of each work group are executed on the processing elements of the compute unit. Work items of the same work group can communicate through the multiprocessor-shared memory (called local memory in OpenCL). Because the GPU just processes data stored in its memory, the program data must be copied from the host to the global memory of GPU before executing the kernel. In the end the results must be copied back to CPU memory.

In the reported experiments we used an NVIDIA GeForce GTX 295 GPU built as an array of multithreaded streaming multiprocessors. Each one consists of eight scalar processor cores, with a set of registers associated, and a common shared memory area of 16KB (Wong et al., 2010). The GTX 295 has 240 cores (at 1.24 GHz) and 1GB of global memory. It was programmed with OpenCL version 1.1. The host machine is an Intel Core 2 Quad Q9550 at 2.83 GHz with 4 GB of RAM and the operating system Windows 7, 64 bits.

11.4   EXPERIMENTS AND RESULTS

The solution quality and obtained speedups of the implemented GA were assessed using two artificial benchmark functions commonly used for GA analysis: Rosenbrock's and Griewangk's functions [1]. Since there was observed no statistically significant difference between the quality of the solutions in the tested implementations our analysis will be exclusively focused on the execution times and consequent speedups.

The parameters of the GA were kept constant in the presented experiments. The minimization problems were transformed into maximization through the transformation $f = c/(c + J)$, where J is the objective function to minimize and $c>0$ a proper regularization constant. The employed selection operator was tournament with two contenders. All the population individuals were subjected to the blend crossover (BLX-alpha) operator chosen due to its suitability to a real-valued chromosome encoding [5]. The probability of mutation was equal to 0.025. It was used an elitist approach maintaining the first and second best solutions.

The population's sizes were varied in a geometric progression from 256 to 262,144 with a constant ratio of four and the number of genes was varied from 2 to 16 with a constant ratio of two. Results were obtained as the average of 20 runs with 500 iterations each.

11.4.1 SEQUENTIAL WORKLOAD

Prior to build a parallel version of the canonical genetic algorithm for GPU, the sequential execution was characterized. This is done, calculating the execution time percentages in CPU for each main step of the algorithm. The sequential implementation is briefly outlined in Algorithm 1.

In Figures 11.3 and 11.4 the graphs with the relative execution time in CPU for the Rosenbrock and Griewangk functions, considering a population of 262144 individuals and varying the dimensionality of the problem for 2, 4, 8 and 16 genes, are depicted.

Algorithm 1 – Sequential implementation of the canonical GA
1 – Initialise and evaluate the first population
2 – Repeat until convergence
2.1 – Selection (Tournament)
2.2 – Crossover
2.3 – Mutation
2.4 – Compute the two best elements (Best)
2.5 – Evaluation

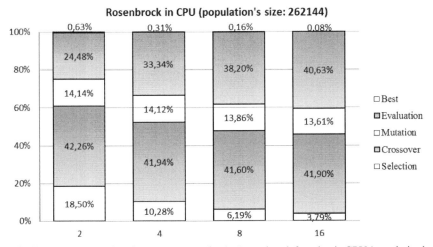

FIGURE 11.3 Execution time percentages for the Rosenbrock function in CPU (population's size = 262,144) varying the number of genes.

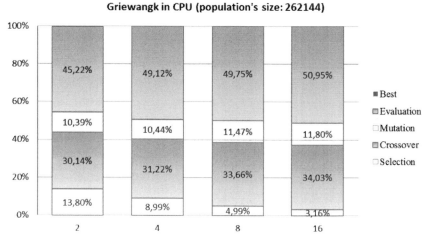

FIGURE 11.4   Execution time percentages for the Griewangk function in CPU (population's size = 262144) varying the number of genes.

As can be seen, in the sequential implementation for the Rosenbrock function, most of the execution time is spent with the crossover operator. It corresponds to about 42% of the time for all the considered dimensionalities and also for all the studied populations' sizes. Only the graphs with the results for the largest population are presented because for the others populations the results are similar. The evaluation comes in second place with percentages varying from about 24% (2 genes) to about 41% (16 genes) in all the population's sizes. The mutation operator spends about 13% of the time for all the cases and the selection operation decreases from about 18% for 2 genes until about 3% of the time for 16 genes. Computing the best values just depends on the population's size, thus its absolute execution time remains the same when the number of genes is increased. In sequential versions for CPU, the time spent in computing the best values is less than about 0.6% of the time for all the studied cases, being difficult to observe its value in the graphics.

Considering the sequential implementation for the Griewangk function, the more noticeable alteration is the exchange of positions between the crossover operator and the evaluation. The evaluation is now the most time consuming operation, which is justifiable by the Griewangk function to be more complex than the Rosenbrock function. Now, the time spending in evaluation varies from about 45% (2 genes) to about 52% (16 genes).

Crossover comes second with about 32% of the time and mutation spends about 11% of the time. Selection varies now from about 13% (2 genes) to about 3% (16 genes).

In summary, when the absolute execution time for evaluation increases, the portion of time spent in each of the other operators decreases because its absolute time remains approximately the same. The absolute execution times for the sequential and parallel versions can be observed in Tables 11.1 (Rosenbrock) and 11.2 (Griewangk) for the largest population. From the results obtained with other population's sizes, it can also be concluded that the time portion spent in each operator is almost independent of the population's size.

11.4.2 PARALLEL WORKLOAD

In the parallel algorithm outlined in Algorithm 2, four kernels (K1, ... K4) are considered, one for each genetic operation. In the parallel implementation, the four kernels are totally executed in parallel. The calculation of the two best values used in the elitist approach is done in CPU since, according to our tests, it is faster to compute those values in CPU rather than in GPU.

TABLE 11.1 Single Iteration Execution Times (Milliseconds) for the Rosenbrock Function in CPU and GPU for a Population Size of 262,144 Individuals*

Rosenbrock function, population's size = 262,144, sequential execution times in CPU (ms)						
Genes	*Selection*	*Crossover*	*Mutation*	*Evaluation*	*Best*	*Total*
2	44,06	100,65	33,68	58,29	1,49	238,17
4	49,34	201,34	67,78	160,04	1,51	480,02
8	59,71	401,52	133,73	368,63	1,51	965,10
16	73,48	811,57	263,60	786,95	1,52	1937,12
Parallel execution times in GPU (ms)						
Genes	*Selection*	*Crossover*	*Mutation*	*Evaluation*	*Best*	*Total*
2	1,23	0,25	0,13	0,58	1,85	4,05
4	2,37	0,80	0,19	1,68	1,91	6,96
8	4,82	2,60	0,31	4,00	1,90	13,64
16	9,86	4,28	0,55	8,60	1,88	25,17

*Times are shown by section (kernel) of code and by the number of genes.

TABLE 11.2 Single Iteration Execution Times (Milliseconds) for the Griewangk Function in CPU and GPU for a Population Size of 262,144 Individuals*

Griewangk function, population's size= 262,144, sequential execution times in CPU						
Genes	Selection	Crossover	Mutation	Evaluation	Best	Total
2	47,09	102,86	35,46	154,30	1,53	341,23
4	59,21	205,66	68,79	323,58	1,53	658,77
8	60,60	408,68	139,27	604,07	1,52	1214,14
16	74,68	803,96	278,67	1203,77	1,52	2362,60
Parallel execution times in GPU (ms)						
Genes	Selection	Crossover	Mutation	Evaluation	Best	Total
2	1,23	0,25	0,13	0,51	1,84	3,96
4	2,38	0,79	0,19	0,97	1,92	6,25
8	4,82	2,58	0,31	1,91	1,85	11,47
16	9,87	4,29	0,55	4,96	2,28	21,94

*Times are shown by section (kernel) of code and by the number of genes.

Algorithm 2 – Parallel implementation in GPU

1 – Initialize, evaluate and copy to GPU the first population

2 – Repeat until convergence

    2.0 – Copy the two best elements to GPU

    2.1 – Launch kernels

        K1 – Selection [roulette | tournament]

        K2 – Crossover

        K3 – Mutation

        K4 – Evaluation

    2.2 – Copy the fitness vector from GPU to CPU

    2.3 – Compute the two best elements

The Figures 11.5 and 11.6 present the relative parallel execution times in GPU for the Rosenbrock and Griewangk functions respectively, considering the largest population.

Now, as can be seen from these figures, the relative importance of the best values computation (Best) has increased since the execution time is almost the same as before, but the four kernels are much faster than in the sequential implementation. According to Tables 11.1 and 11.2, the time to compute the two best elements in the parallel version is slightly higher than

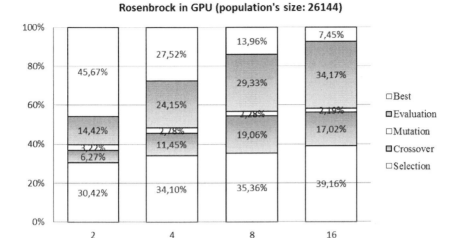

FIGURE 11.5   Execution time percentages for the Rosenbrock function in GPU (population's size = 262,144) varying the number of genes.

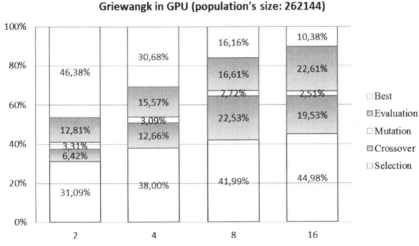

FIGURE 11.6   Execution time percentages for the Griewangk function in GPU (population's size = 262,144) varying the number of genes.

in the sequential version, this happens because the time to copy the fitness vector from GPU to CPU is now included in the "Best" column.

In the parallel version the more time consuming operations are the selection and computing the best values. For large populations, computing the best values is the most time consuming operation when the number of genes is the smallest (2) and selection is the most time consuming operation when the number of genes is bigger. Evaluation, crossover and mutation operations show a huge advantage in being parallelized. From these three operations, evaluation is nearly always the operation that spends the highest percentage of time, followed by crossover and mutation for the last.

When comparing CPU and GPU times for the biggest population size, a speedup (sequential time/parallel time) of around one hundred times for the Griewangk function, and a speedup of around 60 times for the Rosenbrock function can be observed (see Figures 11.7 and 11.8). For a population of

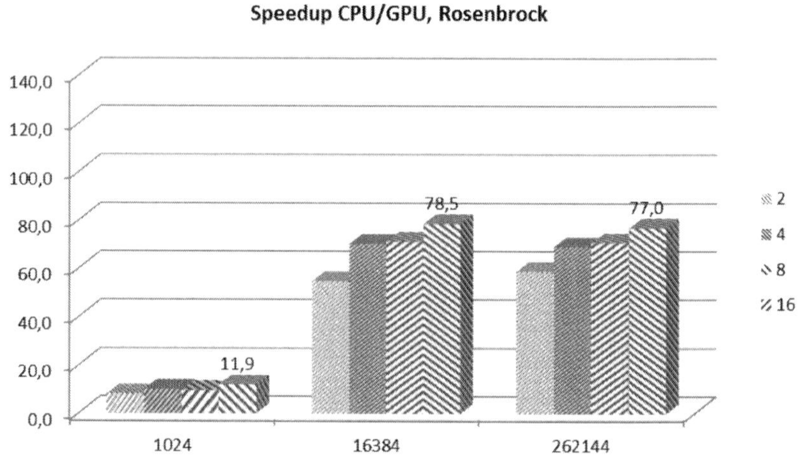

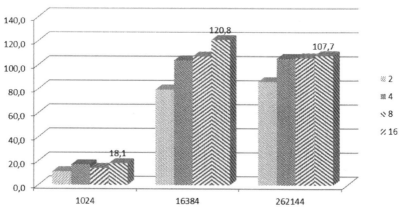

FIGURE 11.7 Speedups (sequential CPU/Parallel GPU) for the Rosenbrock function depending on the population's size and the number of genes.

FIGURE 11.8 Speedups (sequential CPU/Parallel GPU) for the Griewangk function depending on the population's size and the number of genes.

1024 individuals the speedup is less than 18 times for Griewangk function and less than 12 times for the Rosenbrok function. It can also be seen that in most of the cases the speedup grows when the number of genes increases.

11.4.3   USING A TWO-DIMENSIONAL INDEX SPACE

The previous parallel implementation of the kernels considered a one-dimensional index space. This means that, one thread operates over a chromosome. By using multi-dimensional index spaces, instead of having a single work item per individual, it is possible to have one work item for each gene of each individual. This can be applied to the crossover and mutation operators. As for big populations the mutation operator just corresponds to about 2 or 3% of the execution time, the use of a bi-dimensional index space is studied for the crossover operation. Thus, the kernel for crossover is defined using a bi-dimensional index-space, where one dimension is the number of genes and the second dimension is the population's size. The speedups obtained with this version are shown in Figures 11.9 and 11.10.

As can be seen, in the parallel version with a bi-dimensional crossover kernel (called GPU 2D) the speedups are smaller than in the one-dimensional version for 2 and 4 genes, but the speedups are bigger for 8 and 16 genes. That is, the parallelization at the gene level is just worthwhile for large dimensionalities.

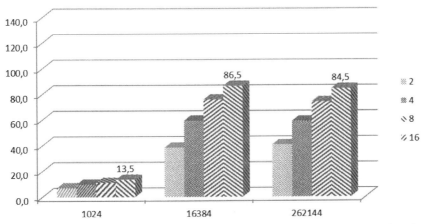

FIGURE 11.9   Speedups (sequential CPU/Parallel GPU 2D) for the Rosenbrock function depending on the population's size and the number of genes.

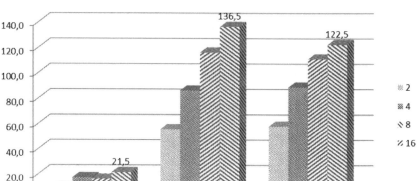

FIGURE 11.10 Speedups (sequential CPU/Parallel GPU 2D) for the Griewangk function depending on the population's size and the number of genes.

11.5   CONCLUSION

In this work a set of evolutionary optimization schemes was set up in an attempt to obtain a better understanding of the putative gains resulting from the adoption of a GPU architecture when tackling optimization problems solved by population-based metaheuristic methods. In our implementation the different granularity levels subsumed in the different phases of a canonical GA were considered. The crossover and mutation operators were implemented as 2D operators allowing the parallel treatment of not only each chromosome but also each gene inside it.

For the cases of heavy workloads the results show that a proper design can result in significant speedup gains. The reported experiments also show that the two-dimensional granularity offered by the GPU architecture is an asset for the implementation of operators presenting population and inter-genotype independence, both functional and data wise.

Although the presented results were obtained with two particular benchmark functions, the conclusions stimulate both the study of other parallelization models, as well as a theoretical analysis on the relationship between the performance gains and the critical parameters population's cardinality and chromosome's length. The goal being to determine *a priori* the expected speedup, if any, for a given unforeseen optimization problem.

Another interesting observation, heighten by the fact that we used a real representation for the chromosomes, is that the parallelization at the gene level is only justifiable for problems requiring a large chromosome length. Nevertheless, this suggests that for a wide range of optimization problems, e.g., feature selection in the analysis of DNA microarrays or biometric facial recognition, the bi-dimensional parallelization is the right design choice.

ACKNOWLEDGMENT

This work was partly supported by Fundação para a Ciência e Tecnologia (FCT) under the project UID/EEA/50008/2013.

KEYWORDS

- **data parallelism**
- **GPGPU**
- **OpenCL**
- **parallel genetic algorithms**

REFERENCES

1. Bäck, T., Fogel, D., & Michalewicz, Z., *Handbook of Evolutionary Computation*, Institute of Physics Publishing Ltd, Bristol and Oxford Univ. Press, New York, 1997.
2. Davis, L. *Handbook of genetic algorithms*. Van Nostrand Reinold, 1991.
3. Deb, K. *Multiobjective Optimization Using Evolutionary Algorithms*, John Wiley & Sons, New York, USA, 2001.
4. de Veronese, L., & Krohling, R., *Differential evolution algorithm on the GPU with C-CUDA*, IEEE Congress on Evolutionary Computation, CEC, 2010, pp. 1–7.
5. Eshelman, L., & Schaffer, J., Real-coded genetic algorithms and interval-schemata, *FOGA, volume 3*, Morgan Kaufmann, San Mateo, CA, 1993, pp. 187–202.
6. Fujimoto, N., & Tsutsui, S., Parallelizing a Genetic Operator for GPUs. IEEE Congress on Evolutionary Computation June 20–23, Cancún, México, 2013, pp. 1271–1277.
7. Geva, A. Hierarchical unsupervised fuzzy clustering. *IEEE Trans. Fuzzy Systems*, 1999, 7(6), 723–733.
8. Goldberg, D., *Genetic Algorithms in search, optimization and machine learning*. Addison-Wesley, 1989.

9. Holland, J., *Adaptation in Natural and Artificial Systems*, University of Michigan Press, Ann Arbor, 1975.

10. Johar, F., Azmin, F., Suaidi, M., Shibghatullah, A., Ahmad, B., Salleh, S., Aziz, M., & Shukor, M., A Review of Genetic Algorithms and Parallel Genetic Algorithms. Graphics Processing Unit (GPU), IEEE International Conference on Control System, Computing and Engineering, 29 Nov. – 1 Dec. Penang, Malaysia, 2013, pp 264–269.

11. Khronos group: OpenCl – The Open Standard Parallel Computing for Heterogeneous Devices, http://www.khronos.org/opencl/ (accessed in March, 2015).

12. Khronos Group: The OpenCL Specification Version: 1.1, Khronos OpenCL Working Group, editor: Aaftab Munshi, 385 pages, 2011.

13. Lorentz, I., Andonie, R., & Malita, M., An Implementation of Evolutionary Computation Operators in OpenCL. Chapter of Intelligent Distributed Computing V, Studies in Computational Intelligence, Springer-Verlag, 2012, Volume 382, pp. 103–113.

14. NVIDIA Corporation: NVIDIA CUDA Programming guide, version 4.0, 2011.

15. Oliveira, J. V., Semantic constraints for membership function optimization, *IEEE Trans. on Systems, Man, and Cybernetics, Part A: Systems and Man,* 1999, 29(1), 128–38.

16. Pospichal, P., Jaros, J., & Schwarz, J., *Parallel Genetic Algorithm on the CUDA Architecture*, Di Chio et al. (Eds.): EvoApplications, Part I, LNCS vol. 6024, Springer, Heidelberg, 2010, pp. 442–451.

17. Prata, P., Fazendeiro, P., Sequeira, P., & Padole, C., *A Comment on Bio-Inspired Optimization via GPU Architecture: The Genetic Algorithm Workload.* B.K. Panigrahi et al. (Eds.): SEMCCO 2012, LNCS 7677, Springer-Verlag Berlin Heidelberg. 2012, pp. 670–678.

18. Shah, R., Narayanan, P., & Kothapalli, K., GPU-Accelerated Genetic Algorithms. Workshop on Parallel Architectures for Bio-inspired Algorithms, 2010.

19. Tsutsui, S., Parallelization of an evolutionary algorithm on a platform with multi-core processors. 9th International Conference on Artificial Evolution (EA'09), Springer-Verlag, Berlin, Heidelberg, 2009, pp. 61–73.

20. Wahib, M., Munawar, A., Munetomo, M., & Akama, K., Optimization of parallel Genetic Algorithms for nVidia GPUs. IEEE Congress on Evolutionary Computation (CEC), New Orleans, LA, 5–8 June 2011, pp. 803–811.

21. Wong, H., Papadopoulou, M., Sadooghi-Alvandi, M., & Moshovos, A., Demystifying GPU Microarchitecture through Microbenchmarking, IEEE International Symposium on Performance Analysis of Systems & Software, 2010, pp. 236–246.

CHAPTER 12

AN EFFICIENT APPROACH FOR POPULATING DEEP WEB REPOSITORIES USING SFLA

SHIKHA MEHTA[1] and HEMA BANATI[2]

[1]*Jaypee Institute of Information Technology, Noida, India*

[2]*Dyal Singh College, University of Delhi, Delhi*

E-mail: [1]mehtshikha@gmail.com; [2]banatihema@hotmail.com

CONTENTS

ABSTRACT

Since its inception, search engine websites like Google, Yahoo!, AltaVista, etc. are serving the information needs of the billions of users. However, these general purpose search engines employ generic crawlers that have limited capabilities to distinguish the web pages as static/surface web page or deep web pages. Hence, the number of deep web urls retrieved (for a particular query) by these search engines is very less to satisfy the users. This derives the need to create separate repositories for both the surface and deep web to serve deep web information requirements of the users. This chapter proposes a distinct meta-crawler approach to locate and populate the deep web urls for populating the contextual deep web repositories. Development of deep-net meta-crawler eliminates the need of performing broad search for crawling deep web urls. The main contributions in the development of deep net meta-crawler (DNM) include: (i) an approach to articulate deep web semantic thematic queries, (ii) development of rule based classifier/filter to recognize the web pages with deep search interfaces, and (iii) evolution of deep web semantic queries using Shuffled Frog Leaping Algorithm (SFLA) to augment crawler performance. Experimental studies were performed over varied contextual domains such as 'computers,' 'vehicles,' 'health,' and 'consumer electronics.' Results established that formulation of deep web semantic terms improved precision of DNM up to 15% as compared to non-semantic deep web queries. Subsequent optimization of deep web semantic terms using SFLA, accelerated precision of DNM to more than 90%. Presented deep web classifier is able to achieve high classification accuracy of up to 98%. Evaluation of SFLA based DNM with respect to Genetic algorithm based DNM further substantiated the efficacy of proposed approach. SFLADNM significantly outperformed GADNM by retaining upto 22% more diverse optimal species. Thus presented SFLA based DNM may serve as good choice for populating the deep web repositories in order to serve the deep web requirements of the users.

12.1   INTRODUCTION

The current size of World Wide Web is approximately 50 billion pages and continues to grow. The unprecedented growth of Internet has resulted in considerable focus on Web crawling/spidering techniques. Crawlers are defined as "software programs that traverse the World Wide Web information space by following hypertext links and retrieve web documents by standard HTTP protocol" [9]. Crawlers are used to populate the local repositories with web pages to be indexed later using various ranking algorithms [10]. Prevalent search engines use generic web crawlers that automatically traverse the web by downloading documents and following links from page to page. Thus, these search engines cover only the surface web or "publicly indexable Web" [22] whereas large portion of the WWW is dynamically generated. This covert side of the WWW is popularly known as the Hidden [13] or Invisible Web [33]. Bergman [6] pointed that the invisible web contains 400 to 550 times more information as compared to the surface web and is often stored in specialized databases [23]. Deep web constitutes 307,000 sites, 450,000 databases, and 1,258,000 interfaces [15]. The volume of the hidden web is estimated to increase more rapidly with the advancements in web technology trends. As more web applications are developed using server-side scripts like java server pages (JSP), active server pages (ASP/ASP.net) and hypertext pre-processor (PHP) etc. and databases like MySql, Oracle etc., rather than just HTML, the amount of hidden web content will grow more. Despite of this importance, very limited studies have been performed to populate the repositories with deep web urls. Ntoulas et al. [29] proposed a strategy for automatic query generation to excavate the high quality gold mine of deep web data. Jianguo et al. [18] presented a sampling technique to select the appropriate queries to be submitted to search interface. The approach mainly focuses on formulation of queries in order to minimize the retrieval of duplicate data. Lu et al. [24] employed linguistic, statistics and HTML features to formulate the queries for crawling the deep web. An approach to automatically fill the forms was proposed Caverlee et al. [7], Xiang et al. [37] and by Banati & Mehta [2]. Alvarez et al. [1] presented an approach to automatically fill and submit only the domain specific forms. Similarly Madhavan et al. [25] also presented techniques for indexing deep websites with multiple search interfaces. All these studies focus on crawling the deep web data from individual deep web websites via automatic query generation and submission of forms. The various techniques discussed above work with the assumption that deep websites are already recognized.

On the contrary, Raghavan and Garcia-Molina [31] introduced 'HiWeb' architecture, a task-specific approach to crawl the hidden web pages. However, the architecture had limited capability to recognize and respond to dependencies between multiple search interfaces and for processing the forms with partial values. Lage et al. [21] presented a technique for collecting hidden web pages for data extraction based on web wrappers. However, this technique was limited to access only the web sites having common navigation patterns. Cope et al. [11] developed a decision tree based technique to automatically discover the web pages with search interfaces. The algorithm incorporated long rules and large number of features in training samples, which may lead to over-fitting. Fontes and Silva [14] developed SmartCrawl strategy to automatically generate the queries to crawl the deep web pages. However, Smartcrawl had limited capabilities to analyze the web pages, thus it also indexed the web pages with errors/no results. Barbosa & Freire [5] presented Form Focused Crawler (FFC) to crawl the web pages with search forms. Nevertheless, FFC requires substantial manual tuning including identification of features. Besides, the set of forms retrieved by FFC are highly heterogeneous and it could be time consuming to train the link classifier. Umara et al. [35] presented an ontology-based approach for automatic classification of deep web databases, which can be used for building focused information retrieval systems. Tim et al. [34] presented OPAL—a comprehensive approach to understand the forms via form labeling and form interpretation. Heidi et al. [16] formulated eight rules to classify the web query interfaces and evaluated them using various machine-learning techniques. It was observed that J48 algorithm performed the best with 99% accuracy.

The various studies discussed above focus either on optimization of queries to fetch data or content from individual deep web sites or to recognize the web pages with deep web search interfaces. This chapter presents a comprehensive approach to develop the contextual deep web repositories along with developing the simple yet efficient approach to recognize the deep web query interfaces. The approach involves formulation of semantic queries to crawl the deep web urls via search portals. These deep web semantic queries are submitted to the search engines for crawling the deep web urls. The work also contributes a simple and effective rule based classifier to recognize the deep web search interfaces. Subsequently deep web semantic queries are evolved using Shuffled Frog Leaping Algorithm (SFLA) to augment the retrieval of deep web urls. The rest of the chapter is organized as follows. Section 12.2 presents the methodology for formulation of semantic contextual queries and their evolution using SFLA. This is followed by experiments and discussion of results in Section 12.3. Section 12.4 outlines the future prospects and conclusion.

12.2   SHUFFLED FROG LEAPING ALGORITHM BASED APPROACH TO POPULATE CONTEXTUAL DEEP WEB REPOSITORIES

The proposed approach to populate the contextual deep web repositories involves development of deep web meta-crawler to crawl the deep web urls. The working of Deep Net Meta-Crawler (DNM) involves automatic formulation of semantic deep web queries by associating domain specific terms with keywords, which semantically refer to deep web repositories/ databases etc. These semantic deep web queries are submitted to state-of-the-art search engines to retrieve the deep web urls. Subsequently, semantic deep web terms are evolved using SFLA to boost the retrieval of deep web urls from prevalent search engines.

The detailed approach to populate the deep web repositories is depicted in Figure 12.1. It begins with the formulation of deep web semantic contextual/thematic terms. These terms are submitted to the search engine to fetch the context relevant deep web urls from the WWW. From the retrieved results, deep web urls are identified using the proposed deep web classifier/filter. The filtered deep web urls are subsequently stored in the repositories.

Thereafter, fitness of the formulated semantic terms is computed based on their capability to retrieve the deep web urls. Subsequently, the thematic terms are evolved using SFLA and the process continues till the convergence criterion is satisfied. Thus, the efficacy of presented deep web meta-crawler is determined mainly by the techniques used to develop the following approaches:

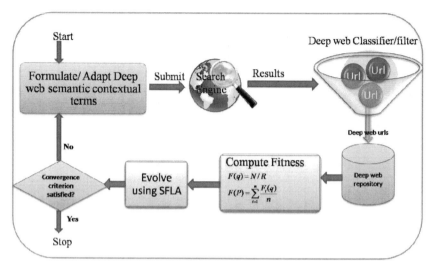

FIGURE 12.1   Deep Web Meta-crawler Approach.

- Deep web search interface pages classification/filtering;
- Formulation of deep web semantic contextual/thematic terms;
- Evolution of deep web semantic contextual terms using SFLA.

12.2.1 DEEP WEB SEARCH INTERFACE CLASSIFICATION/ FILTERING

Deep web classifier is a tool to filter the deep web urls from the other urls. To develop the filter, a comprehensive study of approx 150 different dynamic web applications was performed. The study revealed that all dynamic web applications have a common feature; presence of 'form' tag in the HTML source code. Further analysis exposed that there are two types of forms— search forms and other/non search forms. Search forms precisely refer to those forms in web pages which retrieve website content from the databases on submitting a query. They may be simple search forms with single textbox or complex search forms having multiple drop-down boxes with different selection criterion as shown in Figures 12.2 and 12.3, respectively.

As highlighted in the figures, all the search forms contain terms like "search" one or more times either in the entire <form> tag or in its surrounding code. The keyword "search" is also present in form id, value of submit button, label, action url or in the form of comments. Besides, there are other forms as shown in Figure 12.4, such as the custom search form, login form, registration form and subscription form etc. The login form

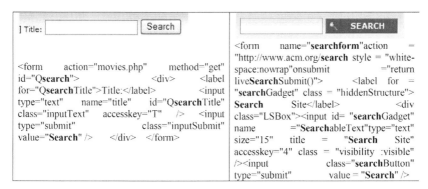

FIGURE 12.2   Single Text Box based Search Forms with varied ways to use 'search' keyword in web pages.

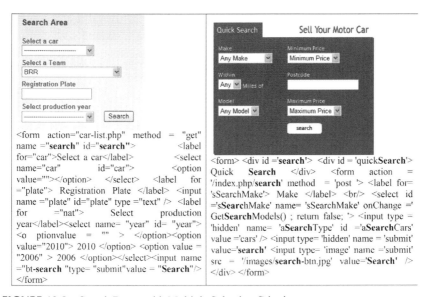

FIGURE 12.3   Search Boxes with Multiple Selection Criteria.

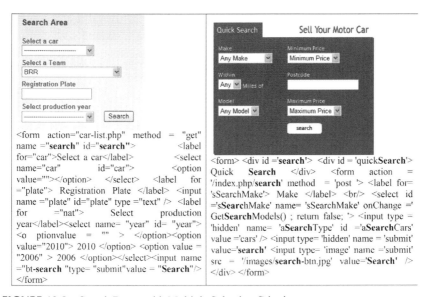

FIGURE 12.4   'Custom' search and 'Login' Forms.

contains keywords like "login," "sign in," "password" etc. Many websites also provide search forms to search the web pages from the world-wide web e.g. various websites include "Custom Search Engine" or "site search" in their pages in order to facilitate the users while browsing their website. These search forms are identified through strings such as "id='cse-search-box'," "Search this site," "Google search," "Google custom search," "Google CSE Search Box," "Search Google," "Search Bing," "cse-search," "20 search," "alltheweb," "altavista," "aolsearch," "askjeeves," "ebay," "excite," "iwon," "jocant," "lycos," "msn," "netscape," "dmoz," "webcrawler," "yahoo," "bing," etc. Since such search forms are not deep web forms, they are categorized as web pages with conventional search engines. The presented deep web classifier uses all these features to filter and extract 'hidden' web page urls from the other web pages with textboxes as shown in Figure 12.5. Since hidden web belongs to dynamic part of the web, hidden web classifier begins with input urls and primarily filters the static web pages from the dynamic web pages. In the second stage, dynamic web pages are classified and filtered into hidden web pages, web pages with login, registration forms, conventional search engine etc. Thus, based on this information four rules are formulated, which are used by the proposed hidden web classifier are discussed in the next paragraph.

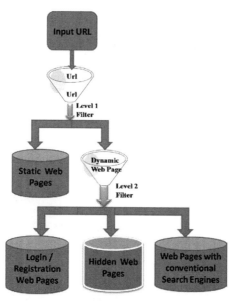

FIGURE 12.5   Multi-level Filtering by Hidden Web Classifier.

Proposed Set of Rules to be Employed by the Hidden Web Classifier:

Rule 1: *Check the presence of the <form> tag in the HTML document. If Exists, classify the document as a dynamic web page else classify it as a static web page.*

Apply rule 2 to 4 for all dynamic web pages, which contain more than one form

For a dynamic web page:

Rule 2: *Check if form id in search form contains any values of V where V={"cse-search-box'," "Search this site," "Google search," "Google custom search," "Google CSE Search Box," "Search Google," "Search Bing," "cse-search," "20 search," "alltheweb," "altavista," "aolsearch," "askjeeves," "ebay," "excite," "iwon," "jocant," "lycos," "msn," "netscape," "dmoz," "webcrawler," "yahoo," "bing" etc.}, then classify it as a form with conventional search engine interface; else classify it as a deep web interface.*

Rule 3: *If web page is dynamic, check HTML code present within the opening <form> and closing </form> tags contains the keyword "search" in any of values (U) where U= {"form id,"" value of submit button," "label" and" action url"}. If exists classify it as a deep web search form.*

Rule 4: *Check if the HTML code section within the opening <form> and closing </form> tags contains the keyword "login," "sign in," "password" in any of values (W) where W= {"form id,"" value of submit button," "label" and" action url"} then classify it as a registration or login form.*

12.2.2   FORMULATION OF DEEP WEB SEMANTIC CONTEXTUAL/THEMATIC TERMS

Schlein [27] pointed that the usage of additional terms such as database; repository or archive, etc., along with the query keywords may boost the retrieval of deep web resources from the existing search engines. For example to get the list of science databases, use the search term "science + database or repository or archive." However, authors did not validate their hypothesis. Thus, to corroborate this hypothesis, a query "science database" was submitted to Google, Yahoo and Bing search engines (Figure 12.6). An analysis of the results obtained in Figure 12.6 revealed that majority of the urls obtained by these major search engines belonged to the deep web repositories.

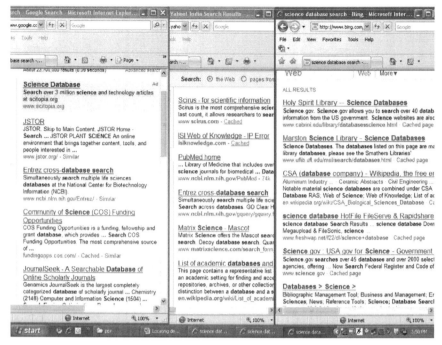

FIGURE 12.6　Results of Semantic Query "Science Database."

Thus, the usages of keywords like 'archive,' 'repository,' 'collection,' 'databases,' 'portal' and 'articles,' etc., which semantically refer to the collection/repositories/databases, etc., facilitate in filtering out the unnecessary web urls from the vast collection of web. To further establish this observation, an experiment was carried out to evaluate the effectiveness of semantic thematic terms in extracting deep web urls. Experiment involved random generation of group of ten thematic queries from two different domains – consumer electronics and computers. These thematic queries were associated with varied deep web semantic terms. Subsequently, for each semantic query submitted to the Google search engine, top 10 results were extracted and used to evaluate the strength of both the semantic and non-semantic/normal queries in extracting deep web urls. Efficiency of these semantic thematic terms was measured using the commonly used decision support accuracy metric; precision, which is defined as the total number of deep web urls retrieved out of the top 10 urls extracted on submitting the query.

The confusion matrix used to compute the precision for deep web classifier is as follows:

TABLE 12.1   Confusion Matrix

True		Actual	
		True	
Predicted	Positive	True Positive (TP)	False Positive (FP)
	Negative	False Negative (FN)	True Negative (TN)

$$Precision = \frac{TP}{(TP + FP)} \tag{1}$$

Accordingly, precision (Eq. 1) is defined as the positive predictive value or the total number of deep web pages, which were actually deep web urls and are also classified as deep web urls by the classifier out of the total number true positives and false positives. Figure 12.7 and 12.8 demonstrate the results obtained from 20 independent runs for both non-semantic query and deep web semantic queries formulated by associating the various deep web semantic terms. It can be observed from the results that precision range of non-semantic query terms varies from 22% to 30% whereas precision of deep web semantic terms varies in the range of 40–45%. These results also demonstrate the limited capabilities of prevalent search engines to satisfy the deep web information needs of the users. Thus, the approach to associate deep web semantic terms with thematic terms although fostered precision by approximately 15%; the maximum average precision of approx 45% is not adequate for developing the deep web repositories. To further enhance the precision, deep web semantic thematic terms are subsequently optimized using shuffled frog leaping algorithm as discussed in the next section.

12.2.3 EVOLUTION OF DEEP WEB SEMANTIC CONTEXTUAL TERMS USING SHUFFLED FROG LEAPING ALGORITHM (SFLA)

The low values of precision attained in Section 12.2.2 indicate that the vocabulary used to generate the thematic terms determines the quality of the resources collected. Prevalent search engines exploit only a fixed number of initial query terms to retrieve the results and ignore subsequent terms. Thus, articulation of appropriate thematic terms to retrieve the deep web urls forms a big challenge and is observed as an optimization problem. Yang & Korfhage [38] investigated the utility of genetic algorithm (GA)

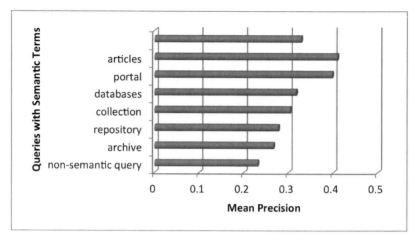

FIGURE 12.7   Precision of Deep Web Semantic vs. Non-Semantic Terms in 'Consumer Electronics' domain.

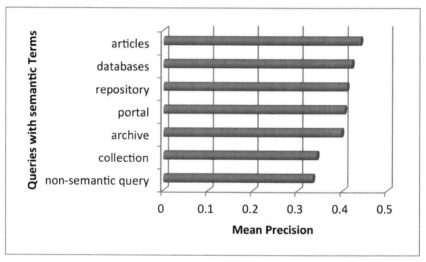

FIGURE 12.8   Precision of Deep Web Semantic vs. Non-Semantic Terms in 'Computer' domain.

for query modification in order to improve the performance of information retrieval systems. Owais [30], Cecchini et al. [8] and Hogenboom et al. [17] employed genetic algorithm for optimization of Boolean queries, topical queries and RDF-chain queries, respectively. Nasiraghdam et al. [28] employed hybrid genetic algorithm for query optimization in distributed databases. Banati and Mehta [4] employed SFLA to evolve the queries for populating contextual web repositories. It was established that SFLA performs better than GA in evolving contextual queries. Banati & Mehta [4] evaluated the performance of genetic algorithm, memetic algorithm, particle swarm optimization and shuffled frog leaping algorithm over benchmark test functions for global optimization. These studies confirmed that SFLA has better chance of attaining global optimal results. It was also established that context aware filtering using shuffled frog leaping algorithm depicts better results as compared to GA [4]. Hence, in the presented work, shuffled frog leaping algorithm is used for evolving the contextual queries to optimize the performance of deep net meta-crawler for populating the contextual deep web repositories.

SFLA is a memetic meta-heuristic for combinatorial optimization developed by Eusuff and Lansey [12]. It is a random search algorithm inspired by natural memetics. A simple and high performance algorithm, SFLA integrates the benefits of both the Memetic Algorithm (MA) [27] and the Particle Swarm Optimization (PSO) [20] algorithm. It is based on the evolution of memes carried by interactive individuals and global exchange of information within the population. Population in SFLA consists of a set of frogs divided into diverse clusters known as memeplexes. Each frog in the memeplex represents a viable solution to an optimization problem. Within each memeplex, every frog holds beliefs that can be influenced by the ideas of other frogs and cultivates through a process of memetic evolution known as local search. After a number of memetic evolutionary steps, all the memeplexes are shuffled together leading to global evolution. The local search and the shuffling processes continue until the defined convergence criteria are satisfied. The precise steps of SFLA are enumerated as the pseudo-code in Figure 12.9. In SFLA, an initial population P of frogs (solutions) F_k is generated randomly. After computing the fitness of all initial solutions, whole population is sorted in the descending order of their fitness and global best solution is determined. Subsequently all solutions (F_k) are distributed into 'm' memeplexes M^1, M^2, $M^3 \ldots M^m$ as follows:

$$M^d = \{F_k^d \mid F_k^d = F_{d-m(k-1)}\} \qquad (2)$$

In Eq. (2), $k = 1, 2 \ldots n$ represent the solution number in the population and $d=1, 2, 3 \ldots m$ refers to the memeplex number of solution k.

In Figure 12.9, D_{max} and D_{min} are respectively the maximum and minimum allowed changes in a frog's position and *rand()* function generates a random number between 0 and 1. Within each memeplex, the fitness of

Begin;

Generate random population of P solutions;

For each individual i in P : calculate fitness f(i);

Sort the population P in descending order of their fitness;

Determine the fitness of global best solution as X^{gb}

Divide P into m memeplexes;

         For each memeplex m

Determine the fitness of local best solution as X^{lb}

Determine the fitness of local worst solution as X^{w}

         For each iteration k

       For each dimension d in individual i

     Improve the position of worst solution using Eq.3 with respect to

         local best solution

New Position $X_d^w = X_d^w + D_d$;

         Compute fitness of Xw;

         If fitness improves

New Position $X_d^w = X_d^w + D_d$;

           Else

     Improve the position of worst solution using Eq.4 with respect to

         global best solution

New Position $X_d^w = X_d^w + D_d$;

         Compute fitness of Xw;

         If fitness improves

New Position $X_d^w = X_d^w + D_d$;

           Else

New Position $X^w = rand() * D_{max} + rand() * (D_{min})$;

         End;

End;

End;

Combine the evolved memeplexes;

Sort the population P in descending order of their fitness;

         Check if termination is true;

End;

FIGURE 12.9   Pseudo-code for Shuffled Frog Leaping Algorithm.

worst solution is improved by adjusting the fitness landscape according the local and global best solutions as per the Eqs. (3) and (4), respectively.

$$Change\ in\ frog\ position(D_d) = rand(\ )*(X_d^{lb} - X_d^{w}) \tag{3}$$

$$(D)_{max} \geq D_d \geq D_{min}）$$

$$Change\ in\ frog\ position(D_d) = rand(\ )*(X_d^{gb} - X_d^{w}) \tag{4}$$

$$(D)_{max} \geq D_d \geq D_{min}）$$

Accordingly, if the fitness of solution improves, it replaces the worst solution else a new solution is generated at random to replace the least fitted solution. The SFLA parameters adapted to evolve the quality of deep web semantic contextual terms are as follows:

12.2.3.1 Initialization

The initial population *(P)* of solutions consists of semantic domain specific set of terms that can be submitted to a search engine. Each individual in the population refers to the deep web semantic query. This constitutes a list of terms extracted from the thematic context associated with the term obtained from the deep web semantic term set. The number of terms in each of the initial queries is random with fixed upper and lower bound on the query size.

12.2.3.2 Objective Function

For every individual *i P,* fitness is computed to evaluate its quality. Quality of the deep web semantic query is computed as its ability to retrieve deep web urls when submitted to a search engine. The relevance of these urls is being assessed manually to verify the results. Thus the precision/fitness *F(q)* of an individual query *q* is defined as number of deep web urls *(N)* retrieved out of the total number of resultant urls *(R)* as shown in equation 5 and fitness of the whole population of size *n* is computed as mean fitness of all the queries as given equation 6

$$Precision\ (P(q) = \frac{N}{R} \tag{5}$$

$$Average\ Precision(P) = \sum_{i=1}^{n} \frac{P_i(q)}{n} \tag{6}$$

12.2.3.3   Convergence Criteria

The convergence criterion was fixed to number of generations to stop the evolution of deep web semantic terms using shuffled frog leaping algorithm.

12.3   EVALUATION OF SFLA BASED DEEP NET META-CRAWLER (SFLADNM)

To demonstrate the effectiveness of presented SFLA based deep web contextual meta-crawler, experiments were performed on Core2 Duo 1.67 GHz processor and 3-GB RAM computer. The algorithm was implemented in Java using Jdk1.6 toolkit. Meta-crawler takes an incremental approach to evolve high quality terms for retrieving context relevant web resources. It starts by generating an initial population of queries using terms extracted from a thematic context. Among the various prevailing search engines, Google search engine has the biggest repository of web urls. Thus, thematic terms were submitted to the Google search engine for retrieving the context relevant web urls. Due to the practical limitations, only the snippets (or webpage summary) of the top ten results returned by the search engine were used to evaluate the fitness of the queries. The queries were then evolved incrementally based on their ability to retrieve the context relevant web urls. For experiments, thematic contexts were created from Concept hierarchy of Open Directory Project popularly known as DMOZ (www.DMOZ.org).

The strength of SFLA based deep web meta-crawler was evaluated with respect to GA as it is the most common algorithm being used for query optimization. For SFLA the whole population was divided into 5 memeplexes of equal size such that Population = m × n (m is number of memeplexes = 5 and n is number of individual solutions in each memeplexes = 20). Each experiment was performed for 20 independent runs for the population of 100 queries. The strength of the presented SFLA based contextual deep web meta-crawlers was evaluated using Eq. (6). The studies were performed to:

- Compute the Classification Accuracy of Hidden Web Classifier employed by Deep-Net Meta-Crawler
- Assess the Effect of Evolving Deep Net Meta-Crawler with Semantic and Non-semantic Queries using SFLA
- Compare the performance of SFLADNM with GA based Deep Net Meta-Crawler (GADNM)

• Evaluate the effectiveness of SFLADNM in maintaining Diversity of Optimal Species as compared to GADNM

12.3.1   COMPUTE THE CLASSIFICATION ACCURACY OF HIDDEN WEB CLASSIFIER

This study validates the classification accuracy of rule based hidden web classifier employed in deep net meta-crawler. Experiment involved random generation of 10 queries for the domains – 'Vehicles,' "Book," "Science," and 'Health.' Subsequently, for each query submitted to the Google search engines top 10 results were extracted and used to evaluate the strength of deep web classifier. The results retrieved by the deep net meta-crawler on submitting these queries were tested manually to compute their actual classification accuracy. The classification accuracy was computed using the commonly used decision support accuracy metrics; recall, precision and f1 measure. The confusion matrix used to compute the precision, recall and f1 measure for deep web classifier is shown in Table 12.1.

$$Recall = \frac{TP}{(TP + FN)} \qquad (7)$$

$$F1 = \frac{2 * Precision * Recall}{(Precision + Recall)} \qquad (8)$$

Equations (1), (7), and (8) depict the formulas for precision, recall and F1 measure respectively. Accordingly, precision is defined as the positive predictive value or the total number of deep web pages, which were actually deep web urls and are also classified as deep web urls by the classifier out of the total number true positives and false positives. Similarly recall is defined as the true positive rate of the classifier. F1 measure is the harmonic mean of precision and recall.

Figure 12.10 displays the precision, recall and f1 measure for all the domains. Results portray that for all the presented domains, classifier is able to achieve high classification accuracy with precision up to 98%, recall 97% and f1 measure up to 97%. Table 12.2 depicts the results of various conventional approaches used for developing the classifier. Assuming that authors have given the best results of their work, it can be observed that

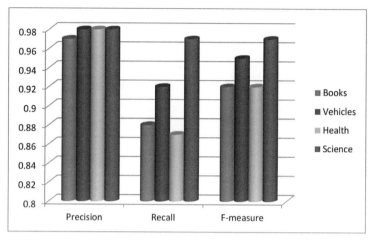

FIGURE 12.10   Performance of Deep Web Classifier various domains.

TABLE 12.2   Different Approaches for Hidden Web Classifier

Reference	Technique	Accuracy (%)
Cope et al. [11]	Decision tree algorithm with more than 500 features generated from HTML tags	85
Barbosa and Freire [5]	Decision tree algorithm with 14 features generated from HTML tags	90
Kabisch et al. [19]	Based on nine domain independent general rules	87.5
Heidi et al. [16]	Based on Form elements and Eight rules for classification	99
Proposed	Form elements and four rules for classification	98

classification accuracy of proposed approach is better than the first three approaches. Although accuracy of presented approach is slightly less than Heidi et al. [16]; this slight difference is negligible due to the fact that our approach employs only four rules and Heidi et al. [16] classifier is based on eight rules which increase the time complexity of the classifier. Moreover the rules formulated by Heidi et al. [16] ignore the case where forms with conventional search engines are embedded within the HTML documents. Such forms search WWW rather than deep web and should be considered non-searchable forms. These results establish the effectiveness of presented rule based classifier for filtering deep web urls. Thus the approach presents a simplistic methodology to obtain promising results.

12.3.2 ASSESS THE EFFECT OF EVOLVING DEEP NET META-CRAWLER WITH SEMANTIC AND NON-SEMANTIC QUERIES USING SFLA

This experiment assesses the effect of optimizing the performance of deep web meta-crawler with semantic thematic queries and non-semantic thematic queries. Semantic queries were formulated by associating the randomly generated domain specific keywords with the terms, which semantically refer to deep web repositories/databases. Figures 12.11–12.14 illustrate the Mean

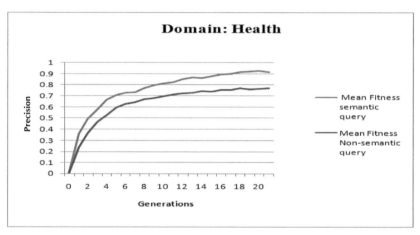

FIGURE 12.11   Semantic vs. Non-semantic Query Optimization using SFLA over 'Health' domain.

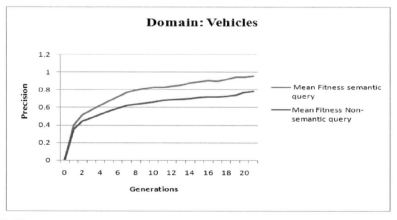

FIGURE 12.12   Semantic vs. Non-semantic Query Optimization using SFLA over 'Vehicles' domain.

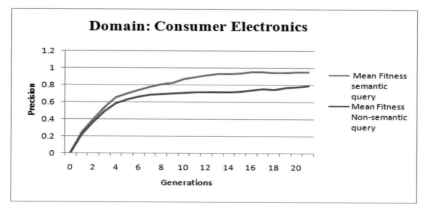

FIGURE 12.13 Semantic vs. Non-semantic Query Optimization using SFLA over 'Consumer Electronics' domain.

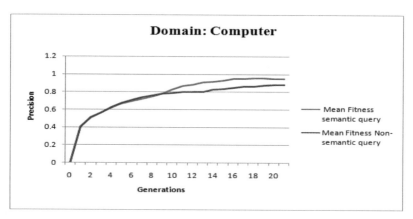

FIGURE 12.14 Semantic vs. Non-semantic Query Optimization using SFLA over 'Computer' domain.

Fitness/Mean Precision (as given in Eq. 5) of population after every generation for both the deep web semantic query and the non-semantic query for the domains – computer, health, vehicles and consumer electronics.

All these results precisely exhibit that over the generations, using SFLA, deep web semantic meta-crawler evolve faster and more to attain global maxima as compared to non-semantic meta-crawler. Tables 12.3 and 12.4 depict the percentage improvement in precision obtained due to evolution of meta-crawler using semantic terms as compared to non-semantic terms.

Bold values in Table 12.3 illustrate that using deep web semantic terms, SFLA based meta-crawler converges faster and achieves gain in precision

TABLE 12.3 Enhancement in Precision of SFLA-Based Semantic Deep Net Meta-Crawler

Domains	Precision of Deep Net Meta-Crawler		
	Non-Semantic SFLA	Semantic SFLA	Gain (%)
Consumer Electronics	0.79	0.95	16
Health	0.77	0.92	15
Vehicles	0.78	0.95	17
Compute	0.88	0.95	7

TABLE 12.4 Enhancement in Precision of SFLA-Based Semantic Deep Net Meta-Crawler

Domains	Precision of Deep Net Meta-Crawler		
	Without SFLA	With SFLA	Gain (%)
Consumer Electronics	0.25	0.95	70
Health	0.36	0.92	56
Vehicles	0.4	0.95	55
Computer	0.4	0.95	55

up to 16%. Table 12.4 establishes that optimization of meta-crawler using SFLA brings significant improvement in the precision of deep web meta-crawler. Shuffled frog leaping algorithm enhances the performance of deep net meta-crawler by up to 70% in consumer electronics domain and more than 50% in all the other three contextual domains. Thus the presented SFLA based semantic deep web metacrawler enhances the precision more than 90% as compared to FFC [5], whose average precision is only 16% for the given set of domains. FFC employs broad search technique to crawl the deep web. Barbosa and Freire [5], developed Adaptive Crawler for Hidden Web Entries (ACHE) to address the limitations of their previous work. Authors work did not provide information about the exact precision attained. An analysis of their performance graphs (in published work) depicts that although the performance is improving over time, ACHE is able to retrieve on an average 1100 forms after crawling 100,000 web pages from the various domains. These results substantiate the strength of presented approaches to develop the deep web meta-crawler.

12.3.3   COMPARISON OF SFLA-BASED DEEP NET META-CRAWLER (SFLADNM) WITH GENETIC ALGORITHM (GA)-BASED DEEP NET META-CRAWLER (GADNM)

This study compares the convergence speed of SFLA with respect to GA for deep net meta-crawler. GA was implemented on same lines as SFLA. It was initialized with the population of 100 queries. For GA, crossover probability was set to 0.7 and mutation probability of 0.03. The experiment involves optimization of deep web semantic thematic queries using SFLA and GA for deep net meta-crawler. Results as depicted in Figures 12.15–12.18 establish that SFLA based meta-crawler performs significantly better than GA in evolving deep web semantic queries for all the four domains – Computer, Vehicle, Health and Consumer Electronics. SFLA based DNM depicts steady improvement whereas GA based DNM is fluctuating and stagnates early. Thus, SFLA based DNM has better chance to attain global optima.

12.3.4   EVALUATE THE EFFECTIVENESS OF SFLADNM VS. GADNM IN MAINTAINING DIVERSITY OF OPTIMAL SPECIES

It is very important for the meta-crawler to populate the repositories with diverse web urls in order to serve the varied information needs of the users.

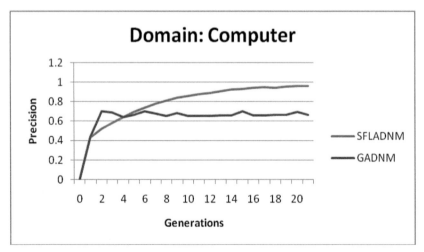

FIGURE 12.15   Comparison of Convergence Speed for SFLADNM vs. GADNM over 'Computer' domain.

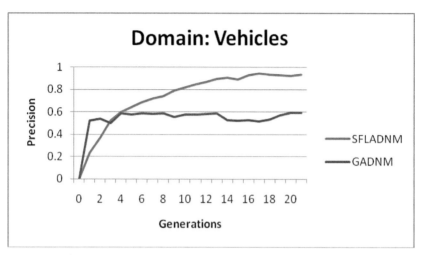

FIGURE 12.16   Comparison of Convergence Speed for SFLADNM vs. GADNM over 'Vehicles' domain.

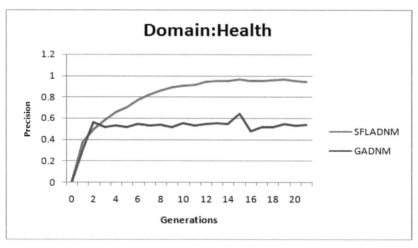

FIGURE 12.17   Comparison of Convergence Speed for SFLADNM vs. GADNM over 'Health' domain.

Thus, the efficiency of SFLADNM and GADNM to retain varied optimal species is evaluated for all the four domains on the basis of:

- Percentage of Diverse Species/Queries Retained in Optimized Population;
- Percentage of Diverse Deep Web Urls Retrieved by the Optimal Species/Queries.

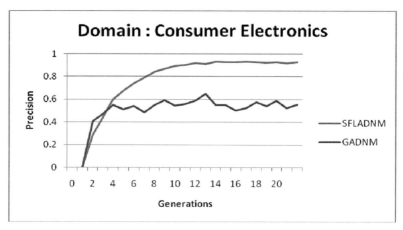

FIGURE 12.18 Comparison of Convergence Speed for SFLADNM vs. GADNM over 'Consumer Electronics' domain.

For SFLADNM results depicted in Table 12.5 were obtained after 20 generations when mean fitness of the population reached to more than 90% as illustrated in previous experiment. Since for GA, convergence was quite slow for DNM, only the solutions with 90% or more fitness are considered. As shown in Table 12.6, presented SFLADNM is able to enhance retention of varied optimal queries up to 27 percent as compared to GADNM. Results also establish that SFLADNM is able to generate up to 22% of more unique urls as compared GADNM. These results substantiate that SFLADNM may serve as better choice for populating the contextual deep web repositories.

12.4 CONCLUSION

The chapter presented SFLA-based deep net meta-crawler to populate the deep web repositories. The approach involved formulation of deep web semantic thematic terms and their submission to a search engine. From the retrieved results, web pages with deep web interfaces are filtered

TABLE 12.5   Percentage of Diverse Queries and Urls Generated by SFLADNM and GADNM

	Diverse Queries (%)		Diverse Urls (%)	
Domain	**SFLADNM**	**GADNM**	**SFLADNM**	**GADNM**
Consumer Electronics	53	26	45	23
Vehicles	32	18	34	18
Computer	43	35	41	27
Health	36	28	27	12

TABLE 12.6   Percentage Enhancement in Diversity of SFLADNM vs. GADNM

	Enhancement of Diverse Queries (%)	**Enhancement of Diverse urls (%)**
Domain	**SFLADNM vs. GADNM**	**SFLADNM vs. GADNM**
Consumer Electronics	27	22
Vehicles	14	16
Computer	8	14
Health	8	15

using proposed rule based classifier. Subsequently, based on their ability to retrieve the deep web urls, deep web semantic terms are evolved using SFLA. Experimental evaluation demonstrated a significant increase in precision of meta-crawler (from 45% to 90%) through formulation of semantic deep web queries and their subsequent optimization using SFLA. SFLA based meta-crawler also performed better than Genetic algorithm based approach. Hence, the deep web repositories populated using proposed deep net meta-crawler may provide better search experience to the user.

KEYWORDS

- **deep web classifier**
- **deep web meta-crawler**
- **deep web semantic query optimization**
- **SFLA based query optimization**
- **shuffled frog leaping algorithm**

REFERENCES

1. Alvarez, M., Raposo, J., Pan, A., Cacheda, P., Bellas, F., & Carneiro V.: Crawling the Content Hidden Behind Web Forms, Proceedings ICCSA, LNCS 4706(II),. 2007, 322–333.
2. Banati, H., & Mehta, S.: VibrantSearch: A BDI Model of Contextual Search Engine, Next Generation Web Services Practices (NWeSP), 2010, 623–629.
3. Banati, H., & Mehta, S.: SEVO: Bio-inspired Analytical Tool for Uni-modal and Multimodal Optimization, Proceedings of the International Conference on Soft Computing for Problem Solving (SocProS 2011), Advances in Intelligent and Soft Computing, Springer 2011, 130, 557–566.
4. Banati, H., & Mehta, S.: Evolution of Contextual queries using Shuffled Frog Leaping Algorithm, 2010 International Conference on Advances in Communication, Network, and Computing, IEEE, 2010b, 360–365.
5. Barbosa, L., Freire, J.: An Adaptive Crawler for Locating Hidden-Web Entry Points, International Proceedings of the 16th international conference on World Wide Web,. 2007, 441–450.
6. Bergman, M. K.: The Deep Web: Surfacing Hidden Value. The Journal of Electronic Publishing, 2001, 7(1), 4.
7. Caverlee, J., Liu, L., & David, B.: Probe, Cluster and discover: focused extraction of qa-pagelets from the deep web. Proceeding of the 20th International Conference of Data Engineering, 2004, 103–114.
8. Cecchini, R. L., Lorenzetti, C. M., Maguitman, A. G., & Brignole, N.B.: Using genetic algorithms to evolve a population of topical queries, Information processing and Management, 2008, 44, 1863–1878.
9. Cheong, F. C.: Internet Agents: Spiders, Wanderers, Brokers, and Bots. New Riders Publishing, Indianapolis, Indiana, USA (1996).
10. Cho, J., & Garcia-Molina, H.: The Evolution of the Web and Implications for an Incremental Crawler, In Proceedings of the 26th International Conference on Very Large Databases, 2000, 200–209.
11. Cope, J., Craswell, N., & Hawking, D.: Automated discovery of search interfaces on the web. Fourteenth Australasian Database Conference, 2003, 181–189.

12. Eusuff, M. M., & Lansey, K. E.: Optimization of water distribution network design using the shuffled frog leaping algorithm. Journal of Water Resources Planning and Management, 2003, 129(3), 210–225.

13. Florescu, D., Levy, A., & Mendelzon, A.: Database techniques for the World-Wide Web: a survey. ACM SIGMOD Record, 1998, 27(3), 59–74.

14. Fontes, A. C., & Silva, F. S.: SmartCrawl: A New Strategy for the Exploration of the Hidden Web, Proceedings of the 6th annual ACM international workshop on Web information and data management, 2004, 9–15.

15. He, B., Patel, M., Zhang, Z., & Chang, K. C. C.: Accessing the deep web: Attempting to locate and quantify material on the Web that is hidden from typical search engines, Communications of the ACM, 50(5),. 2007, 95–101.

16. Heidy M. Marin-Castro, Victor J. Sosa-Sosa, Jose, F. Martinez-Trinidad, & Ivon Lopez-Arevalo: Automatic discovery of web Query Interfaces using Machine Learning techniques. Journal of Intelligent Information Systems, 2013, 40, 85–108

17. Hogenboom, A., Milea, V., Frasincar, F., Kaymak, U., RCQ-GA: RDF Chain query optimization using genetic algorithms, Proceedings of the Tenth International Conference on E-Commerce and Web Technologies (EC-Web), 2009, 181–192.

18. Jianguo, L., Yan, W., Jie, L., Jessica, C.: An Approach to Deep Web Crawling by Sampling 2008 IEEE/WIC/ACM International Conference on Web Intelligence and Intelligent Agent Technology, 2008, 718–724.

19. Kabisch, T., Dragut, E. C., Yu, C. T., & Leser U.: A hierarchical approach to model web query interfaces for web source integration. Proceedings, Very Large Databases, 2009, 2(1), 325–336.

20. Kennedy, J., & Eberhart, R. C. Particle Swarm Optimization. In: Proceedings of the IEEE International Joint Conference on Neural Networks, 1995, 1942–1948.

21. Lage, J. P., Da, Silva, A. S., Golgher, P. B., & Laender, A. H. F.: Collecting hidden web pages for data extraction, Proceedings of the 4th international workshop on Web Information and Data Management, 2002, 69–75.

22. Lawrence, S., & Giles, C. L.: Accessibility of information on the web. Nature 400 (6740): 107. DOI: 10.1038/21987. PMID 10428673 (1999).

23. Lin, K., & Chen, H.: Automatic Information Discovery from the "Invisible Web." Proceedings of the International Conference on Information Technology: Coding and Computing, IEEE, 2002, 332–337.

24. Lu J., Zhaohui W., Qinghua Z., & Jun L.: Learning Deep Web Crawling with Diverse Features, 2009 IEEE/WIC/ACM International Conference on Web Intelligence and Intelligent Agent Technology – Workshops, 2009, 572–575.

25. Madhavan, J., Ko, D., Kot, L., Ganapathy, V., Rasmussen, A., & Halevy, A.: Google's Deep-Web Crawl, Proceedings of VLDB Endowment, 2008, 1(2), 1241–1252.

26. Mehta, S., & Banati, H., "Context aware Filtering Using Social Behavior of Frogs," Swarm and Evolutionary Computation, Volume 17, August 2014, Pages 25–36, Elsevier.

27. Moscato, P., & Cotta, C.: A Gentle Introduction to Memetic Algorithms. In: Handbook of Meta- heuristics, Kluwer, Dordrecht, 1999, 1–56.

28. Nasiraghdam, M., Lotfi, S., & Rashidy, R.: Query optimization in distributed database using hybrid evolutionary algorithm, International Conference on Information Retrieval & Knowledge Management, (CAMP), 2010, 125–130.

29. Ntoulas, A., Zerfos, P., & Cho, J.: Downloading textual hidden web content through keyword queries. Proceedings of the 5th ACM/IEEE-CS joint conference on Digital libraries, 2005, 100–109.

30. Owais, S. S. J.: Optimization of Boolean Queries in Information Retrieval Systems Using Genetic Algorithms – Genetic Programming and Fuzzy Logic, Journal of Digital Information and Management, 2006, 4(1), 249–255.

31. Raghavan, S., & Garcia-Molina, H.: Crawling the hidden web, Proceedings of the 27th International Conference on Very Large Data Bases, 2001, 129–138.

32. Schlein, A. M.: Find it online: The complete guide to online research. 3rd ed., Tempe, AZ: Facts on Demand Press, 2002, 122–131.

33. Sherman, C., & Price, G.: The Invisible Web: Uncovering Information Sources Search Engines Can't See. Thomas H. Hogan, Sr. 4th edition (2003).

34. Tim F., Georg G., Giovanni G., Xiaonan G., Giorgio O., Christian S.: OPAL: Automated Form Understanding for the Deep Web, WWW 2012, 829–838.

35. Umara N., Zahid R., & Azhar R.: TODWEB: Training-less Ontology based Deep Web Source Classification. iiWAS'11-The 13th International Conference on Information Integration and Web-based Applications and Services, 2011, 190–197.

36. Wang, Y., Li, H., Zuo, W., He, F., Wang, X., & Chen, K.: Research on discovering deep web entries. Computer Science and Information Systems, 2011, 8(3), 779–799.

37. Xiang, P., Tian, K., Huang, Q.: A Framework of Deep Web Crawler. Proceedings of the 27th Chinese Control Conference, 2008, 582–586.

38. Yang, J., & Korfhage, R.: Query Optimization in Information Retrieval using Genetic Algorithms, Proceedings of the 5th International Conference on Genetic Algorithms, Morgan Kaufmann Publishers Inc., 1993, 603–613.

CHAPTER 13

CLOSED LOOP SIMULATION OF QUADRUPLE TANK PROCESS USING ADAPTIVE MULTI-LOOP FRACTIONAL ORDER PID CONTROLLER OPTIMIZED USING BAT ALGORITHM

U. SABURA BANU

Professor, Department of Electronics and Instrumentation Engineering, BS Abdur Rahman University, Vandalur, Chennai – 600048, Tamilnadu, India

CONTENTS

ABSTRACT

Quadruple Tank process is a two-input-four-output process. Control of the Quadruple tank is a mind-boggling problems. Using law of conservation, the mathematical modeling of the MIMO process is computed. Linear model, the State space model is obtained using Jacobian method. State space model is converted into transfer function matrices. Reduced order FOPDT model is obtained from the fourth order transfer function model. Interaction study is performed using Relative Gain Array (RGA) and the input and output are paired. Steady state gain matrix is used for determining illness of MIMO system. Singular value decomposition technique is used to determine condition number to avoid sensitivity problem, which arises due to small change of process output. In proposed method, multiloop decentralized fractional order PID controller is designed for minimum phase Quadruple Tank process. The fractional order PID controller parameters are obtained using Bat Algorithm. Simulation studies show the likelihood of the proposed method for the computational analysis of the nonlinear minimum phase interacting process. The experimental results indicate that the developed control schemes work well under servo, regulatory and servo-regulatory conditions.

13.1   INTRODUCTION

The control of liquid level in tank process is basic control problem in many process industries. Nonlinear multi input multi output process with high interaction are complex to control. Such multivariable system can be controlled either by centralized multivariable control scheme or decentralized multi loop control scheme. In many process industries multi-loop PID control scheme is used for it advantage. The main advantage of multi-loop scheme is that if any loop fails, the control engineer can easily identify and replace controller, easy to implement and tune. In this technique multi input multi output (MIMO) process decomposed into single input single output process. Condition numbers are found to determine the worst-case deviation of the open loop transfer function. Some linear systems, a small change in one of the values of the coefficient matrix or the right-hand side vector causes a large change in the solution vector. A solution of some linear system more sensitive to small change in error. That is small error in process can change large error in control action, therefore such system can to be decomposed. The ill conditioned plant estimated in terms of certain singular values and condition numbers. Gain matrix with large condition number said to be ill-conditioned [1]. Many methods has been proposed for tuning decentralized PID controller such as auto-tuning method from step tests [2], effective transfer function based methods [3], automatic tuning method [4], relay feedback tuning method [5], detuning method, sequential loop closing method [6]. The steps involved in tuning of multi-loop PID are tedious. From simulation point of view, the mathematical model of the process needs to be computed. The linear region should be chosen from input output characteristics and state space and transfer function for each linear need to be computed. The influence of manipulated variable and controlled variable should be analyzed by interaction study [7, 8]. Process interaction plays vital role in design of optimal controller, such interaction can be compensated by feed forward element called decoupler [9].

Many of MIMO process are higher order dynamics because of sensors and final control element. For decentralized control design, higher order models approximated into reduced first order model plus dead time or second order model plus dead time. All the real world system is not an exact integer order system. In earlier time, all the real world processes are approximated into integer order process due to lack of method availability. But now, emerging numerical techniques used to convert approximated integer order process into accurate fractional order process [10].

The fractional calculus was 300 years old topic in the branch of mathematics. Mathematician was privileged to deal with fractional order calculus. Now the engineering start to explore fractional order technique in engineering application. The physical meaning of fractional order system elaborated and identified in recent times. For last three decades, the occurrence and importance of fractional calculus techniques in control engineering has been increased [11, 12]. PID controllers have been widely used in all kind of processing industries because of simplicity and lower percentage overshoot, robustness [13]. The conventional PID controller flexibility, robustness and accuracy improved by additional two extra degree of freedom called integral order λ and derivative order μ. [14, 15]. The reason for using Fractional order PID control is that it produces satisfactory control performance than integer order controllers.

Tuning of five different parameters of controller is difficult and challenging task. Recently various swarm intelligence techniques like particle swarm optimization, bee colony optimization, bacterial foraging, particle swarm optimization, bat algorithm [16, 17], differential evolution (DE) [18], Genetic algorithm and some hybrid optimization [19–21], etc., has gained popularity in the field of automation. Bat algorithm is a new metaheuristic method to find global optimum value. The micro bats find their prey by its natural echolocation behavior. The global optimum value of fractional order controller parameters is found by bat algorithm [22, 23]. In the proposed work, multiloop fractional order PID controller parameters are brought out by using the optimal values obtained from Bat algorithm for quadruple tank process.

In this chapter, multi loop fractional order PID control is used for quadruple tank process. The chapter has been organized as follows: Section 13.2 discusses the quadruple tank process, Section 13.3 elaborates the bat algorithm in general, Section 13.4 gives an insight on the fractional calculus, Section 13.5 details multiloop fractional order PID controller tuning, section 6 discusses result analysis and discussion and finally Conclusion.

13.2 QUADRUPLE TANK PROCESS DESCRIPTION

The schematic diagram for quadruple tank process is shown in Figure 13.1. Johannson [25] proposed laboratory Quadruple tank process, which consists of four interconnected water tanks with two pumps v_1 and v_2. Pump v_1 is connected to tanks 1 and 4 with distribution valve γ_1. Similarly pump 2 is connected to tanks 2 and 3 with distribution valve γ_2. By adjusting the valve position γ_1 and γ_2, the dynamic of process can be changed by introducing transmission poles and interaction. If the sum of distribution valve γ_1, γ_2 is between 1 and 2, then the system is minimum phase with transmission zero

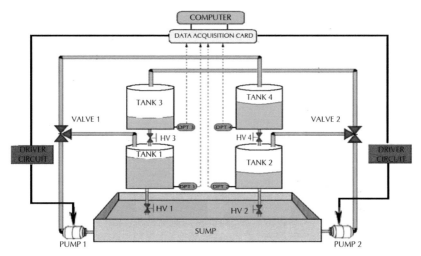

FIGURE 13.1   Schematic diagram of Quadruple tank process.

in left half. If the sum is between 0 and 1, then the system is non-minimum phase with transmission zero in right half, which makes the process unstable. If the sum of distribution valve constant is 1, then transmission zero will be at origin. The flow of tank is the product of pump gain, ratio of valve opening and voltage applied to the pump. The voltage to the pump is manipulated variable, which is applied pump driver circuit. The inflow water to Tank 1 is $\gamma_1 k_1 v_1$ and flow to Tank 4 is $(1-\gamma_1)k_1 v_1$, Similarly for Tanks 2 and 3. The control objective is to maintain the level in the lower two tanks 1 and 2 with two pumps.

13.2.1   MATHEMATICAL MODELING

Mathematical models of system were developed for many reasons. They may be constructed to assist in the interpretation of experimental data, to predict the consequences of changes of system input or operating condition, to deduce optimal system or operating conditions and for control purposes. But the main problem in modeling is the process dynamics should be captured otherwise no use in modeling the process. The dynamic model of the process has been derived from the application of fundamental physical and chemical principles to the system, using a conventional mathematical modeling approach.

$$\frac{dh_1}{dt} = -\frac{a_1}{A_1}\sqrt{2gh_1} + \frac{a_3}{A_1}\sqrt{2gh_3} + \frac{\gamma_1}{A_1}k_1v_1 \tag{1}$$

$$\frac{dh_2}{dt} = -\frac{a_2}{A_2}\sqrt{2gh_2} + \frac{a_4}{A_2}\sqrt{2gh_4} + \frac{\gamma_2}{A_2}k_2v_2 \tag{2}$$

$$\frac{dh_3}{dt} = -\frac{a_3}{A_3}\sqrt{2gh_3} + \frac{(1-\gamma_2)}{A_3}k_2v_2 \tag{3}$$

$$\frac{dh_4}{dt} = -\frac{a_4}{A_4}\sqrt{2gh_4} + \frac{(1-\gamma_1)}{A_3}k_1v_1 \tag{4}$$

The nominal values of the parameters and variables are tabulated in Table 13.1.

The open loop data was generated in the Quadruple tank system by varying the inflow rate by V_1 and V_2 open loop data generated. Operating points found out from the input output characteristics.

13.2.2   STATE SPACE MODEL

State space model for the quadruple tank process is obtained by linearizing the mathematical model using Jacobian approximation and substituting the operating conditions. In Table 13.2, State space model and transfer function matrix is shown.

The linearized state space model represented as

TABLE 13.1   Nominal Values of the Parameters Used

Parameter	Description	Value
A_1, A_3	Cross sectional area of tanks 1 and 3	28 cm²
A_2, A_4	Cross sectional area of tanks 2 and 4	32 cm²
a_1, a_3	Cross-sectional area of the outlet hole for tanks 1 and 3	0.071 cm²
a_2, a_4	Cross-sectional area of the outlet hole for tanks 1 and 3	0.057 cm²
G	Gravitational constant	981 cm/s²
γ_1, γ_2	Flow distribution constant	0.70, 0.60

TABLE 13.2 Operating Conditions and the Conventional State Space and Transfer Function Model of the Quadruple Tank Process

Operating points

$V_{1s}=3.15; V_{2s}=3.15, h_{1s}=21.67, h_{2s}=22.54, h_{3s}=13.12, h_{4s}=14.12$

State space model

$$A = \begin{bmatrix} -0.11 & 0 & 0 & 0.11 \\ 0 & -0.08 & 0.08 & 0 \\ 0 & 0 & -0.08 & 0 \\ 0 & 0 & 0 & -0.11 \end{bmatrix} . B = \begin{bmatrix} 0.38 & 0 \\ 0 & 0.27 \\ 0.44 & 0 \\ 0 & 0.59 \end{bmatrix} \begin{matrix} C = \begin{bmatrix} 1 & 0 & 0 & 0 \\ 0 & 1 & 0 & 0 \end{bmatrix}; \\ D = \begin{bmatrix} 0 & 0 \\ 0 & 0 \end{bmatrix} \end{matrix}$$

Transfer Function Matrix G(s)

$$\begin{bmatrix} \dfrac{0.08s^3+0.007s^2+0.0001s+1.3\times10^{-6}}{s^4+0.1s^3+0.003s^2+5.3\times10^{-5}s+2.5\times10^{-7}} & \dfrac{0.001s^2+6.2\times10^{-5}s+7.3\times10^{-7}}{s^4+0.1s^3+0.003s^2+5.3\times10^{-5}s+2.5\times10^{-7}} \\ \dfrac{0.002s^2+9.2\times10^{-5}s+7.6\times10^{-7}}{s^4+0.1s^3+0.003s^2+5.3\times10^{-5}s+2.5\times10^{-7}} & \dfrac{0.06s^3+0.005s^2+0.001s+1.47\times10^{-6}}{s^4+0.1s^3+0.003s^2+5.3\times10^{-5}s+2.5\times10^{-7}} \end{bmatrix} .$$

$$x = \begin{bmatrix} \dfrac{-1}{T_1} & 0 & \dfrac{A_3}{A_1T_3} & 0 \\ 0 & \dfrac{-1}{T_2} & 0 & \dfrac{A_4}{A_3T_4} \\ 0 & 0 & \dfrac{-1}{T_3} & 0 \\ 0 & 0 & 0 & \dfrac{-1}{T_4} \end{bmatrix} x + \begin{bmatrix} \dfrac{\gamma_1 K_1}{A_1} & 0 \\ 0 & \dfrac{\gamma_2 K_2}{A_2} \\ 0 & \dfrac{(1-\gamma_2)K_2}{A_3} \\ \dfrac{(1-\gamma_1)K_1}{A_4} & 0 \end{bmatrix} u \quad (5)$$

$$y = \begin{bmatrix} 1 & 0 & 0 & 0 \\ 0 & 1 & 0 & 0 \end{bmatrix} x \quad (6)$$

where $T_i = \dfrac{A_i}{a_i}\sqrt{\dfrac{2h_i}{g}}, i = 0,1,2,3,4.$

13.2.3 APPROXIMATION OF THE HIGHER ORDER TRANSFER FUNCTION TO FIRST ORDER PLUS DEAD TIME MODEL FOR THE QUADRUPLE TANK PROCESS

Any higher order process can be approximated to first order plus dead time. Analysis and controller design will be easy if higher order system is approximated to a FOPDT system. An attempt has been made to approximate the second order process to FOPDT system.

$$G(s) = \begin{bmatrix} \dfrac{5.192}{62.07\,s + 1} & \dfrac{2.836}{126.61\,s + 1} e^{-0.2s} \\ \dfrac{2.986}{89.6\,s + 1} e^{-0.2s} & \dfrac{5.705}{91.75\,s + 1} \end{bmatrix}. \tag{7}$$

13.3 INTERACTION ANALYSIS

13.3.1 CONDITION NUMBER

Gain matrix for the entire process is computed from Transfer function matrix. Normally Gain matrix is used to determine illness of MIMO process and also whether system can be decoupled or not. The Singular value decomposition is a matrix technique that determines if a system is able to be decoupled or not. The eigenvalues for the system are obtained from gain matrix. s_1 and s_2 are the positive square roots of the respective eigenvalues. The condition number is the ratio of the larger value to the smaller value. The plant is said to be ill-conditioned when the condition number has higher value. The CN indicates how close a matrix is to singularity. A system with high value of condition number indicates how the linear dependent system is based on Singular value decomposition of the gain matrix. When dealing with condition number concept in control system, if it is below 50 then the system can be easily decoupled.

$$b = g_{11}^2 + g_{12}^2 \tag{8}$$

$$c = g_{11}g_{21} + g_{12}g_{22} \tag{9}$$

$$d = g_{21}^2 + g_{22}^2 \tag{10}$$

$$\sigma_1 = s_1^2 = \frac{(b+d) + \sqrt{(b-d)^2 + 4c^2}}{2} \qquad (11)$$

$$\sigma_2 = s_2^2 = \frac{bd - c^2}{s_1^2} \qquad (12)$$

$$\Sigma = \begin{bmatrix} s_1 & 0 \\ 0 & s_2 \end{bmatrix}$$

$$\text{Condition number } CN = \frac{s_1}{s_2} = \frac{\sigma_{max\ value}}{\sigma_{min\ value}} \qquad (13)$$

where σ_{max} and σ_{min} are the maximum and minimum singular values of Gain matrix.

13.3.2 CALCULATING RGA WITH STEADY STATE GAIN MATRIX

Relative gain array is formed. RGA is a normalized form of the gain matrix. RGA has been widely used for the multi-loop structure design, such as a ratio of an open-loop gain to a closed-loop gain. It is used to find the influence of input to output. Form the transfer function matrix dynamic relative gain arrays are formulated. Variable pairing is done by selecting the values of relative gain close to value 1 (Table 13.3).

13.3.3 NI ANALYSIS WITH RGA

The stability of control loop pairing is analyzed by Niederlinski index. Stability analysis of control loop pairing is calculated by Niederlinski Index.

TABLE 13.3 Condition Number For Operating Range

Gain matrix	Condition No.	RGA
$G(0) = \begin{bmatrix} 5.192 & 2.836 \\ 2.986 & 5.705 \end{bmatrix}$	C.N=3.3	$\begin{bmatrix} 1.4 & -0.4 \\ -0.4 & 1.4 \end{bmatrix}$

This stability analysis is perfectly applicable for two input two output system. Negative value of NI indicates the instability of pairing [26].

$$NI = \frac{|G|}{\prod_{i=1}^{n} G_{ii}} \tag{14}$$

$$G(0) = \begin{bmatrix} 5.192 & 2.836 \\ 2.986 & 5.705 \end{bmatrix}$$

NI can be calculated by

$$NI = \frac{(5.192 \times 5.705) - (2.986 \times 2.836)}{(5.192 \times 5.705)} = +0.68$$

13.4   BAT ALGORITHM

Swarm Intelligence techniques, such as particle swarm optimization, ant colony optimization, bee colony optimization are gaining popularity in the recent past in the control engineering areas to solve complex problems. In the proposed research work, Bat algorithm is used for the optimization purpose. Bats have echolocation property [1, 2] based on SONAR effects to identify prey, shun obstruction, to find gap in the dark. Loud sound emanates from Bats, which is bounced back by the surrounding objects. The resultant signal bandwidth is correlated to the various species. The duration of the pulse is a few thousandths of a second (approx. 8 to 10 ms), with a constant frequency in the range of 25kHz to 150kHz. During hunting, the pulse rate is about 200 pulses per second nearer to their prey, specifying the signal processing capability of the bats. The bat ear has an integration time around 300 to 400 s. Speed of sound in air is typically $v = 340$ m/s, the wavelength λ of the ultrasonic sound bursts with a constant frequency f is given by $\lambda = v/f$, which is in the range of 2 mm to 14 mm for the typical frequency range from 25kHz to 150 kHz. The loudness varies from loudest to quietest from searching to nearing the prey. They even avoid hinderance in the size of hair. The time dealy between emission and reception gives three-dimensional view of the scene Bats have the capability to identify the location of the prey, its moving speed and its direction. The sight, smell properties and the Doppler effect helps detect their prey with easiness. Steps involved for the Bat algorithm is given below:

1. Echolocation a doppler effect is used to locate the prey and obstacles.
2. Bats start at position x_i with a velocity v_i and fixed frequency f_{min} and varying wavelength λ and loudness A_0 to search for prey. They vary the pulse emission based on the location of the target.
3. Loudness changes from a large (positive) A_0 to a minimum constant value A_{min}. 3D view is not generated which is available in nature. Also, the range of the wavelength and the frequency and the detectable range are chosen arbitrarily and not fixed.

13.4.1 PROCEDURE FOR BAT ALGORITHM

Step 1: Bat population x_i ($i = 1,2,...., n$) for the ten parameters K_{p1}, K_{i1}, K_{d1}, λ_1, μ_1, K_{p2}, K_{i2}, K_{d2}, λ_2, μ_2 (Controller parameters for the two loops considering the interactions) and their corresponding v_i are initialized.

Step 2: Pulse frequency f_i at x_i, pulse rate r_i and loudness A_i are defined.

Step 3: The population x_i to the Multiloop PI controller are applied and the multiobjective optimal function are computed by providing 50% weightage given to the ITAE of the first loop and 50% weightage given to the ITAE of the second loop.

Step 4: New solutions are generated by adjusting frequency, Updating velocities and locations/solution. If rand $>r_i$
A solution is selected among the best solution.
A local solution is generated around the selected best solution.

Step 5: A new solution is generated by flying randomly.

Step 6: If rand $<A_i$ and $f(x_i) < f(x_+)$.
The new solutions are accepted. The parameter r_i is increased and A_i is reduced.

Step 7: The bats are ranked and the current best x_+ is found.

Step 8: The results are post processed and visualized.

Step 9: Step 3 is proceeded till the maximum number of iteration is reached or the sopping criteria met.

13.5 BASICS OF FRACTIONAL CALCULUS

The differ-integral operator, $_aD_t^q$, is a combined differentiation-integration operator commonly used in fractional calculus. This operator is a notation for taking both the fractional derivative and the fractional integral in a single expression and is defined by

$$_aD_t^q = \begin{cases} \dfrac{d^q}{dt^q} & q>0 \\ 1 & q=0 \\ \displaystyle\int_a^t (d\tau)^{-q} & q<0 \end{cases} \tag{15}$$

where q is the fractional order which can be a complex number and a and t are the limits of the operation. There are some definitions for fractional derivatives. The commonly used definitions are Grunwald-Letnikov, Riemann-Liouville and Caputo definitions. The Grunwald-Letnikov definition is given by

$$_aD_t^q = \frac{d^q f(t)}{d(t-a)^q} = \lim_{N\to\inf} \left[\frac{\text{t-a}}{\text{N}}\right]^{-q} \sum_{j=0}^{N-1} (-1)^j \binom{q}{j} f\left(\text{t-j}\left[\frac{\text{t-a}}{\text{N}}\right]\right) \tag{16}$$

The Riemann-Liouville definition is the simplest and easiest definition to use. This definition is given by

$$_aD_t^q f(t) = \frac{d^q f(t)}{d(t-a)^q} = \frac{1}{\Gamma(n-q)} \frac{d^n}{dt^n} \int_0^t (t-\tau)^{n-q-1} f(\tau)d\tau \tag{17}$$

where n is the first integer which is not less than q, i.e., $n-1\leq q<n$ and Γ is the Gamma function.

$$\Gamma(z) = \int_0^{\inf} t^{z-1} e^{-t} dt \tag{18}$$

For functions f(t) having n continuous derivatives for $t\geq 0$ where $n-1\leq q<n$, the Grunwald-Letnikov and the Riemann-Liouville definitions are equivalent. The Laplace transforms of the Riemann-Liouville fractional integral and derivative are given as follows:

$$L\{_aD_t^q f(t)\} = s^q F(s) - \sum_{k=0}^{n-1} s^k \, _0D_t^{q-k-1} f(0); \qquad n-1<q<n \tag{19}$$

The Riemann-Liouville fractional derivative appears unsuitable to be treated by the Laplace transform technique because it requires the knowledge of the non-integer order derivatives of the function at t=0.

This problem does not exist in the Caputo definition that is sometimes referred as smooth fractional derivative in literature. This definition of derivative is defined by

$$
{}_aD_t^q f(t) = \begin{cases} \dfrac{1}{\Gamma(m-q)} \displaystyle\int_0^t \dfrac{f^{(m)}(\tau)}{(t-\tau)^{q+1-m}}\, d\tau\ ; & m\text{-}1{<}q{<}m \\[4mm] \dfrac{d^m}{dt^m} f(t) & ;\qquad q{=}m \end{cases}
\tag{20}
$$

where m is the first integer larger than q. It is found that the equations with Riemann-Liouville operators are equivalent to those with Caputo operators by homogeneous initial conditions assumption. The Laplace transform of the Caputo fractional derivative is

$$
L\{{}_0D_t^q f(t)\} = s^q F(s) - \sum_0^{n-1} s^{q-k-1} f^{(k)}(0); n-1 < q < n
\tag{21}
$$

Contrary to the Laplace transform of the Riemann-Liouville fractional derivative, only integer order derivatives of function f are appeared in the Laplace transform of the Caputo fractional derivative. For zero initial conditions, previous equation reduces to

$$
L\{{}_0D_t^q f(t)\} = s^q F(s)
\tag{22}
$$

The numerical simulation of a fractional differential equation is not simple as that of an ordinary differential equation. Since fractional order differential equations do not have exact analytic solutions, approximations and numerical techniques are used. The approximation method, Oustaloup filter is given by

$$
s^q = k \prod_{n=1}^{N} \dfrac{1+\dfrac{s}{\omega_{zn}}}{1+\dfrac{s}{\omega_{pn}}}\ ;\ q{>}0
\tag{23}
$$

The approximation is valid in the frequency range $[\omega_l, \omega_h]$; gain k is adjusted so that the approximation shall have unit gain at 1 rad/sec; the number of poles and zeros N is chosen beforehand (low values resulting in simpler approximations but also causing the appearance of a ripple in both gain and phase behaviors); frequencies of poles and zeros are given by

$$\alpha = \left(\frac{\omega_h}{\omega_l} \right)^{\frac{q}{N}}$$

$$\eta = \left(\frac{\omega_h}{\omega_l} \right)^{\frac{1-q}{N}}$$

$$\omega_{zn} = \omega_{p,n-1}\eta, \quad n=2,....,N$$

$$\omega_{pn} = \omega_{z,n-1}\alpha, \quad n=1,....,N$$

13.6   MULTILOOP FRACTIONAL ORDER PID CONTROLLER TUNING USING BAT ALGORITHM

Multi loop single input single output controller widely used in all multivariable process. Whole MIMO process is treated into many SISO process to make control strategy simple lower order control can be improved by decentralized control. In spite of many advanced control strategies, multi loop control strategies widely used in process industry due to reasonable performance and easy implementation, easiness to handle loop failure. Multi loop control design involves two methods. First one is decentralized single input single output system and second one is decoupler to eliminate the interaction effect.
m × m process output can be described as

$$Y_1(s) = G_{11}(s) \cdot U_1(s) + \sum_{i=2}^{m} G_{1i}(s) \cdot U_i(s) \tag{26}$$

2 × 2 process output can be described as

$$Y_1(s) = G_{11}(s) \cdot U_1(s) + G_{12}(s) \cdot U_2(s) \tag{27}$$

Input output pairing is very essential in multi loop control design. Interaction study such as Relative gain array RGA are used to find the interaction between input and output, so maximum interaction between input and output at some conditions is selected for pairing. So desired output can be controlled by manipulating the paired input. Paired input and output is U_1, Y_1 so $G_{12}(s) \cdot U_2(s)$ is a disturbance for output Y_1 due to interaction effect.

The interaction affects the SISO PI/PID controller performance. So tuning of PID is very difficult due to interaction. Many detuning methods available such as sequential loop closing method, Relay auto tuning method, independent loop method. Generally, the improvement of high-level control is dependents on performance of low level PID loops. Decentralized PID controller with decoupler is widely used to get satisfactory performance.

2×2 process represented by,

$$G(s) = \begin{bmatrix} g_{11}(s) & g_{12}(s) \\ g_{21}(s) & g_{22}(s) \end{bmatrix} \tag{28}$$

A multivariable controller,

$$K_{ij} = kp_{ij}\left(1 + \frac{1}{Ti_{ij}\gamma s^{\lambda_{ij}}} + Td_{ij}\gamma s^{\mu_{ij}}\right)$$

where Kp_{ij} is the proportional gain, Ti_{ij} is integral time constant or reset time (mins/repeat), Td_{ij} derivative time, λ_{ij} integrator order and μ_{ij} derivative order. The control law of PID rewritten as

$$K_{ij}(s) = kp_{ij} + Ki_{ij} \cdot \frac{1}{s^{\lambda_{ij}}} + kd_{ij} \cdot s^{\mu_{ij}} \quad i,j = \{1,2,....,n\} \tag{30}$$

where integral gain $ki_{ij} = kp_{ij}/Ti_{ij}$ and derivative gain.

In the proposed method, decentralized control of MIMO interaction of one input on another output consider as disturbance. But normally disturbance are independent on input, however designing multi loop fractional order PID will reject all disturbances by optimal tuning methods. Here interactions are considered as disturbance for tuning purpose alone. The multi objective optimization function of Bat algorithm is weighted percentage of ITAE of the two loops. Table 13.4 shows the FOPID parameters from the proposed scheme. Figure 13.2 shows the block diagram of the multiloop fractional order PID controller tuned using Bat Algorithm. The objective function is selected to be the Integral Time Absolute Error (ITAE). The aim

TABLE 13.4 FOPID Parameters for the Proposed Scheme

Technique	Loop 1	Loop 2
Controller tuned using BAT algorithm	$K_{p1} = 6.8152$	$K_{p2} = 0.6626$
	$K_{i1} = 0.82$	$K_{i2} = 2.1056$
	$K_{d1} = 0.7184$	$K_{d2} = 1.611$
	$\lambda_1 = 0.9686$	$\lambda_2 = 0.7788$
	$\mu_1 = 0.5313$	$\mu_2 = 0.4235$

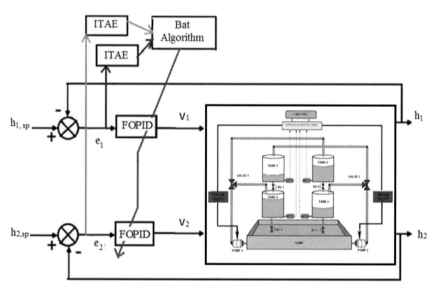

FIGURE 13.2 Block diagram of the Multiloop fractional order PID controller tuned using Bat Algorithm for two interacting conical tank process.

of the controller is to tune the FOPID parameters for the two loops considering the interactions. Normally, a multiloop PID controller requires various steps, such as determination of the relative gain array and find the interaction effects, then tune fractional order PID controllers for the loop with higher interaction effects. Then detune the controller parameters so that the effect of the other manipulated variable is taken into account leading to a complex and lengthy procedure. Whereas, in the proposed technique, the closed loop process is selected and fractional order PID controller parameters were computed for minimum Integral time Absolute error.

13.7   ANALYSIS

13.7.1   SERVO RESPONSE

Figure 13.3 shows the servo response for quadruple tank process. Variable set point is given at an interval of 2×10^3 sec and the controller has complete

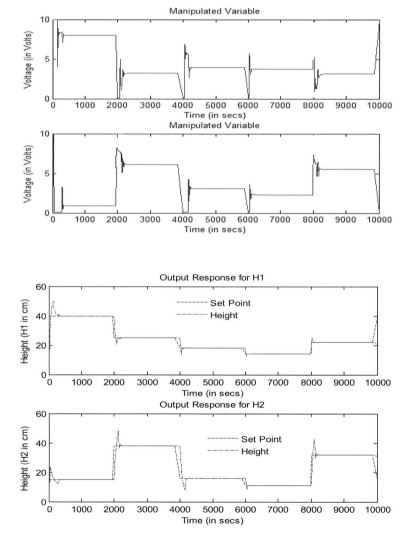

FIGURE 13.3   Servo response of FOPID controller 1,2.

control over the process and the controller tracks the set point. The ISE, IAE, ITAE value of Fractional order PID controller 1,2 under servo problem is tabulated in Table 13.5.

13.7.2   REGULATORY RESPONSE

Figure 13.4 shows the regulatory response of the proposed control scheme for quadruple tank process. Disturbance is given to the process at an interval of 1000 sec. By adjusting the manipulated variable, the process output is maintained constant irrespective of the disturbance applied within the range of ±10%. The response of disturbance rejection at steady sate is shown in Figure 13.4. The ISE, IAE, ITAE value of Fractional order PID controller 1,2 under regulatory problems is tabulated in Table 13.6.

13.7.3   SERVO REGULATORY RESPONSE

Figure 13.5 shows the servo regulatory response of the proposed controller. A variable step inputs are applied and a disturbance is applied after the response reaches the steady state value. Even though both the set point and disturbance are applied simultaneously, the proposed Fractional order PID controller scheme is capable of providing efficient control action. The ISE,IAE,ITAE value of Fractional order PID controller 1,2 under regulatory problems is tabulated in Table 13.7.

TABLE 13.5   Performance Indices Under Servo Problems

Servo	IAE	ISE	ITAE
Loop 1	3375	38480	5.727e6
Loop 2	4522	3.949e4	1.016e6

TABLE 13.6   Performance Indices Under Regulatory Problems

Servo	IAE	ISE	ITAE
Loop1	1588	9904	3.783e6
Loop2	1113	7753	5.907e5

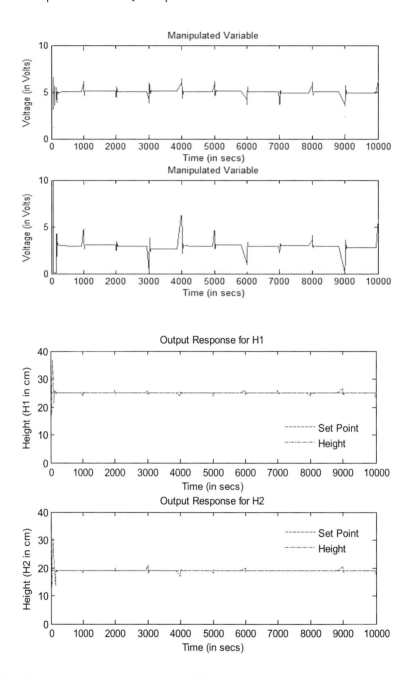

FIGURE 13.4 Regulatory Response of FOPID controller 1,2.

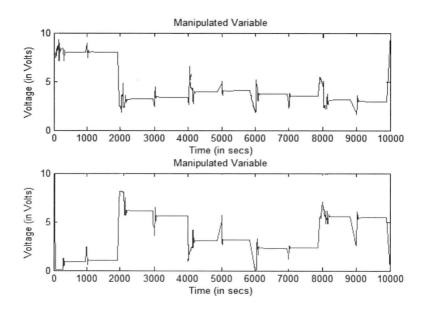

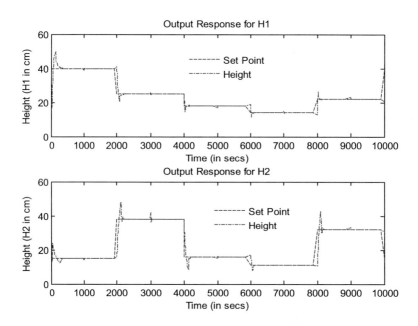

FIGURE 13.5   Servo and Regulatory Response of FOPID controller 1,2.

TABLE 13.7   Performance Indices under Servo-Regulatory Problems

Servo Regulatory Response	FOPID optimized using Bat Algorithm			FOPID optimized using Genetic Algorithm			FOPID optimized using Particle Swarm Optimization		
	IAE	ISE	ITAE	IAE	ISE	ITAE	IAE	ISE	ITAE
Loop 1	3483	39130	6.228e6	4549	45271	1.87e7	3980	42920	8.912e6
Loop 2	4237	3.38e4	1.112e6	5879	4.8e4	1.98e6	4980	3.9e4	1.78e6

13.8   CONCLUSION

In the proposed work, multi-loop fractional order PID controller is optimally tuned using Bat algorithm for quadruple tank process. Gain matrix was computed to measure the interaction of process at steady state. The sensitivity problem due to interaction is determined from SVD conditional number. The stability of control loop pairing is analyzed by Niederlinski index. Multi-loop fractional order PID parameters are computed using Bat algorithm minimizing Integral Time Absolute Error. The proposed controller is validated for servo, regulatory and servo-regulatory problems and the result shows that the scheme will result in a simple design of the multi-loop fractional order PID controller for quadruple tank process. Performance Indices shows that the parameters optimized using Bat algorithm is far better than parameters optimized using Genetic Algorithm and Particle Swarm Optimization technique.

KEYWORDS

- **Bat**
- **Condition number**
- **Multi loop fractional order PID control**
- **Nonlinear process**
- **Quadruple tank process**
- **Relative gain array**

REFERENCES

1. Jie Chen, James S. Freudenberg, & Carl N. Nett: The role of the condition number and the relative gain array in robustness analysis. Automatica, 1994, 30(6), 1029–1035.
2. Wang, Q.-G., Huang, B., & Guo, X., Auto tuning of TITO decoupling controllers from step test. ISA Transaction, 2000, 39, 407–418.
3. Xiong, Q., Cai, W.-J., & He, M.-J., Equivalent transfer function method for PI/PID controller design of MIMO processes. Journal of Process Control,. 2007, 13, 665–673.
4. Palmor, Z. J., Halevi, Y., & Kransney, N., Automatic tuning of decentralized PID controllers for TITO process. Automatica, 1995, 31(7), 1001–1010.
5. Wang, Q.-G., Zou, B., Lee, T. H., & Bi, Q., Auto tuning of multivariable PID controllers from decentralized relay feedback. Automatica, 1997, 33(3), 319–330.
6. Hovd, M., & Skogestad, S., Sequential design of decentralized controllers. Automatica, 1994, 30, 1601–1607.
7. Xiong, Q., Cai, W. J., & He, M. J. A practical loop pairing criterion for multivariable processes. Journal of Process Control, 2005, 15, 741–747.
8. Bristol, E. H. On a new measure of interaction for multivariable process control. IEEE Transactions on Automatic Control, 1966, 11 (1), 133–134.
9. Nordfeldt, P., & Hagglund, T. "Decoupler and PID controller design of TITO systems," Journal of Process Control, 2006, 16, 923–933.
10. Petras, I. The fractional order controllers: Methods for their synthesis and application. Journal of Electrical Engineering, 1999, 284–288.
11. Manabe, S. The non-integer integral and its application to control systems. ETJ of Japan, 1961, 6(3/4), 83–87.
12. Podlubny, "Fractional-Order Systems and Controller," IEEE Transaction on Automatic Control, 1999, 44(1), 208–214.
13. Astrom, K., & Hagglund, T. "The Future of PID Control," Control Engineering Practice, 2001, 9, 1163–1175.
14. Fabrizio Padula, VisioliA, Tuning rules for optimal PID and fractional-order PID controllers. Journal of Process Control, 2011, 21, 69–81.
15. Truong Nguyen Luan Vu, & Moonyong Lee: Analytical design of fractional-order proportional-integral controllers for time-delay processes. ISA Transactions, 2013, 52, 583–591.
16. Zamani, M., Karimi-Ghartemani, M., Sadati, N., & Parniani, M. Design of a fractional order PID controller for an AVR using particle swarm optimization. Control Engineering Practice, 2009, 17, 1380–1387.
17. Mendes, R., Kennedy, J., & Neves, J. The fully informed particle swarm: simpler, maybe better. IEEE Transaction on Evolutionary Computation, 2004, 8, 204–21.
18. Arijit Biswas, Swagatam Das, Ajith Abraham, & Sambarta Dasgupta: Design of fractional order $PI^\lambda D^\mu$ controller with improved differential evolution. Engineering Applications of Artificial Intelligence, 2009, 22, 343–350.
19. Duarte Valerio, & Jose Sada Costa: Tuning of fractional PID controllers with Ziegler–Nichols-type rules, Signal Processing, 2006, 86(10), 2771–2784.
20. Concepción A. Monje, Blas M. Vinagre, Vicente Feliu, & Yang Quan Chen: "Tuning and auto-tuning of fractional order controllers for industry applications," Control Engineering Practice, 2008, 16(7), 798–812.
21. Fabrizio Padula, & Antonio Visioli: "Tuning rules for optimal PID and fractional order PID controllers," Journal of Process Control, 2011, 21, 69–81.

22. Xin-She Yang, A New Metaheuristic Bat-Inspired Algorithm. Studies in Computational Intelligence, 2010, 284, 65–74.

23. Xin She Yang, A. H. Gandomi, Bat algorithm: a novel approach for global engineering optimization, Engineering Computation, 2012, 29 (5), 464–483.

24. Johansson, K. H. The Quadruple-Tank Process: A Multivariable Laboratory Process with an Adjustable Zero. IEEE Transactions on Control Systems Technology, 2000, 8(3), 456–465.

25. Niederlinski, A., A heuristic approach to the design of linear multivariable interacting control systems. Automatica, 1971, 7, 691–701.

PART III

THEORY AND APPLICATIONS OF SINGLE AND MULTIOBJECTIVE OPTIMIZATION STUDIES

CHAPTER 14

A PRACTICAL APPROACH TOWARDS MULTIOBJECTIVE SHAPE OPTIMIZATION

G. N. SASHI KUMAR

Scientific Officer, Computational Studies Section, Machine Dynamics Division, Bhabha Atomic Research Centre, Trombay, Mumbai, India–400 085, Tel.: +91-22-2559-3611; E-mail: gnsk@barc.gov.in

CONTENTS

14.1   INTRODUCTION TO SHAPE OPTIMIZATION

Shape optimization problem is unlike other optimization problems involving in finding the optimal set of parameters. These sets of parameters, which represent the shape of a body should optimize the required objective functions. Thus, the parameters to be optimized are indirectly related to the objective functions, through parameterization and flow solution. The first challenge in shape optimization is to find appropriate parameterization, so that the feasible set P is best represented. The bounds for these parameters are to be specified by the designer, which is sometimes, a difficult task. Ant Colony Optimization (ACO) resolves this difficulty, as it has the capability to reach an optimum, which is outside the designer specified bounds. The shape optimization problems are generally governed by partial differential equation leading to multi-modal objective functions (OF). Multi-modal objective functions have numerous local optima. The evolutionary algorithms with ability to search global optimum are best suited for shape optimization problems.

The computational cost (time and effort in obtaining a Computational Fluid Dynamics (CFD) solution) involved in the evaluation of OF is high. Any evolutionary approach converging in less number of function evaluations is preferred [15]. Most of the engineering applications in shape optimization involve more than one objective function [3, 5, 18, 19]. For example in the optimization of a process plant, such as a multistage flash desalination plant, the production of potable water is of prime concern yet the uniformity of flashing in various stages is also a necessary criterion for enhanced life of the equipment [18]. Such an optimization needs multiobjective optimization (MOO) approach. Figure 14.1 shows the schematic of multiobjective shape optimization approach with ACO.

Since the doctoral work of Tiihonen on shape optimization in 1987 [22], numerous approaches have evolved in this area with majority of the applications in aerodynamics and engineering [9]. Shape optimization has been traditionally practiced by evaluating the gradients using adjoint/ADIFOR solver. Automatic differentiation, using tools such as TAPENEDE (http://www-sop.inria.fr/) is expected to ease out the difficulties in gradient based approaches. These Newton based solvers converge to the nearest local optimum (LO) and do not have a mechanism to neither escape LO nor get out of it, when trapped in LO. With the development of evolutionary techniques many researchers have successfully adapted them for shape optimization.

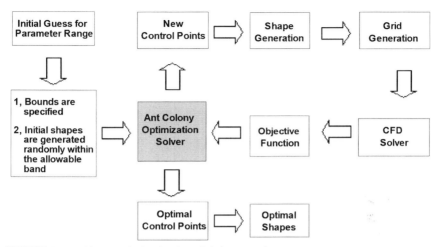

FIGURE 14.1 Shape optimization in MOO framework.

Evolutionary technique using a population, with its heuristics has mechanisms that help the algorithm avoid convergence to LO. A few of the evolutionary algorithms are also known for their capability to attain the optimum in minimum number of function evaluations.

The process of developing an efficient shape optimizer involves: (i) Appropriate parameterization of the shape, (ii) Selection of appropriate global optimizer like Genetic Algorithm (GA), ACO, etc., (iii) Objective function evaluators like CFD solver or surrogate model, etc., and (iv) Selection of MOO variant. Thus, the development of an efficient shape optimizer involves many disciplines.

14.2 PARAMETERIZATION

The shape Γ which is defined by a small number of design variables x_c are considered for optimization as parameters. Parameterization should be able to represent exhaustive family of complex geometries, including constraints and singularities. The number of parameters should be as minimum as possible and there should be a mechanism for controlling the smoothness of the geometry. There are numerous approaches used in parameterization of shape [2]. Most popular three approaches are: (i) computer aided design (CAD) based, (ii) polynomial representation, and (iii) free form deformation approach. Use of CAD tools cuts down the development time of optimization solver. The CAD representation uses boundary represented (B-rep) solids.

It is more versatile and many of the current large-scale multi disciplinary optimization solvers are based on CAD parameterization. In polynomial approach the commonly used methods are: i) Polynomial representation, ii) Bezier curve, iii) B-spline, and iv) NURBS. Parameterization $P(x)$ can be represented as follows.

$$P(x) = \sum_{i=0}^{n-1} A_i x^i = \sum_{i=1}^{n} A_i B_{i,p}(x) = \sum_{i=1}^{n} A_i N_{i,p}(x) = \frac{\sum_{i=1}^{n} A_i W_i N_{i,p}(x)}{\sum_{i=1}^{n} W_i N_{i,p}(x)} \tag{1}$$

where $A_i, B_{i,p}(x), N_{i,p}(x), W_i$ are control points, Bezier, B-spline of degree p and weights respectively. However, these methods are suitable for two-dimensional optimization and simple 3 dimensional geometries. The free form deformation originates from computer graphics [20]. It does not depend on the surface/curve definition. The displacement caused due to deformation (δq) can be defined by a third order Bezier tensor product.

$$\delta q = \sum_{i=0}^{n_i} \sum_{j=0}^{n_j} \sum_{k=0}^{n_k} B_{i,n_i}(\eta_q) B_{j,n_j}(\xi_q) B_{k,n_k}(\zeta_q) \Delta P_{ijk}^{'} \tag{2}$$

where $B_{i,n_i}, B_{j,n_j}, B_{k,n_k}$ are Bernstein polynomials of order n_i, n_j, n_k, respectively. ΔP_{ijk} are control point displacements.

14.3   A PRACTICAL APPROACH FOR MULTI OBJECTIVE OPTIMIZATION

Multi objective optimization (MOO) problem is defined as finding the optimum parameters, $\overline{x}^* = \left[x_1^*, x_2^*, \ldots, x_n^* \right]^T$, $\overline{x}^* \in X \subset \mathbb{R}^n$ such that the objective functions $f_i(\overline{x}) \in F \subset \mathbb{R}^m$ are simultaneously minimized. Here $f_i(\overline{x}) : \mathbb{D} \Rightarrow \mathbb{R}$, $i = 1, 2, \ldots m$, are m continuous and bounded functions and $\mathbb{D}$ is the search space of $\overline{x}$. The solution approach to MOO problem can be classified as cooperative and non-cooperative strategies. The prominent non-cooperative approach, Nash Equilibrium has been applied to aerodynamic shape /design optimization problems [21]. Bi-level approach, a type of Stackelberg strategy has also been applied for shape optimization [10]. The Pareto-optimal solution is the popular cooperative strategy. Pareto-optimal solutions consist of a set of non-dominated design points $\overline{x}^{(1)}$ and $\overline{x}^{(2)}$ represented as $\overline{x}^{(1)} \not\prec \overline{x}^{(2)}$ and $\overline{x}^{(2)} \not\prec \overline{x}^{(1)}$.

In many of the engineering problems the decision maker has an a priori knowledge of the target/reference values of the objective functions. There has been a class of algorithms based on these target values. There are three approaches depending on, when the decision maker intervenes, namely, (i) prior to the search (a priori), (ii) during the search (interactive), and (iii) after the search (a posteriori). The goal programming (a priori approach) requires the target vector $\bar{T}$ to be supplied before the search. The objective function is defined as Minimization of $\sum_{i=1}^{m}\left|f_i(\bar{x}) - T_i\right|$. This is referred as goal programming [11]. The reference point method proposed by Wierzbicki [23] is an interactive MOO technique based on achievement scalarization function (ASF). When a MOO problem is posed, the decision maker (DM) supplies a reference point (not target value). If the optimum is not within the expected bounds of DM a new reference point is supplied and the process continues till the DM attains a satisfactory solution. The ASF s: $Q \times F \rightarrow R^l$, where reference vector $q(\bar{x}) \in Q \subset R^m$, the min-max scalarization can be defined as $s = min\left[max\left\{\left|f_1(\bar{x}) - q_1\right|, \left|f_2(\bar{x}) - q_2\right|, \ldots, \left|f_m(\bar{x}) - q_m\right|\right\}\right]$. The posteriori approach common in MOO is the weighted metrics method. The objective function is defined as minimization of scalarization function, L_p.

$$L_p = \left(\sum_{i=1}^{m} w_i \left|f_i(x) - q_i\right|^p\right)^{1/p} \tag{3}$$

This approach becomes linear combination of weights when $p = 1$ and becomes a Tchebycheff problem if $p = \infty$. Miettinen [14] shows that the weighted metrics is Pareto optimal, if (i) the optimal solution is unique, and (ii) $w_i > 0$ for $1 \leq p < \infty$.

The method of determining ranking is based on either (i) distance from the reference point or (ii) based on ε-dominance [19] defined by the following equation (see Figure 14.2).

$$[\vec{f}* - \vec{f}_{ref}] \prec [\vec{f} - \vec{f}_{ref}] :\Leftrightarrow \forall_{i \in \{1,2,\ldots,m\}} ([f_i(\bar{x}*) - f_{i,ref}] \leq [f_i(\bar{x}) - f_{i,ref}]) \wedge$$
$$\exists_{i \in \{1,2,\ldots,m\}} (L_p[f_i(\bar{x}*) - f_{i,ref}] < L_p[f_i(\bar{x}) - f_{i,ref}] - \varepsilon_i); \ \varepsilon_i = \varsigma_i(f_i^{max} - f_i^{min}), \varsigma_i \in (0,1) \tag{4}$$

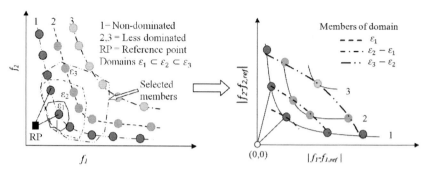

FIGURE 14.2   The scalarization approach leads to a Pareto optimal point.

where the values of $\vec{\varsigma}$ determines the region considered for further evaluations. A larger $\vec{\varsigma}$ would require longer time for optimization, at same time a smaller region may miss the global optimum of the problem.

14.4   ANT COLONY OPTIMIZATION FOR MOO

The popular shape optimizers based on evolutionary algorithms use GA, ACO or swarm colony optimization [7, 15]. GA has been applied to shape optimization since early 1990's [13]. The multi objective concept of NSGA-II by Deb et al. [6] has yielded a reliable multiobjective optimizer. GA based search is limited to the bounds of the parametric space specified by the designer. ACO has been preferred in this work for its advantages over GA, namely (i) The parametric search space is not restricted to that of the designer defined bounds, (ii) ACO takes less number of function evaluations to attain the optimum compared to GA [19]. The multi objective optimization based on ant colony optimization is discussed in this chapter. The concept of goal vector has been used for solving multi objective optimization (MOO) problems.

Ant Colony optimization a population based technique, has proved itself in field of combinatorial optimization [8]. Abbaspour et al. [1] has successfully extended it to parametric optimization using the route of inverse modeling. The author has earlier demonstrated shape optimization using ACO coupled with computational fluid dynamics (CFD) solver [15, 16, 17].

In multiobjective ACO (MACO) algorithm the search space of each parameter i ($i= 1,.. n$) with bounds $\left[\beta_i^{min}, \beta_i^{max}\right]$ is discretized into finite number of levels, l_i ($i= 1,.. n$). Ant path is defined as the set obtained by choosing one level from each parameter (see Figure 14.5). So, there will be $M = l_1 \times l_1 \times .. \times l_n$ exhaustive ant paths (or exhaustive search space, ESS). Each ant path requires a function evaluation for knowing it's worth. In order to minimize the number of function calls a subset of M is chosen randomly, named as random search space (RSS). RSS is typically $> 0.2 \times$ESS. The author from his experience with numerous problems arrived at a formula for optimum RSS (Equation 5). This value is less than Abbaspour's recommendation [1] when $n > 3$ and more when $n \leq 3$.

$$RSS = \frac{l_1 \times l_2 \times .... \times l_n}{2^{n-1}} \qquad (5)$$

Each of the ant pathway determines a heuristic for that iteration. The intensity of trial $(\tau_u(i))$ on each pathway u for the iteration i as defined by Abbaspour et al. [1] is

$$\tau_u(i) = \begin{cases} \exp\left[\ln(100)\left(\dfrac{L_{p,u} - L_{p,cr}}{L_{p,\min} - L_{p,cr}}\right)\right], & L_{p,u} \leq L_{p,cr} \\ 0 & L_{p,u} > L_{p,cr} \end{cases} \qquad (6)$$

where $L_{p,u}$ is value of the scalarization function for the uth pathway (Eq. 3), $L_{p,cr}$ is the critical value above which τ_u is forced to be zero. $L_{p,\min}$ is the minimum of all L_p's. The trail share for each stratum $\beta_{ij}, (j = 1,2,..l_j)$ from each path can be summed to yield Φ_{ij}. Scores (S_{ij}) are calculated for each stratum β_{ij}.

$$\Phi_{ij} = \sum_{u \in crossing\ pathways} \tau_u \quad \text{and} \quad S_{ij} = \frac{(\Phi_{ij})^A (\sigma_{ij})^N}{\sum_i \sum_j (\Phi_{ij})^A (\sigma_{ij})^N} \qquad (7)$$

Crossing pathways are all the ant paths that cross stratum β_{ij}, σ_{ij} is the standard deviation of all values of the scalarization function involving β_{ij}. The values of A, N, T as given by Abbaspour et al. [1] and later modified by author [19] are $A = 1$, $N = \eta_{ij}/\mu_g$ and $T = L_{p,\min} + \sigma_g$. μ_{ij}, η_{ij} are mean and standard deviation of trail for β_{ij} stratum. μ_g, σ_g are mean and standard deviation of scalarization function. A pseudo code of ACO is given in the appendix

The steps involving (i) descritization of parametric space, (ii) ant path selection, (iii) evaluation of scalarization function, and (iv) calculation of scores for each stratum constitute one iteration. The low scoring strata have been dropped and higher scored strata are retained. The ACO has a constraint, the new bounds obtained from the retained strata should be continuous. There can be two situations arising while truncating the bounds of parameters, namely, (i) the maximum scoring stratum is within the previous defined bounds (see Figure 14.3a), or (ii) the maximum scoring stratum is at the edge of the bound (*see* Figure 14.3b). In such a case the domain is extended in the direction to accommodate the optimum (*see* Figure 14.3b). This facilitates the searching of optimum, which is outside parametric bounds defined by the designer. The bounds thus obtained forms the search space for next iteration. The iterations are continued till the convergence criterion is satisfied.

Pseudo code

> *Initialize the domain for all parameters (Iteration = 0)*
>
> 10 *The entire range of all the parameters is discretized into levels*
>
> *Iteration = Iteration + 1*
>
> *Do i = 1, Number of ants (RSS)*
>
> > *Choose a path for ant i by selecting one level from each parameter randomly*
> >
> > *Construct the shape of the body defined by the path of the ant*
> >
> > > *If (Are you using a Meshless solver. eq. yes) then*
> > >
> > > > *Regenerate connectivity in the zone near to the body*
> > >
> > > *else*
>
> *Generate the Grid with new shape*
>
> *End If*
>
> *Run the CFD solver for prescribed convergence*
>
> *Evaluate the Scalarization function with solution from CFD solver*
>
> *End Do*
>
> *Evaluate the scores for each level in all of the parameters*
>
> *Apply cut-off criteria, which shrinks the range for the parameters*
>
> *New range for the parameters are thus obtained*
>
> *If (does not meet the convergence criteria) go to 10*
>
> *Print the best optimum obtained*

The methodology of MACO is further elaborated using a test case in Section 14.5.

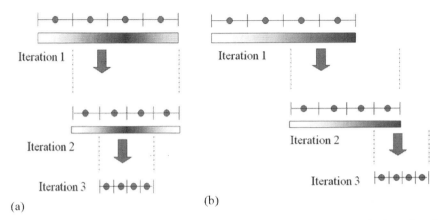

FIGURE 14.3 Change in bounds of a parameter using MACO algorithm (the intensity of gray defines the score value).

14.5 APPLICATION TEST CASES

In order to illustrate the methodology of MACO shape optimization, a quasi 1D supersonic nozzle with a shock has been discussed in detail. Case 1 has four parameters, which represent the shape of a supersonic nozzle. Nozzle is of 3 units length and 1 square unit area at the throat. Throat is at the middle, i.e., 1.5 units length. The target of optimization in case 1 is to determine the shape of the nozzle that would produce a shock in the nozzle with downstream pressure at 0.6784. This shock should occur at distance of 2 units from inlet and velocity of the gas before shock should be 3 Mach. Case 2 demonstrates the same problem in case 1 with an additional parameter. The backpressure which was fixed in case 1 was assumed to be a parameter. Thus these two cases would emphasize the importance of correct parameterization in shape optimization problems.

14.5.1 CASE 1

Step 1: Parameterization: The nozzle is assumed to be axi-symmetric. A Bezier curve is used for constructing the surface of nozzle (i.e., radius Vs length). Four control points have been chosen (see Figure 14.4) at various positions along the length of nozzle. The x-coordinates of control points remain fixed, while the y-coordinates are variables. The area of throat is a fixed value (=1). Two Bezier curves are used to represent the subsonic and supersonic sections independently. A Bezier curve representing the nozzle shape (with b grid nodes) is given as.

$$x_i = (1-t)^2 x_{C1} + 2(1-t)tx_{C2} + t^2 x_{C3}; t = (i-1)/(1-b)$$

$$y_i = (1-t)^2 y_{C1} + 2(1-t)ty_{C2} + t^2 y_{C3} \tag{8}$$

Here $(x_{Cj}, y_{Cj}), j = 1,2,3$ are control points. Two Bezier curves require 6 control points. The x_{Ci}, $i = 1,2,3,4,5,6$ are fixed while values y_{Ci}, $i = 1,2,5,6$ are function of parameters (α_i, $i=1,..4$) that are to be optimized. $y_{C3}(= y_{C4})$ is radius of throat. The only constraint on parameters is that $\alpha_3 > 0.6$ (i.e., should be > throat radius). The domain for optimization for these 4 parameters is defined (see Figure 14.4). At end of optimization these bounds shrink to represent the optimized nozzle shape.

Step 2: Discretization of parameters: The 4 parameters, α_i, $i=1,..4$ are discretized into 5 levels each (see Figure 14.5). An ant chooses any one level from each parameter, which forms its path. This path represents a unique nozzle shape. There could be $5^4 = 625$ ant paths (ESS) from which a RSS of 80 has been chosen ($=5^4/2^3$ using Eq. 5). These 80 ant paths are randomly chosen to form the candidate shapes in each iteration.

Step 3: Objective functions and its evaluation: The two objective functions chosen are (i) the position, f_1 and (ii) Mach number (f_2) at which shock occurs. The reference point chosen, $q = (2,3)$, that is we are interested in arriving at a nozzle shape where a shock is seated in the supersonic zone at length = 2

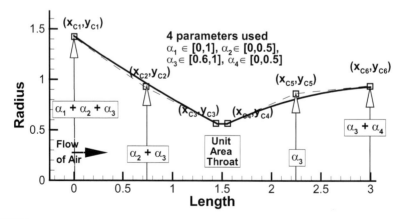

FIGURE 14.4 Parameterization of Nozzle.

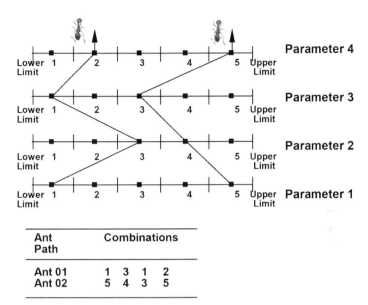

Ant Path	Combinations			
Ant 01	1	3	1	2
Ant 02	5	4	3	5

FIGURE 14.5   Parameters and their discretization.

and the shock should be occurring at Mach = 3 with a fixed downstream pressure of 0.6784 ($P_{upstream}$=1). The scalarization function has been taken as.

$$L_p(f,w,q) = \left(\sum_{i=1}^{m} w_i \left| f_i(x) - q_i \right|^p \right)^{1/p} \text{ with } w_i = \frac{1}{M}, m = 2, p = 2 \quad (9)$$

Varying the weights one can generate various points on the Pareto front [11]. It is authors observation that $w_i=1/m=1/2$, optimizes to a Pareto point nearest to the reference point [19].

An explicit MacCormack technique [4] with quasi one-dimensional flow equations was solved for determining the objective functions. The codes accompanying this chapter has the complete implementation in ForTran (nozshk.for).

Step 4: Ranking of the solutions: Ordering of the scalarization function is performed such that the equation

$$L_2 (f,w,q) \big|_1 \leq L_2 (f,w,q) \big|_2 \leq \ldots \ldots \leq L_2 (f,w,q) \big|_n \quad (10)$$

is satisfied. The first few members (= $RSS \times \xi$, $\xi \in (0,1)$; typical value is 0.1) are considered for ranking. The ranking is based on the ε- dominance criterion defined in Section 14.2.

Step 5: Arriving at bounds for next iteration: Each of the ant pathways determines a heuristic for that iteration. The scores for each stratum are evaluated. Depending on the criterion for cut-off, (see Table 14.1) the levels with scores $> (\text{Score}_{min} \times 1.5)$ are retained. These form the bounds for the next iteration.

Step 6: Looping and Termination: The steps 1 to 5 constitute of one iteration in shape optimization problem. The steps are repeated till the convergence criterion is satisfied. Convergence criterion could be: (i) the bounds of the parameters shrink to <5% of initial bounds, or (ii) the best individual obtained within designers tolerance.

The bounds arrived after 5th iteration for α_1, α_2 and α_4 are outside the initial specified limits (see Table 14.2). The domain of α_1 after 10th iteration is larger than the initial guess. Such a situation arises when all the levels in a parameter are performing equally good or equally bad. This emphasizes that the optimization problem has been ill-posed. Figure 14.6 shows the change of nozzle shape with iterations. Any experienced designer would identify that arrived optimum shape of nozzle is unusual/unphysical.

The optimization process has been stopped after 20 iterations. Table 14.2 (see Figure 14.9) shows that the improvement in the scalarization function with iteration is negligible. A possible reason could be: (i) insufficient number of parameters used in optimization, or (ii) bad parameterization.

TABLE 14.1   The Details of 1st Iteration

Parameter	Initial guess	Scores after 1st iteration	New bounds
α_1	[0, 1.0]	0.021 0.017 0.017 **0.100 0.055**	[0.6, 1.0]
α_2	[0, 0.5]	**0.050 0.090 0.061 0.000 0.017**	[0.0 0.5]
α_3	[0.6, 1.0]	**0.210 0.056** 0.000 0.000 0.000	[0.6, 0.76]
α_4	[0, 0.5]	0.000 0.000 0.000 **0.204 0.065**	[0.3, 0.5]

Criterion for cut-off: Levels with scores $> (\text{Score}_{min} \times 1.5)$ are retained.

TABLE 14.2   The Progress of Shape Optimization Using MACO

Parameter	Initial guess	After 5th Iteration	After 10th Iteration
α_1	[0, 1.0]	[1.3296, 1.9440]	[0.1745, 1.7671]
α_2	[0, 0.5]	[0.3944, 1.1720]	[0.8608, 1.1927]
α_3	[0.6, 1.0]	[0.6538, 07183]	[0.6188, 0.7545]
α_4	[0, 0.5]	[0.4424, 0.5720]	[0.5344, 0.5523]
Best SF	0.778	0.736	0.713

SF = Scalarization function.

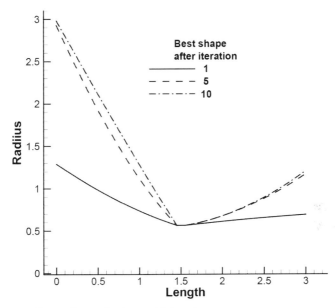

FIGURE 14.6   Change in nozzle shape with iteration.

An additional parameter, the back-pressure of the nozzle has been added to see whether MACO succeeds in attaining a better optimum. This parameter (back pressure) was fixed at 0.6784 in case 1. This extended case 1 with one additional parameter will be discussed as case 2.

14.5.2   CASE 2

The shape parameterization of case 1 has been used unaltered. 195 ant paths are used to represent the RSS ($=5^5/2^4$ using Eq. 5). The addition of new process parameter has reduced the Scalarization function value from 0.7 (in case 1) to 0.08 (see Figure 14.9). This shows the importance of appropriate parameterization for obtaining the optimal shape. Readers attention is brought to another test case that use 3 control points in super-sonic zone (instead of 2 control points as in case 1 and 2) improves the scalarization function to 0.002 [19]. Figures 14.7 and 14.8 show the convergence and change in nozzle shape up to 10th iterations. Thus, it is very important in shape optimization problems to: (i) incorporate the right controlling variables as parameters to be optimized, and (ii) to use appropriate shape definition.

14.5.3 CASE 3

Third test case demonstrated is shape optimization of a bump in a closed channel flow. The motive fluid is Nitrogen gas with trace amounts of Oxygen. This problem addresses the transportation/ flow of binary gas in a viscous dominated closed channel. The target of the study is to obtain a bump shape with given constraints of length and height, so that there is minimum change in minor component composition across the shape. The free stream Mach number of inlet Nitrogen gas (N_2=99.5%, O_2=0.5%) has been fixed as 0.3 (N_{Re}=1,000) at temperature of 273 K and density 1.08×10^{-2} kg/m³. The bump shape is to be optimized for a fixed length and height. The shape of the bump is constructed using three parameters $\{\alpha_1, \alpha_2, x_{mean}\}$. The upward and downward y-coordinates of bump shape are parameterized as follows.

$$y_{upward}=0.2(\cos(x-x_{mean}))^{\alpha 1}; \; y_{downward}=0.2(\cos(x-x_{mean}))^{\alpha 2} \qquad (11)$$

The initial domains of these three parameters are discretized into 5 levels each.

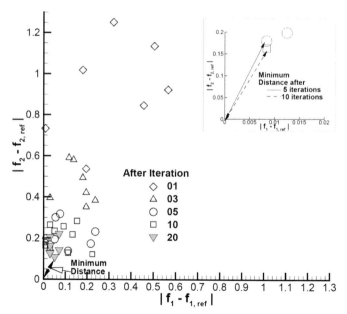

FIGURE 14.7   Progress of optimization with iteration.

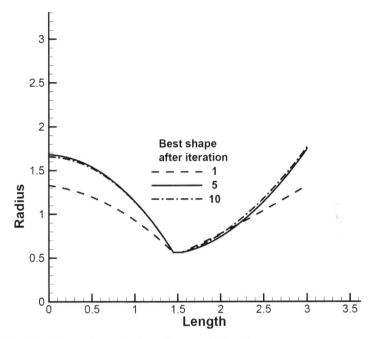

FIGURE 14.8 Change in nozzle shape change with iteration.

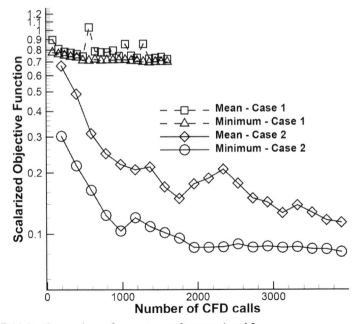

FIGURE 14.9 Comparison of convergence for cases 1 and 2.

RSS of 31 ($=5^3/2^2$ using Eq. 5) has been used. The two objectives (functions) in this problem are: (i) the deviation of concentration of O_2 (W_A) from inlet value at any location inside the channel should be minimum, that is, $f_1 = \underset{i \in Nodes}{\overset{Maximum}{}} |W_{A,i} - 0.005|$, and (ii) pressure drop due to the bump should be minimum, that is, $f_2 = \Delta P$. Both these optimization functions have contradicting influence on the shape of the bump. Each objective function evaluation requires a CFD solution. For this purpose a Meshless flow solver has been used. The kinetic flux vector splitting based SLKNS solver [12] was used in evaluation of the flow. It is a compressible flow solver, which is inherently stable due to its upwinding at molecular level [12]. The modified split stencil approach proposed by Mahendra et al. makes the solver accurate for viscous solutions [12, 19]. For detailed discussion on the update equations and boundary conditions, please refer [12, 15]. The flow simulation predicts the pressure drop, bulk velocity and temperatures. As the heavier component of Oxygen is of small quantity dilute mixture assumption has been used in evaluation of diffusive flux of the heavier species. There are two important contributions that lead to the solution of this problem. Firstly, the use of meshless CFD solver which does not require grid generation after every shape change. Figure 14.11 shows the procedure adopted in meshless solver by which grid generation step is bye-passed. In meshless flow solver each node is defined by its location and its connectivity set (*see* Figure 14.10). The least squares formula is used within the connectivity to obtain the gradients of state vectors [12].

Gradient of flux g at node p with respect to x, y can be written as

$$g_x = \frac{\sum \Delta y_i^2 \sum \Delta x_i \Delta g_i - \sum \Delta x_i \Delta y_i \sum \Delta y_i \Delta g_i}{\sum \Delta x_i^2 \sum \Delta y_i^2 - \left(\sum \Delta x_i \Delta y_i\right)^2} ;$$

$$g_y = \frac{\sum \Delta x_i^2 \sum \Delta y_i \Delta g_i - \sum \Delta x_i \Delta y_i \sum \Delta x_i \Delta g_i}{\sum \Delta x_i^2 \sum \Delta y_i^2 - \left(\sum \Delta x_i \Delta y_i\right)^2} \qquad (12)$$

where $\sum$ stands for $\sum_{i=1}^{connectivity}$. The procedure is as follows: A background grid is generated. The changed shape is constructed from the control points. This shape is appended to the background grid.

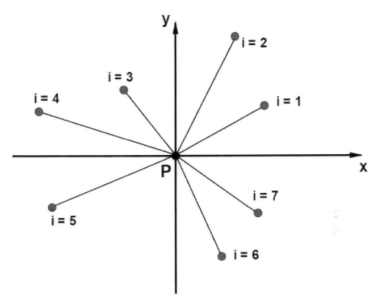

FIGURE 14.10 The connectivity set for node *p*.

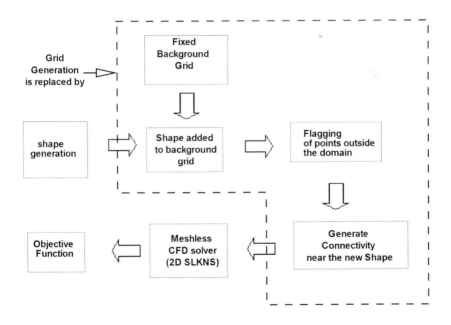

FIGURE 14.11 Process flow sheet "How is grid generation step avoided?".

A few boundary layer nodes are also added along the surface. Next nodes that fall outside the domain (bump) are flagged, but not removed. The changed connectivity set is generated in the vicinity of the bump shape. Now the new grid is ready to be used by the CFD solver. In this process, the generation of connectivity near the changed shape is less time consuming than generation of full grid at each shape change [17]. Figure 14.12 shows the addition of bump nodes on the background grid and flagging of nodes. An elaborate discussion on the use of meshless CFD solver in shape optimization problems can be found in reference [17].

The second contribution is the implementation of kinetic based splitting in the diffusion equation. The generalized Stefan-Maxwell equation for solution of concentration profile in flow field reduces to

$$\frac{\partial}{\partial t}(\rho_A) + \frac{\partial}{\partial x}(\rho_A u) + \frac{\partial}{\partial y}(\rho_A v) + \frac{\partial}{\partial x}(J_x) + \frac{\partial}{\partial y}(J_y) = 0 \qquad (13)$$

when a binary species with very low concentration of heavier species such that $(1-W_A) \rightarrow 1$, W_A is mole fraction of species A and

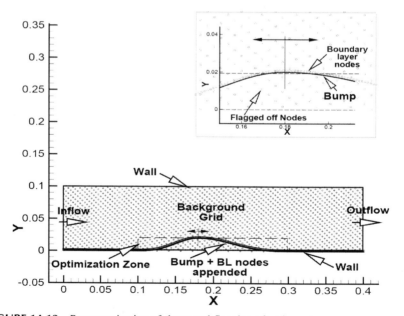

FIGURE 14.12   Parameterization of shape and flagging of nodes.

$$J_x = -\left(\frac{C^2}{\rho}\right)^2 M_A M_B D_{AB} \left[\frac{\partial}{\partial x}(W_A) + W_A\left(\frac{1}{p} - \frac{M_A}{\rho RT}\right)\frac{\partial}{\partial x}(p)\right]$$

$$J_y = -\left(\frac{C^2}{\rho}\right)^2 M_A M_B D_{AB} \left[\frac{\partial}{\partial y}(W_A) + W_A\left(\frac{1}{p} - \frac{M_A}{\rho RT}\right)\frac{\partial}{\partial y}(p)\right] \quad (14)$$

where C is the concentration in kmole/m³, p is pressure in N/m², ρ_A is density of species A in kg/m³, D_{AB} is diffusion coefficient in m²/s. The derivatives of diffusive fluxes J_x and J_y are evaluated using full stencil connectivity. Traditionally the derivative of convective fluxes ρ_{Au} and ρ_{Av} are evaluated using streamline upwinding. The approach of kinetic flux vector splitting has been used to derive the expressions for upwind stencils. These use CIR splitting at molecular level implemented by stencil subdivision at macroscopic level [15]. The derivatives can be written as

$$\frac{\partial}{\partial x}(\rho_A u)|_{US} = \frac{\partial}{\partial x}(G_1^+)|_{N_1} + \frac{\partial}{\partial x}(G_1^-)|_{N_2};$$

$$\frac{\partial}{\partial y}(\rho_A v)|_{US} = \frac{\partial}{\partial y}(G_2^+)|_{N_3} + \frac{\partial}{\partial y}(G_1^-)|_{N_4} \quad (15)$$

where $G_1^\pm, G_2^\pm$ are the positive and negative split fluxes in x and y directions, respectively. N_1 and N_2 are positive and negative stencil in x-direction. Similarly, N_3 and N_4 are positive and negative stencil in y-direction.

$$G_1^\pm = \rho_A u\left[\frac{1 \pm erf\left(u\sqrt{\beta}\right)}{2}\right] \pm \frac{\rho_A}{\sqrt{\pi\beta}}\frac{e^{-u^2\beta}}{2}$$

$$G_2^\pm = \rho_A v\left[\frac{1 \pm erf\left(v\sqrt{\beta}\right)}{2}\right] \pm \frac{\rho_A}{\sqrt{\pi\beta}}\frac{e^{-v^2\beta}}{2} \quad (16)$$

$\beta = 1/2RT$, Wall Boundary condition is implemented by equating the gradient of J along surface normal to zero.

For validation and elaborate discussion on the topic the user is encouraged to download the author's thesis [15]. The advantages of using kinetic based approaches are (i) high robustness, and (ii) negative concentration are never encountered. A typical Mach profile of the shape is shown in Figure 14.13. The change in shape of bump as optimization progressed for the reference point q(x) = (0.001, 20 Pa) is shown in Figure 14.13.

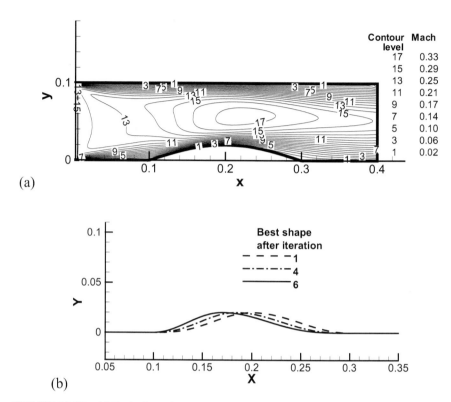

FIGURE 14.13   (a) Typical Mach contours; (b) Change in shape with iteration.

14.6   CONCLUSIONS

The reference point multi objective shape optimization approach has been demonstrated as a practical way of attaining solutions to engineering problems. The combination of reference point strategy along with parametric ant colony optimization using with ε-dominance based ranking was successful in capturing the members of the Pareto front, which are closest to the reference point.

The optimizer has capability to search the domain outside the initial parametric space specified by designer. Cases studies in this chapter have shown that shape optimization problem is a strong function of parameterization. The meshless solver minimizes on grid generation for every change in shape of the body.

Optimization tools require the use of robust CFD solvers, meshless SLKNS solver along with Kinetic diffusion solver is one such combination of robust solver. The present strategy (MACO) of shape optimization is a powerful tool for engineering applications.

AUTHOR'S CONTRIBUTIONS

Author has been working in area of development and applications of compressible CFD solvers, which are based on Kinetic Flux Vector Splitting and meshless discretization [15, 17, 19]. He has been working in the development of shape optimization solvers based on GA and ACO. Involved in applying the tools on various process/ shape optimization problems [15, 16, 18].

ACKNOWLEDGMENTS

Author is deeply indebted to Dr. A.K. Mahendra, B.A.R.C., Trombay, Mumbai, India for his guidance and for many invaluable discussions on optimization and meshless compressible flow solvers. The author acknowledges Prof. S.V. Raghurama Rao, Aerospace Department, I.I.Sc., Bangalore, India for introducing him to ant colony optimization.

KEYWORDS

- ant colony optimization
- computational fluid dynamics
- convergent divergent nozzle
- kinetic diffusion solver
- MACO
- meshless
- multi objective optimization
- shape optimization
- SLKNS

REFERENCES

1. Abbaspour, K. C., Schulin, R., & Genuchten, M.Th. van., Estimating unsaturated oil hydraulic parameters using ant colony optimization, *Adv. Water Res.*, 2001, 24, 827–841.

2. Abou El Majd, B., & Desideri, J-A., Duvigneau, R., Shape design in Aerodynamics: Parameterization and sensitivity, *Eu. J. Comp. Mech.*, 2008, 17(1–2).

3. Ali Elhama, Michel, J. L., & van Toorenb, Winglet multiobjective shape optimization, *Aerospace Science and Technology*, 2014, 37, 93–109.

4. Anderson, J. D. Jr., *Compuational Fluid Dynamics, The Basics with Applications*, McGraw-Hill, Inc., 1st Ed., 1995.

5. Arias-Montaño, A., Coello Coello Carlos, A., & Mezura-Montes Efrén, Evolutionary Algorithms Applied to Multiobjective Aerodynamic Shape Optimization, In Koziel, S. Yang, Xin-She, ed., *Computational Optimization, Methods and Algorithms, Studies in Computational Intelligence,* Springer Berlin Heidelberg, 2011, 211–240.

6. Deb, K., *Multiobjective Optimization Using Evolutionary Algorithms,* John Wiley and Sons, 2001.

7. Deb, K., Sundar, J., Bhaskara, U., & Chaudhuri, S., Reference Point Based Multiobjective Optimization Using Evolutionary Algorithms, *ISSN Intl. J. Comp. Intel.Res.*, 2006, 2(3), 273–286.

8. Dorigo, M., & Stutzle, T., *Ant Colony Optimization*, Prentice-Hall of India Pvt. Ltd., 2005.

9. Haslinger, J., Mäkinen, R. A. E., *Introduction to Shape Optimization, Theory, Approximation and Computation,* SIAM, 2003.

10. Herskovits, J., Leontiev, A., Dias, G., & Santos, G., Contact shape optimization: a bilevel programming approach. *Int. J. of Struc. and Multidisc. Optim.,* 2000, 20, 214–221.

11. Ignizio, J., *Goal Programming and Extensions*, DC Heath Lexington, Massachusetts, USA, 1976.

12. A. K. Mahendra, R. K. Singh, and G. Gouthaman. Meshless kinetic upwind method for compressible viscous rotating flows. Computers & Fluids, 46:325–332, 2011.

13. Mohammadi, B., & Pironneau, O., *Applied Shape Optimization for Fluids,* Oxford University Press, 2nd Edn., 2009.

14. Miettinen, K. M., *Nonlinear Multiobjective Optimization*, Kluwer Academic Publishers, Boston, Massachusetts, USA, 1998.

15. Sashi Kumar, G. N., *Shape Optimization using a Meshless Flow Solver and Modern Optimization Techniques,"* MSc(Engg.) thesis, Indian Institute of Science, Bangalore, India, 2006.

16. Sashi Kumar, G. N., Mahendra, A. K., & Raghurama Rao, S. V., AIAA paper No.. 2007–3830, 18th AIAA-CFD Conference Miami, FL, 25th–28th June,. 2007.

17. Sashi Kumar, G. N., Mahendra, A. K., & Deshpande, S. M., Optimization using Genetic Algorithm and Grid-Free CFD solver, *Intl. Soc. of CFD/CFD J.*, 2008, 16(4), 425–433.

18. Sashi Kumar, G. N., Mahendra, A. K., Sanyal, A., & Gouthaman, G., Genetic algorithm-based optimization of a multi-stage flash desalination plant, *Desal. and Water Treat.*, 2009, 1(1–3), 88–106.

19. Sashi Kumar, G. N., Mahendra, A. K., & Gouthaman, G., Multiobjective shape optimization using ant colony coupled computational fluid dynamics solver, *Comp. and Fluids*, 2011, 46(1), 298–305.

20. Sederberg, T., & Parry, S., Free form deformation of solid geometric models, *Comp. Grap.*, 1986, 20(4), 151–160.

21. Tang, Z., Désidéri, J. A., & Périau, J., Multicriterion Aerodynamic Shape Design Optimization and Inverse Problems Using Control Theory and Nash Games, *J. Opti. Th. and Appl.,*. 2007, 135(3), 599–622.
22. Tiihonen, T., *Shape Optimization and Unilateral Boundary Value Problems*, PhD Thesis, U. of Jyvaskyla, Finland, 1987.
23. Wierzbicki, A. P., The use of reference objectives in multiobjective optimization, In Fandel, G., Gal, T., ed., *Multiple Criteria Decision Making Theory and Applications*, Springer-Verlag, 1980, p.468.
24. Zitzler, E., Deb, K., & Thiele, L., Comparison of multiobjective evolutionary algorithms: Empirical results, *Evol. Comp.,* 2000, 8, 173–195.

APPENDIX

An analytical MOO test case (cited as ZDT2 [24]) is illustrated here for better understanding to beginners in MOO. Each of the design parameters $x_i \in (0,1), i = 1,...5$ were discretized into 5 levels and 195 ($=5^5/2^4$ using Equation 5) ant paths were chosen randomly. The two objective functions were scalarized with respect to various reference points. It was observed that one arrives to a value on the Pareto front nearest to the reference point (see Table A1).

$$f_1(x) = x_1; \quad f_2(x) = g(x)[1 - (x_1/g(x))^2];$$

$$g(x) = 1 + 9(\sum_{i=2}^{5} x_i)/(n-1); \quad x_i \in [0,1]; \quad i = 1,2,...5 \quad \text{(A.1)}$$

TABLE A1 Optimized Results for Various Reference Points

Reference point	(0.2, 0.8)	(0.1, 0.5)	(0.6, 0.2)
Initial Min. distance	1.459	1.766	2.045
Final distance	0.141	0.452	0.194
# function evaluations	1950	975	4095
Best solution	(0.35, 0.89)	(0.63, 0.75)	(0.86, 0.27)

Data for test case 3: The properties of the Nitrogen used at 875 Pa (i)Thermal conductivity, $k = 1.68 \times 10^{-4}$ W/(mK), (ii) Viscosity, $\mu = 2.21 \times 10^{-5}$ kg/(m.s) at 20°C and (iii) N_2-O_2, $D_{AB} = 1.13 \times 10^{-5}$ $T^{1.724}$ atm-cm²/s

Codes included with this chapter

(1) MACO solver for a test function ZDT2.

(2) MACO solver coupled with nozzle solver for multi objective optimization of a supersonic nozzle. (i) Case 1 and (ii) Case 2 discussed in this chapter.

Pseudo code of ACO

1	*Initialize the domain for all parameters (Iteration = 0)*
2 10	*The entire range of all the parameters is discretized into levels*
3	*Iteration = Iteration + 1*
4	*Do i = 1, Number of ants (RSS)*
5	*Choose a path for ant i by selecting one level from each parameter randomly*
6	*Evaluate the Objective function for each ant pat*
7	*End Do*
8	*Calculate intensity of trial ($\tau_u(i)$) on each pathway u for the iteration i*

$$\tau_u(i) = \begin{cases} \exp\left(\ln(100)\left(\dfrac{L_{p,u} - L_{p,cr}}{L_{p,min} - L_{p,cr}} \right) \right), & L_{p,u} \leq L_{p,cr} \\ 0 & L_{p,u} > L_{p,cr} \end{cases}$$

where $L_{p,u}$ is value of the OF for the uth pathway, $L_{p,cr}$ is the critical value above which τ_u is forced to be zero. $L_{p,min}$ is the minimum of all L_p's

9   The trail share for each stratum β_{ij} (j = 1,2,...l) from each path can be summed to yield Φ_{ij}.

$$\Phi_{ij} = \sum_{u \in crossing\ pathways} \tau_u$$

Crossing pathways are all the ant paths that cross stratum β_{ij}

10   Evaluate the scores (S_{ij}) for each level in all of the parameters

$$S_{ij} = \frac{\left(\Phi_{ij}\right)^A \left(\sigma_{ij}\right)^N}{\sum_i \sum_j \left(\Phi_{ij}\right)^A \left(\sigma_{ij}\right)^N} \quad g_{cr} = T$$

σ_{ij} is the standard deviation of all values of the OF involving β_{ij}.

$A = 1$, $N = \eta_{ij}/\mu_g$ and $T = L_{p,min} + \sigma_g$.

μ_{ij}, η_{ij} are mean and standard deviation of trail for β_{ij} stratum.

μ_g, σ_g are mean and standard deviation of OF

11 Apply cut-off criteria (for example, $S_{ij} < S_{ij,Max}/2$ will be removed), which shrinks the range for the parameters

12 New range for the parameters are thus obtained

13 If (optimization meets convergence criteria) else goto 10

14 Print the best optimum obtained

CHAPTER 15

NATURE-INSPIRED COMPUTING TECHNIQUES FOR INTEGER FACTORIZATION

MOHIT MISHRA,[1] S. K. PAL,[2] and R. V. YAMPOLSKIY[3]

[1]*Department of Computer Science and Engineering, Indian Institute of Technology (Banaras Hindu University), Varanasi, India*

[2]*Senior Research Scientist, Scientific Analysis Group, Defence Research and Development Organization, Ministry of Defence, Govt. of India, New Delhi, India*

[3]*Associate Professor, Department of Computer Engineering and Computer Science, University of Louisville, KY, USA*

CONTENTS

15.1   THE PROBLEM OF INTEGER FACTORIZATION

Integer Factorization is a vital number theoretic problem often used in field of cryptography and security. Resolving a semi-prime number into its component prime factors is one of the most important arithmetic task in mathematics and computer science.

Integer factorization can be formally defined as the decomposition or resolving a composite number into its component prime factors, which are nontrivial. The most difficult instances of integer factorization problems are the task of figuring out the prime factors of a *semi-prime number.*

A semiprime number is a composite number comprising *two* non-trivial divisors which are prime and when multiplied together yield the semi-prime number under consideration. For instance, 77 is a semi prime number since it is composed of two prime factors, 7 and 11. The prime factorization task becomes even harder when:

1. The two prime numbers are randomly chosen, and are very large (of the order more than thousands of bits).
2. The two prime numbers are of similar sizes in terms of number of digits.

For such problems, no efficient polynomial time algorithm exists on non-quantum computers. However, Shor [2] introduced a quantum computer based algorithm that solves it. The reason is absolutely understandable. The integer factorization function is a one-way mathematical function [3]. Given two prime numbers, it is easy to multiply the numbers to produce a semi-prime number. However, given a semi-prime number, it is hard to compute the prime factors of the number which when multiplied together would yield the same semi-prime number. Because of the computational intractability, integer factorization is widely used in many public-key cryptographic protocols such as RSA encryption [1].

The state-of-the-art of the problem records the largest semi-prime known to have been factored is RSA-768 (232 digits) reportedly in December 2009 that Lenstra et al. [4]. This task was accomplished using a much-optimized version of the general number sieve algorithm (GNFS) [5], which has the best-known asymptotic running time complexity for a b-bit semi-prime number:

$$O\left(exp\left(\left(\frac{64}{9}b \right)^{\frac{2}{3}} \right)(\log b)^{\frac{2}{3}} \right) \tag{1}$$

It is essential that we discuss about the complexity nature of integer factorization problem, which can provide us with a deep understanding of the designing of functions that can help compute the prime factors of semiprime numbers. We discuss about the complexity classes involved in integer factorization in Section 15.2.

A number of approaches have been taken to solve the problem of integer factorization. Broadly and categorically, they can be classified into three different forms:

1. general;
2. special; and
3. alternative.

The general form aims at solving any type of number, in contrast to the special form where the algorithm takes advantage of the specific structure or form of the number involved, which results in better efficiency as compared to general forms. Alternative forms provide an altogether different ways of approaching the problem like genetic programming, swarm intelligence, neural networks, etc.

15.2   COMPLEXITY CLASSES OF THE PROBLEM

Integer Factorization can be represented in two different forms, which lead to different categories of complexity classes defining the problem's complexity [5]:

1. *Function Problem*: The function problem version can be stated: Given an integer N (a semi-prime number), find an integer/integers (prime number(s)) that divide N. This problem lies in NP, that is, it can be solved by a non-deterministic Turing machine in polynomial time. However, we cannot suggest if it does lie in FP, that is, if the problem is solvable by a deterministic Turing machine in polynomial time. This problem is more complex than a decision problem since it does not just involves a YES or NO as its output but involves figuring out the prime factors of the number.
2. *Decision Problem*: The decision problem for integer factorization can be defined as: Given an integer N (a semi-prime) and an integer. M such that $M \in [1,N]$, does R have a factor f such that $f(1,M)$? Thus the output is basically either YES or NO. This representation is useful because most well-studied complexity classes are defined as classes

of decision problems. In combination with a binary search algorithm, a solution-function to a decision version of Integer Factorization can solve the general case of Integer Factorization in logarithmic number of queries [15]. However, the determination of classifying the decision version into a certain specific complexity class still remains an open question.

Integer Factorization is known to belong to both nondeterministic polynomial time (NP) and co-NP classes, because both 'yes' and 'no' decisions can be verified given the prime factors either via polynomial time primality test such as AKS primality test [16, 17] or via simple multiplication of divisors.

15.3 STATE-OF-ART OF THE PROBLEM

In the quest for solving the integer factorization problem, we have seen a remarkable progress in the computationally difficult task, which directly affects the security systems based on the computationally intractability of integer factorization in polynomial times for non-quantum computers.

As discussed before in Section 15.1, overall, the approaches to solving the problem of integer factorization can be classified into three groups, namely Special, General and Alternative. Table 15.1 (Source: www.crypto-world.com [23]) presents a brief chronological order of the Integer Factorization by record and year using GNFS, MPQS, SNFS and QS.

Table 15.2 presents a brief chronological order of the top 10 Integer Factorization records using Elliptic Curve Method [24] (Source: Zimmerma [25]).

TABLE 15.1   Integer Factorization Records Using GNFS, SNFS, MPQS and QS*

Number	Digits	Algorithm	Date Completed
RSA-768	232	GNFS	Dec, 2009 [23]
RSA-200	200	GNFS	May, 2005 [23]
c176	176	GNFS	May, 2005 [23]
RSA-568	174	GNFS	Dec, 2003 [23]
RSA-160	160	GNFS	Mar, 2003 [23]
c158	158	GNFS	Jan, 2002 [23]
RSA-155	155	GNFS	Aug, 1999 [23]
RSA-140	140	GNFS	Feb, 1999 [23]
RSA-130	130	GNFS	April, 1996 [23]

TABLE 15.1   Continued

Number	Digits	Algorithm	Date Completed
P13171	119	GNFS	Nov, 1994 [23]
RSA-129	129	MPQS	April, 1994 [23]
RSA-120	120	MPQS	June, 1993 [23]
RSA-110	110	MPQS	April, 1992 [23]
RSA-100	100	MPQS	April, 1991 [23]
c116	116	MPQS	1990 [23]
$2^{1061}-1$	320	SNFS	Aug, 2012 [23]
$2^{1039}-1$	313	SNFS	May,. 2007 [23]
$(6^{353}-1)/5$	274	SNFS	Jan, 2006 [23]
$32633^{41}-1$	186	SNFS	Sept, 1998 [23]
$7^{352}+1$	128	GNFS+FPGA+DAPDNA-2	Sept, 2006 [23]
$2^{481}+2^{241}+1$	116	QS	June, 1996 [23]

*Legends: MPQS – multiple polynomial quadratic sieve (a variation of quadratic sieve (QS), GNFS – general number field sieve, FPGA – field programmable gate array, SNFS – special number field sieve. RSA – RSA security challenge number. c – Cunningham number co-factor.

TABLE 15.2   Top 10 Integer Factorization Records Using ECM

	Factor's digits	Number	B1/B2	Sigma	Date
1	83	$7^{337}+1$	7.6e9	3882127693	2013-Sep-07
2	79	$11^{306}+1$	800e6	3648110021	2012-Aug-12
3	77	N188	1e9	366329389	2013-Jun-15
4	75	$11^{304}+1$	1e9	3885593015	2012-Aug-02
5	75	EM47	850e6	2224648366	2012-Sep-11
6	74	$12^{284}+1$	600e6	2396540755	2014-Oct-26
7	73	$2^{1181}-1$	3e9	4000027779	2010-Mar-06
8	73	$2^{1163}-1$	3e9	3000085158	2010-Apr-18
9	72	$3^{713}-1$	6e8	1631036890	2012-Jan-01
10	72	$3^{560}-2$	29e8	2677823895	2013-Jan-02

Source: [25].

15.4   INTEGER FACTORIZATION AS DISCRETE OPTIMIZATION TASK

We can formulate the problem of integer factorization as discrete optimization task in the form of instances of integer programming problems in two ways:

<u>Form 1:</u> $minimize\ f(x) = (x^2 - y^2)(mod\ N)$

$constraints:\ (x - y)(mod\ N) \neq 0$

$$x, y \in \left[2, \frac{(N-1)}{2} \right]$$
(2)

Having found a pair *(x, y)* that satisfies the above equation, we can find then find the prime factors of the semi-prime number *N* by determining the GCD between either *(x-y, N)* or *(x+y, N)*.

<u>Form 2:</u> $minimize\ f(x) = N(mod\ x)$ (3)

$$constraint: x \in \left[2, \sqrt{N} \right]$$

Form 1 is a two-variable integer-programming problem. The output is a pair of integers *(x, y)* which satisfies *f(x, y)*. There are multiple solutions to this discrete optimization problem, which are almost symmetrical in nature (Figures 15.1 and 15.2), and thus we are at a privilege to just determine one such pair satisfying the problem statement. However, domain of the problem becomes enormously large with increasing *N*.

The problem of integer factorization can also be represented as Form 2, a one-variable integer-programming problem with a smaller domain as compared to Form 1 problem formulation. However, unlike Form 1 where we had a number of possible combinations of solutions, in case of Form 2, there

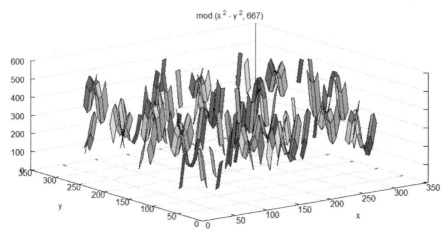

FIGURE 15.1   Surface plot of $(x^2 - y^2)$ mod(667) (produced by Octave).

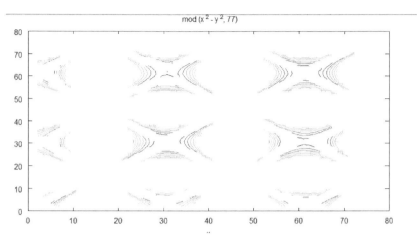

FIGURE 15.2   Contour plot of $(x^2 - y^2)$ mod(77) with domain [2, 76] (produced by Octave).

is exactly one solution. This form is a combination of the function problem and the decision problem. As a function problem, we need to figure x that would divide N. At the same time, as a decision problem, we need to determine if there exists a factor f such that $f \in (1, \sqrt{N})$.

Figure 15.1 shows the surface plot of Eq. (2) for N = 667. Same has been depicted as a contour in Figure 15.2 but for value of N as 77 (for better understanding). Further, if we make use of the symmetry observed in Figure 15.2 by scaling down the domain to [2, (N-1)/2], we obtain the contour plot as shown in Figure 15.3. Figure 15.4 presents a function plot for Eq. (3). As observed, there are multiple steep slopes and peaks, which can make convergence suboptimal.

It is vital to understand that the design of an objective function in case of integer factorization is high importance since the objective function drives the entire evolutionary or swarm computing process. Form 1 and Form 2 are exact mathematical functions that satisfy the property of a semi-prime number. We will also discuss about an alternative approach based on the degree of similarity of digits of the given semi-prime number and the number formed from the product of evolving prime numbers in Section 15.1.

15.5   TOWARDS NATURE-INSPIRED COMPUTING

The exact algorithms often incur huge memory and runtime expenses as we have seen in many such algorithms applied to various NP-Hard problems.

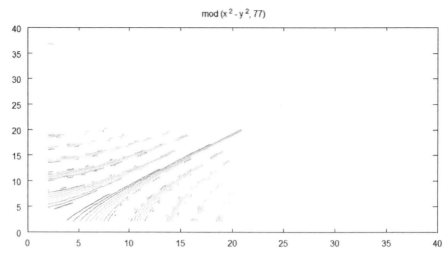

FIGURE 15.3   Contour plot of (x^2-y^2) mod(77) with domain [2, 38] (produced by Octave).

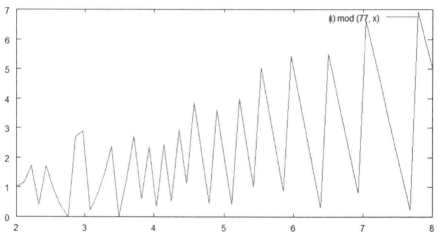

FIGURE 15.4   Function plot for 77 mod (x) with domain in Refs. [2, 25] (produced by Octave).

For a simple instance, dynamic programming applied to Traveling Salesman Problem leads to enormous memory and runtime cost.

On the other hand, approximation algorithms and soft computing techniques may not guarantee the exactness or accuracy of the solutions obtained. However, they may provide reasonably acceptable and approximate solutions in polynomial times with less memory overhead. A lot of optimization problems can solved using these algorithms where exact algorithms incur huge overhead upon time and memory.

In recent years, researchers have adapted nature-inspired computing techniques to solve a number of problems for whom no polynomial time algorithms exist. These techniques have proven to be highly successful and easy to implement with less memory and runtime complexity. Modeling natural phenomenon for our computationally difficult tasks provide a great deal of insight into working towards resolving solutions for our problems.

There have been such alternative approaches to solving integer factorization found in Refs. [7–12] which suggest the growing interest of researchers across the academia towards nature-inspired computing for integer factorization. We discuss about some recent research work done in this field using Genetic Algorithms [12], Firefly Algorithm [11] and Molecular Geometry Optimization Algorithm [21] as a part of the authors' contribution to this field.

15.5.1   GENETIC ALGORITHM

Genetic Algorithms [22] have been shown to solve a number of linear and nonlinear optimization problems. However, some optimization problems contain numerous local optima, which may be difficult to distinguish from the global maximum. This can result in suboptimal solutions. As a consequence, several population diversity mechanisms have been proposed to "delay" or "counteract" the convergence of the population by maintaining a diverse population of members throughout its search.

Evolution inspired algorithms face in solving problems in which the only measure of fitness is a binary – correct/incorrect result, since there is no possibility for the algorithm to converge on a solution via hill climbing. On one hand, one can assume that the integer factorization problem exhibits such binary information (a number is a factor or is not a factor), however, it is actually not the case as can be seen from the following trivial example, which demonstrates a series of partial solutions with gradual increase in fitness value of factor-approximating numbers.

The proposed GA is generational [6]:

- a population of N possible solutions is created
- the fitness value of each individual is determined
- repeat the following steps N/2 times to create the next generation a choose two parents using tournament selection b with probability p_c, crossover the parents to create two children, otherwise simply pass parents to the next generation c with probability p_m for each child, mutate that child d place the two new children into the next generation

- repeat new generation creation until a satisfactory solution is found or the search time is exhausted.

Yampolskiy in his paper [12] described a way to factorize a semi-prime composed of prime numbers p and q of equal number of digits. The measure of fitness considers the degree of similarity between the number being factored and the product of evolving factors p and q in terms of placement and value of constituting numbers as well as overall properties of the numbers such as size and parity.

The implementation involves assumption of a chromosome to be a string consisting of the factors (say p and q) of semi-prime N. We assume that the number of digits in the factors is same since we are interested in solving the most difficult cases of integer factorization.

In a typical genetic algorithm, there are two major operations, namely crossover and mutation. The crossover operation, in our case, is accomplished by exchanging subsets of varying size of numbers between individuals selected to be "parents." This is done according due the superiority of the fitness values. Mutation operation transfers a chosen digit to another digit, making sure that the legitimacy is not violated. For example, it is absolutely sure that the least significant digit cannot be even and the most significant digit cannot be zero.

Following, we present the algorithm used for integer factorization. For extensive details, readers are advised to refer to Ref. [12].

Algorithm:

1. Initialize chromosomes. Each chromosome is a string concatenation of p and q, the prime factors to be figured out.
2. Evaluate the fitness values by comparing the digits of the product of p and q with the given value of semi-prime number.
3. While (termination condition is not met).
4. Perform selection operation.
5. Perform cross-over between the parents selected to produce offspring.
6. Perform mutation on offspring by changing a digit to a legitimate digit.
7. End while.

It is interesting to note that this approach results in a number of suboptimal solutions because of presence of local extrema. This leads to leakage of information partially about the relationship between semi-prime number given to be factored, and product of candidates solutions for p and q. The example provided in Table 15.2 given a number $N = 4885944577$ [12] depicts such

TABLE 15.2 Partial Solutions to a Sample Factorization Problem With Increasing Fitness

Potential solution	Approximation of N	Fitness: number of matching digits
$p = 14531, q = 73341$	1065718071	1
$p = 54511, q = 43607$	2377061177	2
$p = 84621, q = 43637$	3692606577	3
$p = 84621, q = 41637$	3523364577	4
$p = 84621, q = 51637$	4369574577	5
$p = 94621, q = 51637$	4885944577	10

relationship. Observe that as the size of N increases the degree of inter-influence of digits of N located at a distance from each other further reduces. This results in more independent evaluation of partial solutions. In semi-prime numbers like RSA numbers the local maxima points come from numbers which are also a product of at least two integers and which match the number to be factored in terms of its constituting digits to a certain degree.

Such local maxima are frequent in the IF domain; in fact, Yampolskiy [12] showed that given any semi-prime number N with n decimal digits there are exactly $2 \times 10^{n-1}$ unique pairs of numbers p_i and q_i up to n digits each, which if multiplied, have a product matching all n digits of N precisely. For example for $N = 77$ ($n = 2$), that number is $2 \times 10^{2-1} = 2 \times 10^1 = 20$, or to list explicitly: (01 × 77 = 77), (03 × 59 = 177), (07 × 11 = 77), (13 × 29 = 377), (09 × 53 = 477), (21 × 37 = 777), (17 × 81 = 1377), (27 × 51 = 1377), (19 × 83 = 1577), (39 × 43 = 1677), (31×67 = 2077), (23 × 99 = 2277), (57×61 = 3477), (49 × 73 = 3577), (41 × 97 = 3977), (91 × 47 = 4277), (63 × 79 = 4977), (33 × 69 = 5577), (87 × 71 = 6177), and (93 × 89 = 8277).

The best result reported by Yampolskiy's approach [12] was a factorization of a 12 digit semi-prime (103694293567 = 143509 × 722563). This result took a little over 6 h on a Intel 2 core 1.86 GHz processor with 2 GB of RAM and was achieved with a population consisting of 500 individuals, two point crossover, mutation rate of 0.3% and genome represented via decimal digits.

15.5.2 FIREFLY ALGORITHM

The Firefly Algorithm [11] is a recent addition to the family of swarm-based metaheuristics. It is inspired by the flashing behavior of the fireflies.

The less bright fireflies get attracted towards brighter fireflies taking into account the media around the problem domain. Three assumptions are made while designing the algorithm:

1. All fireflies are unisex, such that each firefly gets attracted to all other fireflies.
2. The attractiveness factor is proportional to the brightness of the firefly. A less bright firefly will get attracted to a brighter firefly. Also, as the distance between two fireflies increase, the attractiveness factor between them reduces since the intensity of brightness reduces, which is inversely proportional to the square of the distance between the fireflies.
3. Landscape of the problem domain also affects the movement of the fireflies. That is, a media with a large absorption coefficient will reduce the intensity of brightness of the fireflies, thereby hampering movement.

The firefly algorithm has been exploited to solve a number of hard optimization problems. Suppose there are N fireflies, that is, N candidate solutions. Each candidate solution x_i has a cost associated with it, which is obtained from the objective function of the problem under consideration. The objective function determines the brightness function I_i of each firefly:

$$I_i = f(x_i) \qquad (4)$$

Based on the computed fitness values, the less bright firefly gets attracted and moves towards the brighter one. At the source, the brightness is higher than any other point in the environment. This brightness decreases with increase in the distance between any two fireflies. At the same time, surrounding media absorbs some amount of light determined by an absorption coefficient, γ. For two fireflies i and j, the fitness values vary inversely with distance between the two fireflies r_{ij}. Since light intensity follows the inverse square law, we have the light intensity as a function of the distance between any two fireflies at distance r from each other as:

$$I(r) = I_0 / r_{ij}^2 \qquad (5)$$

where I_0 is the light intensity at the source.

By incorporating the media absorption coefficient γ, light intensity can be represented in the Gaussian Form as:

$$I(r) = I_0 \exp(-\gamma r_{ij}^2) \tag{6}$$

The attractiveness function thus becomes:

$$\beta(r) = \beta_0 \exp(-\gamma r_{ij}^2) \tag{7}$$

where β_0 is the attractiveness value at r = 0.

The movement of a firefly in the search domain consists of two essential components:

a. attraction component (involving β), and
b. randomization component (involving α).

A firefly moves towards a brighter firefly, and while doing so, a small radius of randomization is maintained to exploit the search domain around the point where the firefly has reached while moving towards the brighter firefly, thus applying the principle of exploration and exploitation.

In the classical firefly algorithm, when firefly i is less bright than the firefly j, the movement of the firefly i follows the equation:

$$x_{ik} = x_{ik} + \beta_0 e^{-\gamma r_{ij}^2}\left(x_{ik} - x_{jk}\right) + \alpha S_k \left(rand_{ik} - 0.5\right) \tag{8}$$

where k = 1,2,3,…D (D is the dimension of the problem in consideration). The term with α is called the randomization parameter. S_k is the scaling parameter over the domain of the problem.

$$S_k = |u_k - l_k| \tag{9}$$

where u_k and l_k are the upper and lower bounds of x_{ik}. Distance between two fireflies i and j are calculated by the Cartesian distance between them:

$$I(r) = I_0 \exp(-\gamma r_{ij}^2) \tag{10}$$

Mishra et al. [11] adapted a multithreaded bound varying chaotic firefly algorithm for factorizing semi-prime numbers. The objective function used was Form-1 function as described in Section 15.3:

$$f(x) = N \ (mod \ x) \tag{11}$$

where N is the semi-prime to be factored out, and $x \in (lower\_bound,$ square_root $(N))$. The lower_bound is defined as

$$lower\_bound = 10^{d-1} \tag{12}$$

where d is the number of digits in the square root of N floored to an integer value. Since the above objective function is dependent on a single independent variable, we can parallelize our function and introduce multithreading. In this case, each thread works on a domain size of 10^{d-1}. Each thread will return local minima, out of which one will be the global minimum when the objective function evaluates to zero, giving us one factor of our semi-prime in consideration. The other factor will then be easily computed.

Further, Mishra et al. [11] introduce chaos in the algorithm parameters α and γ using Logistic maps [18] and [19], and update them after each iteration as follows:

$$\alpha_{t+1} = \mu_1 . \alpha_{t.} (1 - \alpha_t) \tag{13}$$

$$\gamma_{t+1} = \mu_2 . \gamma_{t.} (1 - \gamma_t) \tag{14}$$

where t is the sample, μ_1 and μ_2 are control   parameters,

$$0 \leq \mu_1, \mu_2 \leq 4.$$

The best result reported was factorization of 14 digit semi-prime $51790308404911 (= 5581897 \times 9278263)$ in 1880.6 iterations with 500 fireflies.

Algorithm:

1. Initialize the algorithm parameters – $\alpha_0 (\neq \{0.0, 0.25, 0.5, 0.75, 1.0\})$, β_0 and γ.
2. Initialize the firefly population $(x_1, x_2, x_3,..., x_n$. Each firefly is a candidate solution, i.e., it represents the prime factors.
3. While (t < MaxGeneration)
4. for i = 1 : n (all n fireflies)
5. for j = 1 : n (n fireflies)
6. if $(f(x_i) > f(x_j))$, // where f(x) is same as Eq. (2)/Eq. (3)
7. move firefly i towards j;
8. end if
9. Vary attractiveness with distance r via Eq. (6)
10. Evaluate new solutions and update light intensity;

11. end for j
12. end for i
13. Rank fireflies and find the current best;
14. end while
15. Report the best solution
16. end

In our experiments, we initialized α_0 to 0.65, β_0 to 1.0 and γ to 0.6. The firefly population was initialized to 500 and maximum iterations set to 5000. Each test case was run 10 times. The results of the experiments are shown in Table 15.3. For extensive details on this work, readers may refer to our previous work [11].

15.5.3 MOLECULAR GEOMETRY OPTIMIZATION

Molecular Geometry Optimization (MGO) Algorithm [21] is inspired by the computational chemistry behind the arrangement of a group of atoms in space such that the net inter-atomic forces between the atoms are minimized (close to zero). This would lead to minimum surface energy potential which any molecule would try to achieve, that is, the position on the potential energy surface is a stationary point.

For instance, while optimizing the geometry of a water molecule, the aim is to establish the hydrogen-oxygen bond lengths and the hydrogen-oxygen-hydrogen bond angle which would minimize the inter-atomic forces that would otherwise be either pulling or pushing the atoms apart.

However, it should be noted that not always the molecular geometry optimization process seeks to obtain global or local minimum energy, but may optimize to a saddle point on the potential energy surface (transition state), or to fix certain coordinates as may be the case.

TABLE 15.3   Experiment Results and Observations with Firefly Algorithm

Test Case	Size in Bits	Success Rate (%)	Mean Iterations
5325280633	33	100	22.4
42336478013	35	100	298.8
272903119607	38	100	378.0
11683458677563	44	80	1976.5
51790308404911	46	80	1880.6

The positions of atoms are represented by position vectors which in turn can be represented by Cartesian Coordinates. With such representation, one can introduce energy function as a function of the position vector r, $E(r)$. With such energy function, the molecular geometry optimization essentially becomes a mathematical optimization task, in which the aim is to find the position vectors of the atoms such that it $E(r)$ is at a local minimum. That means, the derivative of the energy function with respect to the position of the atoms $\partial E/\partial r$ is a zero vector $\boldsymbol{0}$. For such an arrangement, the second derivative matrix of the system, $\partial \partial E/\partial r_i \partial r_j$, known as Hessian Matrix, is positive definite, that is, all the Eigen values are positive.

We use the above concept in modeling an algorithm for our problem of integer factorization. Below is the description of the Molecular Geometry Optimization (MGO) Algorithm:

1. Initialize the position of atoms which represent the solutions with values drawn from uniform random distribution. Let n be the number of atoms (solution pool)
2. While a stopping criterion is met, follow steps through 3 to
3. For i=1:n
4. For j=1:n
5. Calculate the force between atoms i and j, $f_{i,j}$.
6. If force > target threshold:
7. Move the atoms by some computed step Δr that is predicted to reduce the force
8. If the new position yields the energy function lower than the current energy function value of atom i, update the energy value with the energy at this new position.
9. Go to step 3.

A. Energy Function

The energy function in the algorithm represents the objective functions defined in Section III.

B. Force between two atoms

The force function plays a vital role in determining the movement of the atoms. Theoretically, the force function is the derivative of the energy function with respect to the position vector of the atom. But in our case, we are dealing with discrete spaces, and hence the energy functions are not differentiable. However, we represent the force function in a way that closely represents a differentiable form. The force function is different for different energy functions and position vectors.

For Form-1 of the objective function in Section 15.3, we adopt the following force function between two atoms i and j, when atom i moves towards j, or towards a better possible solution:

$$force(i,j) = (2x_i\delta r_{i,j} - 2y\delta r_{i,j})(mod\ N) \tag{15}$$

where $\delta r_{i,j} = |r_i - r_j|$

Formulation of force function for Form-2 of the objective function in Section 15.3 is a difficult task, and is open to various formulations as long as it closely represents the force between the two atoms that can provide an effective measure for movement of the atoms around. In our case, we adopt the following force function:

$$force(i,j) = E(r_i)r_i - E(r_j)r_j \tag{16}$$

C. Movement equation

The movement equation involves a computed step Δr which is predicted to reduce the force between atoms i and j. With this in view the movement equation for our problem consists of an attraction term and a randomization term. We formulate the movement equation as follows:

$$r_i = r_i + rand.(r_j - r_i) + randomization\_term \tag{17}$$

where $rand()$ is a function that generates a random number $\in (0,1)$ drawn from a random distribution (uniform in our case). The randomization term computes a random step-size in a way that r_i does not cross the boundary of the search domain. To implement such a randomization term, the following steps are followed:

1. Calculate r_i while excluding the randomization term, i.e., $r_i = r_i + rand().(r_j - r_i)$
2. $u = r_i - lower\_bound$
3. $v = upper\_bound - r_i$
4. $s = rand()-0.5$
5. if $(s<0)$
6. $rand\_factor = -1.\alpha.rand()*u$
7. else

8.   *rand_factor = α.rand().v*
9. $r_i = r_i + rand\_factor$

α is the randomization parameter drawn from Logistic Map [18, 19] as follows:

$$\alpha = \mu. \, \alpha. \, (1 - \alpha) \qquad\qquad (18)$$

where μ is the control parameter which is 4 in our case. α was initialized to 0.65.

Table 15.4 presents the testing of the algorithm with Form-2 type of objective function on semi-prime numbers extending upto 46 bits with 100 atoms (solution vectors) and maximum iterations set as 5000.

15.6   CONCLUSION

Recent advances in nature-inspired computing techniques have proven to possess a great potential in solving some of the hardest problems in computer science in an efficient manner. In this chapter, we explored how such techniques can help solve the problem of integer factorization though we do not assert that these will guarantee the results and if they are scalable or not. We presented three algorithms as part of the authors' contribution in this field, namely Genetic Algorithm, Firefly Algorithm and Molecular Geometry Optimization. The preliminary results presented in Refs. [11], [12] and [21] show that these algorithms provide a deep insight into the problem as well as provide new directions to solve the problem. The research work also talks about the difficulty in solving the problem owing to the designing of appropriate objective function and binary nature of the problem. Readers are

TABLE 15.4   Observations for Form-2 Using MGO Algorithm on Bigger Semi-Primes

Test Case	Size in Bits	Success Rate (%)	Mean Iterations
5325280633	33	100	46.26
42336478013	35	100	168.30
272903119607	38	100	251.48
11683458677563	44	85	1473.00
51790308404911	46	69	2143.35

advised to refer to Ref. [26] for details on the limitations of bio-inspired approaches for problems of binary nature. However, it is a great progress in this field in exploring the power of natural phenomena in optimizing hard computing problems like integer factorization itself.

KEYWORDS

- firefly algorithm
- genetic algorithm
- integer factorization
- molecular geometry optimization
- swarm intelligence

REFERENCES

1. Rivest, R., Shamir, A., & Adleman, L. "A Method for Obtaining Digital Signatures and Public-Key Cryptosystems," Communications of the ACM, Feb. 1978, vol. 21, Issue 2, pp. 120–126.
2. Shor, P. W. "Polynomial time algorithms for prime factorization and discrete logarithms on a quantum computer," SIAM Journal Sci. Statist. Computing, 1997, Vol. 26, pp. 1484–1509.
3. Goldwasser, S., & Bellare, M. "Lecture Notes on Cryptography," July 2008, retrieved from http://cseweb.ucsd.edu/~mihir/papers/gb.pdf in Jan–Feb 2013.
4. Kleinjung, T. et al., "Factorization of a 768-Bit RSA Modulus," Advances in Cryptology – CRYPTO 2010, Lecture Notes in Computer Science, 2010, vol. 62223, pp. 333–350.
5. Pomerance, C. "A Tale of Two Sieves," Notices of the AMS, 1996, Vol. 43, pp. 1473–1485.
6. Yampolskiy, R., Anderson, P., Misic, P., Arney, J., Misic, V., & Clarke, T. "Printer model integrating genetic algorithm for improvement of halftone pattern," Western New York Image Processing Workshop (WNYIPW), IEEE Signal Processing, 2004.
7. Society, Rochester, N. Y., & Chan, D. M. "Automatic generation of prime factorization algorithms using genetic programming," Genetic Algorithms and Genetic Programming at Stanford. pp. 52–57, Stanford Bookstore. Stanford, California, 2002.
8. Meletiou, G., Tasoulis, D. K., & Vrahatis, M. N. "A first study of the neural network approach to the RSA cryptosystem," IASTED 2002 Conference on Artificial Intelligence, Banff, Canada, 2002, pp. 483–488.
9. Jansen, B., & Nakayama, K. "Neural networks following a binary approach applied to the integer prime-factorization problem," IEEE International Joint Conference on Neural Networks (IJCNN), July 2005, pp. 2577–2582.

10. Laskari, E. C., Meletiou, G. C., Tasoulis, D. K., & Vrahatis, M. N. "Studying the performance of Artificial Neural Networks on problems related to cryptography," Nonlinear Analysis: Real World Applications, 2006, Vol. 7, pp. 937–942.

11. Mishra, M., Chaturvedi, U., & Pal, S. K. "A Multithreaded Bound Varying Chaotic Firefly Algorithm for Prime Factorization," IEEE International Advance Computing Conference (IACC), Feb 2014, Gurgaon, pp. 1321–1324.

12. Yampolskiy, R. V. "Application of bio-inspired algoritm to the problem of intger factorization," International Journal of Bio-inspired Computation. 2010, Vol. 2, No. 2, pp. 115–123.

13. "Energy Minimization," Wikipedia, The Free Encyclopedia, Retrieved from http://en.wikipedia.org/wiki/Energy_minimization on Jan. 10, 2014.

14. Elaine Rich, "Automata, computability and complexity: theory and applications," Prentice Hall, 2008.

15. Adi Shamir, E. T. "Factoring large numbers with the TWIRL device," Crypto – The 23rd Annual International Cryptology Conference, Santa Barbara, California, USA, pp. 1–26, 2003.

16. Agrawal, M., Kayal, N., & Saxena, N. 'PRIMES is in P,' Annals of Mathematics 2004, 160 (2), 781–793.

17. Brent, R. P. "Recent progress and prospects for integer factorisation algorithms," Computing and Combinatorics: Sixth Annual International Computing and Combinatorics Conference, Sydney, Australia, pp. 3–22, 2000.

18. May, R. M. "Simple mathematical models with very complicated dynamics," Nature 1976, 261(5560), pp. 459–467.

19. Eric, W. Weisstein, "Logistic Equation," MathWorld. Retrived from http://mathworld.wolfram.com/LogisticEquation.html on Jan. 14, 2014.

20. Laskari, E. C., Meletiou, G. C., & Vrahatis, M. N. "Problems of cryptography as discrete optimiation tasks," Nonlinear Analysis, 2005, Vol. 63, pp. 831–837.

21. Mishra, M., Chaturvedi, U., & Shukla, K. K. "A New Heuristic Algorithm based on Molecular Geometry Optimization Algorithm and its Application to the Integer Factorization Problem," Accepted for publication in the proceedings of the 2014 Intl. Conference on Soft Computing and Machine Intelligence, Sept. 26–27, 2014, New Delhi, India.

22. Goldberg, D. E. "Genetic Algorithms in Search, Optimization and Machine Learning, Reading, MA, Addison-Wesley Professional, 1989.

23. Factorization Announcements, crypto-world. Retrieved from www.crypto-world.com/Factor/FactorAnnoucements.html on July 09, 2015.

24. Zimmermann, P. "Elliptic Curve Method for Factoring," Encyclopedia of Cryptography and Security, Springer, 2005.

25. 50 Largest Factors Found By ECM. Retrieved from www.loria.fr/~zimmerma/records/top50.html on July 09, 2015.

CHAPTER 16

GENETIC ALGORITHM BASED REAL-TIME PARAMETER IDENTIFIER FOR AN ADAPTIVE POWER SYSTEM STABILIZER

WAEL MOHAMED FAYEK[1] and O. P. MALIK[2]

[1]*Assistant Professor, Department of Electrical Engineering, Helwan University, Cairo, Egypt*

[2]*Professor Emeritus, Department of Electrical and Computer Engineering, University of Calgary, Alberta, Canada*

CONTENTS

Electric power systems are complex multi-component dynamic systems. Spread over vast geographical areas and no longer operating as isolated systems, but as interconnected systems, they are subjected to many kinds of disturbances and abnormalities. Their characteristics vary with the variation of loads and generation schedules. The present-day tendency of operating generators with relatively small stability margins has made these systems even more fragile. A good system should have the ability to retain its normal operating condition after a disturbance. As stability is ultimately concerned with the quality of the electric power supply, it is considered as one of the main topics of power systems research.

Most power system elements are highly non-linear and some of them are combinations of electrical and mechanical parts that have very different dynamic behavior. Interaction between electrical and mechanical parts within an individual element and the interactions between the elements results in complicated system dynamics and transient behavior. Consequently, these lead to various kinds of unstable characteristics.

Although there are several sources of positive damping in a power system, there are also sources of negative damping, notably voltage-regulating and speed-governing systems. Furthermore, although the natural positive damping predominates, in some circumstances the net damping is negative.

If the natural damping of a power system is negative, the cure is to add artificial positive damping. The most important use of artificial damping is to make net damping positive. Even if the natural damping is already positive, there is a considerable benefit in increasing it artificially to achieve stronger damping.

Most of the generating units installed in the electric utility systems are equipped with continuously acting voltage regulators. The high gain voltage regulator action has a detrimental impact upon the dynamic stability of the power system. Oscillations of small magnitude and low frequency often persist for long periods of time and in some cases present limitations on power transfer capability. Power System Stabilizers (PSSs) were developed to aid in damping these oscillations via modulation of the generator excitation. This development has involved the use of various tuning techniques and input signals, employing various control strategies, and learning to deal with practical problems such as noise and interaction with turbine-generator shaft torsion modes of vibration.

A combined on-line Genetic Algorithm (GA) identifier and Pole Shift (PS) linear feedback controller as an Adaptive Power System Stabilizer (APSS) is presented in this Chapter. Genetic algorithms have excellent search capabilities and produce good results in tracking the dynamics of a system. The PS control is preferable for its robustness and stability conditions. The hybrid system combines their advantages while avoiding their weaknesses. The identifier goes through two stages of learning, off-line training and on-line update. The off-line training is used to store a priori knowledge about the system. After the off-line training the identifier is further updated every sampling period making it an adaptive approach. The GA population is searched every sampling instant to obtain its best member as the system regression coefficients. The PS control uses these regression coefficients to calculate the closed-loop poles. The unstable poles are moved inside the unit circle in the z-plane and the control is calculated so as to achieve the desired performance. A third order Auto Regressive model with eXogenous signal (ARX) has been used to model the power system dynamics.

The main drawback of GA is the long time required for convergence. A new method, suitable to the problem under study, is proposed to greatly reduce the time required to identify the system parameters in real-time for the on-line identifier. To the authors; knowledge it is the first time that GA gas been used for real-time control.

Simulation studies are carried out on a single-machine infinite bus power system to assess the performance of the APSS. After successful simulation studies, the APSS is further tested on a scaled physical model of a power system. The performance of the APSS shows significant benefits over Conventional Power System Stabilizer (CPSS) in terms of performance improvement and no requirement for parameter tuning.

16.1 POWER SYSTEM STABILIZERS

The basic function of a power system stabilizer is to extend stability limits by modulating generator excitation to damp the oscillations of synchronous machine rotors relative to one another. These oscillations of concern typically occur in the frequency range of approximately 0.2–2.5 Hz, and insufficient damping of these oscillations may limit the ability to transmit power. To provide damping, the stabilizer must produce a component of electrical torque on the rotor, which is in phase with speed variations. The implementation details differ, depending upon the stabilizer input signal and control strategy employed. However, for any input signal the transfer function of the stabilizer must compensate for the gain and phase characteristics of the excitation system, the generator, and the power system, which collectively determine the transfer function from the stabilizer output to the component of electrical torque that can be modulated via excitation control. This transfer function is strongly influenced by voltage regulator gain, generator power level, and AC system strength.

16.1.1 CONVENTIONAL POWER SYSTEM STABILIZERS

The most commonly used PSS, referred to as the Conventional PSS (CPSS), is a fixed parameter lead-lag type of device. The first CPSS, proposed in 1950's, is based on the use of a transfer function designed using the classical control theory [1]. A supplementary stabilizing signal derived from speed deviation, power deviation or accelerating power, and a lead-lag compensating network to compensate for the phase difference from the excitation controller input to the damping torque output, is introduced to the excitation controller. By appropriately tuning the phase and gain characteristics of the compensation network, it is possible to make a system have the desired damping ability. The CPSS is widely used in today's excitation controls and has proved effectiveness in enhancing power system dynamic stability.

The CPSS, however, has its inherent drawbacks. It is designed for a particular operating condition around which a linearized transfer function model of the system is obtained. Usually the operating condition where control is needed most is chosen. The high non-linearity, very wide operating conditions and stochastic properties of the actual power system present the following problems to the CPSS:

- How to choose a proper transfer function for the CPSS that gives satisfactory supplementary stabilizing signal covering all frequency ranges of interest?
- How to effectively tune the PSS parameters?
- How to automatically track the variation of the system operating conditions?
- How to consider the interaction between the various machines?

Extensive research has been carried out to solve the above problems. Different CPSS transfer functions associated with different systems have been proposed [2]. Various tuning techniques have been introduced to effectively tune PSS parameters [3]. Effective placement and mutual cooperation between the PSSs in multi-machine systems are also presented [4]. To solve the parameter-tracking problem, variable structure control theory was introduced to design the CPSS [5]. All this research has resulted in great progress in understanding the operation of the PSS and effectively applying PSS in the power systems. However, it cannot change the basic fact – the CPSS is a fixed-parameter controller designed for a specific operating point, which generally cannot maintain the same quality of performance at other operating points [6]. It is for this reason that adaptive control, control that 'adapts' to changing system characteristics, has so much potential to improve power system performance. This idea has led to the research and development of Adaptive Power System Stabilizers (APSSs).

16.1.2 ADAPTIVE POWER SYSTEM STABILIZERS

The adaptive control theory provides a possible way to solve the above-mentioned problems relating to the CPSS. All adaptive control techniques can be classified in two categories:

1. *Direct Adaptive Control* – In this type of control, the objective is to adjust controller parameters so that the output coordinate of the controlled system agrees with that of a reference model. Usually the value of mismatch between the controlled coordinates (y_p) of the system and the model (y_r) is used to perform parameter adjustment.

$$\|y_p - y_r\| = \varepsilon \tag{1}$$

   where ε is called the mismatch error. This kind of adaptive control is often referred as the model reference adaptive control (Figure 16.1). The performance of this algorithm depends on the choice of a suitable reference model and the derivation of an appropriate learning mechanism.

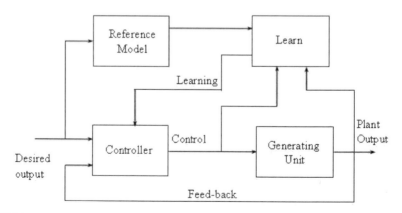

FIGURE 16.1   Direct Adaptive Control.

2. *Indirect Adaptive Control* – In this kind of adaptive control, the objective is to control the system so that its behavior has the given properties. The controller can be thought in terms of two loops. One loop, called the inner loop, consists of the controlled process and an ordinary linear feedback controller. The parameters of the controller are adjusted by the second loop, or outer loop, which is composed of an on-line parameter identifier. This kind of adaptive control is often referred as the self-tuning adaptive control (Figure 16.2).

16.2   SELF-TUNING ADAPTIVE CONTROL

Self-tuning adaptive control is one of the most effective and well-established indirect adaptive control techniques. The self-tuning property lies in its on-line identifier, which is used to estimate the varying parameters of the plant. The control is obtained based on the estimated parameters. The technique has gained recognition because of its flexibility (can be applied to different plants with minimal changes), auto-tuning properties, and ease of implementation using Programmable Logic Controllers (PLCs).

The structure of the self-tuning APSS is shown in Figure 16.2. From this structure, it can be seen that the development of such a device involves the following two parts, an on-line parameter identifier and a controller.

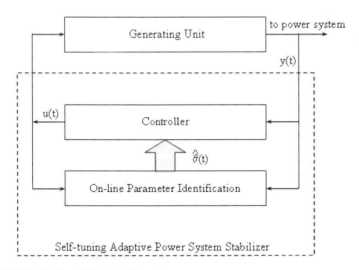

FIGURE 16.2   Indirect Adaptive Control.

16.2.1 ON-LINE PARAMETER IDENTIFIER

This part is the essence of the APSS, which gives the PSS the ability to adapt. At each sampling instant, input and output of the generating unit are sampled, and a mathematical model is obtained by some on-line identification method to represent the dynamic behavior of the generating unit at that instant of time. For a time varying stochastic system, such as a power system, its dynamic behavior varies from time to time. With this on-line identifier, it is expected that the mathematical model obtained each sampling period can track changes in the controlled system. Obviously, the extent to which the identified model fits the dynamics of the actual generating unit determines the failure or success of the APSS. It is for this reason that the on-line identification methods have always been the main subject of research.

A number of identification methods have been proposed, from off-line to on-line methods [7–9]. It is difficult to say which one is the best for all adaptive control applications. Some methods may be more suitable for some specific applications, while others may not be.

16.2.2 CONTROLLER

This part produces the control signal for the generating unit based on the identified model. The control strategy is generally developed by assuming that the identified model is the true mathematical description of the generating unit. However, since the power system is a complex high-order non-linear stochastic continuous system, it is hard for the discrete identified model to precisely describe the dynamic behavior of the power system. Consequently, it is desirable that the control strategy has good tolerance to the errors in the identified model.

The above discussion on identifier and controller highlights two main points:

- Firstly, the on-line identifier should be improved to achieve an identified model, which represents the controlled system as closely as possible.
- Secondly, the control strategy should have the ability to tolerate the identification errors.

With these two parts working together, successful application of the APSS can be achieved.

The identified model of the generating unit is often a Non-Minimum-Phase (NMP) system, which restricts the application of some control strategies in this situation. To overcome the excessive control signal amplitude problem and the unstable feature of the NMP closed-loop system, aGeneralized Minimum Variance (GMV) technique has been proposed and widely used [10–12]. Though it is simple, its closed-loop performance does not often meet the needs of the controller designer. For example, stability, which is an important issue in control design, is not taken into consideration in the GMV control.

Pole Assignment (PA) control strategy is also well known in adaptive control applications. It puts the emphasis on the closed-loop stability rather than the time domain responses. By choosing the closed-loop poles properly, the closed-loop system can be robust to some extent even if the identified model has some errors. However, selection of the proper closed-loop pole locations to meet the need of both the stability and time domain response requirements offers a hard task to the designer.

Pole Shifting (PS) control strategy is a modified version of the PA control strategy [13]. In this strategy, controller parameters are selected on the basis that the closed-loop poles are shifted from the identified open-loop poles in a stabilizing direction (radially inward away from the unit circle in the z-domain). The amount of shift is determined by one parameter, called the pole-shifting factor, which can be adjusted to achieve desired damping. This control strategy simplifies the PA strategy because only one parameter needs to be selected. Obviously, the way to select the pole-shifting factor determines the behavior of the controller.

16.3   COMPUTATIONAL INTELLIGENCE BASED APSS

The last two to three decades have seen a growing interest in applying Artificial Intelligence (AI) techniques such as fuzzy logic, Artificial Neural Networks and evolutionary algorithms for on-line control of power systems to overcome the problems associated with CPSS.

16.3.1   FUZZY LOGIC BASED APSS

The Fuzzy Logic Control (FLC) technique appears to be the most suitable one whenever a well-defined control objective cannot be specified, the system to be controlled is a complex one, or its exact mathematical model is not available. FLCs are robust and have relatively low computation requirements. This decreases the development time and cost.

PSS based on FLC is an active area and satisfactory results have been obtained [14, 15]. Although FLC introduces a good tool to deal with complicated, nonlinear and ill-defined systems, it has the following limitations:

- It is not always easy to construct a rule-base for FLC.
- The selection of membership functions and parameter tuning is a non-trivial task.

The above problems have been overcome using neural networks, self-organizing networks and genetic algorithms to find the rule sets, parameter tuning and optimum number of memberships [16]. However, the complexity of the above models often discourages the user from incorporating the above models in the FLC design process. This limits the application of FLC. Furthermore, the membership functions are decided off-line and then varied during the on-line operation. This sometimes negates the meaning of membership function in FLC.

16.3.2   NEURAL NETWORK BASED APSS

Artificial Neural Networks (ANNs) attempt to achieve good performance via dense interconnection of simple computational elements. ANN structure is based on the present understanding of biological nervous system. The ability to learn is one of the main features of the Neural Networks (NNs) [17]. ANNs can also provide, in principle, significant fault tolerance, since damage to a few links need not significantly impair the overall performance. The massive parallelism, natural fault tolerance and implicit programming of NN computing architectures suggest that they are good candidates for implementing real-time controllers for non-linear dynamic systems, such as power systems.

System identification and control using neural networks was proposed in a pioneering work [18]. Neural networks allow many of the ideas of system identification and adaptive control originally applied to linear and non-linear (or linearized) dynamic systems to be generalized, so as to cope with more severe non-linearities. In various adaptive control techniques (Figures 16.1 and 16.2), ANNs can replace both identifier and controller. Since ANNs have the capability to learn arbitrary non-linearity by adapting their weights they are good candidates for adaptive control applications.

Stability is another important issue in designing ANN based control schemes. Stability of neural network learning algorithm does not necessarily mean stability of the closed-loop system. A detailed review of the neuro-control techniques can be found in Ref. [19].

The disadvantages of the ANN control schemes are:

- The 'black-box' like description of the ANN. It is difficult for an outside user to understand the control process. This discourages the user from applying the control scheme that may yield satisfactory results but cannot describe how the control scheme is obtained.
- ANN may require a long training time to obtain the desired performance. The larger the size of the ANN and the more complicated the mapping to be performed, the longer the training time required.
- The selection of the number of hidden layers and the number of neurons in each layer is not a trivial task. It is, to a large extent, a process of trial and error.

To overcome these problems evolutionary algorithms have been proposed.

16.3.3 GENETIC ALGORITHM BASED APSS

Genetic Algorithms (GAs) are search algorithms, which are based on the genetic processes of biological evolution. GAs work with a population of individuals, each representing a possible solution to a given problem. Each individual is assigned a fitness score according to how well it solves the given problem. The highly adapted individuals will have relatively large number of off-springs. Poorly performing ones will produce few or even no off-springs at all. The combinations of selected individuals produce super-fit off-springs, whose fitness is greater than that of the parents. In this way, the individuals evolve to become more suited to their environment [20]. For these reasons, PSS based on GA is an active area and satisfactory results with off-line training of CPPs have been obtained [21–23].

Although GA is a global optimization method it has the following limitations [20]:

- The optimal solution is determined by going through a number of generations. However, the number of generations necessary to ensure that the most-fit individual is found is a priori unknown.
- Since there are many parameters involved in the algorithm, there is no guarantee that the genetic algorithm can reach a near optimal solution.

If the parameters are not properly selected, it can fall into a local optimal point depending on the topology of the search space.

16.4   SYSTEM CONFIGURATION AND MODEL

The system under study is a single-machine connected to a constant-voltage bus through a double circuit transmission line as shown in Figure 16.3. The sampling frequency is 20 Hz. The mathematical model and parameters of the generator, AVR and governor are given in Appendix A.

The generating unit is identified as a third order ARX discrete model of the form:

$$A(z^{-1})y(t) = B(z^{-1})u(t) + \zeta(t) \tag{2}$$

where $y(t)$, $u(t)$ and $\zeta(t)$ are the system output, system input and white noise, respectively. $A(z^{-1})$ and $B(z^{-1})$ are polynomials in the backward shift operator z^{-1} defined as:

$$A(z^{-1}) = 1 + a_1 z^{-1} + a_2 z^{-2} + a_3 z^{-3} \tag{3}$$

$$B(z^{-1}) = b_1 z^{-1} + b_2 z^{-2} + b_3 z^{-3} \tag{4}$$

The continual re-estimation of the plant model parameters, a_i and b_p, is called recursive parameter estimation such that, at the commencement of each

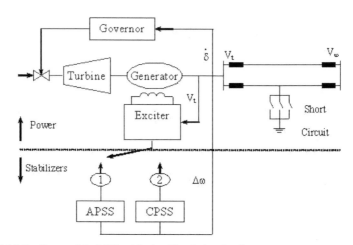

FIGURE 16.3   System Model Used in the Simulation Studies.

sample interval, the estimations obtained during the previous recursion are made available and form a start-up point ready to be updated. A GA based on-line identifier is used for updating these parameter estimates.

The controller synthesis stage can take any one of a number of different forms dependent upon the specified requirements of the controller and overall control objective. A variable PS control algorithm is used.

16.4.1   GENETIC ALGORITHM BASED ON-LINE IDENTIFIER

Referring to Figure 16.2, new control is computed each sampling interval on the assumption that the updated system parameter estimate obtained by the system identifier is 'accurate.' This is carried out in the same way as an off-line procedure. However, in a self-tuning controller it is carried out recursively on-line. This means that while the model parameter estimates are good, the controller output is good, whereas if the model parameter estimates are bad then almost surely the computed control will be bad. This point is particularly important on controller start-up, in that the self-tuning control algorithm will work even if fairly arbitrary initial estimates are entered for the model parameter estimates. This may provide pretty "violent" control; however, the estimates will converge to their true values in a short-space of time.

Different identification quality criteria will result in different identification schemes. Selection of the identification algorithm mainly depends on the mathematical model used. Generally speaking, more sophisticated identification methods will require more computation time. For this reason, when designing an on-line identifier, a compromise must be made between the quality of identification and a reasonable computation time among all possible identification methods.

GA is used as it needs few computations and can be adapted to give better performance. Also, the number of generations needed for convergence is small which is suitable for online identification.

The proposed identifier uses a standard real-valued GA algorithm. The population size is chosen to consist of 99 individuals. The members of the new generation are chosen using the Stochastic Universal Sampling method [24]. At each generation, the member with the best fitness is compared with the best member of the previous population and the fittest to the current measurements is chosen as the best member of the current population.

Rewriting Eq. (2) in the following form suitable for identification:

$$y(t) = \theta^T(t)\varphi(t) + \zeta(t) \tag{5}$$

where $\theta(t)$ is the parameter vector and $\phi(t)$ is the measurement vector defined as

$$\theta(t) = \begin{bmatrix} a_1 & a_2 & a_3 & b_1 & b_2 & b_3 \end{bmatrix}^T \tag{6}$$

$$\varphi(t) = \begin{bmatrix} -y(t-1) & -y(t-2) & -y(t-3) & u(t-1) & u(t-2) & u(t-3) \end{bmatrix}^T \tag{7}$$

Then the predicted value $\hat{y}(t)$ for the system output $y(t)$ is given by

$$\hat{y}(t) = \theta(t) - \varphi(t) \tag{8}$$

The prediction error is defined as

$$\varepsilon(t) = y(t) - \hat{y}(t) \tag{9}$$

The GA objective function is chosen to minimize the square of the identification error defined in Eq. (9). In order to set a one-to-one relationship between the GA genes and the system parameters, each individual member of the GA population represents an assumption for the parameter vector $\theta(t)$ defined in Eq. (6).

Each member is ranked according to its fitness to the current measurements $\varphi(t)$ using the Linear Rank-Based Fitness Assignment method [25]. The termination condition is satisfied when either the number of populations reaches its maximum value of 50 generations or the best member fitness is less than 10^{-6}.

The identifier goes through two stages of training, namely off-line training and on-line update. In off-line training, first the identifier is trained using input-output data for a variety of operating conditions and disturbances. This data is obtained using the system shown in Figure 16.3. The disturbances applied are the voltage reference disturbances, input torque reference disturbances and three phase to ground fault. The training is iteratively done until a pre-specified tolerance of 10^{-4} is met.

After the off-line training, the identifier is further updated on-line. The population of the on-line identifier is selected mostly from best members of the off-line training cases. The rest of the population is selected randomly.

Since the system parameters are varying slowly, the number of members selected to undergo the mutation process is much larger than the number of members selected to undergo the recombination process. This process significantly reduces the time needed to converge making the GA algorithm suitable for the on-line update.

At each sampling instant, the input and the output of the generator are sampled and the input vector to the identifier is formed as in Eq. (7). The GA objective function is to minimize the mean-squared error between the plant output, i.e., desired output, and the identifier estimated output. This process is repeated every sampling period making the on-line update, which in turn results in an adaptive approach to identify a plant.

Typical curves for the convergence of the identified parameters when the system under study undergoes a three phase to ground short circuit fault in the middle of one of the transmission lines, cleared after 100 ms, are shown in Figure 16.4.

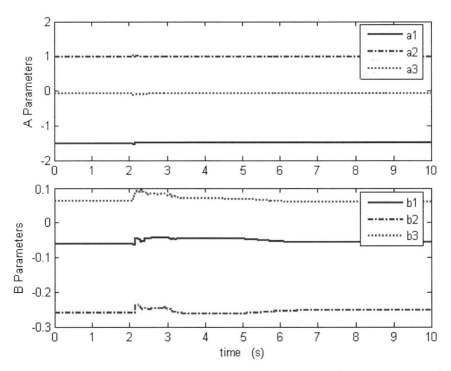

FIGURE 16.4   On-Line Regression Parameters Variation During a Three-Phase to Ground Fault.

16.4.2 CONTROL STRATEGY

Once the system model parameters are identified, the control signal can be calculated based on the ARX discrete model given in Eq. (2). Using PS control algorithm, the characteristic polynomial of the closed-loop system is assumed to have the same form as that of the open-loop system. Also in the closed-loop, the open-loop poles are shifted radially towards the center of the unit circle in the z-plane by a factor α. If the pole-shift factor α is fixed, the PS control algorithm degenerates into a special case of the PA control algorithm. It is evident that the rule determining the pole-shifting factor is very important. For optimum performance α is modified on-line according to the operating conditions of the controlled system [26].

16.4.3 CONTROLLER START-UP

New control is computed each sampling interval on the assumption that the updated system parameter estimate obtained by the system identifier is 'accurate.' This is carried out in the same way as an off-line procedure. However, in a self-tuning controller it is carried out recursively on-line. This means that while the model parameter estimates are good, the controller output is good, whereas if the model parameter estimates are bad then almost surely the computed control will be bad. This point is particularly important on controller start-up, in that the self-tuning control algorithm will work even if fairly arbitrary initial estimates are entered for the model parameter estimates. This may provide pretty "violent" control; however, the estimates will converge to their true values in a short-space of time [27].

This particular point is considered one of the main advantages of the proposed controller as it already has a prior knowledge of the system parameters from the off-line training stored in its initial population which as a result would give a better start up than a controller that has a random estimation of the system parameters.

16.5 SIMULATION STUDIES

Performance of the APSS with the proposed GA-identifier is investigated on a synchronous generator connected to a constant voltage bus through two transmission lines (Figure 16.3). An IEEE Standard 421.5, Type ST1A

AVR and Exciter Model [28] is used in simulation studies. The speed deviation ($\Delta\omega$) is sampled for parameter identification and control computation. Sampling frequencies above 100 Hz provide no practical benefit and the performance deteriorates for sampling rates under 20 Hz. A sampling rate of 20 Hz is chosen to make sure there is enough time available for updating the online GA-identifier and to perform control computations. The absolute physical limits for the control output are ±0.1 pu.

16.5.1 NORMAL LOAD CONDITIONS

The identifier performance is tested with the system working under the full load condition of 0.8pu generated power with a 0.85 power factor lag. The system is subjected to a 0.1pu change in input mechanical power at 1 sec that is removed at 6 s followed by a 0.05pu change in reference voltage at 11 s and removed at 16 s. A three phase to ground fault is applied at 21 s at the middle of one of the transmission lines, the fault is cleared after 100 ms by opening the breakers at both ends of the faulted transmission line. The line is closed back at 26 s.

The speed change under (i) no PSS but with a third order identified model is shown in Figure 16.5, and (ii) PSS with a third-order identified model is shown in Figure 16.6. The results show a good tracking for low frequency oscillations using the GA-identifier for both open-loop and closed-loop control.

In order to verify these results analytically, the residual analysis is applied. Figures 16.7 and 16.8 show the validation results when residual analysis is applied to cases (i) and (ii) described above. The highlighted area denotes the confidence interval. It is clear that these figures give much more precise information about the model quality from an identification point of view.

The high values for the cross correlation during closed-loop control, Figure 16.8, are not due to any deficiency in the identification, but are due to the fact that the system input at time ($t+\tau$) is dependent on the system output at time (t) through the controller transfer function. This creates a correlation between the identification error at time (t) and the system input at time ($t+\tau$).

In order to test the controller as well, a test at full load conditions is conducted. Under these conditions, the CPSS parameters were tuned using the tuning procedure described in Ref. [22]. The parameters of the CPSS

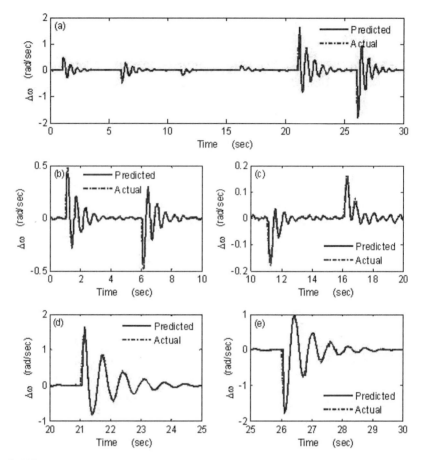

FIGURE 16.5 (a) GA-identifier and plant responses to a 0.1pu step increase in input mechanical power (zoomed in (b)) followed by a 0.03pu step increase in reference voltage (zoomed in (c)) followed by a 3 phase to ground fault at the middle of one of the transmission lines (zoomed in (d)) followed by the re-closure of the faulted transmission line (zoomed in (e)) under full load condition on open-loop control.

were then kept unchanged for all the tests. A disturbance of 0.1 pu step increase in input torque is applied at 1 s, then removed at 6 s. The power angle response of the APSS, CPSS and the open-loop, without stabilizer, are shown in Figure 16.9. It can be seen that CPSS and APSS damp out the oscillations very quickly. In addition, the CPSS and the APSS have a close system response statistics as the CPSS is optimized at this operating condition.

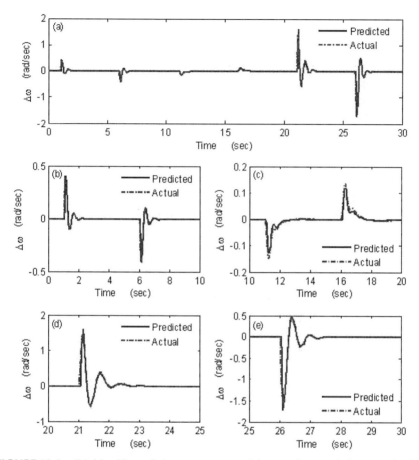

FIGURE 16.6 GA-identifier and plant responses to a 0.1pu step increase in input mechanical power (zoomed in (b)) followed by a 0.03pu step increase in reference voltage (zoomed in (c)) followed by a 3 phase to ground fault at the middle of one of the transmission lines (zoomed in (d)) followed by the re-closure of the faulted transmission line (zoomed in (e)) under full load condition on closed-loop control.

16.5.2 LIGHT LOAD CONDITIONS

In this test, the initial conditions are 0.4 pu power and 0.8 lagging pf. A disturbance of 0.2 pu step increase in input torque is applied at 1 s, then removed at 6 s. The disturbance is large enough to cause the system to operate in anon-linear region. System response for these non-linear conditions is shown in Figure 16.10.

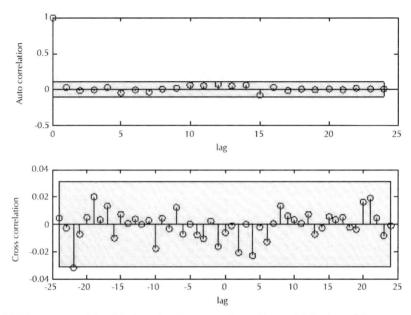

FIGURE 16.7   Model validation when the system is working at full load condition on open-loop control.

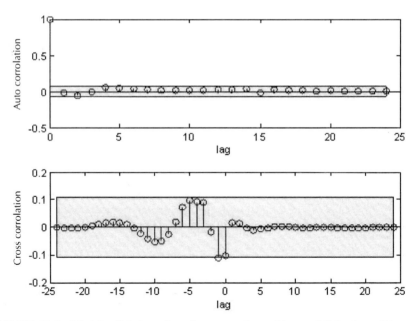

FIGURE 16.8   Model validation when the system is working at full load condition on closed-loop control.

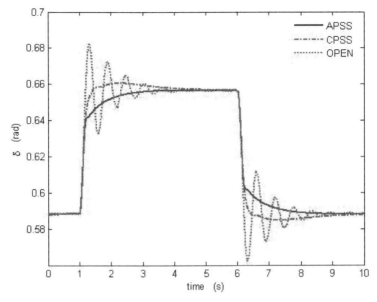

FIGURE 16.9   System Response to a 0.1pu Step Increase in Torque and Return to Initial Conditions Under Full Load.

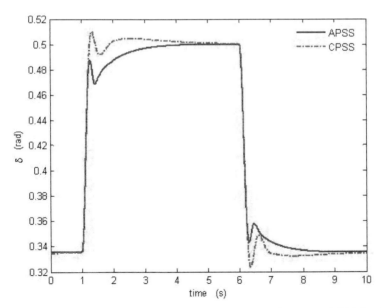

FIGURE 16.10   Response to a 0.2pu Step Increase in Torque and Return to Initial Conditions Under Light Load.

16.5.3   LEADING POWER FACTOR LOAD CONDITIONS

The behavior of the proposed controller is also investigated under leading power factor condition with initial conditions of 0.7 pu power and 0.9 leading pf. It is a difficult situation for the controller because the stability margin is reduced. However, in order to absorb the capacitive charging current in a high voltage power system, it may become necessary to operate the generator at a leading power factor. It is, therefore, desirable that the controller be able to guarantee stable operation of the generator under leading power factor condition.

A disturbance of 0.05 pu step increase in input torque is applied at 1 s then removed at 11 s. Figure 16.11 shows that the oscillation of the system is damped out rapidly and demonstrates the effectiveness of the APSS to control the generator under leading power factor operating conditions.

Although the CPSS might have a better performance from the classical control theory point of view in terms of a quicker response shown as less rising time than the APSS, the APSS has a better performance from the power system point of view as it has much less over-shoot which results in a better damping of the system oscillation and a smoother transition to the new operating point thus improving the dynamic stability properties of the generators.

16.5.4   VOLTAGE REFERENCE CHANGE TEST

In this test, the operating condition is 0.7 pu power, 0.8 pf lagging and 1.092 pu terminal voltage. A 0.05pu decrease in voltage reference is applied at 1 s and removed after 10 s. The power angle and the generator terminal voltage are shown in Figures 16.12 and 16.13, respectively.

In the open loop system without any PSS the severity of the oscillations increases as the reference voltage drops since the system stability margin decreases as the reference voltage drops for a certain active power output.

Figure 16.12 shows that the oscillations are effectively damped by APSS without any over-shoot for both reference voltage increase and decrease which means that the system stability margin is enhanced by using APSS. Although the APSS has a good effect on damping the system oscillations, it has a detrimental effect on the terminal voltage, as seen in Figure 16.13, since it is preferable to have a quick change in terminal voltage.

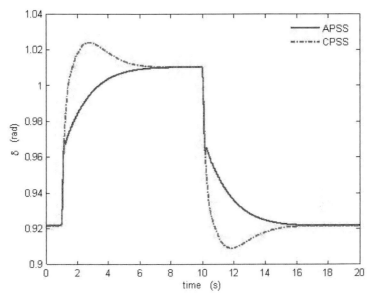

FIGURE 16.11   Response to a 0.05pu step increase in torque and return to initial conditions under leading power factor load.

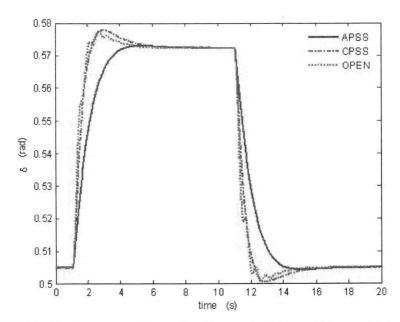

FIGURE 16.12   Response to a 0.05 pu Step Decrease in Reference Voltage and Return to Initial Conditions.

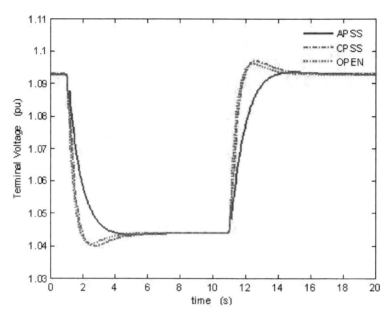

FIGURE 16.13 Generator Terminal Voltage When a 0.05 pu Step Decrease in Reference Voltage is Applied and Return to Initial Conditions.

16.5.5 DYNAMIC STABILITY TEST

In this test the machine is initially working at 1 pu power at unity power factor and the terminal voltage is 1 pu. A ramp of 10–3 pu/s slope is added to the mechanical torque reference at 1 s. Figure 16.14 shows the speed deviation of the generator. It can be seen that the proposed APSS enhances the stability margin of the system. Using the APSS increases the maximum mechanical torque from 1.767 pu (with CPSS) to 1.784 pu.

The simulation results show that the proposed APSS has very good damping characteristics for different operating conditions and disturbances applied on the generating unit. The machine settles to the new operating conditions with very small overshoot and oscillation thus improving the dynamic stability properties of the generator. In addition, the GA population stores a priori knowledge because of off-line training. The genes are further updated on-line adaptively to track the different operating conditions and disturbances. Also, the PS-control uses the on-line updated ARX parameters to calculate the closed-loop poles of the system. The unstable poles are moved inside the unit circle in the z-plane and the control is calculated to

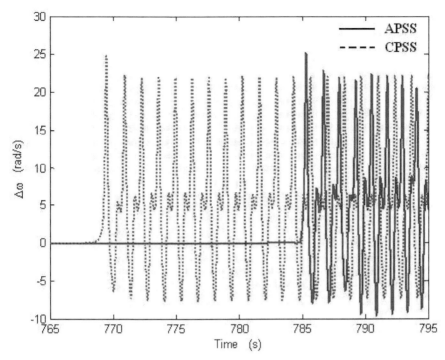

FIGURE 16.14   Generator speed deviation during the dynamic stability test.

optimize the output performance. The PS algorithm assures the stability of the closed-loop system.

16.6   LABORATORY IMPLEMENTATION

In Section 16.5 the theoretical model and simulation results using an APSS consisting of an on-line GA-Identifier and PS-Controller are discussed. The APSS showed enhanced performance compared to a CPSS because of its adaptive parameter tracking capability.

In the computer simulations, the power system was simulated by a set of differential equations, given in Appendix A, and the APSS algorithm was implemented as a sequential algorithm. The numerical model of the power system can only approximate the dynamics of a system to a certain extent. There is always some unexpected dynamic behavior inherent in a system that is not accounted for by any given mathematical model. There exist noise and saturation of elements as well as unexpected disturbances that cause the power system to operate under a continuous small perturbation. Generally, it is necessary to do some laboratory and/or field tests to further assess the evidence from the computer simulation before the installation of the proposed APSS on an actual system. This is especially true for a power system in which the damage could be expensive. For such a system, one solution is to build a scaled physical model of the target system. The scaled physical model is able to emulate the behavior of the actual power plant in the laboratory environment.

16.6.1 POWER SYSTEM MODEL

A physical model of a single machine infinite bus power system is available in the Power System Research Laboratory at the University of Calgary. It consists of a 3-phase, 3 kVA micro-synchronous generator connected to an infinite bus through a double circuit transmission line. An overall schematic diagram of this physical model is given in Figure 16.15. The major units of this model are the turbine model, the generator model, the transmission line model and the AVR.

The turbine is modeled by a 5.5 kW separately excited DC motor. The technical parameters of this DC motor are 220 V, 30 A, 1800 RPM, 7.5 hp.

The generator unit is modeled by a 3-phase, 3 kVA, 220 V micro synchronous generator. The name-plate details of this alternator are 220/127 V, 7.9 A, 3-phase, 60 Hz with a 0.8 power factor. The parameters of this machine are given in Appendix B.

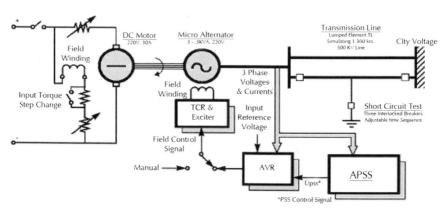

FIGURE 16.15 Structure of the power system model.

The field winding resistance of the synchronous machine, on a per unit basis, is much higher than that of the large machines to be simulated. This results in a comparatively low field transient time constant of 0.765 s, whereas the transient time constant of the large machines is in the neighborhood of 5 s. An electrical device called the Time Constant Regulator (TCR) is used to reduce the effective field resistance and thereby alter the field transient time to the order of that required for the simulation of large machines. With the TCR, the effective field time constant of the micro synchronous generator can be increased up to 10 s.

The transmission line is modeled by a lumped element transmission line. The physical model consists of six 50 km equivalent π sections and gives a frequency response that is close to the actual transmission line response up to 500 Hz. This model simulates the performance of a 300 km long 500 kV, double circuit transmission line connected to an infinite bus as shown in Figure 16.15.

A variety of disturbances can be applied to the system. Using the switch shown in the excitation circuit of the DC motor, Figure 16.15, a step change in input torque of the generator can be applied. Similarly, the input reference voltage of the AVR can be stepped down or up. In addition, different types of faults can be applied to simulate large disturbances. The faults are simulated using relays controlled by short circuit simulation logic (Figure 16.15). The operating condition of the generator, i.e., active power and power factor, can also be changed by changing the armature current of DC motor and terminal voltage of the generator, respectively.

16.6.2   REAL TIME CONTROL ENVIRONMENT

The APSS is implemented using Matlab© Simulink© Real Time Windows© toolbox based model to provide a convenient interface for the operator to monitor the controller variables and parameters. Input/Output (I/O) signals are transferred between the AVR and the controller model using an AT-MIO-16E-2 National Instrument (NI) DSP card through A/D channels.

16.6.3   EXPERIMENTAL STUDIES

Using the physical model of the power system (Figure 16.15) and the real time digital control environment, the experimental studies are conducted.

For comparison purposes results using digital CPSS are also included. All experimental results are saved using the Matlab workspace and plotted later. In the graphs, in order to make the disturbances seem to happen at the same point of time for the different tests, the time axis is adjusted by artificial padding of steady state conditions.

16.6.3.1 *Full Load Tests*

The machine is delivering 0.85 pu power at a 0.9 lagging power factor. At this operating point, two tests are done, a step change in the torque reference and a step change in voltage reference.

In order to make a comparison between CPSS and the proposed APSS under different operating conditions, the parameters of the CPSS are tuned using the Ziegler-Nichols rules for tuning PID controllers [29] to give the best response for the operating conditions of this test. The parameters are given in Appendix B.

Figure 16.16 shows the system response when a 0.1 pu decrease in torque reference is applied at 2 s and removed at 10 s. It can be seen from the figure

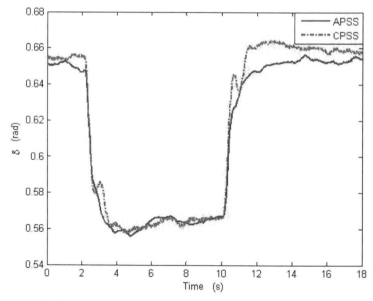

FIGURE 16.16   System response when subjected to a 0.1 pu step change in torque reference under full load condition.

that the performance of both PSSs (the APSS and the CPSS) are close to each other as the CPSS is designed at this operating point.

The system response when a 0.1 pu decrease in voltage reference is applied at 2 s and removed at 10 s is shown in Figure 16.17. It can be seen that the APSS can damp the system oscillation while maintaining the terminal voltage at a reasonable value.

16.6.3.2 Light Load Tests

Three extra tests are applied when the machine was delivering 0.5 pu power at 0.84 lagging power factor.

Figure 16.18 shows the system response when a 0.2 pu step decrease in torque reference is applied at 2 s and removed at 10 s. It can be seen that the APSS can adapt itself for the new operating point and maintain a good damping of the system oscillations whereas the CPSS cannot maintain its previous good performance as it was designed for a different operating point.

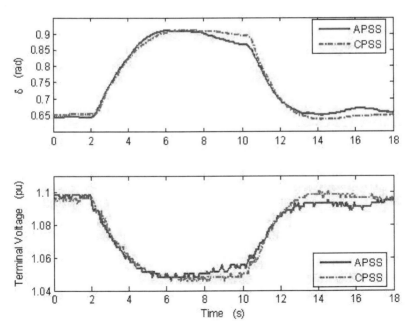

FIGURE 16.17 System response when subjected to a 0.05 pu step change in voltage reference under full load condition.

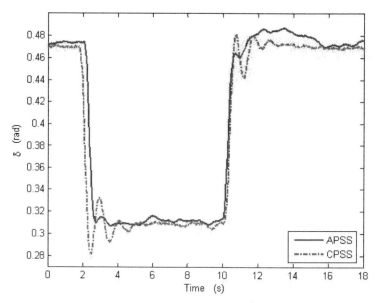

FIGURE 16.18   System response when subjected to a 0.2 pu step change in torque reference under light load condition.

Figure 16.19 shows the system response when a 0.05 pu step decrease in voltage reference is applied at 2 s and removed at 10 s. The results show that both PSSs give a smooth transition to the new voltage settings.

Figure 16.20 shows the system response when a 3 phase to ground short circuit occurs in the middle of one of the transmission lines. The results show that not only the APSS is able to damp the system oscillation effectively but also it can adapt itself to the new operating condition.

16.6.3.3   *Leading PF Load Tests*

The last sets of tests are done when the machine was delivering a 0.6 pu power at 0.9 leading power factor.

During the first test the machine is subjected to a 0.1 pu step decrease in torque reference applied at 2 s and removed at 10 s. Figure 16.21 shows that the response of the APSS is slightly better than the CPSS as the machine is working close to its stability limit.

Figure 16.22 shows the system response during the second test when subjected to a 0.05 pu step decrease in voltage reference applied at 2 s and removed at 10 s. The APSS still can damp the system oscillation while maintaining a reasonable change in the terminal voltage.

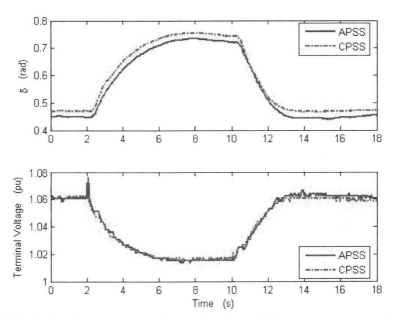

FIGURE 16.19 System response when subjected to a 0.05 pu step change in voltage reference under light load condition.

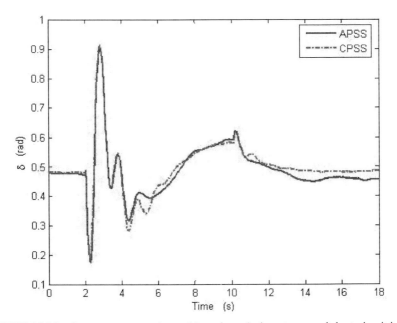

FIGURE 16.20 System response when subjected to a 3 phase to ground short circuit in the middle of one of the transmission lines.

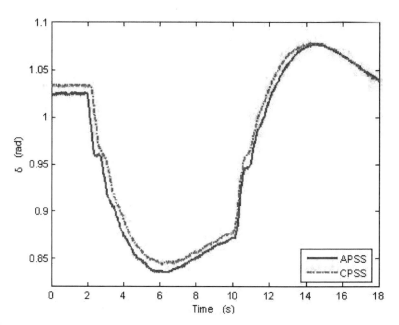

FIGURE 16.21 System response when subjected to a 0.1 pu step change in torque reference under leading pf load condition.

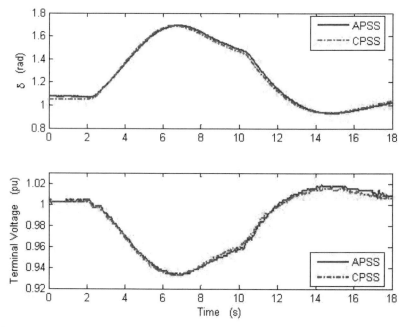

FIGURE 16.22 System response when subjected to a 0.05 pu step change in voltage reference under leading pf load condition.

16.6.4   SUMMARY

Implementation of APSS and CPSS in a laboratory environment and real-time test results on a physical model of power system are presented in this chapter. The APSS was developed as a Matlab© Simulink© Real Time Windows© model with a sample time of 50 ms. The proposed APSS has the following advantages:

- The GA-Identifier is not trained again on-line in the laboratory set-up. Instead, the GA parameters obtained in computer simulation studies in Section 1.5 are directly adopted here. This is favorable from the user point of view as it circumvents the need for on-line training whenever the APSS is used on a new generating unit.
- The identifier is further updated on-line every sampling interval to track the dynamic conditions. The PS-control uses the parameters obtained on-line from the identifiers to compute the control signal.
- Finally the experimental results with the proposed APSS are compared to those of the CPSS. It is demonstrated that the APSS exhibits good performance over a wide range of operating conditions without requiring any further tuning unlike the CPSS.

16.7   CONCLUSIONS

The importance of using PSS in power systems is discussed in this chapter. Due to the non-linear time varying characteristics of the power system, the conventional fixed-parameter controllers (CPSSs) pose operational challenges to the operator. The CPSSs have been found to be inadequate in performance at different operating conditions. The power quality and stability margin is sacrificed when the CPSS is not able to improve the performance at all operating conditions. In addition, the current interconnection requirements such as the need for increasing the line limits by using damping controllers on the system introduce difficulties in terms of retuning the existing PID type controller parameters whenever a new generating unit is added to the system.

This chapter focuses on the following primary issues:

1. The first issue of investigation is how to develop a controller that can draw the attention of the control engineer. Control engineers are often faced with conflicting issues. The conventional PID controllers are inadequate in terms of performance and the required manual tuning

of the parameters. At the same time they are faced with scenarios where the alternative AI control techniques are too complex to use in practice because of the unavailability of information related to their internal working. This chapter addresses this issue by attempting a compromise solution: use the GA for modeling the system dynamics, use the linear feedback controller (such as PS-Control) because of its simplicity and acceptance.

2. The second issue focuses on how to use a GA to represent the dynamic characteristics of the power system. Although the techniques for on-line adaptation are becoming fairly standard, in a real power system environment the demands placed upon adaptive estimation and control techniques can be extremely severe. An ARX methodology is described for problem formulation. The parameters of the linear (PS) controller are obtained by minimizing the square of the identification error. It is possible to achieve the second objective by setting a one-to-one relationship between the GA-genes and the system identified parameters.

3. The third issue focuses on computer simulation results to verify the performance of the APSS. Simulation results show that the proposed APSS is effective for damping the power system oscillations under different operating conditions.

4. The final issue deals with real-time tests with APSS on a scaled physical model of a single-machine connected to an infinite-bus power system. The APSS is implemented using Matlab© Simulink© Real time windows© toolbox. The performance of the APSS suggests significant advantages over the CPSS: performance improvement and no requirement for parameter tuning.

KEYWORDS

- **adaptive power system stabilizer**
- **genetic algorithm**
- **on-line system identification**
- **pole shifting control**
- **power system stabilizer**
- **self tuning control**

REFERENCES

1. Larsen, E. V., & Swann, D. A. Applying Power System Stabilizers, Part I-III. IEEE Trans. on Power Apparatus and Systems, vol. PAS-100, June 1981, pp. 3017–3046.
2. De-Mello, F. P., & Concordia, C. A. Concepts of Synchronous Machine Stability as Affected by Excitation Control. IEEE Trans. on Power Apparatus and Systems, vol. PAS-88, no. 4, April 1969, pp. 316–329.
3. Farmer, R. W. State-of-the-art Technique for System Stabilizer Tuning. IEEE Trans. on Power Apparatus and Systems, vol. PAS-102, 1983, pp. 699–709.
4. Doi, A., & Abe, S. Coordinated Synthesis of Power System Stabilizer in Multi-Machine Power System. IEEE Trans. on Power Apparatus and Systems, vol. PAS-103, no. 6, June 1984, pp. 1473–1479.
5. Chan, W. C., & Hsu, Y. Y. An Optimal Variable Structure Stabilizer for Power System Stabilization. IEEE Trans. on Power Apparatus and Systems, vol. PAS-102, 1983, pp. 1738–1746.
6. Pierre, D. A. A Perspective on Adaptive Control of Power Systems. IEEE Trans. on Power Systems, vol. PWRS-2, no. 5, Sep. 1987, pp. 387–396.
7. Astrom, K. J., Wittenmark, B. Adaptive Control; Addison Wesley Publishing Company, Reading, MA, 1995.
8. Landau, I. D., Lozano, R., & M'Saad, M. Adaptive Control; Springer, London, 1995.
9. Ljung, L. System Identification: Theory for the User; Prentice Hall PTR, Upper Saddle River, New Jersey, 1999.
10. Clark, D. W., & Gawthrop, P. J. A Self-Tuning Controller. IEE Proceedings, 1975, vol. 122, pp. 929–934.
11. Wittenmark, B., & Astrom, K. J. Practical Issues in the Implementation of Self-Tuning Control. Automatica, vol. 20, no. 5, 1984, pp. 595–605.
12. Borrisson, U., & Syding, R. Self-Tuning Control of an Ore Crusher. Automatica, vol. 12, 1976, pp. 1–7.
13. Malik, O. P., Chen, G. P., Hope, G. S., Qin, Y. H., & Xu, G. Y. An Adaptive Self-Optimizing Pole-Shifting Control Algorithm. IEE Proceedings, Part-D, vol. 139, no. 5, 1992, pp. 429–438.
14. El-Metwally, K. A., & Malik, O. P. Fuzzy Logic Power System Stabilizer. IEE Proceedings on Generation, Transmission and Distribution, vol. 143, no. 3, 1996, pp. 263–268.
15. Hariri, A., & Malik, O. P. "Adaptive Network Based Fuzzy Logic Power System Stabilizer," IEEE WESCANEX 95 Proceedings, May 1995, pp. 111–116.
16. Hariri, A., & Malik, O. P. A Fuzzy Logic Based Power System Stabilizer with Learning Ability. IEEE Trans. on Energy Conversion, vol. 11, no. 4, Dec. 1996, pp. 721–727.
17. Irwin, G. W., Warwick, K., & Hunt, K. J. Neural Network Applications in Control; The institution of Electrical Engineers, Herts, U.K., 1995.
18. Zhang, Y., Chen, G. P., Malik, O. P., & Hope, G. S. An Artificial Neural Network Based Adaptive Power System Stabilizer. IEEE Trans. on Energy Conversion, vol. 8, no. 1, 1993, pp. 71–77.
19. Hariri, A., & Malik, O. P. Self-Learning Adaptive Network Based Fuzzy Logic Power System Stabilizer, Proceedings of IEEE Int. Conf. On Intelligent Systems Applications to Power Systems, Florida, vol. 1, no. 4, Jan. 28–Feb. 2 1996, pp. 299–303.
20. Goldberg, D. E. Genetic Algorithms in Search, Optimization, and Machine Learning; Addison Wesley Publishing Company, 1989.

21. Andreoiu, A., & Bhattacharya, K. Lyapunov's Method Based Genetic Algorithm for Multi-Machine PSS Tuning, IEEE Power Engineering Society Winter Meeting, vol. 2, Jan. 27–31, 2002, pp. 1495–1500.

22. Huang, Tsong-Liang; Chang, Chih-Han; Lee, Ju-Lin; Wang, & Hui-Mei. Design of Sliding Mode Power System Stabilizer via Genetic Algorithm, IEEE International Symposium on Computational Intelligence in Robotics and Automation, vol. 2, July 16–20, 2003, pp. 515–520.

23. Yamaguchi, H., Mizutani, Y., Magatani, K., Leelajindakrairerk, M., Okabe, T., Kinoshita, Y., & Aoki, H. A Design Method of Multi-Input PSS by Using Auto-Variable Search Space Type High Speed Genetic Algorithm. IEEE Porto Power Tech Proceedings, vol. 2, Sep. 10–13, 2001, pp. 532–538.

24. Hussein, Wael, & Malik, O.P. GA-identifier and predictive controller for multi-machine power system, IEEE Power India Conference, New Delhi, June 2006.

25. Bäck, T., & Hoffmeister, F. Extended Selection Mechanisms in Genetic Algorithms, Proceedings of the Fourth International Conference on Genetic Algorithms, San Mateo, California, USA: Morgan Kaufmann Publishers, 1991, pp. 92–99.

26. Hussein, Wael, Studies on a Duplicate Adaptive PSS. M.Sc. Thesis, The University of Calgary, Calgary, Alberta, Canada, 2002.

27. Hussein, Wael, & Malik, O. P. Study of System performance with Duplicate Adaptive Power System Stabilizer. Electric Power Components and Systems, vol. 31, No. 9, September 2003, pp. 899–912.

28. IEEE Excitation System Model Working Group, "Excitation System Models for Power System Stability Studies," IEEE Standard P421.5-1992, 1992.

29. Ogata, K. Modern Control Engineering, 4th Ed. Prentice-Hall, Englewood Cliffs, NJ, 2000.

30. Eitzmann, M. Excitation System tuning and Testing for Increased Power System Stability, GE Energy BROC10605, 2004.

APPENDIX A   SINGLE MACHINE POWER SYSTEM

The structural diagram of the single machine infinite bus power system model is shown in Figure 16.3. The generating unit is modeled by seven first order differential equations. The signal to be controlled is the power angle, denoted by δ. The PSS has access to the rotor speed, denoted by ω, or to the generator electric power, denoted by Pe, to produce its control signal, denoted by Upss, which is applied at the input-summing junction of the Automatic Voltage Regulator (AVR) circuit.

A.1   GENERATOR MODEL

The generator is modeled by seven first order differential equations

$$\dot{\delta} = \omega_o \omega \tag{A.1}$$

$$\omega = \frac{1}{2H}(T_m + g + K_d \dot{\delta}_0 - T_e) \tag{A.2}$$

$$\dot{\lambda}_d = e_d + r_a i_d + \omega_o(\omega + 1)\lambda_q \tag{A.3}$$

$$\dot{\lambda}_q = e_q + r_a i_q - \omega_o(\omega + 1)\lambda_d \tag{A.4}$$

$$\dot{\lambda}_f = e_f - r_f i_f \tag{A.5}$$

$$\dot{\lambda}_{kd} = -r_{kd} i_{kd} \tag{A.6}$$

$$\dot{\lambda}_{kq} = -r_{kq} i_{kq} \tag{A.7}$$

where, ω_o: Nominal rotational speed, H: Inertia constant, K_d: damping coefficient, r_a: Armature resistance, r_f: Field resistance, r_{kd}: Direct-axis resistance, and r_{kq}: Quadrature-axis resistance.

The generator parameters are given in Table A.1.

All resistances and reactances are in pu and time constants in seconds

A.2 GOVERNOR MODEL

The governor used in the system has the following transfer function

$$g = \left[a + \frac{b}{1 + sT_g} \right] \dot{\delta} \tag{A.8}$$

The parameters used in the simulation studies are given in Table A.2.

TABLE A.1 Generator Parameters Used in Simulation

$r_a = 0.007$	$H = 3.46$	$r_f = 0.00089$	$r_{kd} = 0.023$
$r_{kq} = 0.023$	$X_d = 1.24$	$X_f = 1.33$	$X_{kd} = 1.15$
$X_{md} = 1.126$	$X_q = 0.743$	$X_{kq} = 0.652$	$X_{mq} = 0.626$
$K_d = -0.0027$			

TABLE A.2   Governor Parameters Used in Simulation.

$a = -0.001328$	$b = -0.17$	$T_g = 0.25$

A.3   AVR AND EXCITER MODELS

The AVR and exciter combination used in the system are from the IEEE
Standard P421.5–1992, Type ST1A shown in Figure A.1 [28]. The parameter values are given in Table A.3.

The AVR control action is determined by the lead lag compensator with
time constants T_B, T_C, T_{BI} and T_{CI}, and by the voltage regulator of proportional integral action of time constant T_A and gain K_A. The local control loop
is closed by the proportional derivative action block time constant T_F and
gain K_F.

Excitation systems with high gain and fast response times greatly aid
transient stability but at the same time tend to reduce small signal stability
[30]. Consequently, to increase the system stability an additional controller
is needed which is called the power system stabilizer.

A.4   IEEE PSS1A CPSS

The CPSS used in the system is from the IEEE Standard P421.5–1992,
PSS1A, shown in Figure A.2. The transfer function of this PSS consists of

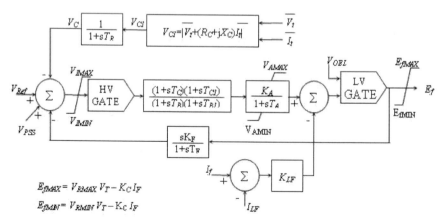

FIGURE A.1   AVR and Excitation Model Type ST1A, IEEE Standard P421.5-1992.

TABLE A.3   AVR and Exciter Parameters Used in Simulation

$R_C = 0.0$	$X_C = 0.0$	$T_R = 0.04$	$K_A = 190$
$K_C = 0.08$	$K_F = 0.0$	$K_{LF} = 0.0$	$I_{LR} = 0$
$T_B = 10.0$	$T_C = 1.00$	$T_A = 0.01$	$T_F = 0.0$
$T_{BI} = 0.0$	$T_{CI} = 0.0$	$V_{OEL} = 999$	$V_{UEL} = -999$
$V_{IMAX} = 999$	$V_{IMIN} = -999$	$V_{AMAX} = 999$	$V_{AMIN} = -999$
$V_{RMAX} = 7.8$	$V_{RMIN} = -6.7$		

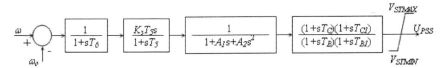

FIGURE A.2   IEEE PSS1A PSS.

a low pass input filter with T_6 time constant, followed by a high pass filter, a derivative type regulator block with time constant T_5 and stabilizer gain K_5. In the third block, a second order torsion filter with A_1 and A_2 parameters is presented. Lastly, a lead-lag compensator with time constants T_1–T_4 is shown in the last block. The lead-lag block provides the appropriate phase-lead characteristics to compensate the phase lag between the exciter input and the generator electrical torque. The parameters used in the simulation studies are given in Table A.4.

APPENDIX B   PHYSICAL MODEL OF POWER SYSTEM

The parameters of the micro-alternator are summarized in Table B.1.

Each transmission line consists of six 50 km equivalent π sections. For each section, the parameters are summarized in Table B.2.

The parameters of the CPSS are summarized in Table B.3.

TABLE A.4   CPSS Parameters Used in Simulation

$T_1 = 0.2$	$T_2 = 0.05$	$T_3 = 0.2$	$T_4 = 0.05$
$T_5 = 2.5$	$T_6 = 0.01$	$A_1 = 0.0$	$A_2 = 0.0$
$V_{STMIN} = -0.1$	$V_{STMAX} = 0.1$	$K_s = 0.11$	

TABLE B.1 Micro Alternator Parameters (All Values Are in pu)

$x_d = 1.20$	$x_q = 1.20$	$r_d = 0.0026$	$r_q = 0.0026$
$x_{md} = 1.129$	$x_{mq} = 1.129$	$x_{kd} = 1.25$	$x_{kq} = 1.25$
$r_{kd} = 0.0083$	$r_{kq} = 0.0083$	$H = 4.75$	
$x_f = 1.27$	$r_f = 0.000747$		

TABLE B.2 Transmission Line Parameters (All Values Are in pu)

$R = 0.036$	$X = 0.0706$	$B = 18.779$

TABLE B.3 CPSS Parameters (All Time Constants Are in Seconds)

$K_s = 0.5$	$T_1 = 0.065$	$T_2 = 0.08$
	$T_3 = 0.065$	$T_4 = 0.08$

CHAPTER 17

APPLIED EVOLUTIONARY COMPUTATION IN FIRE SAFETY UPGRADING

IORDANIS A. NAZIRIS,[1] NIKOS D. LAGAROS,[2] and
KYRIAKOS PAPAIOANNOU[3]

[1]*PhD Candidate, Laboratory of Building Construction and Building Physics, Department of Civil Engineering, Aristotle University of Thessaloniki, 54124, Greece*

[2]*Assistant Professor, Institute of Structural Analysis and Antiseismic Research, School of Civil Engineering, National Technical University of Athens, 15780, Greece*

[3]*Professor Emeritus, Laboratory of Building Construction and Building Physics, Department of Civil Engineering, Aristotle University of Thessaloniki, 54124, Greece*

CONTENTS

17.1   INTRODUCTION

Preserving cultural heritage is compulsory, since historic structures, monuments and works of art are unique and therefore irreplaceable. A major threat for historic buildings is fire hazard; many examples have been reported in the past, where fire caused significant damages to priceless monuments of world cultural heritage [1]. Until recently many reconstruction, materials and fire protection techniques reflected the norms of the time performed, since historic buildings preservation is a relatively modern concept. However, many of these buildings exhibit special features that are not consistent with the requirements of modern fire protection codes. Therefore, fire protection becomes a difficult task that should be based on modern effective techniques, while history and culture should be respected and fire safety interventions should lead to satisfactory safety levels [2]. However, the necessity to preserve the authenticity of historic structures usually leads to expensive interventions, due to the application of special materials and advanced fire protection measures. The main problem is that in such fire safety upgrade actions the available budget is limited and therefore has to be optimally allocated in order to maximize the level of fire safety upgrade.

During the last three decades many numerical methods have been developed to meet the demands of engineering design optimization. These methods can be classified in two categories, gradient-based and derivative-free ones. Mathematical programming methods are among the most popular methods of the first category. Metaheuristic algorithms are nature-inspired or bio-inspired as they have been developed based on processes found in nature like the successful evolutionary behavior of natural systems and these methods belong to the derivative-free category of methods. Metaheuristic algorithms include genetic algorithms [3], simulated annealing [4], particle swarm optimization (PSO) [5], ant colony algorithm [6], artificial bee colony algorithm [7], harmony search [8], cuckoo search algorithm [9], firefly algorithm [10], bat algorithm [11], krill herd [12], and many others. Evolutionary computation algorithms are the most widely used class of metaheuristic algorithms and in particular evolutionary programming [13], genetic algorithms [3, 14], evolution strategies [15, 16] and genetic programming [17]. These methods have been found to be efficient in a variety of engineering problems, both for single and multiobjective formulations. This is the reason that evolutionary computation techniques were adopted in this chapter.

Fire protecting a historic structure can lead in decision-making problems (e.g., selection between two different fire protection measures). The analytic hierarchy process (AHP) is a structured technique used in multiple-criteria decision analysis. It was developed by Thomas L. Saaty in the 1970 s [18, 19] and has been extensively studied and refined since then, having a wide range of applications [20–22]. AHP provides the framework for structuring a decision problem, representing and quantifying its multiple elements, relating those elements to overall goals, and evaluating alternative possible solutions. Specifically the decision problem is decomposed into a hierarchy of sub-problems, each of which can be considered independently. Once the levels of the hierarchy have been identified, the various elements in each level are specified and evaluated with respect to their impact on the elements of the immediate above level. These evaluations are converted to numerical values that can be processed and compared over the entire range of the problem. In the final step of the process, numerical priorities are calculated for each of the decision alternatives.

In this area, methods for multi objective decision making in fire protection of historic buildings has been extensively used. More particular AHP has been incorporated in many previous works in order to facilitate the reduction of fire risk in cultural heritage premises. Shi et al. [23] proposed an improved AHP, based on the coherence of conventional AHP and the Fault Tree Analysis (FTA), which was applied at the Olympic venues in China. Shen et al. [24] discussed the factors of apartment building fire hazards, and used the Analytic Hierarchy Process (AHP) to determine the weights of all fire hazard factors in order to provide a reference for public fire safety assessment. Fera and MacChiaroli [25] using different techniques from the decision support tools, such as the analytic hierarchy process, and through the use of a fire dynamics simulator, suggested a new priority in the classification of the fire-fighting systems in tunnels. Vedrevu et al. [26] designed a participatory multicriteria decision making approach involving Analytical Hierarchy Process to arrive at a decision matrix that identified the important causative factors of fires. Results from this study were quite useful in identifying potential "hotspots" of fire risk, where forest fire protection measures can be taken in advance.

A holistic preservation approach of historic buildings should include, among others, as an inherent part the concern for suitable fire protection design. Such actions, however, are frequently subjected to stringent budget restrictions and specific architectural constraints that lead designers to face challenging decision-making type of problems. Moreover, historic structures, monuments and works of art are subjected to more strict protection policy than modern structures. In this context, the subject of this chapter is composed by two parts where evolutionary computation is applied to real world applications of fire safety upgrading. In the first part, the objective is to provide an integrated systemic scheme, which embodies innovative tools and new technologies for the solution of the budget allocation problem for the fire protection of historic buildings. In particular, we propose a generic selection and resource allocation (S&RA) model for fire safety upgrading of historic buildings; the efficiency of this model is presented for the case of the Mount Athos monasteries. In the second part, we present an AHP hierarchy combined with S&RA for fire safety upgrading of historic buildings; the effectiveness of the proposed approach is assessed for the case of Simonos Petra monastery.

17.2   FIRE PROTECTION MEASURES

The fire safety design can incorporate a number of the available fire protection measures, depending on the special characteristics of the building and the priorities of the designer. For the needs of the present work the measures that can be used in order to improve the fire protection level of a building are subdivided into the following 16 (16) groups. Each one of these can be fully or partially implemented, or be totally absent, depending on the desirable goals, available budget, possible restrictions, etc.:

(i) compartmentation, (ii) fire resistance of structural elements, (iii) control of fire load, (iv) materials (reaction to fire), (v) control of fire spread outside the building, (vi) design of means of escape, (vii) signs and safety lighting, (viii) access of the fire brigade, (ix) detection and alarm, (x) suppression and extinguishing, (xi) smoke control systems, (xii) training of the personnel, (xiii) fire drills – emergency planning, (xiv) management of fire safety, (xv) maintenance of fire safety system, and (xvi) salvage operation.

The existing situation can be characterized by the fact that each class of fire safety measures may already be implemented to a certain degree, which defines the present fire safety level of the building under investigation. The cost of the above measures, in order to be considered, include: capital, manpower-installation, annual running, maintenance and replacement costs. The estimation of the proposed measures' cost is not an easy task and has to be performed for each new project, since it is highly dependent on the particular characteristics of every project.

17.3 EVOLUTIONARY COMPUTATION

Storn and Price [27] proposed a floating-point evolutionary algorithm for global optimization and named it differential evolution (DE), by implementing a special kind operator in order to create offsprings from parent vectors. Several variants of DE have been proposed so far [28]. According to the variant implemented in the current study, a donor vector $v_{i,g+1}$ is generated first:

$$v_{i,g+1} = s_{r_1,g} + F \cdot (s_{r_2,g} - s_{r_3,g}) \tag{1}$$

Integers r_1, r_2 and r_3 are chosen randomly from the interval $[1,NP]$ while $i \neq r_1$, r_2 and r_3. NP is the population size, F is a real constant value, called the mutation factor. In the next step the crossover operator is applied by generating the trial vector $u_{i,g+1}$ which is defined from the elements of $s_{i,g}$ or $v_{i,g+1}$ with probability CR:

$$u_{j,i,g+1} = \begin{cases} v_{j,i,g+1} & \text{if } rand_{j,i} \leq CR \text{ or } j = I_{rand} \\ s_{j,i,g} & \text{if } rand_{j,i} > CR \text{ or } j^1 I_{rand} \end{cases}$$

$$i = 1, 2, ..., NP \text{ and } j = 1, 2, ..., n \tag{2}$$

where $rand_{ji} \sim U[0,1]$, I_{rand} is a random integer from $[1,2,...,n]$ that ensures that $v_{i,g+1} \neq s_{i,g}$. The last step of the generation procedure is the implementation of the selection operator where the vector $s_{i,g}$ is compared to the trial vector $u_{i,g+1}$:

$$s_{i,g+1} = \begin{cases} u_{i,g+1} \text{ if } F(u_{i,g+1}) \geq F(s_{i,g}) \\ s_{i,g} \text{ otherwise} \end{cases}$$

$$i = 1, 2, ..., NP \tag{3}$$

where $F(s)$ is the objective function to be optimized, while without loss of generality the implementation described in Eq. (3) corresponds to maximization.

17.4 ANALYTIC HIERARCHY PROCESS

Analytic hierarchy process (AHP) is a widely used model for dealing with multi-criteria decision making (MCDM) problems, and represents a sub-discipline of operations research which refers to making decisions in the presence of multiple, usually conflicting, criteria. AHP provides a comprehensive and rational framework able to formulate a decision problem, for representing and quantifying its elements, for relating those elements to overall goals, and for evaluating alternative solutions. The basic concept of the hierarchical approach is the decomposition of the problem into multiple levels of hierarchy, usually four or five. The development of a hierarchical approach to fire ranking was initially undertaken at the University of Edinburgh [29–31], in order to facilitate the fire risk assessment on one hand and the fire safety upgrade on the other hand.

Usually there is a need for more than two levels in the hierarchy of fire safety. In this paper four different "decision making levels" have been used (*see* Figure 17.1): (i) Policy (PO) level which represents the general plan for overall fire safety; (ii) Objectives (OB) level which are specific fire safety goals to be achieved; (iii) Strategies (ST) level which are independent fire safety alternatives, each of which contributes wholly or partially to the fulfillment of the fire safety objectives; and (iv) Measures (M), which are components of the fire risk that are determined by direct or indirect measure or estimate.

Once the levels of hierarchy have been identified the corresponding parameters in each level have to be specified. These parameters have to be specified for every type of building, especially those of the lower levels.

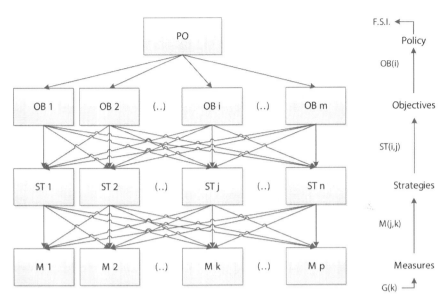

FIGURE 17.1   Typical form of AHP network with 4 levels, like the one used in the present study.

Each one of these parameters has to be expressed numerically in terms of the parameters in the immediate upper level using a weight, which is expressed in a pre-defined scale (in this paper from 1 to 4). The weight coefficients, which are provided by experts, are normalized in order to receive the percentage contribution of a parameter to the parameters of the above level. The parameters of the lowest level are also given an "implementation grade" depending on the extend of implementation of each one (in this paper from 0 to 1). The sum of the products of each parameter with the corresponding parameters of the above level results to a factor, which describes the fire safety level and takes values in the range of 0 to 1 (the larger this factor is, the better in terms of fire safety). In this context, according to the AHP model the fire safety index (FSI) is calculated as follows:

$$FSI = \sum_{i=1}^{m} \sum_{j=1}^{n} \sum_{k=1}^{p} OB(i) \cdot ST(i,j) \cdot M(j,k) \cdot G(k) \qquad (4)$$

where, OB(i) is the weight coefficient of objective OB_i in terms of the policy PO, ST(I,j) is the weight coefficient of strategy ST_j in terms of the objective OB_j, M(j,k) is the weight coefficient of measure M_k in terms of the strategy ST_j and G(k) is the implementation grade of measure M_k. The following expressions have to be fulfilled:

$$\sum_{i=1}^{m} OB(i) = 1,$$

$$\sum_{j=1}^{n} ST(i, j) = 1, \quad \forall \; i = 1, 2, \ldots, m$$

$$\sum_{k=1}^{p} M(j, k) = 1, \quad \forall \; j = 1, 2, \ldots, n \tag{5}$$

In the same context cost is assessed as follows:

$$C = \sum_{k=1}^{p} G(k) \cdot C(k) \cdot A \tag{6}$$

where G(k) is the Implementation grade of measure M_k, C(k) is the Cost per square meter of measure M_k, and A is the area of the building under investigation.

17.5 BUDGET ALLOCATION FOR FIRE SAFETY UPGRADING OF GROUP OF STRUCTURES

17.5.1 PROBLEM OVERVIEW

According to the proposed S&RA model each historic building i is characterized with the implementation level vector $s_{i,initial}$ that denotes the type and implementation grade of the fire protection measures selected. Moreover, each building is assumed to be ranked with reference to its population, the building itself and the significance of its contents. The administration of a group of historic buildings wishes to upgrade their fire safety level to the highest possible level on the basis of existing conditions, building importance and under the available budget constraint; thus, the main objective is to optimally allocate the available budget to the historic buildings for achieving the best overall fire safety upgrade of the group.

17.5.2 MODEL FORMULATION

Optimal fund allocation in this context is defined as follows:

$$\max F(s_{upgrade}) = \sum_{i=1}^{N} \sigma_{pop,i} \cdot \sigma_{build,i} \cdot \sigma_{cont,i} \cdot (s_{upgrade,i} - s_{initial,i}) \tag{7}$$

subject to

$$\sum_{i=1}^{N} K(age_i, size_i, dist_i, extra_i, s_{upgrade,i}, s_{initial,i}) = B_{T\arg et}, \ z_{upgrade,i} \in Z \quad (8)$$

where N is the total number of historic buildings, i represents a historic building (where, $i \in [1, N]$), $\sigma_{pop,I}$ is the importance factor of historic building i related to its *population*, $\sigma_{build,I}$ is the importance factor of historic building i related to its *building value*, $\sigma_{cont,I}$ is the importance factor of historic building i related to its *contents value*, $s_{upgrade,I}$ denotes the anticipated (post-upgrade) fire safety level of building i, $s_{initial,i}$ denotes the current fire safety level of building i, Age_i is the age of historic building i, $Size_i$ is the size of historic building i, $Dist_i$ is the distance of historic building i from the nearest fire service station, $Extra_i$ represents special unquantifiable factors related to fire safety, K is the cost for upgrading historic building i from level $s_{initial,i}$ to level $s_{upgrade,I}$ and B_{Target} is the available budget.

The aim of the objective function given in Eq. (7) is to maximize the fire safety upgrade level of historic buildings; a premium is given to the historic buildings with increased importance rating. The budget constraint given in Eq. (8) implies that cost K per historic building is a function of its characteristics (age, size, distance from fire service station and extra factors) as well as its current and anticipated fire safety level. It should be noted that the aforementioned model is a variant of the well-known knapsack problem [32].

17.5.3 MOUNT ATHOS MONASTERIES

Mount Athos is called a mountain and peninsula situated in Macedonia, Greece. Also known as the Holy Mountain, it is a world heritage site and autonomous polity into the Hellenic Republic, hosting 20 (20) monasteries (see Figure 17.2) under the direct jurisdiction of the Patriarch of Constantinople. Athonite monasteries possess immense deposits of invaluable medieval art treasures, including icons, liturgical vestments and objects (crosses, chalices, etc.), codices and other Christian texts, imperial chrysobulls, holy relics etc. Buildings themselves, their contents referred above, as well as the environment consist a triplet that has to be protected against all possible hazards, including fire, which is one of the major and permanent threats that could cause multiple irreversible damages in many levels, not only economically.

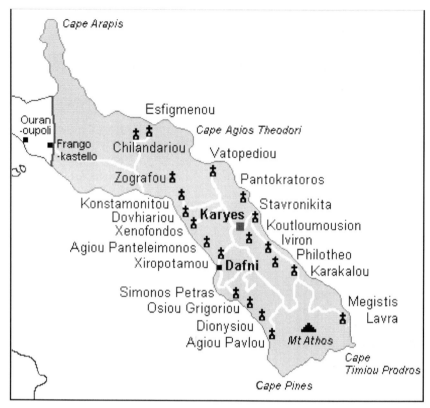

FIGURE 17.2   Map of Mount Athos.

17.5.4   FUND ALLOCATION FOR MOUNT ATHOS MONASTERIES

In this work, we apply the S&RA model to the case of the 20 monasteries of Mount Athos, each one composed of different characteristics that affect the cost of the possible fire protection measures that can be applied to achieve the desirable fire safety upgrading. Namely these parameters are age, size, distance from Karyes and a special extra index that represents some additional specificities, such as: existence of a multi-story building into the territory of one monastery, distance from the forest, complexity of the monastery's layout, etc. The objective function to be maximized is related to three importance factors that characterize each monastery, setting a hierarchy on the priority for application of the fire protection measures. Since human life is always of major significance to be protected, the first importance factor is related to the population of the monastery, including both monks and visitors. The second importance factor is connected to the building value, and for the needs of the present work depends on the Athonite hierarchy, while the third importance factor is connected to the value of the contents of the monastery. All parameters mentioned above for the 20 monasteries are provided in Table 17.1. It should be stated that some of these data are based on observations and personal judgment and may defer from the real situation, since it is extremely difficult to accurately assess some parameters, especially those related to the cultural value.

In this part of the chapter, we apply the proposed S&RA model to the case of the Mount Athos (shown in Figure 17.2) and particularly to its 20 monasteries, each one featured by different age, size and distance from the capital of Mount Athos Karyes (where the a fire service station is situated) along with additional special characteristics. In particular, eight different test cases are examined with varying target budget referring to the available budget for fire protection measures upgrading, aiming to achieve the highest fire safety level for all monasteries: 10.0, 40.0, 160.0 and 640.0 million given in monetary units (MU, corresponding to Dollars or Euros). The parameters used for the implementation of DE algorithm for solving the following budget allocation problem are as follows: the population size NP = 110, the probability CR = 0.125, the constant F = 0.340, while the control variable $\lambda = 0.733$ based on the parameter study presented in Ref. [33].

The budget requirements for upgrading the i-th monastery from its existing fire safety level to different (higher) levels are given in Table 17.1. The estimation of the cost was achieved combining the fire protection measures given in previous section of this chapter. In order to take into account the additional cost requirements imposed by the monastery age for the i-th monastery parameter σ_{age} is used. This parameter corresponds to the ratio of the budget for upgrading, to specific fire safety level, a monastery of the same size and type to that of the same monastery constructed today. Further, in order to consider the requirements for upgrading monasteries of different parameter σ_{size} is used; σ_{size} corresponds to the ratio of the budget for upgrading a monastery, to specific fire safety level, to that of the lower size category monastery. Additionally, the distance from Karyes which is the Capital of Mount Athos, where a fire service station is situated, is taken into account with σ_{dist}, which expresses the additional cost imposed by the

TABLE 17.1   Monasteries of Mount Athos-Characteristics

Monastery	Age	Size	Distance from Karyes	Extra Factor*	σ_{pop}	σ_{build}	σ_{cont}	s_{init}
Great Lavra	Very Old	Very Big	Near	N	1.05	1.20	1.01	3
Vatopedi	Old	Very Big	Far	A1	0.80	1.18	0.99	2
Iviron	Old	Very Big	Near	N	1.03	1.16	1.01	1
Helandariou	New	Big	Very Far	N	0.98	1.14	1.08	2
Dionysiou	Very New	Small	Far	N	1.03	1.12	1.14	3
Koutloumousiou	New	Small	Very Near	A1	0.99	1.09	1.11	4
Pantokratoros	Very New	Small	Near	A2	1.15	1.07	1.09	1
Xiropotamou	Old	Small	Near	B1	1.15	1.05	1.06	2
Zografou	New	Big	Far	N	1.09	1.03	1.03	2
Dochiariou	Old	Small	Far	A1	1.20	1.01	1.07	2
Karakalou	Old	Small	Far	N	1.06	0.99	1.07	4
Filotheou	Old	Small	Near	N	1.07	0.97	1.04	2
Simonos Petras	Very New	Small	Far	A2	0.97	0.95	1.06	2
Agiou Pavlou	Old	Big	Very Far	N	0.93	0.93	0.98	1
Stavronikita	Very New	Very Small	Near	N	1.10	0.91	1.05	3
Xenophontos	Old	Big	Far	B1	0.90	0.88	0.95	4
Osiou Grigoriou	Very New	Small	Far	B1	0.94	0.86	1.01	2
Esphigmenou	Old	Big	Very Far	N	0.91	0.84	0.94	1
Agiou Panteleimonos	New	Big	Far	N	0.99	0.82	0.89	3
Konstamonitou	Old	Small	Far	N	1.10	0.80	0.98	3

* N: Neutral, A1: 1 aggravation factor, A2: 2 aggravation factors, B1: 1 Benefit Factor.

access time of the fire brigade, corresponding to the ratio of the budget for upgrading a monastery, to specific fire safety level, to that of the nearest distance category monastery. Finally some other unquantifiable factors or characteristics of monasteries, which have a significant positive or negative impact on their fire safety level and are not included in the previously mentioned parameters, are represented by σ_{extra}, a parameter which penalizes some special requirements of certain monasteries (e.g., the fact that Monastery of Simonos Petras is a multi-story building imposes additional cost for the implementation of specific fire safety measures, compared to a single story one), or accounts for some benefit factors of other monasteries in order to reduce the total cost (e.g., some monasteries have simple architectural forms/features, that reduces the application cost of specific fire protection measures). Further information about the parameters described above can be found in Ref. [33].

The list of the 20 monasteries along with the age level, size, distance from Karyes, additional factors, population importance factor (related to the number of monks and visitors), building importance (related to the building fabric value), contents importance (related to the value of the artifacts) and the initial fire safety level of the monasteries is provided in Table 17.1.

Figure 17.3 depicts the solutions obtained for the optimal fund allocation problem for the four target budgets examined in this work (10.0, 40.0, 160 and 640.0 million MUs). The figure presents the fire safety level of the 20 monasteries before (initial fire safety level, before upgrading) and after the fire safety upgrading (final fire safety level). It can be seen that for the maximum target budget of 640.0 million MUs the highest fire safety level is achieved for all monasteries, while worth mentioning that in all test cases the available budget is allocated satisfying the constraint considered.

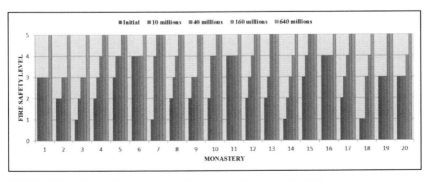

FIGURE 17.3   Fire safety levels for different target budgets (in MU).

17.6 FIRE SAFETY UPGRADING OF A SINGLE BUILDING

17.6.1 PROBLEM OVERVIEW

According to the proposed model, the fire safety level of the building under investigation is expressed by means of the fire safety index. FSI depends on the implementation grade of the fire protection measures selected and ranges between zero (0.0), for total absence of all measures, to monad (1.0), for the case where all measures all fully implemented in the building. The contribution of each fire protection measure to FSI is determined through the AHP network, described in previous section, according to the selected weights that connect the elements of the adjacent levels. The administration of the specific building wishes to upgrade the fire safety level of it to the highest possible level on the basis of the proposed fire protection measures under the available budget constraint; thus, the main objective is to optimally allocate the available budget to the fire protection measures for best overall fire safety upgrade of the building.

17.6.2 MODEL FORMULATION

The optimal fund allocation is defined as a nonlinear programming problem formulated as follows:

$$\max \ FSI(\boldsymbol{G})$$
$$s.t. \ \ C(\boldsymbol{G}) = C_{\text{Target}} \tag{9}$$

In the AHP framework described previously the following criteria and parameters where adopted to describe the fire protection problem of a historic building. On top of the hierarchy, as the main policy, is placed the reduction of the fire risk, representing the fire risk index. On second level the objectives that were chosen are: OB1 – protection of the people (occupants and visitors), OB2 – protection of the building fabric, OB3 – protection of the cultural contents, OB4 – protection of the environment, OB5 – protection of the firemen, and OB6 – safeguard continuity of activity. On next level, which represents the strategies, the following seven parameters were selected: ST1 – reduce the probability of fire start, ST2 – limit fire development in the fire compartment, ST3 – limit fire propagation out of the fire compartment. ST4 – facilitate egress, ST5 – facilitate fire fighting and rescue operations, ST6 – limit the effects of fire products, and ST7 – protection from forest fires. Finally on last level, where the most specific parameters are placed, the following 16 measures were taken into account: M1 – compartmentation, M2 – fire resistance of structural elements, M3 – control of fire load, M4 – materials (reaction to fire), M5 – control of fire spread outside the building, M6 – design of means of escape, M7 – signs and safety lighting, M8 – access of the fire brigade, M9 – detection and alarm, M10 – suppression and extinguishing, M11 – smoke control systems, M12 – training of the personnel, M13 – fire drills and emergency planning, M14 – management of fire safety, M15 – maintenance of fire safety system, and M16 – salvage operation. The weight coefficients of the above elements, required form the AHP network as it was described previously, can be found in Ref. [33].

17.6.3   SIMONOPETRA MONASTERY

The monastery of Simonos Petra, also known as Simonopetra, is probably the most impressive monastery of Mount Athos, founded on a 333 m height steep rock, located in the south-western side of Holy Mountain (see Figure 17.4). The main building complex gives the impression of hovering between sky and earth, and makes its view staggering. Constructed of unreinforced bearing masonry, with the walls on its foundation reaching thickness of 2.0 m, it is a wondrous architectural achievement, described as the "most bold construction of the peninsula" by the Greek Ministry of Culture. The 7.000 sq.m. complex has seven floors with many wooden balconies. The monastery was protected from several invaders due to its inaccessible position, but many times was almost destroyed by several fires such as the ones in 1580, 1622 and 1891 when the Catholicon and the library were burnt down. After the fire of 1891 the monastery was rebuilt to its current form. The last disaster that Simonopetra suffered from, occurred in the summer of 1990, when a big forest fire approached the building complex causing several damages.

Among the monastery's heirlooms, the most important is a piece of Saviour's Cross, holy reliquaries and others. The library, after the last fire, possesses only a few modern manuscripts and books. Currently the monastery is inhabited by a brotherhood of approximately 60 monks, which originates from the Holy Monastery of Great Meteoron in Meteora as in 1973 the Athonite community headed by Archimandrite Emilianos decided to repopulate the almost abandoned monastery. Moreover the monastery can accommodate about 30 visitors.

FIGURE 17.4   The main building complex of Simonopetra Monastery.

17.6.4   MAXIMIZING FIRE SAFETY

In this part of the chapter, we apply the proposed AHP model to the case of the monastery of Simonos Petra (Simonopetra) of Mount Athos. In particular, the formulation described by Eq. (9) is used in order to maximize fire safety level for a given budget. Fire safety level is expressed in terms of FSI, which is an indicator of the fire risk for the building under investigation. FSI directly expresses the policy which is on the top of AHP hierarchy described in previous section of the present study and depends on the implementation grades G(k) of the fire protection measures that are adopted, the approximate cost of which is given in Table 17.2. In fact, and referring to the present situation, most of these measures are partially implemented up to a specific grade, which is given also in Table 17.2 and thus the budget available is invested on the extension of their implementation, which drives to the overall fire safety upgrade of the building.

For comparative reasons two different formulations with reference to the constraint have been implemented while for both ones the same method has been adopted for handling the constraint. In particular, in constraint type 1 (C.t.1) the following inequality constraint has been considered:

$$C \leq C_{\text{Target}} \tag{10}$$

while in the constraint type 2 (C.t.2) the equality constraint, given in the formulation of Eq. (9) is transformed into an inequality one according to the following implementation:

$$|C - C_{\text{Target}}| \leq 0.01 \cdot C_{\text{Target}} \tag{11}$$

The simple yet effective, multiple linear segment penalty function [34] is used in this study for handling the constraints. According to this technique if no violation is detected, then no penalty is imposed on the objective function. If any of the constraints is violated, a penalty, relative to the maximum degree of constraints' violation, is applied to the objective function.

Table 17.2 summarizes the solutions of the proposed formulation for the 1st constraint type. Five different test cases are examined with varying target budget referring to the available budget for the fire protection measures upgrade: 300,000, 600,000, 900,000, 1,200,000 and 1,500,000 Monetary Units (MU). Figure 17.5 depict the percentage distribution of the total budget on the measures available for 2 out of total 5 cases examined, in particular the solutions for the cases of 600,000 and 1,200,000 MU and only for the 1st constraint type.

TABLE 17.2 Optimized Implementation Grades for the 16 Fire Protection Measures for 5 Different Budgets Available

Measure	COST (MU/sq.m.)	0 (initial)	300,000	600,000	900,000	1,200,000	1,500,000
M 1	40	0.30	0.38	0.33	0.78	1.00	1.00
M 2	40	0.30	0.43	0.45	0.50	1.00	1.00
M 3	10	0.60	1.00	1.00	1.00	1.00	1.00
M 4	30	0.60	0.69	0.98	1.00	1.00	1.00
M 5	10	0.50	0.98	0.95	1.00	1.00	1.00
M 6	30	0.60	0.71	0.90	0.89	1.00	1.00
M 7	20	0.30	0.30	0.43	0.30	1.00	1.00
M 8	10	0.20	0.81	1.00	1.00	1.00	1.00
M 9	30	0.60	0.64	0.99	1.00	1.00	1.00
M 10	40	0.50	0.54	0.75	1.00	1.00	1.00
M 11	40	0.00	0.00	0.00	0.10	0.03	1.00
M 12	20	0.90	0.97	0.95	0.98	1.00	1.00
M 13	10	0.50	0.81	0.82	1.00	0.97	1.00
M 14	10	0.90	0.97	1.00	1.00	1.00	1.00
M 15	20	0.60	0.81	1.00	1.00	1.00	1.00
M 16	20	0.40	0.47	0.61	1.00	0.98	1.00
FSI		*0.53*	*0.717*	*0.827*	*0.904*	*0.957*	*1.000*

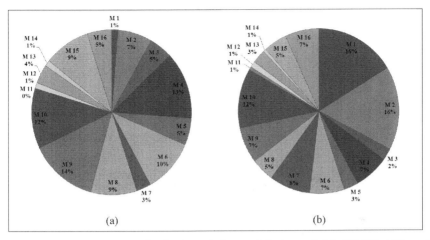

FIGURE 17.5 Percentage distribution of the total budget of (a) 600,000 MU and (b) 1,200,000 MU for C.t.1.

17.7 CONCLUSIONS

In this chapter two real world applications of evolutionary computation are presented in the framework of fire safety upgrading. In particular, we present an evolutionary computation based optimization approach for solving the budget allocation problem of the fire safety upgrading of a group of buildings. The proposed generic selection and resource allocation (S&RA) model was successfully applied to a real world case, the fire protection measures upgrade of the 20 monasteries of the Mount Athos, yielding feasible fund allocation solutions for different budget scenarios. Our proposed approach can successfully lead to optimal selection and resource allocation for any network of buildings.

In the second part, we solve the fire safety-upgrading problem for a single building, which is formulated as a multi-criteria decision making problem based on analytic hierarchy process by means of evolutionary computation. The AHP network, which was adopted for the problem incorporates four hierarchy levels, while the metaheuristic optimization based approach for solving the S&RA model involves differential evolution search algorithms. The proposed model was successfully applied to a real world case, the fire protection measures upgrade of the Mount Athos monastery of Simonos Petra, and can be well implemented in any project that is related to the fire protection design of a building. It should be noted that in both models the results obtained correspond to the indicative parameter values considered and can significantly differ depending to the designer's judgment and priorities.

ACKNOWLEDGMENTS

The first author acknowledges the financial support of the "Anastasios Anastasiou" endowment.

KEYWORDS

- **analytic hierarchy process**
- **cultural heritage**
- **evolutionary computation**
- **fire safety**
- **selection and resource allocation**

REFERENCES

1. Vandevelde, P., & Streuve, E., Fire Risk Evaluation To European Cultural Heritage (FiRE-TECH), Project Proceedings, Laboratorium voor Aanwending der Brandstoffen en Warmteoverdacht, Department of Flow, Heat and Combustion Mechanics, Gent, 2004.
2. Watts, J. M., Jr. Fire Protection Performance Evaluation for Historic Buildings, *J. Fire Prot. Eng.*, 2001, Vol. 11, 197–208.
3. Holland, J., Adaptation in natural and artificial systems. University of Michigan Press, Ann Arbor, Michigan, 1975.
4. Kirkpatrick, S., Jr. Gelatt, C. D., & Vecchi, M. P., Optimization by simulated annealing. *Science*, 1983, 220, 671–680.
5. Kennedy, J., & Eberhart, R., Particle swarm optimization. Proceedings of IEEE International Conference on Neural Networks. IV, 1995.
6. Dorigo, M., & Stützle, T., Ant Colony Optimization, The MIT Press, 2004.
7. Bozorg Haddad, O., Afshar, A., & Mariño, M. A., Honey bees mating optimization algorithm: a new heuristic approach for engineering optimization, Proceeding of the First International Conference on Modelling, Simulation and Applied Optimization (ICMSA0/05), 2005.
8. Geem, Z. W., Kim, J. H., & Loganathan, G. V., A new heuristic optimization algorithm: harmony search. *Simulation*, 2001, 76, 60–68.

9. Yang, X. S., & Deb, S., Engineering optimization by cuckoo search, *Int. J. Math. Model. Num. Optim.*, 2010, 1(4), 330–343.

10. Yang, X. S., Nature-Inspired Metaheuristic Algorithms. Frome: Luniver Press, 2008.

11. Yang, X. S., & Gandomi, A. H., Bat Algorithm: A novel approach for global engineering optimization, *Eng. Computation*, 2012, 29(5).

12. Gandomi, A. H., & Alavi, A. H., Krill Herd: A New Bio-Inspired Optimization Algorithm, Commun. Nonlinear Sci., 2012.

13. Fogel, D. B., Evolving artificial intelligence. PhD thesis, University of California, San Diego, 1992.

14. Goldberg, D. E., Genetic Algorithms in Search Optimization And Machine Learning. Addison Wesley, 1989.

15. Rechenberg, I., Evolution strategy: optimization of technical systems according to the principles of biological evolution. (in German), Frommann-Holzboog, Stuttgart, 1973.

16. Schwefel, H. P., Numerical optimization for computer models. Wiley & Sons, Chichester, UK, 1981.

17. Koza, J. R., Genetic Programming/On the Programming of Computers by Means of Natural Selection, The MIT Press, 1992.

18. Saaty, T. L., A Scaling Method for Priorities in Hierarchical Structures, *J. Math. Psychol.*, 1977, 15: 57–68.

19. Saaty T. L., The Analytic Hierarchy Process, McGraw Hill International, New York, NY, U.S.A., 1980.

20. Forman, Ernest H., & Saul I. Gass. "The analytical hierarchy process – an exposition." *Oper. Research,* July 2001, 49 (4): 469–487.

21. Omkarprasad Vaidya, S., & Sushil Kumar "Analytic hierarchy process: An overview of applications." *Eur. J. Oper. Research*, 2006, 169, 1–29.

22. Thomas Saaty, L., & Luis G. Vargas "Models, Methods, Concepts & Applications of the Analytic Hierarchy Process" (International Series in Operations Research & Management Science) by ISBN-13: 978–1461435969 ISBN-10: 146143596X Edition: 2nd ed., 2012.

23. Shi, L., Zhang, R., Xie, Q., & Fu, L., Improving analytic hierarchy process applied to fire risk analysis of public building, *Chinese Sci. Bull.*, 2009, 54 (8), pp. 1442–1450.

24. Shen, T.-I., Kao, S.-F., Wu, Y.-S., Huang, C.-J., Chang, K.-Y., & Chang, C.-H., Study on fire hazard factors of apartment building using analytic hierarchy process, *J. Applied Fire Sci.*, 2009, 19 (3), pp. 265–274.

25. Fera, M., & MacChiaroli, R., Use of analytic hierarchy process and fire dynamics simulator to assess the fire protection systems in a tunnel on fire, *I. J. R. A. M.*, 2010, 14 (6), pp. 504–529.

26. Vadrevu, K. P., Eaturu, A., & Badarinath, K. V. S., Fire risk evaluation using multicriteria analysis—a case study, *Environ. Monit. Assess.*, 2010, 166 (1–4), pp. 223–239.

27. Storn, R., & Price, K., Differential Evolution – A Simple and Efficient Heuristic for Global Optimization over Continuous Spaces, *J. Global Optim.*, Kluwer Academic Publishers, 1997, Vol. 11 341–359.

28. Das, S., & Suganthan, P. N., Differential evolution: A survey of the state-of-the-art. *IEEE T. Evolut. Comput.*, 2011, 15(1), 4–31.

29. Marchant, E. W., "Fire Safety Evaluation (Points) Scheme for Patient Areas Within Hospitals," Department of Fire Safety Engineering, University of Edinburgh, 1982.

30. Marchant, E. W., Fire Safety Engineering – A Quantified Analysis, Fire Prevention, June 1988, No. 210, pp. 34–38.

31. Stollard, P., "The Development of a Points Scheme to Assess Fire Safety in Hospitals," *Fire Safety J.*, 1984, Vol. 7, No. 2, pp. 145–153.

32. Winston, W., Operations Research: Applications and Algorithms, 4th edition, Duxbury Press, Belmont, CA, 2003.

33. Naziris, I. A., Optimum Performance-Based Fire Protection of Historic Buildings, PhD Thesis, Aristotle University of Thessaloniki, Greece, 2016.

34. Lagaros, N. D., & Papadrakakis, M., Applied soft computing for optimum design of structures, *Struct. Multidiscip. O.*, 2012, 45, 787–799.

CHAPTER 18

ELITIST MULTIOBJECTIVE EVOLUTIONARY ALGORITHMS FOR VOLTAGE AND REACTIVE POWER OPTIMIZATION IN POWER SYSTEMS

S. B. D. V. P. S. ANAUTH, and ROBERT T. F. AH KING

Department of Electrical and Electronic Engineering, University of Mauritius, Reduit 80837, Mauritius

CONTENTS

ABSTRACT

This study focuses on the development and comparative application of elitist multiobjective evolutionary algorithms (MOEAs) for voltage and reactive power optimization in power systems. A multiobjective Bacterial Foraging Optimization Algorithm (MOBFA) has been developed based on the natural foraging behavior of the Escherichia coli bacteria and its performance compared with two contemporary stochastic optimization techniques; an improved Strength Pareto Evolutionary Algorithm (SPEA2) and the Nondominated Sorting Genetic Algorithm-II (NSGA-II).

Simulation results considering different combination of the objective functions on a sample test power system, showed that the three elitist MOEAs were able to locate a whole set of well distributed Pareto-optimal solutions with good diversity in a single run. Fuzzy set theory was successfully applied to select a best compromise operating point from the set of solutions obtained.

A statistical analysis revealed that SPEA2 achieved the best convergence and diversity of solutions among the three MOEAs under study. However, the adaptive chemotaxis foraging behavior modeled in MOBFA allowed the algorithm to refine its search and explore the fitness space more efficiently in lesser computational time thereby producing more extended trade-off solutions curve than NSGA-II and SPEA2.

The proposed MOBFA was found to be a viable tool for handling constrained and conflicting multiobjective optimization problems and decision making, and can be easily applied to any other practical situations where computation cost is crucial.

18.1   INTRODUCTION

This study focuses on the development and comparative application of MOEAs for voltage and reactive power optimization in power systems. Voltage and reactive power control are crucial for power systems to improve their return on investment and to maintain a secure voltage profile while operating closer to their operational limits. This can be achieved by optimal adjustments in voltage controllers such as the field excitation of generators, switchable VAR sources and transformer tap settings.

To achieve a correct balance between conflicting economic and voltage security concerns in power systems and address the main weaknesses of the MOEA in [1], Anauth and Ah King [2] initially applied two stochastic optimization techniques; SPEA2 developed by Zitzler et al. [3] and the NSGA-II developed by Deb et al. [4], to locate a whole set of well distributed Pareto-optimal solutions with good diversity in a single run.

Jeyadevi et al. [5] presented a modified NSGA-II by incorporating controlled elitism and dynamic crowding distance strategies in NSGA-II to multiobjective optimal reactive power dispatch problem by minimizing real power loss and maximizing the system voltage stability. TOPSIS technique is used to determine best compromise solution from the obtained nondominated solutions. Karush–Kuhn–Tucker conditions are applied on the obtained nondominated solutions to verify optimality. Ramesh et al. [6] applied a modified Nondominated Sorting Genetic Algorithm-II to multiobjective Reactive Power Planning problem where three objectives considered are minimization of combined operating and VAR allocation cost, bus voltage profile improvement and voltage stability enhancement. A dynamic crowding distance procedure is implemented in NSGA-II to maintain good diversity in nondominated solutions.

A hybrid fuzzy multiobjective evolutionary algorithm based approach for solving optimal reactive power dispatch considering voltage stability was presented by Saraswat and Saini [7]. A fuzzy logic controller is used to vary dynamically the crossover and mutation probabilities. Based on expert knowledge, the fuzzy logic controller enhances the overall stochastic search capability for generating better pareto-optimal solution. A recent approach using Multi Objective Differential Evolution (MODE) algorithm to solve the Voltage Stability Constrained Reactive Power Planning problem has been proposed by Roselyn et al. [8] where in addition to the minimization of total cost of energy loss and reactive power production cost of capacitors, maximization of voltage stability margin is also considered. MODE emphasizes the non dominated solutions and simultaneously maintains diversity in the non dominated solutions.

Besides, a Bacterial Foraging Optimization approach to multiobjective optimization has been proposed by Niu et al. [9]. Their idea is based on the integration of health sorting approach and pareto dominance mechanism to improve the search to the Pareto-optimal set. Keeping a certain unfeasible border solutions based on a given probability helped to improve the diversity of individuals.

In this chapter, a MOBFA has been developed based on the natural foraging behavior of the *Escherichia coli* (*E.coli*) bacteria present in human intestines. The objective of such organism is to minimize its energy consumption when searching food while considering its physiological and environmental constraints. In accordance to Darwin's theory of evolution by natural selection, bacteria having efficient foraging strategies will survive to pass on their genes to the next generation. These biological concepts and evolutionary principles were modeled in the MOBFA such that it can be applied for optimizing the stated conflicting objectives in power systems. A statistical analysis will be done to evaluate the performance of the proposed MOBFA with NSGA-II and SPEA2 when applied to the voltage and reactive power optimization problem in power systems. Fuzzy set theory [10] will be applied to extract a best compromise operating point from the set of solutions obtained with the three MOEAs under study.

18.2   PROBLEM FORMULATION

The voltage and reactive power optimization problem can be formulated as a nonlinear constrained multiobjective optimization problem where real power loss, the load bus voltage deviations and the allocation cost of additional VAR sources are to be minimized simultaneously.

18.2.1 *OBJECTIVE FUNCTIONS*

18.2.1.1 Real Power Loss

The real power loss in transmission lines can be expressed as follows [1]:

$$P_{loss} = \sum_{\substack{k \in N_{br} \\ k \in (i,j)}} g_k \left[V_i^2 + V_j^2 - 2V_i V_j \cos\left(\delta_i - \delta_j\right) \right] \tag{1}$$

where N_{br} is the set of numbers of transmission lines in the system; g_k is the conductance of the kth transmission line between busses i and j; $v_i \angle \delta_i$ is the voltage at bus i.

18.2.1.2 Voltage Deviation

This objective is to minimize the sum of the magnitude of the load bus voltage deviations that can be expressed as follows [1]:

$$V_d = \sum_{i \in N_{PQ}} \left| V_i - 1.0 \right| \tag{2}$$

where N_{PQ} is the set of numbers of load busses.

18.2.1.3 Investment Cost

The allocation cost of additional VAR sources consists of a fixed installment cost and a variable purchase cost [11] as follows:

$$I_c = \sum_{i \in N_C} \left(C_{fi} + C_{ci} Q_{Ci} \right) \tag{3}$$

where N_C is the set of numbers of load busses for the installation of compensators; C_{fi} is the fixed installation cost of the compensator at bus i (\$); C_{ci} is the per unit cost of the compensator at bus i (\$/MVAR); Q_{ci} is the compensation at bus i (MVAR).

18.2.2 *CONSTRAINTS*

The optimization problem is bounded are discussed in the following subsections.

18.2.2.1 Equality Constraints

The following constraints represent typical load flow equations, which are solved by the Newton-Raphson load flow method:

$$P_{Gi} - P_{Di} - V_i \sum_{j \in N_B} V_j \left[G_{ij} \cos \delta_{ij} + B_{ij} \sin \delta_{ij} \right] = 0 \, i \in N_{B-1} \tag{4}$$

$$Q_{Gi} - Q_{Di} - V_i \sum_{j \in N_B} V_j \left[G_{ij} \sin \delta_{ij} - B_{ij} \cos \delta_{ij} \right] = 0 \, i \in N_{PQ} \tag{5}$$

where N_B is the set of numbers of total busses; N_{B-1} is the set of numbers of total busses excluding the slack bus; P_G and Q_G are the generator real and reactive power, respectively; P_D and Q_D are the load real and reactive power, respectively; G_{ij} and B_{ij} are the transfer conductance and susceptance between bus i and bus j, respectively.

18.2.2.2 Inequality Constraints

The control and state variables are bounded as follows:

Control Variables

a) Generator voltage limits

$$V_{Gi}^{min} \leq V_{Gi} \leq V_{Gi}^{max} \, i \in N_G \tag{6}$$

b) Transformer tap setting limits

$$T_i^{min} \leq T_i \leq T_i^{max} \, i \in N_T \tag{7}$$

c) Reactive power injection limits

$$Q_{Ci}^{min} \leq Q_{Ci} \leq Q_{Ci}^{max} \, i \in N_C \tag{8}$$

State Variables

a) Reactive power generation limits

$$Q_{Gi}^{min} \leq Q_{Gi} \leq Q_{Gi}^{max} \, i \in N_G \tag{9}$$

b) Load bus voltage limits

$$V_{Li}^{min} \leq V_{Li} \leq V_{Li}^{max} \quad i \in N_{PQ} \tag{10}$$

c) Transmission line flow limit

$$S_k \leq S_k^{max} \quad k \in N_{br} \tag{11}$$

18.2.3   MULTIOBJECTIVE FORMULATION

The voltage and reactive power optimization problem can be formulated mathematically as a nonlinear constrained multiobjective optimization problem by aggregating the objective functions and constraints as follows:

Minimize $$\left[P_{loss}, V_d, I_c \right] \tag{12}$$

Subject to: Equality constraints $h(x, y) = 0$ (13)

Inequality constraints $g(x, y) \geq 0$ (14)

where

$x^T = \left[Q_{Gi \, i \in N_G}, V_{Lj \, j \in N_{PQ}}, S_{kk \in N_{br}} \right]$ is the vector of dependent state variables and

$y^T = \left[V_{Gi \, i \in N_G}, T_{j \, j \in N_T}, Q_{Cn \, n \in N_C} \right]$ is the vector of control variables.

18.3   THE BACTERIAL FORAGING ALGORITHM

Passino [12] proposed a single objective evolutionary algorithm known as the Bacterial Foraging Optimization Algorithm (BFOA), which is based on the social foraging behavior of the E.coli bacteria present in the human intestines. The foraging strategy of the E.coli bacteria is governed by four processes, namely – chemotaxis, swarming, reproduction and elimination dispersal. These processes have been analyzed in this section before being adapted to MOBFA for multiobjective optimization.

18.3.1   CHEMOTAXIS

Chemotaxis is the foraging behavior that has been formulated as an optimization problem whereby the E.coli bacterium travels up a positive nutrient gradient, always seeking higher nutrient concentrations and avoiding noxious substances. The movement of the ith bacterium at the jth chemotactic, k^{th} reproductive, and l^{th} elimination dispersal step can be mathematically expressed as follows [12]

$$\theta^i\left(j+1,k,l\right)=\theta^i\left(j,k,l\right)+C\left(i\right)\frac{\Delta(i)}{\sqrt{\Delta^T\left(i\right)\Delta\left(i\right)}} \tag{15}$$

The scalar quantity $C(i)$ indicates the size of the step taken in the random direction specified by the unit length vector. $\Delta(i)$ If the cost at $\theta^i\left(j+1,k,l\right)$ is better than that at the preceding position $\theta^i\left(j,k,l\right)$, then the bacterium will keep taking successive steps of size $C(i)$ in the same direction $\Delta(i)$ otherwise it will tumble in another direction.

18.3.2   SWARMING

Swarming makes the bacteria gather together and move as concentric patterns of swarms. When an *E.coli* bacterium has searched the optimum nutrient gradient, it releases an attractant to signal other nearby bacteria to swarm together. The cost of a position is affected by the cell-to-cell signaling effects $J_{cc}(\theta,P\left(j,k,l\right))$ in E.coli swarms as follows [12]:

$$J\left(i,j,k,l\right)=J\left(i,j,k,l\right)+J_{cc}\left(\theta,P\left(j,k,l\right)\right) \tag{16}$$

$$J_{cc}\left(\theta,P\left(j,k,l\right)\right)=\sum_{i=1}^{S}J_{cc}\left(\theta,\theta^i\left(j,k,l\right)\right) \tag{17}$$

$$=\sum_{i=1}^{S}\left[-d_{attract}\,exp\left(-w_{attract}\sum_{m=1}^{p}\left(\theta_m-\theta^i_m\right)^2\right)\right]$$

$$+\sum_{i=1}^{S}\left[h_{repellent}\,exp\left(-w_{repellent}\sum_{m=1}^{p}\left(\theta_m-\theta^i_m\right)^2\right)\right]$$

where S is the bacteria population size; p is number of control variables; θ^i is a point in the p-dimensional search space for the ith bacterium; $d_{attract}$, $w_{attract}$, $h_{repellent}$ and $w_{repellent}$ determine the intensity of the inter cell attractive and repulsive signals.

18.3.3   REPRODUCTION

The population is sorted in ascending order of fitness or health of each bacterium. The lower half least healthy bacteria die and are replaced when the remaining healthiest ones split into two identical ones, such that each child bacteria undergoes chemotaxis from the same position. Passino [12] defined the health of a bacterium as the cost accumulated during the chemotactic steps. However, if the health is defined like this, reproduction may unknowingly eliminate the best bacterium. A better process has been implemented in MOBFA whereby the minimum cost found by the bacterium during the chemotactic steps is taken as its health.

18.3.4   ELIMINATION – DISPERSAL

Some bacteria are simply dispersed to a random location on the optimization domain, with a very small probability, after some reproduction steps to prevent them from being trapped in local minima. Elimination-dispersal events may help chemotaxis if ever the dispersed bacteria find unexplored higher nutrient concentrations in the search place. However, in most of the cases the chemotactic process is affected due to loss of good solutions. To address the main problem of the single objective BFOA proposed by Passino in Ref. [12], an elitism-preservation mechanism must be implemented in MOBFA to preserve the healthiest bacteria.

18.4   THE PROPOSED MULTIOBJECTIVE BACTERIAL FORAGING ALGORITHM

18.4.1   DESCRIPTION OF MOBFA

Step 1: A random population P_O is created by initializing the parameters

$$S, p, N_f, N_{ch}, N_S, N_{ed}, P_{ed}, C(i)(i = 1,2,...,S) \text{ and } \theta^i$$

where S is the bacteria population; p is number of control variables; N_f is the number of objective functions; N_{ch} is the number of chemotactic steps; N_s is the swim length; N_{ed} is the number of elimination-dispersal events; P_{ed} is the probability of elimination-dispersal events; $C(i)$ is the size of the step taken in the random direction specified by the tumble; θ^i is a point in the p-dimensional search space for the ith bacterium.

Step 2: Elimination-dispersal loop. $l = l + 1$.

Step 3: Chemotaxis loop. $j = j + 1$.

 a) For ($i = 1,2,...,S$), take a chemotactic step for the ith bacterium as follows:

 b) Compute fitness function $J^i(f,j,l)$ using (16) for all objective functions $(f = 1,2,...,N_f)$.

 c) Let $Jlast(f) = J^i(f,j,l)$ since a better cost may be found during the chemtotaxis.

 d) Tumble: Generate a random vector $\Delta(i) \in \mathfrak{R}^p$ with each element $\Delta_n(i) \in [-1,1]$ $(n = 1,2,...,p)$.

 e) Move with a step of size $C(i)$ in the direction of the tumble for the ith bacterium using (15).

$$\theta^i(j+1,l) = \theta^i(j,l) + C(i)\frac{\Delta(i)}{\sqrt{\Delta^T(i)\Delta(i)}} \tag{18}$$

 f) Compute $J^i(f,j,l)$ for all objective functions $(f = 1,2,...,N_f)$.

 g) Swim.

  (i) Let $m = 0$ (counter for the swim length).

  (ii) While $m < N_s$ (limit the length of swim).

   1) Let $m = m + 1$.

   2) If $J^i(f, j+1,l) \leq_d Jlast(f)$ (if dominated), let $Jlast(f) = J^i(f, j+1,l)$ and allow the bacterium to further swim as follows

$$\theta^i(j+1,l) = \theta^i(j+1,l) + C(i)\frac{\Delta(i)}{\sqrt{\Delta^T(i)\Delta(i)}} \tag{19}$$

   and compute the new $J^i(f,j + 1,l)$ for this $\theta^i(j + 1,l)$ as in sub step (f).

3) Else let $m = N_S$ to prevent the bacterium from swimming further.

h) If $i \neq S$, go to sub step (b) to process the next bacterium.

Step 4: Nondominated sorting to select the better-ranked solutions for the next iteration.

a) Combine the new solutions with the old ones in the mating pool as follows:

$$R = P_j \cup P_{j+1} \tag{20}$$

b) Perform a nondominated sorting to R and identify the different fronts: $F_t t = 1,2,...,etc.$

c) A new population $P_{j+1} = \emptyset$ of size S is created and the counter t is set to 1. Until $|P_{j+1}| + |F_t| < S$, perform $P_{j+1} = P_{j+1} \cup F_t$ and $t = t + 1$.

d) Perform the crowding-sort ($F_t, <_c$) procedure as outlined in [4] and include the most widely spread $\left(S - |P_{j+1}|\right)$ solutions by using the crowding distance values in the sorted F_t to P_{j+1}.

Step 5: If $j < N_{ch}$, go to step 3 to continue the chemotaxis.

Step 6: Elimination-dispersal: For each i ($i = 1,2,...,S$), the ith bacterium is dispersed to a random location on the optimization domain with a small probability P_{ed}.

Step 7: Evolution: The step size $C(i)$ of the ith bacterium is decreased as follows:

$$C(i) = \frac{C(i)}{l+1} \tag{21}$$

Step 8: If $l < N_{ed}$, then go to step 2, otherwise end.

18.4.2 CONSTRAINT HANDLING

To handle the constrained multiobjective optimization problem using the proposed MOBFA, this study adopts the constrain-domination principle proposed by Deb et al. [4]. In this approach, two solutions i and j are picked from the population and the better one is chosen. Since, each solution can be either feasible or infeasible the following constrain-domination principle can be used effectively to discriminate between solutions. The solution i is said to constrained-dominate a solution j if any of the following conditions is true [4]:

1) Solution i is feasible and solution j is not.
2) Solutions i and j are both infeasible, but solution i has a smaller overall constraint violation.
3) Solutions i and j are feasible and solution i dominates solution j.

In case both solutions are feasible, the following domination principle is applied [4]:

The solution i is said to dominate the solution j if both of the following conditions are true:

1) Solution i is no worse than solution j in all objective functions.
2) Solution i is better than j in at least one objective function.

However, when the tournament takes place between two infeasible solutions, the infeasible solution with a smaller constraint violation is chosen.

18.4.3   ADAPTIVE CHEMOTAXIS IN MOBFA

Bacterial foraging with a fixed step size $C(i)$ suffers from two main problems:

1) If the step size is too high, then the bacteria will reach near the true Pareto-optimal front quickly but they will not be able to swim further to improve the accuracy of the solutions.
2) If the step size is too small, then the bacteria will take many chemotactic steps to reach the true Pareto-optimal front. However, if the bacteria find a local Pareto-optimal front, they may get trapped into it as the small step size will prevent them from deviating or tumbling too much in order to escape the local optima.

The step size of each bacterium is the main influential factor for both the speed and accuracy of convergence towards the true Pareto-optimal front. The solution to this problem is to use an adaptive chemotactic behavior during the optimization process. In the initial phase, the bacteria are allowed to search the whole solution space with a large step size. During the next elimination-dispersal stage, the step size is decreased to limit the search such that the healthier (nondominated) bacteria can exploit the rich nutrient regions.

This guides the search towards the true Pareto-optimal front with increasing accuracy. The proposed MOBFA uses an initial step size of 0.1, which is decreased during the four elimination and dispersal events, as shown in Figure 18.1. The computational flow of the developed MOBFA is depicted in Figure 18.2.

18.5 SIMULATION RESULTS AND DISCUSSION

18.5.1 IEEE 30-BUS TEST SYSTEM

The three elitist MOEAs were applied to the standard IEEE 30-bus test system [2] [13]. The system consists of 6 generator buses, 24 load buses, 41 transmission lines of which four branches are in-phase transformers with assumed tapping ranges of 10% and 2 installed shunt capacitor banks at bus 10 and bus 24 [2]. The candidate buses for reactive power compensation are 10, 12, 15, 17, 20, 21, 23, 24 and 29 [2]. The lower voltage magnitude limits at all buses are 0.95 p.u while the upper limits are 1.1 p.u for PV buses and 1.05 p.u for load buses and the slack bus [2].

18.5.2 SETTINGS OF THE PROPOSED APPROACH

All three evolutionary algorithms considered in this study were coded in C language on an Intel Core 2 Duo 2.80 GHz processor having 1GB of

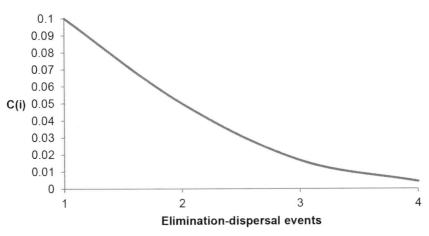

FIGURE 18.1   Evolution in Chemotaxis.

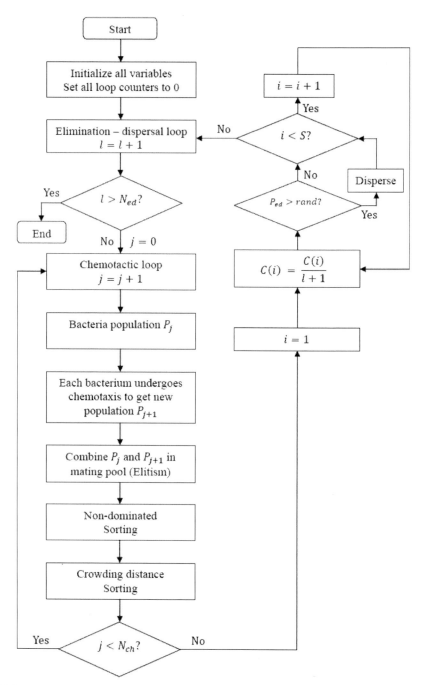

FIGURE 18.2 Computational flow of the developed MOBFA.

RAM [2]. The population size of the MOEAs was chosen as 100. The number of chemotactic steps, number of elimination-dispersal steps and the elimination-dispersal probability in MOBFA were set to 125, 4, and 0.1, respectively. The maximum length of swim of the bacteria was set to 2. The crossover and mutation probabilities in NGSA-II and SPEA2 were set to 0.99 and 0.01, respectively [2]. The distribution index for crossover and mutation were set at 5 and 50, respectively and the simulations were run for 500 generations [2]. The annual energy loss cost (E_c) and total cost (T_c) were computed using the cost settings given in Ref. [11].

18.5.2.1   Case 1: P_{loss} and V_d Minimization

The first study is a bi-objective optimization case whereby the real power loss and the load bus voltage deviations were simultaneously minimized using the three MOEAs. All three MOEAs were able to locate well-distributed Pareto-optimal solutions as shown in Figures 18.3–18.5. The results display the conflicting nature of the two objective functions with operation cost increasing considerably when trying to secure the voltage profile of the power system. This is because conduction loss can be reduced significantly by increasing generator voltage [2]. The ability of the proposed MOBFA and the other MOEAs to produce a set of Pareto-optimal solutions make them an excellent tool for decision making and selecting an operating point with such conflicting objective functions in power systems.

The extreme best solutions obtained using the proposed MOBFA are compared to that of SPEA [1], NSGA-II [2] and SPEA2 [2] in Table 18.1. A best compromise operating point for each MOEA was also extracted using the Fuzzy set theory in [10]. The results show that MOBFA, NSGA-II and SPEA2 produced better solutions than that obtained using SPEA [1]. SPEA does not guarantee the preservation of the boundary solutions with its clustering based diversity preservation technique and is easily outperformed by MOBFA, NSGA-II and SPEA2. The proposed MOBFA and the two other MOEAs implement elitism to prevent the loss of good solutions after the chemotactic and variation operators are applied in the evolutionary algorithms, respectively.

Moreover, MOBFA and SPEA2 produced more extended trade-off solutions than NSGA-II. Hence, MOBFA and SPEA2 could refine their search by exploring better nondominated solutions at the extremes of the trade-off curves than NSGA-II. SPEA2 produced both the best real power loss and the best voltage deviation solutions.

For comparison purposes, the problem was treated as a single objective optimization problem by combining the two objective functions using the weighted-sum or parameterized method as in Eq. (22).

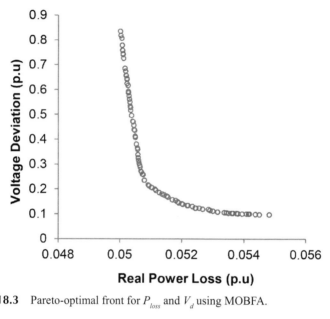

FIGURE 18.3 Pareto-optimal front for P_{loss} and V_d using MOBFA.

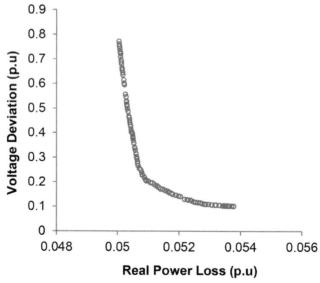

FIGURE 18.4 Pareto-optimal front for P_{loss} and V_d using NSGA-II.

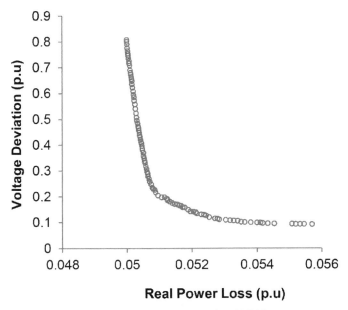

FIGURE 18.5 Pareto-optimal front for P_{loss} and V_d using SPEA2.

$$Minimise \frac{w \times P_{loss}}{\left(P_{loss}^{max} - P_{loss}^{min}\right)} + \frac{(1-w) \times V_d}{\left(V_d^{max} - V_d^{min}\right)} \qquad (22)$$

where w is a weighting factor. The objective function values are normalized to avoid the unnecessary use of a scaling factor as in [1]. The set of nondominated solutions was obtained using real Genetic Algorithm (GA) by linearly varying w from 0 to 1 with an increment of 0.01 for each run. The population size was chosen as 100 and a maximum iteration of 500 was used. The Pareto-optimal front obtained is shown in Figure 18.6.

It can be seen that the proposed MOBFA and the other MOEAs can explore the solution space and find trade-offs between multiple conflicting objectives in one single run while a single objective algorithm such as GA requires multiple runs. The diversity of the solutions obtained using the weighted-sum approach is worst when compared to those obtained using MOEAs. This is because MOEAs use diversity preservation mechanisms and Pareto-dominance based fitness assignment strategies to achieve better diversity and spread of solutions along the Pareto-optimal front.

TABLE 18.1 Best Solutions for P_{loss} and V_d Minimization

Control Variables & Functions	SPEA [1]			NSGA-II [2]			SPEA2 [2]			MOBFA		
	Best P_{loss}	Best V_d	Best Compromise	Best P_{loss}	Best V_d	Best Compromise	Best P_{loss}	Best V_d	Best Compromise	Best P_{loss}	Best V_d	Best Compromise
V_{G1} (p.u)	1.050	1.037	1.050	1.0500	1.0294	1.0500	1.0500	1.0173	1.0500	1.0500	1.0287	1.0499
V_{G2} (p.u)	1.044	1.027	1.044	1.0441	1.0188	1.0416	1.0436	1.0147	1.0421	1.0446	1.0154	1.0415
V_{G5} (p.u)	1.024	1.013	1.023	1.0240	1.0106	1.0178	1.0222	1.0164	1.0182	1.0245	1.0152	1.0170
V_{G8} (p.u)	1.026	1.008	1.022	1.0259	1.0003	1.0195	1.0244	1.0019	1.0192	1.0271	1.0032	1.0192
V_{G11} (p.u)	1.093	1.030	1.042	1.0016	0.9976	0.9905	1.0255	1.0366	0.9992	1.0355	1.0088	0.9945
V_{G13} (p.u)	1.085	1.007	1.043	1.0693	0.9997	1.0210	1.0621	0.9975	1.0178	1.0696	0.9903	1.0332
T_{11} (p.u)	1.078	1.054	1.090	1.0183	1.0123	1.0273	1.0622	1.0601	1.0322	0.9890	1.0263	1.0316
T_{12} (p.u)	0.906	0.907	0.905	0.9538	0.9796	1.0267	0.9036	0.9442	1.0307	0.9954	0.9926	1.0083
T_{15} (p.u)	1.007	0.928	1.020	1.0060	0.9674	1.0168	0.9932	0.9627	1.0130	0.9998	0.9532	1.0158
T_{36} (p.u)	0.959	0.945	0.964	0.9713	0.9657	0.9889	0.9654	0.9688	0.9872	0.9680	0.9774	0.9900
Q_{C10} (MVAR)	-	-	-	2.6596	2.1411	2.5512	0.3166	0.3722	0.3579	0.5287	3.1589	0.5922
Q_{C12} (MVAR)	-	-	-	4.7275	3.4963	4.5494	4.8860	2.9229	4.5836	2.3690	3.0348	1.6216
Q_{C15} (MVAR)	-	-	-	4.2528	4.3367	4.3803	4.3508	4.6194	4.4233	3.6175	4.9802	4.2750
Q_{C17} (MVAR)	-	-	-	4.8447	2.3082	4.7809	4.8409	0.1863	4.6823	3.8689	3.7329	3.9452
Q_{C20} (MVAR)	-	-	-	4.4019	4.3109	4.5742	4.4046	4.9585	4.5780	3.8235	4.9749	4.1836
Q_{C21} (MVAR)	-	-	-	4.9861	3.0570	4.9837	4.8982	0.7859	4.9468	4.9969	4.0524	4.7490
Q_{C23} (MVAR)	-	-	-	2.4735	3.4600	2.5032	2.1037	3.2606	1.8817	2.2256	4.5653	1.7840
Q_{C24} (MVAR)	-	-	-	4.4827	4.7657	4.3712	4.0412	4.7077	4.6059	2.9959	1.8238	2.5745
Q_{C29} (MVAR)	-	-	-	2.5913	2.2054	2.4808	2.2803	2.3344	2.3569	3.1393	4.3235	2.8769
P_{loss} (MW)	5.1065	5.5161	5.1995	5.0045	5.3771	5.0912	**4.9996**	5.5707	5.0963	5.0021	5.4817	5.0928
V_d (p.u)	0.7126	0.1477	0.2512	0.7715	0.1008	0.2075	0.8070	**0.0909**	0.2031	0.8346	0.0968	0.2149

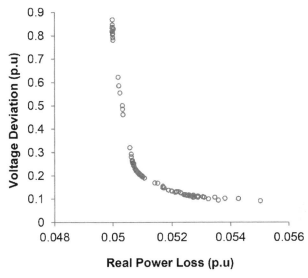

FIGURE 18.6   Pareto-optimal front for weighted-sum approach using GA.

18.5.2.2   Case 2: P_{loss} and I_c Minimization

The second study is a bi-objective optimization case whereby the real power loss and the investment cost were simultaneously minimized using the three MOEAs in order to assess the cost effectiveness of the approach. The trade-off curves obtained are shown in Figures 18.7–18.9. Table 18.2 shows that the results obtained using MOBFA, NSGA-II and SPEA2 are better than that obtained using EP in [11]. Hence, it is better to minimize the real power loss and investment cost simultaneously with the proposed MOBFA and the other MOEAs rather than combining them linearly and using single objective algorithm [2].

Figure 18.7 shows that MOBFA produced a more extended trade-off curve compared to that obtained by NSGA-II and SPEA2 in Figures 18.8 and 18.9, respectively. The results show that MOBFA produced both the best real power loss and the best investment cost while SPEA2 produced the best total cost due to the excellent diversity of its solutions but failed to span its search over the entire trade-off curve as compared to MOBFA.

The results display the conflicting nature of the two objective functions with investment cost increasing considerably when trying to reduce the operational energy lost cost on real power loss. This is because the reactive power compensation brought by additional VAR sources improves the system power factor and reduces the reactive component of the current, thus reducing the Ohmic energy losses [2].

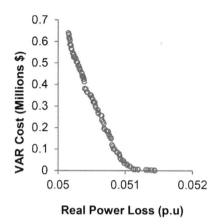

FIGURE 18.7   Pareto-optimal front for P_{loss} and I_c using MOBFA.

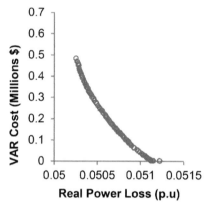

FIGURE 18.8   Pareto-optimal front for P_{loss} and I_c using NSGA-II.

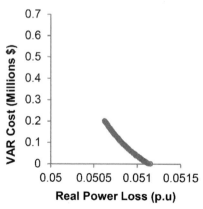

FIGURE 18.9   Pareto-optimal front for P_{loss} and I_c using SPEA2.

TABLE 18.2 Best Solutions for P_{loss} and V_d Minimization

Control Variables & Functions	EP [11] Best T_C	NSGA-II [2] Best P_{loss}	Best I_C	Best Compromise	SPEA2 [2] Best P_{loss}	Best I_C	Best Compromise	MOBFA Best P_{loss}	Best I_C	Best Compromise
V_{G1} (p.u)	1.050	1.0500	1.0498	1.0500	1.0500	1.0500	1.0500	1.0500	1.0500	1.0500
V_{G2} (p.u)	1.044	1.0443	1.0427	1.0439	1.0432	1.0433	1.0436	1.0449	1.0446	1.0441
V_{G5} (p.u)	1.023	1.0238	1.0230	1.0236	1.0220	1.0218	1.0220	1.0242	1.0224	1.0243
V_{G8} (p.u)	1.025	1.0258	1.0234	1.0254	1.0240	1.0236	1.0240	1.0262	1.0300	1.0253
V_{G11} (p.u)	1.050	1.0745	1.0838	1.0849	1.0811	1.0848	1.0814	1.0509	1.0814	1.0765
V_{G13} (p.u)	1.050	1.0633	1.0646	1.0679	1.0720	1.0737	1.0737	1.0693	1.0687	1.0776
T_{11} (p.u)	0.950	1.0310	1.0288	1.0283	0.9963	0.9945	0.9968	1.0106	1.0296	0.9991
T_{12} (p.u)	1.100	0.9524	0.9177	0.9329	0.9831	0.9836	0.9694	0.9649	0.9645	0.9566
T_{15} (p.u)	1.025	0.9806	0.9696	0.9797	0.9857	0.9808	0.9863	0.9949	0.9864	0.9960
T_{36} (p.u)	1.050	0.9531	0.9463	0.9525	0.9530	0.9483	0.9489	0.9590	0.9734	0.9483
Q_{C10} (MVAR)	0	0.0354	0.0005	0.0025	0.0001	0.0008	0.0004	0.7587	0.0000	0.0995
Q_{C12} (MVAR)	0	0.0638	0.0006	0.0009	0.0002	0.0000	0.0001	1.3489	0.0000	0.0041
Q_{C15} (MVAR)	0	2.8224	0.0006	0.4900	0.0007	0.0005	0.0007	3.1834	0.0000	0.3408
Q_{C17} (MVAR)	0	1.5333	0.0004	0.0258	0.0007	0.0006	0.0003	4.2994	0.0000	0.1059
Q_{C20} (MVAR)	0	2.5528	0.0002	0.8987	1.2644	0.0006	0.0006	2.2456	0.0000	0.0336
Q_{C21} (MVAR)	0	3.0724	0.0019	0.0008	0.0194	0.0083	0.0153	3.7541	0.0000	0.4652
Q_{C23} (MVAR)	0	2.1942	0.0004	0.8094	2.0594	0.0002	1.0808	1.7059	0.0000	1.0379
Q_{C24} (MVAR)	0	2.6892	0.0002	2.2240	2.6069	0.0005	1.6213	1.3098	0.0000	0.7271

TABLE 18.2 Continued

Control Variables & Functions	EP [11] Best T_c	NSGA-II [2] Best P_{loss}	Best I_c	Best Compromise	SPEA2 [2] Best P_{loss}	Best I_c	Best Compromise	MOBFA Best P_{loss}	Best I_c	Best Compromise
Q_{C29} (MVAR)	0	0.8736	0.0003	0.6021	0.5893	0.0006	0.3292	2.3972	0.0000	0.2841
P_{loss} (MW)	5.159	5.0252	5.1222	5.0698	5.0621	5.1151	5.0828	**5.0153**	5.1441	5.0822
E_c (×10⁶ $)	2.7116	2.6412	2.6922	2.6647	2.6606	2.6885	2.6715	**2.6361**	2.7037	2.6712
I_c (×10⁶ $)	0	0.4841	0.0011	0.1586	0.2012	0.0012	0.0954	0.6391	**0**	0.1019
T_c (×10⁶ $)	2.7116	3.1253	2.6933	2.8233	2.8618	**2.6898**	2.7669	3.2751	2.7037	2.7731
% Saving	3.64	-11.06	4.29	-0.33	-1.70	**4.42**	1.67	-16.39	3.92	1.45

18.5.2.3   Case 3: P_{loss}, V_d and I_c minimization

The third study is a tri-objective optimization case whereby all the three conflicting objective functions were simultaneously minimized using the three MOEAs. Figures 18.10–18.11 and Table 18.3 show that SPEA2 provided better diversity of Pareto-optimal solutions than MOBFA and NSGA-II and also found the best total cost solution. This is because SPEA2 considers density information during fitness assignment to achieve a better spread of solutions [2]. However, the proposed MOBFA was able to refine its search to find both the best voltage deviation and the best investment cost while NSGA-II could only produce the best real power loss.

18.6   PERFORMANCE ANALYSIS

The two main goals in a multiobjective optimization are to minimize the generation distance of the solutions to the true Pareto-optimal set and to maximize the diversity of the solutions along the Pareto-front [2]. The following performance metrics evaluate the performance of the algorithms based on a reference Pareto-optimal front obtained by selecting the best nondominated solutions from the combined Pareto-optimal solutions of the three MOEAs obtained for the 21 independent runs [2].

18.6.1   GENERATION DISTANCE AND SPREAD METRICS

The generation distance metric [14] evaluates the closeness of the nondominated set obtained by an algorithm to the reference Pareto-optimal front while the spread metric [4] evaluates how evenly the nondominated solutions are distributed in the objective space [2]. Table 18.4 shows that performance of the proposed MOBFA was almost comparable to that of NSGA-II. Moreover, SPEA2 produces better convergence and diversity of solutions than MOBFA and NSGA-II as supplemented by the smallest generation distance and spread metrics.

18.6.2   STATISTICAL ANALYSIS

A statistical analysis was performed using the Mostats5 toolbox [15], which superimposes and samples the attainment surfaces of the three MOEAs

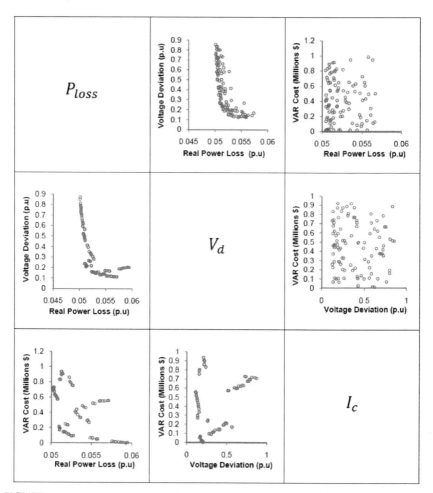

FIGURE 18.10 Pareto-optimal solutions for P_{loss}, V_d and I_c minimization. The upper diagonal plots are for SPEA2 and lower diagonal plots are for MOBFA.

throughout the fitness space to determine the percentage by which each algorithm outperforms the others [2].

The results in Table 18.5 show that SPEA2 was unbeaten in 75% of the fitness space covered by the three algorithms while in 52.5% of the fitness space it outperformed MOBFA and NSGA-II. The proposed MOBFA is the second best as it outperformed NSGA-II and SPEA2 in 20% of the fitness space and was unbeaten 32.5% of the fitness space. NSGA-II did well in part of the fitness space but could not outperform MOBFA and SPEA2 at the 95% confidence level.

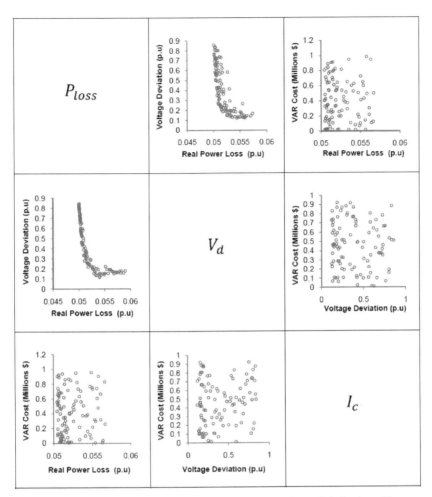

FIGURE 18.11 Pareto-optimal solutions for P_{loss}, V_d and I_c minimization. The upper diagonal plots are for SPEA2 and lower diagonal plots are for NSGA-II.

18.6.3 COMPUTATIONAL COMPLEXITY

Table 18.6 shows the computational time of the three MOEAs during the different simulation cases considered. SPEA2 took about twice as much time as NSGA-II for the bi-objective simulation cases and about 3.5 times as much time for the tri-objective simulation case. This is because the truncation operator used in SPEA2 is more computationally expensive than the nondominated sort technique in NSGA-II [2]. Moreover, the proposed MOBFA is the second best in terms of computational time, as it adopts the same nondominated sort mechanism as NSGA-II.

TABLE 18.3 Best Solutions for P_{loss}, V_d and I_c Minimization

Control Variables & Functions	NSGA-II [2]				SPEA2 [2]				MOBFA			
	Best P_{loss}	Best V_d	Best I_c	Best Compromise	Best P_{loss}	Best V_d	Best I_c	Best Compromise	Best P_{loss}	Best V_d	Best I_c	Best Compromise
V_{G1} (p.u)	1.0500	1.0291	1.0253	1.0489	1.0500	1.0470	1.0487	1.0486	1.0500	1.0173	1.0175	1.0495
V_{G2} (p.u)	1.0449	1.0271	1.0152	1.0425	1.0435	1.0329	1.0448	1.0451	1.0443	1.0122	1.0089	1.0359
V_{G5} (p.u)	1.0240	1.0057	0.9976	1.0218	1.0212	1.0113	1.0201	1.0202	1.0238	1.0177	1.0194	1.0177
V_{G8} (p.u)	1.0247	0.9963	0.9973	1.0192	1.0244	1.0001	1.0240	1.0218	1.0258	1.0087	0.9908	1.0055
V_{G11} (p.u)	1.0326	1.0627	1.0024	1.0179	1.0453	0.9857	1.0834	1.0568	1.0589	1.0009	1.0749	1.0774
V_{G13} (p.u)	1.0682	0.9974	1.0106	1.0378	1.0688	0.9915	1.0624	1.0376	1.0631	0.9988	1.0727	1.0407
T_{11} (p.u)	1.0345	1.0906	0.9993	1.0355	1.0055	1.0049	1.0371	1.0360	1.0224	1.0188	1.0854	1.0363
T_{12} (p.u)	0.9366	0.9354	0.9012	0.9348	0.9826	0.9473	0.9321	0.9876	0.9668	0.9449	1.0035	0.9575
T_{15} (p.u)	0.9950	0.9478	0.9293	1.0039	0.9936	0.9313	0.9886	0.9896	0.9806	0.9366	0.9503	0.9745
T_{36} (p.u)	0.9646	0.9541	0.9286	0.9707	0.9584	0.9643	0.9450	0.9602	0.9662	0.9738	0.9472	0.9797
Q_{C10} (MVAR)	1.5438	0.1400	0.0131	0.1948	3.8616	0.8059	0.0027	0.0280	1.3393	0.0497	0.0000	0.1298
Q_{C12} (MVAR)	0.3628	0.2166	0.0007	0.1030	0.0529	1.1700	0.0636	0.0408	0.0672	0.2011	0.0000	0.0083
Q_{C15} (MVAR)	4.5874	2.9780	0.0004	0.8357	3.9489	0.7247	0.1096	0.2171	2.8710	2.5762	0.0000	0.1109
Q_{C17} (MVAR)	4.8370	1.0568	0.0004	0.2636	4.6507	0.5351	0.0239	0.0751	4.3757	1.8405	0.0000	0.0011
Q_{C20} (MVAR)	3.0088	3.6961	0.0029	0.2829	2.0008	1.9937	0.0664	0.0657	2.2286	3.0231	0.0000	0.0565
Q_{C21} (MVAR)	4.9354	1.3326	0.0011	0.2160	4.0790	3.5579	0.0057	0.0076	4.9668	2.4952	0.0000	0.0262
Q_{C23} (MVAR)	1.6366	3.3982	0.0005	0.3096	1.2413	4.0209	0.0053	0.0100	1.0639	2.0878	0.0000	0.5543
Q_{C24} (MVAR)	4.6964	2.7807	0.0068	0.1426	4.7691	4.9744	0.0333	0.0078	2.8983	2.2396	0.0000	0.5611

TABLE 18.3 continued

Control Variables & Functions	NSGA-II [2]				SPEA2 [2]				MOBFA			
	Best P_{loss}	Best V_d	Best I_c	Best Compromise	Best P_{loss}	Best V_d	Best I_c	Best Compromise	Best P_{loss}	Best V_d	Best I_c	Best Compromise
Q_{C29} (MVAR)	2.3491	0.9378	0.0006	0.5791	2.0106	1.9579	0.0123	0.0283	3.7673	3.7721	0.0000	1.4014
P_{loss} (MW)	**5.0003**	5.4028	5.6022	5.2023	5.0085	5.4410	5.1444	5.2125	5.0147	5.7003	5.9286	5.2782
V_d (p.u)	0.8374	0.1149	0.1940	0.2441	0.8559	0.1220	0.6590	0.2790	0.8509	**0.1110**	0.2033	0.2811
E_C (× 10^6 $)	**2.6282**	2.8397	2.9445	2.7343	2.6325	2.8598	2.7039	2.7397	2.6357	2.9961	3.1161	2.7742
I_C (× 10^6 $)	0.8477	0.5051	0.0098	0.0968	0.8075	0.6012	0.0187	0.0234	0.7163	0.5576	0	0.0945
T_C (× 10^6 $)	3.4759	3.3448	2.9543	2.8312	3.4399	3.4610	**2.7226**	2.7631	3.3521	3.5536	3.1161	2.8687

TABLE 18.4 Generation Distance and Spread Metrics

	Generation Distance Metric			Spread Metric		
	MOBFA	NSGA-II	SPEA2	MOBFA	NSGA-II	SPEA2
Mean	0.001983	0.001162	**0.000755**	0.487836	0.455250	**0.386633**
Variance	3.057E-07	3.692E-07	**2.511E-07**	0.000464	0.000222	**0.000699**

TABLE 18.5 Statistical Analysis

	MOBFA	NSGA-II	SPEA2
Unbeaten (%)	32.5	27.5	75
Beats all (%)	20	0	**52.5**

TABLE 18.6 Computational Time of MOBFA, NSGA-II and SPEA2

	Computational Time (seconds)								
	Case 1			Case 2			Case 3		
	MOBFA	NSGA-II	SPEA2	MOBFA	NSGA-II	SPEA2	MOBFA	NSGA-II	SPEA2
Mean	140.233	**94.911**	195.702	134.054	**81.729**	158.317	159.102	**97.783**	345.043
STD. Deviation	1.808	1.299	11.725	3.050	2.337	14.001	0.445	1.187	11.292

18.7 CONCLUSIONS

A comparative application MOBFA with NSGA-II and SPEA2 was provided for voltage and reactive power optimization in power systems. The proposed MOBFA is an extension of the single objective BFOA [12], which was adapted to handle the constrained tri-objective optimization problem. The speed and accuracy of convergence of BFOA was improved by introducing adaptive step size during chemotaxis. Moreover, elitism was introduced in MOBFA to prevent the loss of good solutions when evolving to the next generation.

The optimization cases considered using different combinations of the objective functions, show that the proposed MOBFA and the other two MOEAs were able to locate a whole set of well distributed Pareto-optimal solutions with good diversity in a single run and outperformed those obtained with SPEA in [1] and EP in Ref. [11]. Fuzzy logic theory was successfully applied to select a best compromise operating point from the trade-off solutions obtained during the different simulation cases considered using the three MOEAs.

The simulation results and statistical analysis showed that SPEA2 found better convergence and spread of solutions than MOBFA and NSGA-II. However, the adaptive chemotaxis foraging behavior modeled in MOBFA allowed the algorithm to refine its search and explore the fitness space more efficiently with lesser computational time thereby producing more extended trade-off solutions than NSGA-II and SPEA2.

The main advantages of the proposed MOBFA are that it is easy to implement, capable of exploring better extreme solutions than NSGA-II and requires exceptionally lesser computational time than SPEA2.

Hence, it can be concluded that MOBFA is a viable tool for handling constrained and conflicting multiobjective optimization problems and decision making, and can be easily applied to any other practical situations where computation cost is crucial.

KEYWORDS

- **Elitist multiobjective evolutionary algorithms**
- **fuzzy logic theory**
- **multiobjective bacterial foraging algorithm**
- **optimal VAR dispatch**

REFERENCES

1. Abido, M. A., & Bakhashwain, J. M. Optimal VAR Dispatch Using a Multiobjective Evolutionary Algorithm. International Journal of Electrical Power & Energy Systems. 2005, 27(1), 13–20.

2. Anauth, S. B. D. V. P. S., & Ah King, R. T. F. Comparative Application of Multiobjective Evolutionary Algorithms to the Voltage and Reactive Power optimization Problem in Power Systems. In proceeding of: Simulated Evolution and Learning – 8th International Conference. SEAL 2010. Springer. Lecture Notes in Computer Science. 2010, 6457, 424–434.

3. Zitzler, E., Laumanns, M., & Thiele, L. SPEA2: Improving the Strength Pareto Evolutionary Algorithm for Multiobjective Optimization. In: Giannakoglou, K., et al. (eds.) Evolutionary Methods for Design, Optimization and Control with Application to Industrial Problems (EUROGEN 2001). International Center for Numerical Methods in Engineering (CIMNE). 2002, 95–100.

4. Deb, K., Pratap, A., Agrawal, S., & Meyarivan, T. A Fast and Elitist Multiobjective Genetic Algorithm: NSGA-II. IEEE Transactions on Evolutionary Computation. 2002, 6(2), 182–197.

5. Jeyadevi, S; Baskar, S., Babulal, C. K., & Iruthayarajan M. W. Solving multiobjective optimal reactive power dispatch using modified NSGA-II. Electrical Power and Energy Systems. 2011, 33, 219–228.

6. Ramesh, S., Kannan, S., & Baskar, S. Application of modified NSGA-II algorithm to multiobjective reactive power planning. Applied Soft Computing. 2012, 12, 741–753.

7. Saraswat, A., & Saini, A. Multiobjective optimal reactive power dispatch considering voltage stability in power systems using HFMOEA. Engineering Applications of Artificial Intelligence. 2013, 26, 390–404.

8. Roselyn J. P., Devaraj, D., & Dash, S. S. Multi Objective Differential Evolution approach for voltage stability constrained reactive power planning problem. Electrical Power and Energy Systems. 2014, 59, 155–165.

9. Niu, B., Wang, H., Wang, J., & Tan, L. Multiobjective bacterial foraging optimization. Neurocomputing. 2013, 116, 336–345.

10. Dhillon, J. S., Parti, S. C., & Khotari, D. P. Stochastic Economic Load Dispatch. Electric Power Systems Research. 1993, 26, 179–186.

11. Lai, L. L., & Ma, J. T. Evolutionary Programming Approach to Reactive Power Planning. IEE Proceedings-Generation Transmission Distribution. 1996, 143(4), 365–370.

12. Passino, K. M. Biomimicry of bacterial foraging for distributed optimization and control. IEEE Control Systems Magazine. 2002, 22(3), 52–67.

13. Alsac, O., & Scott, B. Optimal Load Flow with Steady-State Security. IEEE Transactions on Power Apparatus and Systems. 1974, 93, 745–751.

14. Veldhuizen, D. Multiobjective Evolutionary Algorithms: Classifications, Analyses, and New Innovation. PhD thesis. Department of Electrical Engineering and Computer Engineering. Airforce Institute of Technology. Ohio. 1999.

15. Knowles, D., & Corne, W. Approximating the nondominated front using the Pareto archived evolution strategy. J. Evolutionary Computation. 2000. 8(2), 149–172.

CHAPTER 19

EVALUATION OF SIMULATED ANNEALING, DIFFERENTIAL EVOLUTION, AND PARTICLE SWARM OPTIMIZATION FOR SOLVING POOLING PROBLEMS

YING CHUAN ONG, SHIVOM SHARMA, and G. P. RANGAIAH*

Department of Chemical and Biomolecular Engineering, National University of Singapore, Engineering Drive 4, Singapore 117585, Republic of Singapore, E-mail: gprangaiah@gmail.com

CONTENTS

ABSTRACT

In the recent past, several stochastic optimization methods such as simulated annealing (SA), differential evolution (DE) and particle swarm optimization (PSO) have been improved for solving global optimization problems. Pooling problems are a particular type of blending problems involving intermediate storages. Research on its formulation and solution by global optimization is important due to economic and environmental factors. However, SA, DE and PSO have not been applied to pooling problems. Therefore, in this chapter, improved variants of these methods, namely, Modified SA, Very Fast SA, Integrated DE and Unified Bare-Bones PSO are first tested on several benchmark mathematical problems, to establish their relative efficacy. Then, these methods are applied and evaluated for solving pooling problems using r-formulation developed by Zhang and Rangaiah [35]. The results show that IDE is generally more efficient and reliable on benchmark mathematical and pooling problems tested.

19.1   INTRODUCTION

The pooling problem, which merges the features of both the network flow and blending problems, is an important global optimization problem for achieving significant energy and cost savings [17, 24, 25]. A pooling network contains a number of process/source streams flowing to a set of intermediate pools (i.e., storage tanks), which are further connected to a set of final product tanks (Figure 19.1). For example, several source streams emerge from the process units in a petroleum refinery such as distillation columns, reformers and catalytic crackers, to be blended with additives in one or more intermediate pools before being channeled to product tanks [14]. It is possible to send source streams directly from the process units to product tanks but the process involving pools is commonly used due to its greater operational flexibility. However, the process involving pools is more complex and challenging to optimize.

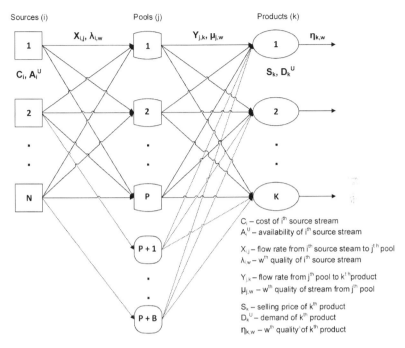

FIGURE 19.1   Schematic of network for the pooling problem formulation; note that each of the pools P+1, ..., P+B has only one inlet stream.

Given the source streams with different costs, composition (e.g., sulfur content) and/or qualities (e.g., octane number), the pooling problem is to determine the optimal flow rates from sources to product tanks through pools, which will minimize the cost while meeting the product demands and specifications. In this optimization problem, product demand, storage capacity and feed availability lead to inequality constraints, and mass balances around each pool are the equality constraints. Pooling problems are categorized into three classes: standard, generalized and extended pooling problems [25]. In the present work, only the standard pooling problems are studied.

Stochastic global optimization (SGO) methods such as genetic algorithms (GA), simulated annealing (SA), differential evolution (DE) and particle swarm optimization (PSO) have been proposed and improved in the last two decades [28]. They require little or no assumptions on the characteristics of optimization problems, are easy to implement and apply, and are likely to locate the global optimum within reasonable computational effort. Further, they have been adopted and extensively used for multiobjective optimization [29].

SA was proposed by Kirkpatrick et al. [21], who drew an analogy between the physical process of slow cooling of a molten metal (i.e., annealing) and the combinatorial optimization whose cost function is defined in a discrete domain. It has been extended to optimization in continuous domain [10]. There have been several modifications such as hybridization with the simplex method [10] and direct search [2], and adaptive parameter tuning [20]. However, SA of Corana et al. [13] is chosen and modified in this study because of its many applications in chemical engineering. For example, it was used for correlation of activity coefficients in aqueous solutions of ammonium salts using local composition models [19], liquid-liquid phase equilibrium modeling [16], and calculation of homogenous azeotropes in reactive and non-reactive mixtures [9]. SA of Corana et al. [13] was found to be reliable but less efficient for these applications. Also, it was claimed to be reliable for parameter estimation in vapor-liquid equilibrium modeling compared to very fast SA (VFSA) and direct search SA [10]. VFSA [30] is chosen as a faster variant of SA. In the calculation of phase stability analyzes for non-reactive and reactive mixtures, it is shown to require far fewer number of function evaluations (NFE, which is an indication of computational effort in application problems) with a tradeoff in reliability as compared to SA [5]. Both SA of Corona et al. [13] and VFSA have not been applied to pooling problems.

DE is a population based optimizer proposed by Storn and Price [32] especially for non-linear and non-differentiable continuous functions; more details on DE can be found in Price et al. [26]. It mimics biological evolution by performing mutation, crossover and selection steps (as in GA) to escape from the local minima. DE has fewer parameters compared to other stochastic algorithms, its principle is easy to understand, has relatively faster convergence and high reliability to find the global optimum [3, 31]. It has been successfully applied to phase stability and parameter estimation problems, fed-batch bioreactor, synthesis of cost-optimal heat exchange networks; *see* Chen et al. [11] for an overview of these applications. There are also many studies to improve its performance; most of them are focused on two aspects: adaptation of DE parameters [23, 27], and hybridization with other optimization methods [3, 31, 34]. The algorithm used in the present study is the integrated DE (IDE) developed by Zhang and Rangaiah [35] due to its reliability and efficiency in solving chemical engineering applications such as modeling vapor-liquid equilibrium data, phase equilibrium calculations, reactive phase equilibrium calculation and phase stability problems [36, 37]. IDE integrates three solid strategies, namely, tabu list, self-adaptation of parameters and mutation strategy, and a novel stopping criterion [34]. It was recently applied to pooling problems by Zhang and Rangaiah [34].

PSO is also a population-based method proposed by Eberhart and Kennedy [15]; it is inspired by the social behavior of bird flocks and fish schools. In principle, it explores the search space to identify promising areas having better solutions before exploiting these areas for the best solution. In the classical PSO, the population of potential solutions is called the swarm, and each solution is called a particle. Results from PSO application to phase stability and equilibrium calculations in reactive and non-reactive systems, show that it is a reliable method with good performance [7]. It has been applied to parameter estimation in vapor-liquid equilibrium modeling problems [37] and mean activity coefficients of ionic liquids [5]. A number of studies to improve PSO are focused on a parameter-free PSO, known as bare-bones PSO (BBPSO) and hybridization with other stochastic algorithms. For example, BBPSO-MC-gbest and BBPSO-MC-lbest integrate crossover and mutation operators of DE based on global and local best topologies, and both these are combined into unified BBPSO (UBBPSO) by Zhang et al. [37]. UBBPSO has not been applied to pooling problems.

For modeling and solving standard pooling problems, several formulations have been proposed. p-formulation of Haverly [18] uses stream flows and product qualities as the decision variables, q-formulation employs proportion of stream flows entering pools, instead of stream flows, as the decision variables [4], and pq-formulation combines both p- and q-formulations [33]. The recent r-formulation by Zhang and Rangaiah [34], motivated by the q-formulation, reduces the search space, number of decision variables and constraints. Solution of standard pooling problems using p-, q- and r-formulations suggests that r-formulation is better than the other two [34].

In the present chapter, four SGO methods, namely, Modified SA, VFSA, IDE and UBBPSO are studied and compared for solving pooling problems, for the first time. Another novelty in this systematic study is the use of r-formulation of Zhang and Rangaiah [34] for pooling problems. Modified SA, VFSA, IDE and UBBPSO are first tested on benchmark unconstrained and constrained problems to establish their robustness and efficiency. Then, they are applied to 13 benchmark-pooling problems. The results show that IDE is better than Modified SA, VFSA and UBBBPSO for solving the mathematical and pooling problems tested.

19.2   SELECTED STOCHASTIC GLOBAL OPTIMIZATION METHODS

Consider the following minimization problem with bounds on decision variables, inequality and equality constraints.

Minimize $$f(x)$$ 1(a)

Subject to $$\mathbf{x}^L \leq x \leq \mathbf{x}^U$$ 1(b)

$$g(x) \leq 0$$ 1(c)

$$h(x) = 0$$ 1(d)

Here, x is the vector of D number of decision variables with lower ($\mathbf{x}^L$) and upper ($\mathbf{x}^U$) bounds; $\mathbf{g}$ and $\mathbf{h}$ are the set of inequality and equality constraints, respectively.

The flow chart of Modified SA algorithm is shown in Figure 19.2. In order to minimize the objective function f(x), the algorithm begins with initial temperature (T_0), initial solution vector ($\mathbf{x}_1$, inside the bounds on decision variables) and step length vector for decision variables (**v**). The value of the objective function (f) is calculated at the initial solution, which is tentatively set as the optimum ($\mathbf{x}_{opt}$, f_{opt}). Then, new trial solutions are generated, one by one along each of all decision variables, and the objective function value at each of them is computed. The trial solution is accepted if it has a lower objective function value or it satisfies the Metropolis criterion (which allows acceptance of a solution having higher objective value with some probability). The trial solution with the smaller objective function value is set as the optimum. The above step (i.e., generation and evaluation of trial solutions along all decision variables) is repeated NV times for each value of **v**, and then **v** is adjusted. After NT adjustments of **v**, temperature is decreased. This completes one iteration/generation of SA. The search is terminated when improvement in the objective function value in the latest Nft iterations is lower than a pre-specified value or temperature is adjusted more than MNG times. See Corana et al. [13] for more details on SA. In the Modified SA, if the updated value of v_d is larger than $x_d^U - x_d^L$, then value of v_d is changed to $x_d^U - x_d^L$ to ensure that trial solutions are likely to satisfy bounds on decision variables. Further, value of **v** is set to $\mathbf{x}^U - \mathbf{x}^L$ after each temperature update for greater exploration.

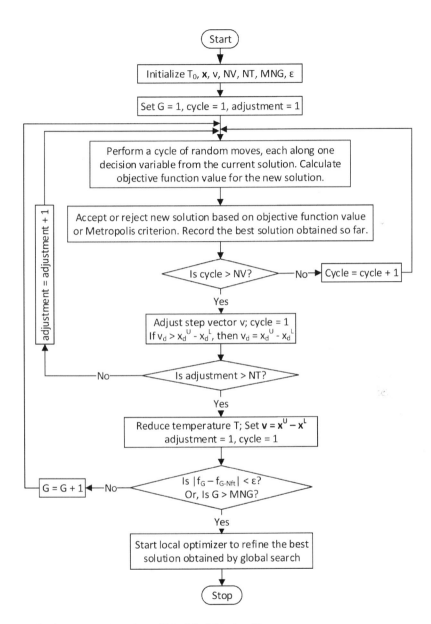

FIGURE 19.2 Flow chart of Modified SA algorithm.

VFSA algorithm uses a different strategy for generation of trial solutions and a different cooling scheme [30] and it does not have the step length parameter **v**. Trial solutions are generated (one by one along each of all decision variables) and evaluated for the objective function value. This is repeated NT times for each value of T, and then T value is modified. Finally, search is terminated based on the termination criteria.

In DE (Figure 19.3), the initial population of NP individuals is generated within the search space using uniformly distributed random numbers. The objective function value for each individual is calculated, and the trial

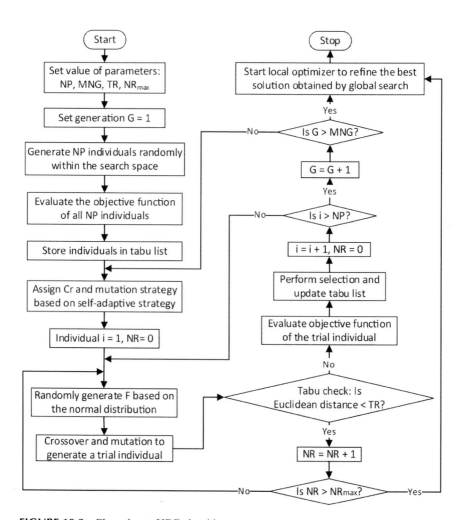

FIGURE 19.3 Flow chart of IDE algorithm.

individual is stored in the tabu list. The mutation, crossover and selection steps of DE are performed on the population in each generation. Before these steps, the self-adaptive strategy is used to assign mutation strategy and crossover rate (Cr) for each generation. The i-th trial individual is then produced according to the assigned mutation strategy, a randomly generated F value (with mean 0.5 and standard deviation 0.3) and Cr value, through mutation and crossover steps of DE. A boundary violation check is performed on the decision variable values of the trial individual generated; if lower/upper bound of any variable is violated, this variable value is generated randomly within its bounds, for the trial individual under consideration.

The trial individual is then compared with the individuals in the tabu list. If it is near to any in the tabu list (as indicated by the Euclidean distance in the decision variables space), the trial individual is rejected and a new trial individual is generated through the mutation and crossover operations. If the number of such rejections (NR) is greater than the given NR_{max} for the same trial individual, then it indicates the algorithm has either converged to the global optimum or trapped at a local optimum, and so the best solution found so far is unlikely to improve in the subsequent generations and the search is terminated. If NR is less than NR_{max} and the generated trial individual is away from all individuals in the tabu list, then the objective function and constraints are calculated.

The evaluated trial individual is stored in the tabu list. Selection between the target and trial individuals is performed based on their objective function values. If the trial individual is selected, it replaces the target individual in the population immediately and may participate in the subsequent mutation and crossover operations to enhance the convergence speed. NR is reset to 0 for generating the trial individual for the next target individual, and the above procedure is repeated until all NP target individuals are covered. This completes one generation in DE. Generations continue for MNG unless the stopping criterion on NR_{max} is already met in the earlier generations. See Zhang and Rangaiah [34] for more details on IDE.

Figure 19.4 shows the flow chart of UBBPSO algorithm. In PSO, a population of NP particles is initialized randomly within the bounds on decision variables. Then, objective function values are calculated for all particles, and personal best and global best are selected. For each particle in the population, a new particle is generated using particle velocity based on its personal best

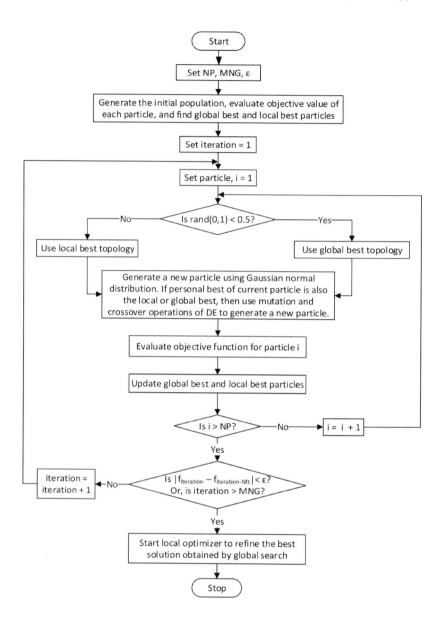

FIGURE 19.4   Flow chart of UBBPSO algorithm.

position and the global best position. Particle velocity and the new particle location should be maintained inside the search space. Value of the objective function is evaluated for each new particle, and personal and global best positions are updated as necessary. These steps are repeated for all particles in the population, to complete one iteration.

In order to eliminate the parameters in PSO, Gaussian normal distribution can be used to generate the new particle, and the resulting algorithm is referred as BBPSO. The search does not progress if personal best of the current particle is also the global best. Hence, the global best index particle is updated using mutation and crossover operations of DE algorithm, which is referred as BBPSO-MC-gbest. UBBPSO algorithm employs both global best and local best (among the particle and its two neighbors) topologies with equal probabilities (Figure 19.4). See Zhang et al. [37] for more details on UBBPSO.

The selected four algorithms are implemented in MATLAB platform. The computer system employed is Intel Core i7 (2630QM CPU @ 2 GHz, 4 GB DDR3 RAM) for which MFlops (million floating point operations per second) for the LINPACK benchmark program (at http://www.netlib.org/) for a matrix of order 500 are 899. In this work, a local optimizer is employed, after the global optimization fulfills its stopping criteria, to obtain a precise solution. MATLAB's inbuilt function: fmincon is used as the local optimizer. There are 4 local methods: trust-region-reflective, interior-point, sequential quadratic programming (SQP) and active-set, in fmincon. Firstly, interior point method is used, which is capable of handling large, sparse and dense problems. Then, SQP and active-set methods are used as they are small scale local methods, and give a precise solution efficiently. This sequence is recommended by MATLAB (http://www.mathworks.com/help/optim/ug/choosing-a-solver.html#bsbwxm7).

Any of the methods in fmincon may give a solution that violates inequality constraints under some circumstances. So, a sanity check on the inequality constraint violation is required after each method. If the constraint violation is more than 0.001, the solution returned by that local optimizer is disregarded and the next local method is continued with the optimal solution obtained by the previous global/local algorithm to find the precise optimum. If all three local optimizers fail to give a solution with constraint violation less than 0.001, then the solution obtained by the global search is considered as the final/optimal solution.

19.3 EVALUATION OF MODIFIED SA, VFSA, IDE, AND UBBPSO

19.3.1 EVALUATION PROCEDURE, CONSTRAINTS HANDLING AND PARAMETER SETTINGS

The performance of the modified SA, VFSA, IDE, and UBBPSO algorithms are tested on eight unconstrained test problems. The ability of global optimizers to handle inequality constraints is evaluated on another 7 benchmark mathematical problems with inequality constraints, given in CEC 2006 [22]. These constrained problems involve 2 to 20 decision variables (with bounds on them) and 2 to 15 inequality constraints. Mathematical characteristics of these benchmark mathematical problems are summarized in the appendix (Tables A1 and A2).

The performance evaluation is based on several criteria such as success rate (SR defined as number of successful runs in 100 trials) after local optimization and mean number of function evaluations (MNFE) after global search, out of 100 trials. Each trial for all the benchmark problems is started with a different random number seed. A successful run means that the algorithm found the objective function value within $\pm 1 \times 10^{-6}$ from the known global optimum value. Hence, SR indicates reliability of the algorithm to find the global optimum. The computational efficiency can be inferred from MNFE instead of computational time because function evaluation is computationally intensive in application problems; also, MNFE is independent of the computer used. In the performance comparison, MEGS is the acronym for the median of errors between the best objective function values found by the global solver and the known global optimum value. Similarly, MEGLS is the median of errors of the best objective function values found by local solvers after global search and the known global optimum value. The difference between MEGS and MEGLS shows the usefulness of local optimization after the global search.

The inequality constraints in all constrained test problems are handled by the penalty function approach, which is commonly used for handling inequality constraints [12]. In this approach, constrained minimization problem is transformed into an unconstrained one by adding a penalty term (which is determined by the extent of constraint violation) to the objective function. In this work, static penalty wherein the penalty factor is kept constant throughout the optimization is employed. The penalty factor is tuned using the difficult constrained test problems (g02 and g18 in Table A2) and pooling problems (Adhya 1, Adhya 2, Adhya 3, Adhya 4, Haverly 2 and Foulds 2 in Table A3). The performance of three penalty factors: 100, 1000 and 10^6 is evaluated, and finally penalty factor of 100 is used in all global optimizers.

It is important to identify suitable values for all parameters in the selected stochastic optimization algorithms. In this study, two different parameter settings are used to solve benchmark mathematical and pooling problems: moderate and high. The moderate parameter values for each and every algorithm are taken from the literature (Table 19.1). If the result of a particular benchmark problem is not satisfactory, then the high parameter settings (Table 19.1) are used to improve the results. High parameter settings require more computational effort.

19.3.2 EVALUATION ON BENCHMARK UNCONSTRAINED PROBLEMS

Figure 19.5 shows the global SR (i.e., average SR of an algorithm on a number of optimization problems) at several different NFE for solving unconstrained test problems using the four algorithms. It can be seen that IDE has the highest SR on all problems tested. UBBPSO has good SR on problems solved using moderate parameter settings (Figure 19.5a) but it has very low SR on Rastrigin and Schwefel problems as can be seen from

TABLE 19.1 Suggested Parameters Values for Modified SA, VFSA, IDE and UBBPSO

Modified SA		VFSA		IDE		UBBPSO	
T_0	10000	T_0	10000	NP	20 (50)	NP	20 (50)
NV	5 (10)			TLs	20	MNG	1500
NT	5 (10)	NT	10	LP	20		
NFEmax	300000 (750000)	NFEmax	300000 (750000)	TR	10^{-3} D (10^{-6} D)		
rT	0.85	CS	0.45	MNG	1500		
v_0	$x^U - x^L$	C_G	1	NRmax	12		
c	2						
Nft	5	Nft	5			Nft	5
ε	10^{-6}	E	10^{-6}			ε	10^{-6}

*High parameter settings for solving difficult problems are given in brackets.

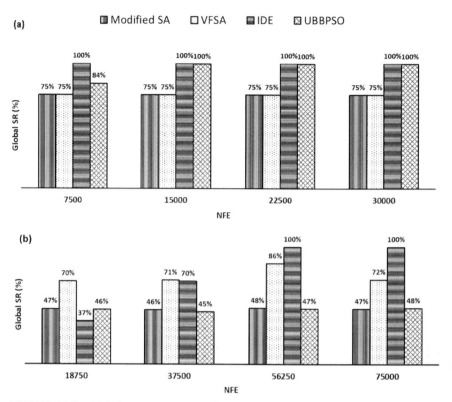

FIGURE 19.5 Global success rate at different NFE for solving (a) Ackley, Camelback, Sphere and Step problems using moderate parameter settings, and (b) Griewank, Rastrigin, Rosenbrock and Schwefel problems using high parameter settings, by Modified SA, VFSA, IDE and UBBPSO.

very high MEGLS in Table 19.2; note that the performance results in this and other similar tables are for the global optimization method using the termination criteria stated in the previous section. Similar performance of UBBPSO is reported in Zhang et al. [37]. For UBBPSO, relatively large MEGS and MEGLS for Rastrigin and Schwefel problems indicate that, unlike other algorithms, the search is trapped in a local minimum, which is away from the global minimum. VFSA has outperformed Modified SA, in terms of efficiency, by a large margin for all test problems; however, its large MEGS, especially on difficult problems (Table 19.2), means that the solution found is far from the known global minimum, and the robust performance of VFSA is mainly due to the local search from the solution found by the global search.

TABLE 19.2 Performance of Modified SA, VFSA, IDE and UBBPSO For Solving Benchmark Unconstrained Problems

Problem	Indicator	Modified SA	VFSA	IDE	UBBPSO
Ackley	MEGS	9.71E-05	**1.94E-08**	1.47E-05*	6.19E-06
	MEGLS	1.90E-08	**1.94E-08**	1.82E-08*	1.86E-08
	MNFE	131273	63718	29239*	**28464**
Camelback	MEGS	4.06E-09	0.001089	0.031695	**8.45E-09**
	MEGLS	2.66E-15	2.66E-15	**0**	2.66E-15
	MNFE	7243	1454	1354	**1274**
Sphere	MEGS	**8.40E-07**	0.106005	0.001273	2.13E-06
	MEGLS	1.77E-14	**5.17E-16**	8.55E-14	4.99E-14
	MNFE	119296	54127	**7305**	20511
Step	MEGS	1.67E-06	4.653095	0.001271	**2.22E-06**
	MEGLS	1.51E-30	**2.16E-30**	1.51E-30	1.49E-30
	MNFE	119723	24658	**7302**	20450
Griewank*	MEGS	0.012321	0.690568	1.38E-11	**2.51E-06**
	MEGLS	0.012320	0.039364	**1.62E-14**	1.52E-14
	MNFE	478531	70999	56382	**39764**
Rastrigin*	MEGS	5.91E-07	1.967297	**7.99E-08**	13.58623
	MEGLS	**0**	**0**	**0**	12.43699
	MNFE	480481	**44530**	72428	75000
Rosenbrock*	MEGS	**0.141767**	423.9303	17.722	79.93874
	MEGLS	1.74E-16	1.74E-16	1.74E-16	1.74E-16
	MNFE	750001	**8911**	75000	75000
Schwefel*	MEGS	1.03E-06	31.1218	**3.64E-11**	1662.781
	MEGLS	**0**	**0**	**0**	1662.688
	MNFE	477841	**18112**	70796	74809

* These problems are solved using the high parameter settings.

**Best values are in bold font.

Overall, IDE is the better than other algorithms tested as it has close to perfect SR (Figure 19.5) and has relatively low MNFE (Table 19.2). It also has the highest consistency as it can reliably handle all test problems unlike Modified SA, VFSA and UBBPSO, which have shown relatively low SR on some unconstrained problems. Many times, SR decreases slightly with the progress of search; for example, SR of Modified SA has decreased from 47% after 18,750 function evaluations to 46% after 37,500 function evaluations in Figure 19.5b. This anomaly is because the solution obtained by the global optimization is not in the vicinity of the global optimum, leading the local optimizer to the nearby local optimum. Recall that, after the global optimization algorithm, a local optimizer is used to find the solution precisely and then compare it with the known global solution to determine the success of the run.

19.3.3   EVALUATION ON BENCHMARK CONSTRAINED PROBLEMS

Figure 19.6 shows the global SR at several different NFE for solving all constrained test problems using the four algorithms. Modified SA, IDE and UBBPSO algorithms with moderate parameter settings have nearly same global SR after 15,000 function evaluations for 5 of 7 constrained problems tested. All algorithms have generally a low SR on g02 problem. This observation agrees with Zhang and Rangaiah [34], who have reported a low SR for IDE. The g18 problem requires high parameter settings, and Modified SA, IDE and UBBPSO have SR of around 80% compared to 52% by VFSA for this problem (Figure 19.6c).

Similar to unconstrained test problems, IDE requires fewer MNFE to reach the global optimum compared to the other algorithms (Table 19.3). Modified SA is found to be less efficient but more reliable than VFSA. Overall, the former is less efficient and reliable compared to both IDE and UBBPSO on the constrained problems tested. VFSA has the largest MEGS and MEGLS on all constrained problems, which is similar to its performance on the unconstrained problems. Among four algorithms, IDE is the most reliable and efficient algorithm for the constrained problems tested.

19.4   POOLING PROBLEM FORMULATION

Figure 19.1 shows the general process network for producing D units of desired K products (with η qualities) from N source streams via P pools and B bypasses. To simplify the formulation, each bypass stream is considered as an additional pool.

19.4.1   p- AND q-FORMULATIONS

The bi-linear programming formulation, known as the p-formulation, was proposed by Haverly [18] to describe pooling problems mathematically. As

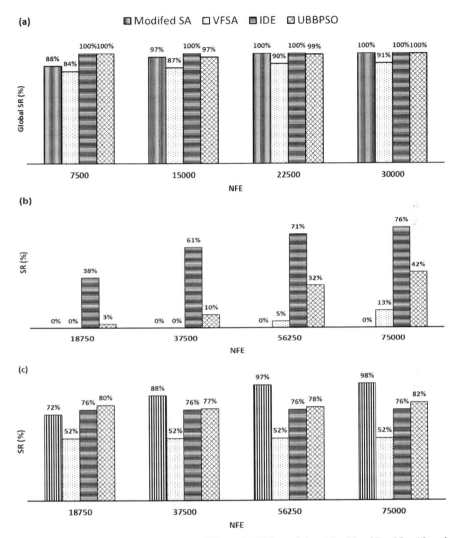

FIGURE 19.6   Global success rate at different NFE for solving (a) g01, g07, g08, g19 and g24 problems using moderate parameter settings, (b) g02 problem using high parameter settings, and (c) g18 problem using high parameter settings, by Modified SA, VFSA, IDE and UBBPSO.

shown in Figure 19.1, the source streams with flow rates ($X_{i,j}$ from i-th source stream to j-th pool), qualities ($\lambda_{i,w}$ for w-th quality of i-th source stream) and costs (C_i of i-th source stream), are channeled to one or more pools (including those representing bypasses) for blending. In a typical pooling network, not all sources are connected to every pool because of bypass streams and process requirements. Hence, $X_{i,j}$ will be zero for these unconnected streams

TABLE 19.3 Performance of Modified SA, VFSA, IDE and UBBPSO For Solving Benchmark Constrained Problems

Problem	Indicator	Modified SA	VFSA	IDE	UBBPSO
g01	MEGS	0.000214657	0.084094	0.099827	**1.2E-06**
	MEGLS	0	0	0	0
	MNFE	51150	37839	**5705**	12546
g07	MEGS	0.259666	13.32076	**0.104197**	0.870323
	MEGLS	**1.05E-10**	1.09E-10	1.06E-10	1.1E-10
	MNFE	32389	**6178**	10513	29294
g08	MEGS	**2.54E-09**	0.061144	7.82E-07	3.91E-09
	MEGLS	0	0	0	0
	MNFE	31776	**365**	799	1363
g19	MEGS	5.368844	118.4109	**3.689978**	9.316079
	MEGLS	0	7.82E-14	0	9.95E-14
	MNFE	53169	**16101**	20642	29921
g24	MEGS	6.81E-05	0.110728	0.002028	**5.95E-06**
	MEGLS	**0**	**0**	**0**	7.11E-15
	MNFE	6618	2009	**1255**	8786
g02*	MEGS	0.010304	0.052804	**0.000779**	0.011236
	MEGLS	0.008957	0.025422	**4.21E-13**	0.009746
	MNFE	299101	160115	**75000**	**75000**
g18*	MEGS	0.000361	0.437	**1.57E-09**	0.001183
	MEGLS	**1.2E-12**	0.191044	**1.2E-12**	2.86E-12
	MNFE	132195	**3228**	73390	62357

*These problems are solved using the high parameter setting.

*Best values are in bold font.

from sources to pools. Constraints to limit the availability of i-th source stream, A_i^U and more than one quality variable can be the extra requirements in some cases. Next, outlet streams from the pools with flow rate ($Y_{j,k}$ from j-th pool to k-th product) and quality ($\mu_{j,w}$) are channeled to product tanks to be blended for the second time. Similar to source streams, not all pool outlet streams go to each product tank and so $Y_{j,k}$ will be zero for such unconnected pool outlet streams. Unlike the availability constraints for source streams, requirements on product demand, D_k^U and product quality ($\eta_{k,w}$) are imposed and have to be satisfied for producing the final products with their respective selling price, S_k.

The objective of a pooling problem is to maximize the profit defined as the difference between the revenue from selling the products and the cost of raw materials, or equivalently to minimize the negative of the profit [see Eq. (2)]. Equations (3) and (4) are equality constraints arising from, respectively, mass and quality balances about each pool (including those for bypasses). In Eq. (5), there are N inequality constraints as total flow rates of the source stream cannot exceed the available limit. In Eq. (6), there are K inequality constraints as total flow rate into each product tank cannot exceed its requirements on the product demand. There are K×L inequality constraints in Eq. (7) in order to meet the quality requirements of all products.

$$\text{Objective Function}\left[\ \min \sum_{j=1}^{P+B}\sum_{i=1}^{N}\left(C_i X_{i,j}\right)-\sum_{k=1}^{K}S_k\left(\sum_{j=1}^{P+B}Y_{j,k}\right)\right. \tag{2}$$

$$\text{Mass Balance about Pools}\left[\ \sum_{i=1}^{N}X_{i,j}-\sum_{k=1}^{K}Y_{j,k}=0 \text{ for } j=1,2,\ldots,P+B\right. \tag{3}$$

$$\text{Quality Balances about Pools}\left[\ \sum_{i=1}^{N}\left(\lambda_{i,w}X_{i,j}\right)-\mu_{j,w}\left(\sum_{k=1}^{K}Y_{j,k}\right)=0\right.$$
$$\text{for } j=1,2,\ldots,P+B;\, w=1,2,\ldots,L \tag{4}$$

$$\text{Feed Availability}\left[\ \sum_{j=1}^{P+B}X_{i,j}-A_i^{U}\le 0 \text{ for } i=1,2,\ldots,N\right. \tag{5}$$

$$\text{Product Demand}\left[\ \sum_{j=1}^{P+B}Y_{j,k}-D_k^{U}\le 0 \text{ for } k=1,2,\ldots,K\right. \tag{6}$$

$$\text{Product Quality}\left[\ \sum_{j=1}^{P+B}\left(\mu_{j,w}Y_{j,k}\right)-\eta_{k,w}\left(\sum_{j=1}^{P+B}Y_{j,k}\right)\le 0\right.$$
$$\text{for } k=1,2,\ldots,K;\, w=1,2,\ldots,L \tag{7}$$

The bounds for decision variables X, Y and μ are as follows.

$$0\le X_{i,j}\le \min\left\{A_i^{U},\sum_{k=1}^{K}D_k^{U}\right\}\text{ for } i=1,2,\ldots,N;\, j=1,2,\ldots,P+B \tag{8}$$

$$0 \leq Y_{j,k} \leq \min\left\{ D_k^U, \sum_{i=1}^{N} A_i^U \right\} \text{ for } j=1,2,\ldots,P+B; k=1,2,\ldots,K \qquad (9)$$

$$\min_{i}\left\{\lambda_{i,w}\right\} \leq \mu_{j,w} \leq \max_{i}\left\{\lambda_{i,w}\right\} \text{ for } j=1,2,\ldots,P+B; w=1,2,\ldots,L \qquad (10)$$

In the q-formulation proposed by Ben-Tal et al. (1994), the decision variable, $X_{i,j}$ is replaced by a new decision variable ($q_{i,j}$) defined as the fraction of total flow from a pool to all connected product tanks. Hence, $X_{i,j}$ is given by

$$X_{i,j} = q_{i,j} \sum_{k=1}^{K} Y_{j,k} \text{ for } i=1,2,\ldots,N; j=1,2,\ldots,P+B \qquad (11)$$

The mass balance about pools is given by the following equation (instead of Eq. (3)).

$$\sum_{i=1}^{N} q_{i,j} = 1 \text{ for } j=1,2,\ldots,P \qquad (12)$$

For each bypass ($j = P+1, \ldots, P+B$), only one $q_{i,j}$ is 1.0 for its particular source stream and all other $q_{i,j}$ are 0.0. Hence, Eq. (12) does not include bypasses.

19.4.2 r-FORMULATION

Motivated by q-formulation, two new decision variables, $r_{i,j}$ and $R_{j,k}$ are introduced in r-formulation to reduce the search space, number of decision variables and constraints [34]. $r_{i,j}$ is the fraction of the maximum possible value for $q_{i,j}$ (used to calculate $X_{i,j}$ via Eq. (11)), as follows.

$$q_{1,j} = r_{1,j} \text{ for } j=1,2,\ldots, P \qquad (13a)$$

$$q_{i,j} = r_{i,j}\left(1 - \sum_{m=1}^{i-1} q_{m,j}\right) \text{ for } i=2,3,\ldots,N-1; j=1,2,\ldots, P \qquad (13b)$$

$$q_{N,j} = 1 - \sum_{m=1}^{N-1} q_{m,j} \text{ for } j=1,2,\ldots, P \qquad (13c)$$

Equation (13b) takes into account the previous flow rates from the first to $(i-1)$th sources which are connected to j-th pool. The flow rate of the last stream from N-th source to j-th pool is calculated using Eq. (13c). Equation (13) eliminates the equality constraints arising from the mass balances for pools, and also implicitly satisfies Eq. (12). For each pool representing a bypass ($j = P+1, \ldots, P+B$), only one $q_{i,j}$ is 1.0 for its particular source stream and all other $q_{i,j}$ are 0.0. Hence, $r_{i,j}$ variables are not required for $j = P+1, \ldots, P+B$.

Similar to $r_{i,j}$, $R_{j,k}$ is the fraction of the maximum possible value for $Y_{j,k}$, as follows.

$$Y_{1,k} = R_{1,k}\, D_k^U \text{ for } k = 1,\, 2,\ldots,K \qquad (14a)$$

$$Y_{j,k} = R_{j,k}\left(D_k^U - \sum_{m=1}^{j-1} Y_{m,k} \right) \text{ for } j = 2,\, 3,\ldots,P+B;\ k = 1,\, 2,\ldots,\ K \qquad (14b)$$

In Eq. (14b), the upper bound of the product demand and the flow rates from all previous $j-1$ pools into k-th product tank are taken into account. The decision variables in r-formulation are r's and R's which are different to those in p- and q- formulations. Flow rates in the pooling network (X and Y) can be calculated explicitly using Eqs. (11), (13) and (14), and so they are the dependent variables.

The objective function, constraints, decision variables and their bounds in r-formulation are as follows.

$$\text{Objective Function}\left[\min \sum_{i=1}^{N}\sum_{j=1}^{P+B}\sum_{k=1}^{K}(C_i q_{i,j} Y_{j,k}) - \sum_{k=1}^{K} S_k\left(\sum_{j=1}^{P+B} Y_{j,k} \right) \right. \qquad (15)$$

$$\text{Feed Availability}\left[\sum_{j=1}^{P+B}\sum_{k=1}^{K}(q_{i,j} Y_{j,k}) - A_i^U \le 0 \text{ for } i = 1,\, 2,\ldots,N \right. \qquad (16)$$

$$\text{Product Quality}\left[\sum_{i=1}^{N}\sum_{j=1}^{P+B}(\lambda_{i,w} q_{i,j} Y_{j,k}) - \eta_{k,w} \sum_{j=1}^{P+B} Y_{j,k} \le 0 \right.$$

$$\text{for } k = 1,\, 2,\ldots,\ K;\ w = 1,\, 2,\ldots,L \qquad (17)$$

Variables and their Bounds

$$0 \le r_{i,j} \le 1 \text{ for } i = 1,\, 2,\ldots,N-1;\ j = 1,\, 2,\ldots,\ P+B \qquad (18)$$

$$0 \leq R_{j,k} \leq 1 \text{ for } j=1, 2,\ldots, P+B; k=1, 2,\ldots, K \qquad (19)$$

In the Eqs. (15)–(17), q's are calculated via Eq. (13). As in the p-formulation, if a source stream is not connected to a pool, the corresponding $r_{i,j}$ will be zero and can be excluded from the list of decision variables. Similarly, if a pool outlet stream is not connected to a product tank, the corresponding $R_{i,j}$ will be zero and can be excluded from the list of decision variables. The number of decision variables is thus reduced; also, the decision variables are intrinsically normalized [Eqs. (18) and (19)].

Maximum numbers of decision variables, inequality and equality constraints in the p-, q- and r-formulations are summarized in Table 19.4. It is clear from this table that the number of decision variables and number of constraints are fewer in r-formulation than those in p- and q-formulations. More importantly, r-formulation does not have any equality constraints, which are very difficult to handle in stochastic global optimization. On the other hand, the objective function and inequality constraints in r-formulation are more nonlinear than those in other formulation but stochastic global optimization methods are not much affected by nonlinearity.

19.4.3   RESULTS AND DISCUSSION

In this work, 13 benchmark pooling problems are solved using Modified SA, VFSA, IDE and UBBPSO algorithms. The basic details of these pooling problems including the number of qualities and global optimum are summarized in Table A3 in the appendix; see Adhya et al. [1] for flow charts and other details of these pooling problems. In the present study, Adhya 3 and Haverly

TABLE 19.4   Problem Characteristics Obtained Using p-, q- and r-Formulations

Problem characteristic	p-formulation	q-formulation	r-formulation
Maximum number of decision variables	$N \times (P+B) + P \times$ $L + (P+B) \times K$	$N \times (P+B) +$ $(P+B) \times K$	$(N-1) \times (P+B) +$ $(P+B) \times K$
Maximum number of inequality constraints	$(L+1) \times K + AC^*$	$(L+1) \times K + AC^*$	$L \times K + AC^*$
Maximum number of equality constraints	$B + (L+1) \times P$	P	Nil

* AC is the number of additional constraints for availability of source streams; it will be zero if source streams are available without any limit.

2 pooling problems are solved using high parameter settings, whereas the remaining 11 problems are solved using moderate parameter settings.

Figure 19.7 presents global SR at several different NFE for solving the 13 benchmark pooling problems using different algorithms. All algorithms gave nearly same SR on 11 benchmark pooling problems solved using moderate parameter settings. IDE has better SR on Adhya 3 problem, whereas

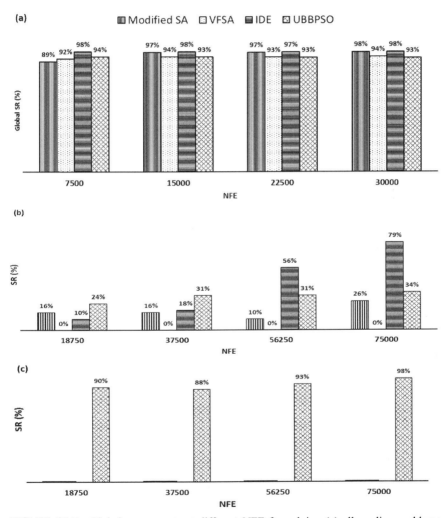

FIGURE 19.7 Global success rate at different NFE for solving (a) all pooling problems (except Adhya 3 and Haverly 2) using moderate parameter settings, (b) Adhya 3 problem using high parameter settings, and (c) Haverly 2 problem using high parameter settings, by Modified SA, VFSA, IDE and UBBPSO.

UBBPSO outperformed other algorithms on Haverly 2 problem. VFSA has very low reliability of close to zero SR on these two pooling problems solved using high parameter settings. Further, it gave largest MEGS on most problems, except Adhya 2, Foulds 4, Foulds 5 and Haverly 2. This is similar to the observation made in the testing of the benchmark constrained problems. IDE is the most effective because it generally shows higher efficiency and satisfactory reliability on most pooling problems tested. UBBPSO has consistent reliability on all pooling problems, but it requires significantly more MNFE compared to IDE on all pooling problems (Table 19.5). Modified SA is the least efficient algorithm on pooling problems solved using moderate parameter settings.

TABLE 19.5   Performance of Modified SA, VFSA, IDE and UBBPSO For Solving Benchmark Pooling Problems

Problem	Indicator	Modified SA	VFSA	IDE	UBBPSO
Adhya 1	MEGS	2.822448	3.618286	1.651335	**0.551008**
	MEGLS	0	0	0	0
	MNFE	30925	11706	**10127**	29231
Adhya 2	MEGS	2.925723	2.231498	2.273026	**0.337097***
	MEGLS	0	0	0	0*
	MNFE	30504	30504	**9041**	74367*
Adhya 4	MEGS	0.019845	6.428764	4.619536	**0.003248**
	MEGLS	0	0	0	0
	MNFE	59349	36667	**7068**	29587
BenTal 4	MEGS	13.73799	68.78063	2.707946	**1.58E-06**
	MEGLS	0	0	0	0
	MNFE	18390	**3423**	4980	22860
BenTal 5	MEGS	83.88828	232.5963	210.7711	**45.04492**
	MEGLS	0	0	0	0
	MNFE	107759	118088	**29350**	30000
Foulds 2	MEGS	95.04999	206.2892	104.8188	**0.362153**
	MEGLS	0	0	0	0
	MNFE	62493	38940	**20583**	30000
Foulds 3	MEGS	**0.297243**	1.595881	1.315962	0.754774
	MEGLS	**1.82E-08**	1.94E-08	2.49E-08	2.75E-08
	MNFE	300201	300961	**21679**	30000

TABLE 19.5   Continued

Problem	Indicator	Modified SA	VFSA	IDE	UBBPSO
Foulds 4	MEGS	**0.36694**	3.58874	4.7397	6.068965
	MEGLS	**9.65E-09**	2.54E-08	1.31E-08	1.19E-08
	MNFE	300201	300961	**24547**	30000
Foulds 5	MEGS	**0.074927**	1.356551	2.049346	2.59537
	MEGLS	1.42E-14	**0**	-2.8E-14	-2.8E-14
	MNFE	301301	300841	**26755**	30000
Haverly 1	MEGS	7.289876	50.93795	2.277262	**1.48E-07**
	MEGLS	0	0	0	0
	MNFE	14782	**2572**	3999	16390
Haverly 3	MEGS	0.029277	22.47323	0.231756	**5.08E-08**
	MEGLS	0	0	0	0
	MNFE	17516	**2624**	3268	8952
Adhya 3*	MEGS	5.273889	8.313221	7.122064	**1.611929**
	MEGLS	**0**	1.429303	**0**	1.429302
	MNFE	269825	**67692**	75000	75000
Haverly 2*	MEGS	200.138824	255.374095	200.002931	**200.00**
	MEGLS	200	200	200	**0**
	MNFE	60071	**1879**	22267	34119

*These problems are solved using the high parameter settings.

**Best values are in bold font.

19.5   CONCLUSIONS

In this study, performance of Modified SA, VFSA, IDE and UBBPSO is compared on the benchmark unconstrained, constrained and pooling problems. For this, pooling problems are represented using r-formulation, which reduces the search space, number of decision variables and number of inequality constraints, and also eliminates all equality constraints, all of which greatly facilitate solution of pooling problems by stochastic global optimization methods. Results show that IDE is the most consistent, reliable and reasonably efficient optimization algorithm compared to Modified SA, VFSA and UBBPSO, on constrained and unconstrained test problems. Further, it is efficient and reliable for solving most of the benchmark pooling problems tested. Hence, IDE is recommended over Modified SA, VFSA and

UBBPSO. The results for unconstrained, constrained and pooling problems tested indicate that Modified SA is generally the least efficient and VFSA is generally the least reliable.

KEYWORDS

- **differential evolution**
- **global optimization**
- **particle swarm optimization**
- **pooling problems**
- **simulated annealing**

REFERENCES

1. Adhya, N., Tawarmalani, M., & Sahinidis, N. V. A Lagrangian Approach to the Pooling Problem, *Ind. Eng. Chem. Res.* 1999, 38, 1956–1972.
2. Ali, M. M., Törn, A., & Viitanen, S. A Direct Search Variant of the Simulated Annealing Algorithm for Optimization Involving Continuous Variables. *Computers & Operations Research* 2002, 29(1), 87–102.
3. Babu, B. V., & Munawar, S. A. Differential Evolution Strategies for Optimal Design of Shell-and-Tube Heat Exchangers. *Chemical Engineering Science.* 2007, 62(14), 3720–3739.
4. Ben-Tal, A., Eiger, G., Gershovitz, V., & Israel, T. Global Minimization by Reducing the Duality Gap. *Mathematical Programming* 1994, 63, 193–212.
5. Bonilla-Petriciolet, A., Fateen, S. E. K., & Rangaiah, G. P. Assessment of Capabilities and Limitations of Stochastic Global Optimization Methods for Modeling Mean Activity Coefficients of Ionic Liquids. *Fluid Phase Equilibria* 2013, 340, 15–26.
6. Bonilla-Petriciolet, A., & Segovia-Hernández, J. G. Particle Swarm Optimization for Phase Stability. *19th European Symposium on Computer Aided Process Engineering* 2009, 635–640.
7. Bonilla-Petriciolet, A; Iglesias-Silva, G. A., & Hall, K. R. Calculation of Homogeneous Azeotropes in Reactive and Non-Reactive Mixtures Using a Stochastic Optimization Approach. *Fluid Phase Equilibria* 2009, 281(1), 22–31.
8. Bonilla-Petriciolet, A., Rangaiah, G. P., & Segovia-Hernández, J. G. Evaluation of Stochastic Global Optimization Methods for Modeling Vapor–Liquid Equilibrium Data. *Fluid Phase Equilibria* 2010, 287(2), 111–125.
9. Bonilla-Petriciolet, A., Vazquez-Roma, R., Iglesias-Silva, G. A., & Hall, K. R. Performance of Stochastic Global Optimization Methods in the Calculation of Phase Stability Analyses for Nonreactive and Reactive Mixtures. *Ind. Eng. Chem. Res.* 2006, 45(13), 4764–4772.

10. Cardoso, M. F., Salcedo, R. L., & Feyo de Azevedo, S. The Simplex-Simulated Annealing Approach to Continuous Non-Linear Optimization. *Computers & Chemical Engineering* 1996, 20(9), 1065–1080.

11. Chen S. Q., Rangaiah, G. P., & Srinivas, M. Differential Evolution: Methods, Development and Chemical Engineering Applications. In Rangaiah G. P. (editor) *Stochastic Global Optimization: Techniques and Applications in Chemical Engineering*, World Scientific 2010, 57–109.

12. Coello Coello, C. A. Theoretical and Numerical Constraint-Handling Techniques Used with Evolutionary Algorithms: A Survey of the State of the Art. *Computer Methods in Applied Mechanics and Engineering* 2002, 191(11–12), 1245–1287.

13. Corana, A., Marchesi, M., Martini, C., & Ridella, S. Minimizing Multimodal Functions of Continuous Variables with the Simulated Annealing Algorithm. *ACM Transactions on Mathematical Software* 1987, 13(3), 262–280.

14. Dewitt, C. W., Lasdon, L. S., Waren, A. D., Brenner, D. A., & Melhem S. A. An Improved Gasoline Blending System for Texaco, *Interfaces* 1989, 19(1), 85–101.

15. Eberhart, R., & Kennedy, J. A New Optimizer Using Particle Swarm Theory. *Proceedings of 6th International Symposium on Micro Machine and Human Science* 1995, 39–43.

16. Ferrari, J. C., Nagatani, G., Corazza, F. C., Oliveira, J. V., & Corazza, M. L. Application of Stochastic Algorithms for Parameter Estimation in the Liquid–Liquid Phase Equilibrium Modeling. *Fluid Phase Equilibria* 2009, 280(1–2), 110–119.

17. Foulds, L. R., Haugland, D., & Jörnsten, K. A Bilinear Approach to the Pooling Problem. *Optimization* 1992, 24(1–2), 165–180.

18. Haverly, C. A. Studies of the Behavior of Recursion for the Pooling Problem. ACM SIGMAP Bulletin 1978, 25, 19–28.

19. Jaime-Leal, J. E., & Bonilla-Petriciolet, A. Correlation of Activity Coefficients in Aqueous Solutions of Ammonium Salts Using Local Composition Models and Stochastic Optimization Methods. *Chemical Product and Process Modeling* 2008, 3(1), Article 38.

20. Jezowski, J. M., Poplewski, G., & Bochenek, R. Adaptive Random Search and Simulated Annealing Optimizers : Algorithms and Application Issues. In Rangaiah G. P. (editor) *Stochastic Global Optimization: Techniques and Applications in Chemical Engineering*, World Scientific 2010, 57–109.

21. Kirkpatrick, S., Gelatt, C. D., & Vecchi, J. M. P. Optimization by Simulated Annealing, *Science* 1983, 220, 671–680.

22. Liang, J. J., Runarsson, T. P., Mezura-Montes, E. Clerc, M., Suganthan, P. N., Coello Coello, C. A., & Deb, K. Problem Definitions and Evaluation Criteria for the CEC 2006 Special Session on Constrained Real-Parameter Optimization Problem 2006, 1–24.

23. Mernik, M., Brest, J., & Zumer, V. Self-adapting Control Parameters in Differential Evolution : A Comparative Study on Numerical Benchmark Problems. *IEEE Transactions on Evolutionary Computation* 2006, 10(6), 646–657.

24. Misener, R., & Floudas, C. A. Global Optimization of Large-Scale Generalized Pooling Problems: Quadratically Constrained MINLP Models. *Ind. Eng. Chem. Res.* 2010, 49(11), 5424–5438.

25. Misener, R., Thompson, J. P., & Floudas, C. A. Global Optimization of Standard, Generalized, and Extended Pooling Problems via Linear and Logarithmic Partitioning Schemes. *Computers & Chemical Engineering* 2011, 35(5), 876–892.

26. Price, K., Storn, R. M., & Lampinen, J. A. *Differential Evolution: A Practical Approach to Global Optimization*, Springer 2005.

27. Qin, A. K., Huang, V. L., & Suganthan, P. N. Differential Evolution Algorithm with Strategy Adaptation for Global Numerical Optimization. *IEEE Transactions on Evolutionary Computation* 2009, 13(2), 398–417.
28. Rangaiah G. P. *Stochastic Global Optimization: Techniques and Applications in Chemical Engineering*, World Scientific 2010.
29. Rangaiah G. P., & Bonilla-Petriciolet A. (Eds.). *Multiobjective Optimization in Chemical Engineering: Developments and Applications,* Wiley 2013.
30. Sen, M. K., & Stoffa P. L. *Global Optimization Methods in Geophysical Inversion*, Elsevier Publisher 1995.
31. Srinivas, M., & Rangaiah, G. P. Differential Evolution with Taboo List for Solving Nonlinear and Mixed-Integer Nonlinear Programming Problems. *Ind. Eng. Chem. Res..* 2007, 46, 7126–7135.
32. Storn, R., & Price, K. Differential Evolution – A Simple and Efficient Adaptive Scheme for Global Optimization Over Continuous Spaces, Technical Report TR-95-012, ICSI, March 1995.
33. Tawarmalani, M., & Sahinidis, N. *Convexification and Global Optimizationi Continuous Andmixed-Integer Nonlinear Programming: Theory, Algorithms, Software, and Applications*, Kluwer Academic: Dordrecht 2002.
34. Zhang, H., & Rangaiah, G. P. Integrated Differential Evolution for Global Optimization and its Performance for Modeling Vapor-Liquid Equilibrium Data. *Ind. Eng. Chem. Res.* 2011, 50(17), 10047–10061.
35. Zhang, H., & Rangaiah, G. P. Optimization of Pooling Problems for Two Objectives Using ε-Constraint Method. In Rangaiah, G. P., & Bonilla-Petriciolet, A. (editors) *Multiobjective Optimization in Chemical Engineering: Developments and Applications,* Wiley 2013, 17–34.
36. Zhang, H., Kennedy, D. D., Rangaiah, G. P., & Bonilla-Petricolet, A. Novel Bare-Bones Particle Swarm Optimization and its Performance for Modeling Vapor–Liquid Equilibrium Data. *Fluid Phase Equilibria* 2011a, 301(1), 33–45.
37. Zhang, H., Fernández-Vargas, J. A., Rangaiah, G. P., Bonilla-Petriciolet, A., & Segovia-Hernández, J. G. Evaluation of Integrated Differential Evolution and Unified Bare-Bones Particle Swarm Optimization for Phase Equilibrium and Stability Problems. *Fluid Phase Equilibria* 2011b, 310(1–2), 129–141.

APPENDIX

Details of benchmark unconstrained, constrained and pooling problems are given in Tables A1, A2 and A3, respectively.

TABLE A1 Details of Benchmark Unconstrained Problems

Problem	Number of decision variables (n)	Lower and upper bounds	Global minimum
Ackley: $F_{obj}(x) =$ $-20\exp\left(-0.2\exp\left(\sqrt{\frac{1}{n}\sum_{i=1}^{n}x_i^2}\right)\right) - \exp\left(\frac{1}{n}\sum_{i=1}^{n}\cos(2\Pi x_i)\right) + 20 + \exp(1)$	30	$[-32, 32]^n$	0.0 at $x = \{0,\ldots,0\}$
Camelback: $F_{obj}(x) = 4x_1^2 - 2.1x_1^4 + \frac{1}{3}x_1^6 + x_1x_2 - 4x_2^2 + 4x_2^4$	2	$[-5, 5]^n$	-1.0316285 at $x = \{0.0898, -0.7126\}$ or $x = \{-0.0898, 0.7126\}$
Sphere: $F_{obj}(x) = \sum_{i=1}^{n}x_i^2$	30	$[-100, 100]^n$	0.0 at $x = \{0,\ldots,0\}$
Step: $F_{obj}(x) = \sum_{i=1}^{n}(x_i^2 + 0.5)^2$	30	$[-100, 100]^n$	0.0 at $x = \{0,\ldots,0\}$
Griewank*: $F_{obj}(x) = \frac{1}{4000}\sum_{i=1}^{n}x_i^2 - \prod_{i=1}^{n}\cos\left(\frac{x_i}{\sqrt{x_i}}\right) + 1$	30	$[-600, 600]^n$	0.0 at $x = \{0,\ldots,0\}$
Rastrigin*: $F_{obj}(x) = 10n + \sum_{i=1}^{n}[x_i^2 - 10\cos(2\Pi x_i)]$	30	$[-5.12, 5.12]^n$	0.0 at $x = \{0,\ldots,0\}$
Rosenbrock*: $F_{obj}(x) = \sum_{i=1}^{n-1}\left[100(x_{i+1} - x_i^2)^2 + (x_i - 1)^2\right]$	30	$[-30, 30]^n$	0.0 at $x = \{1,\ldots,1\}$
Schwefels*: $F_{obj}(x) = -\sum_{i=1}^{n}x_i\sin\left(\sqrt{|x_i|}\right)$	30	$[-500, 500]^n$	$-418.983n$ at $x = \{420.9687,\ldots,420.9687\}$

* Difficult problems.

TABLE A2   Basic Details of Benchmark Constrained Problems;

Problem	Number of decision variables (n)	Feasible region (% of search space)	Number of inequality constraints	Lower and upper bounds	Global minimum
g01	13	0.0111	9	$x_{1...9,13} = [0,1], x_{10,11,12} = [0,100]$	−15.0
g07	10	0.0003	8	$[−10, 10]^n$	24.306209
g08	2	0.8560	2	$[0, 10]^n$	−0.095825
g19	15	33.4761	5	$[0, 10]^n$	32.655593
g24	2	79.6556	2	$x_1 = [0,3], x_2 = [0,4]$	−5.508013
g02*	20	99.9971	2	$[0, 10]^n$	−0.803619
g18*	9	0.0000	13	$x_{1...8} = [−10,10], x_9 = [0,20]$	−0.866025

*Difficult problems.

**See Liang et al. [22] for more details.

TABLE A3 Basic Details of 13 Benchmark Pooling Problems*

Problem	Source/ input streams	Number of pools + bypasses	Number of products	Number of qualities	Number of decision variables	Number of inequality constraints	Global optimum
Haverly 1	3	1 + 1	2	1	5	2	-400
Haverly 2	3	1 + 1	2	1	5	2	-600
Haverly 3	3	1 + 1	2	1	5	2	-750
Foulds 2	6	2 + 2	4	1	18	4	-1100
Foulds 3	11	8 + 0	16	1	152	16	-8
Foulds 4	11	8 + 0	16	1	152	16	-8
Foulds 5	11	4 + 0	16	1	92	16	-8
Ben–Tal 4	4	1 + 1	2	1	6	3	-450
Ben–Tal 5	5	3 + 1	5	2	29	11	-3500
Adhya 1	5	2 + 0	4	4	11	16	-549.8
Adhya 2	5	2 + 0	4	6	11	24	-549.8
Adhya 3	8	3 + 0	4	6	17	24	-561.05
Adhya 4	8	2 + 0	5	4	16	20	-877.65

*Number of decision variables and number of inequality constraints are in r-formulation.

CHAPTER 20

PERFORMANCE IMPROVEMENT OF NSGA-II ALGORITHM BY MODIFYING CROSSOVER PROBABILITY DISTRIBUTION

K. V. R. B. PRASAD[1] and P. M. SINGRU[2]

[1]*Professor, Department of E.E.E., MITS, P.B. No: 14, Kadiri Road, Angallu (V), Madanapalle – 517325, Chittoor District, Andhra Pradesh, India, E-mail: prasad_brahma@rediffmail.com*

[2]*Associate Professor, Department of M.E., BITS, Pilani – K.K. Birla Goa Campus, NH-17B, Zuarinagar, South Goa District, Goa, India, E-mail: pravinsingru@gmail.com*

CONTENTS

20.1 INTRODUCTION

The optimization is the act of obtaining the best result under the given circumstances. The optimization refers to finding one or more feasible solutions, which correspond to extreme values of one or more objectives. The need for finding such optimal solutions in a problem comes mostly from the extreme purpose of either designing a solution for minimum possible cost of fabrication, or for maximum possible reliability, or others. Because of such extreme properties of optimal solutions, optimization methods are of great importance in practice, particularly in engineering design, scientific experiments and business decision-making. In order to widen the applicability of an optimization algorithm in various different problem domains, natural and physical principles are mimicked to develop robust optimization algorithms like evolutionary algorithms (EAs) and simulated annealing (SA). If a physical system is modeled as an optimization problem with one objective function, the task of finding the optimal solutions is called single-objective optimization. If an optimization problem involves more than one objective function, the task of finding one or more optimum solutions is known as multiobjective optimization [1–3].

The EA mimics nature's evolutionary principles to drive its search towards an optimal solution. The EAs include genetic algorithm (GA), differential evolution (DE), evolutionary strategies (ES), evolutionary programming (EP), genetic programming (GP), etc. The GA is a search and optimization procedure that is motivated by the principles of natural genetics and natural selection. It gives near global population of optimal solutions. It works with a population of solutions and gives multiple optimal solutions in one simulation run. It has two distinct operations such as selection and search. It is flexible enough to be used in a wide variety of problem domains. It has three main operators such as reproduction, crossover and mutation which are playing important role in creating a new population of solutions. The reproduction or selection operator is used to make duplicates of good solutions and eliminate bad solutions in a population, while keeping the population size constant. The crossover and mutation operators are used to create new solutions [1, 2, 4, 5].

20.2 NSGA-II ALGORITHM

In nondominated sorting genetic algorithm (NSGA), the dual objectives of a multiobjective optimization algorithm are maintained by using fitness assignment scheme which prefers nondominated solutions and by using a sharing strategy which preserves diversity among solutions of each nondominated front. The main advantage of an NSGA is the assignment of fitness according to nondominated sets. The NSGA progresses towards the Pareto-optimal region front-wise, by providing systematic emphasis to better nondominated sets. Moreover, performing sharing in the parameter space allows phenotypically diverse solutions to emerge when using NSGAs. If desired, the sharing can also be performed in the objective space. The NSGA has difficulties which include high computational complexity of nondominated sorting, lack of elitism and need for specifying the sharing parameter (σ_{share}) [1].

The elitist nondominated sorting algorithm (NSGA-II) with actual simulated binary crossover (SBX-A), having normal probability distribution, uses elite-preservation strategy along with an explicit diversity-preserving mechanism. The algorithm carries out nondominated sorting of combined parent and offspring population. For the solutions of the last allowed front, a crowding distance-based niching strategy is used to resolve which solutions are to be carried over to the new population. The diversity among nondominated solutions is introduced by using the crowding comparison procedure which is used with the tournament selection and during the population reduction phase. The elitism does not allow the already found Pareto-optimal solutions (POS) to be deleted.

20.3 SBX-A OPERATOR

The SBX-A operator, used in this algorithm, works with two parent solutions and creates two offsprings. This operator simulates the working principle of the single-point crossover operator on binary strings. This operator respects the interval schemata processing, in the sense that common interval schemata between parents are preserved in children [1, 6, 7].

The probability distribution used to create a child solution is obtained from the following relations [7].

$$P(\beta_s) = 0.5(\eta_c + 1)\beta_s^{\eta_c} \quad \text{if } \beta_s \leq 1 \tag{1}$$

$$P(\beta_s) = 0.5(\eta_c + 1)\frac{1}{\beta_s^{\eta_c+2}} \quad \text{otherwise} \tag{2}$$

where β_s is the spread factor and η_c is the crossover index. The value of η_c gives a probability for creating near parent solutions; large value gives a higher probability for creating near parent solutions.

In GA, the reproduction operator makes duplicates of good solutions, to eliminate the bad solutions from the population. The crossover and mutation operators create new solutions, by recombination. The crossover operator is the main search operator in GA. The search power of a crossover operator is defined as a measure of how flexible the operator is to create an arbitrary point in the search space. The role of mutation is to restore lost or unexpected genetic material into population to prevent the premature convergence of GA to suboptimal solutions [1, 7].

20.4 SBX-LN OPERATOR

The SBX-A operator creates children solutions proportional to the difference in parent solutions. In this operator, near parent solutions are more likely to be chosen as children solutions than solutions away from parents. The performance of NSGA-II algorithm is improved by modifying the probability distribution of SBX-A [7, 8].

In this chapter, the probability distribution of SBX-A operator, used in NSGA-II algorithm, is modified with lognormal probability distribution (SBX-LN). In this operator, the probability of creating offspring away from the parents is influenced by the η_c. This possibility decreases with the decrease in η_c and hence the SBX-LN becomes more parent-centric operator. In this operator, both parents are given equal probability of creating offspring in its neighborhood. The variance of intra-member distance increases due to the application of recombination operator, SBX-LN. This operator also assigns children solutions proportional to the spread of parent solutions, thereby making GA with this operator potential to exhibit self-adoption [7].

In single point binary crossover, the functional relationship of the contracting crossover probability distribution is obtained from the following relation

$$C(\beta_s) = \frac{k}{(1 - \beta_s)} \tag{3}$$

where k is constant of probability distribution. In single point binary crossover, the functional relationship of the expanding crossover probability distribution is obtained from the following relation.

$$E(\beta_s) = \frac{k}{\beta_s(\beta_s - 1)} \tag{4}$$

The Eq. (4) is modified by expressing the expanding crossover probability distribution in terms of contracting crossover probability distribution. This is shown in the following relation.

$$E(\beta_s) = \frac{1}{\beta_s^2} C\left(\frac{1}{\beta_s}\right) \tag{5}$$

The lognormal distribution, defined with the probability density function, is obtained from the following relation.

$$f(\beta_s) = \left(\frac{1}{\beta_s \eta_c \sqrt{2\pi}}\right) \exp\left[\frac{-1}{2}\left(\left(\frac{\log \beta_s - \mu}{\eta_c}\right)^2\right)\right] \quad 0 < \beta_s < \infty \tag{6}$$

where μ is the mean of variable's natural logarithm. The β_s is defined as the ratio of the absolute difference in children to that of the parent values. The probability distribution of β_s, using lognormal distribution (with $\mu=0$), is obtained from the following relation.

$$P(\beta_s) = \frac{1}{\beta_s \eta_c \sqrt{2\pi}} \exp\left[\frac{-1}{2}\left(\frac{\log \beta_s}{\eta_c}\right)^2\right] \quad 0 < \beta_s < \eta_c \tag{7}$$

The probability of contracting crossover is more desirable than the expanding crossover. This increases the probability of creating offspring between the parents. This is more parent-centric for small value of η_c.

In order to keep children solutions within the bounds defined for decision variable, the factor C_p is defined with the normal probability distribution. This is shown in the following relation.

$$C_p = \frac{1}{P(Z \leq z)} \tag{8}$$

where Z is the standardized normal value and z is the upper limit of standardized normal value. The upper limit of standard normal variable is obtained from the following relation.

$$z = \frac{\log k}{\eta_c} \tag{9}$$

The factor representing normal probability distribution is obtained from the following relation.

$$\alpha_p = \frac{1}{C_p} \tag{10}$$

The constant of probability distribution is obtained from the parent solutions. This is shown in the following relation.

$$k = \min\left(1 + \frac{2\left(p_i^{max} - p_i^2\right)}{\left(p_i^2 - p_i^1\right)}, 1 + \frac{2\left(p_i^{min} - p_i^1\right)}{\left(p_i^2 - p_i^1\right)}\right) \tag{11}$$

where p_i^{max} is the maximum parent value of i-th real variable, p_i^{min} is the minimum parent value of i-th real variable, p_i^1 is the first parent solution of i-th real variable and p_i^2 is the second parent solution of i-th real variable. The lognormal distribution, after incorporating the factor representing normal probability distribution, is obtained from the following relation.

$$P(\beta_s) = \frac{1}{\left(\alpha_p \beta_s \eta_c \sqrt{2\pi}\right)} \exp\left(\frac{-1}{2}\left(\frac{\log \beta_s}{\eta_c}\right)^2\right) \quad 0 < \beta_s < k \tag{12}$$

From lognormal probability distribution, the ordinate is obtained from the following relations.

$$\beta_{sq} = e^{-z_n u} \quad \text{if } \alpha_p u < 0.5 \tag{13}$$

$$\beta_{sq} = e^{z_n u} \quad \text{otherwise} \tag{14}$$

where z_n is the standard normal value and u is the random number. The children solutions are obtained, after obtaining the ordinate, using the following relations.

$$x_i^{(1, tp+1)} = 0.5\left(\left(1 + \beta_{sq}\right)x_i^{(1, tp)} + \left(1 - \beta_{sq}\right)x_i^{(2, tp)}\right) \tag{15}$$

$$x_i^{(2,tp+1)} = 0.5\left(\left(1-\beta_{sq}\right)x_i^{(1,tp)} + \left(1+\beta_{sq}\right)x_i^{(2,tp)}\right) \tag{16}$$

where $x_i^{(1,tp+1)}$ is the first child solution of i-th real variable at (tp+1)th generation, $x_i^{(1,tp)}$ is the first parent solution of i-th real variable at tp-th generation,
$x_i^{(2,tp)}$ is the second parent solution of i-th real variable at tpth generation and
$x_i^{(2,tp+1)}$ is the second child solution of i-th real variable at (tp+1)th generation.
The use of mutation may destroy already found good information. It is suggested that GAs may work well with a large crossover probability (p_c) and
small mutation probability (p_m). Hence, the mutation index (η_m) is chosen
with a small value.

The probability distribution of contracting and expanding crossover is
shown in Figure 20.1. From Figure 20.1, it is observed that the probability
of contracting crossover is more and non-uniform. The probability distribution of contracting and expanding crossovers for the lognormal distribution
is shown in Figure 20.2. From Figure 20.2, it is observed that the probability
of contracting crossover is more for small value of η_c.

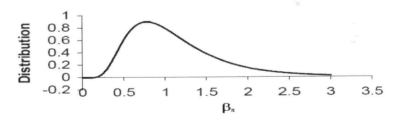

FIGURE 20.1   Probability distribution of contracting and expanding crossover.

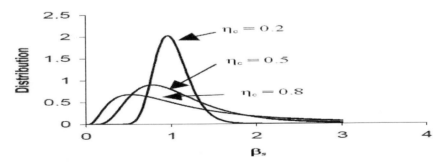

FIGURE 20.2   Probability distribution of contracting and expanding crossovers for the
lognormal distribution.

The lognormal probability distribution of children solutions along with their parents, for different cases, is shown in Figure 20.3. From Figure 20.3, it is observed that this operator is highly parent-centric.

20.5   TEST FUNCTIONS

In this chapter, 20 different multiobjective functions are tested by using the NSGA-II algorithm with SBX-A and SBX-LN crossover probability distributions. These test functions are described in Table 20.1 [1].

20.6   NSGA-II ALGORITHM FOR TEST FUNCTIONS

In this chapter, the NSGA-II algorithm with SBX-A and SBX-LN crossover probability distributions is used to test 20 different multiobjective functions. The variance generational distance (GD) is the performance metric used to measure the performance of SBX-A and SBX-LN operators. The GD of convergence metric finds an average distance of the solutions of optimum

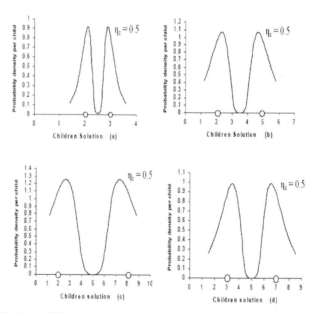

FIGURE 20.3   Probability distribution of children solutions (a) with closely spaced parents at 2 and 3; (b) with parents at 2 and 5 (intermediate case); (c) with distant parents at 2 and 8; (d) with parents at 3 and 7 (intermediate case).

TABLE 20.1 Description of Test Functions

S. No.	Function	No. of variables	Variable bunds	Objective functions (Minimize)	Constraint
1	SCH1	1	$-10 \leq x \leq 10$	$f_1(x) = x^2$	0
				$f_2(x) = (x-2)^2$	
2	SCH2	1	$-5 \leq x \leq 10$	$f_1(x) = \begin{cases} -x & if \ x \leq 1 \\ x-2 & if \ 1 < x \leq 3 \\ 4-x & if \ 3 < x \leq 4 \\ x-4 & if \ x > 4 \end{cases}$	0
				$f_2(x) = (x-5)^2$	
3	FON	5	$-4 \leq x_i \leq 4,$ $i = 1, ..., 5.$	$f_1(x) = 1 - e^{\left(-\sum_{i=1}^{5}\left(x_i - \frac{1}{\sqrt{5}}\right)^2\right)}$	0
				$f_2(x) = 1 - e^{\left(-\sum_{i=1}^{5}\left(x_i + \frac{1}{\sqrt{5}}\right)^2\right)}$	
4	KUR	3	$-5 \leq x_i \leq 5,$ $i = 1, 2, 3.$	$f_1(x) = \sum_{i=1}^{2}\left(-10e^{-0.2\sqrt{x_i^2 + x_{i+1}^2}}\right)$	0
				$f_2(x) = \sum_{i=1}^{3}\left(|x_i|^{0.8} + 5\sin(x_i^3)\right)$	

TABLE 20.1 Continued

S. No.	Function	No. of variables	Variable bounds	Objective functions (Minimize)	Constraint
5	VNT	2	$-3 \leq x_1 \leq 3,$ $-3 \leq x_2 \leq 3.$	$f_1(x) = \left[0.5\left(x_1^2 + x_2^2\right) + \sin\left(x_1^2 + x_2^2\right) \right]$ $f_2(x) = \left[\dfrac{\left(3x_1 - 2x_2 + 4\right)^2}{8} + \dfrac{\left(x_1 - x_2 + 1\right)^2}{27} + 15 \right]$ $f_3(x) = \left[\dfrac{1}{\left(x_1^2 + x_2^2 + 1\right)} - 1.1e^{\left[-\left(x_1^2 + x_2^2\right)\right]} \right]$	0

TABLE 20.1 Continued

S. No.	Function	No. of variables	Variable bunds	Objective functions (Minimize)	Constraint
6	POL	2	$-\pi \leq x_1 \leq \pi,$ $-\pi \leq x_2 \leq \pi.$	$f_1(x) = \left[\begin{array}{l} 1 + (A_1 - B_1)^2 \\ + (A_2 - B_2)^2 \end{array} \right]$ $f_2(x) = \left((x_1 + 3)^2 + (x_2 + 1)^2 \right)$ $A_1 = \left\{ \begin{array}{l} 0.5\sin(1) - 2\cos(1) \\ +\sin(2) - 1.5\cos(2) \end{array} \right\}$ $A_2 = \left\{ \begin{array}{l} 1.5\sin(1) - \cos(1) \\ +2\sin(2) - 0.5\cos(2) \end{array} \right\}$ $B_1 = \left\{ \begin{array}{l} 0.5\sin(x_1) - 2\cos(x_1) \\ +\sin(x_2) - 1.5\cos(x_2) \end{array} \right\}$ $B_2 = \left\{ \begin{array}{l} 1.5\sin(x_1) - \cos(x_1) \\ +2\sin(x_2) - 0.5\cos(x_2) \end{array} \right\}$	0

TABLE 20.1 Continued

S. No.	Function	No. of variables	Variable bounds	Objective functions (Minimize)	Constraint
7	ZDT1	30	$0 \leq x_i \leq 1,$ $i = 1, ..., 30.$	$f_1(x) = x_1$ $f_2(x) = g(x).h(f_1(x), g(x))$ $g(x) = 1 + \dfrac{9}{n-1} \sum_{i=2}^{n} x_i$ $h(f_1(x), g(x)) = 1 - \sqrt{x_1 / g(x)}$	0
8	ZDT2	30	$0 \leq x_i \leq 1,$ $i = 1, ..., 30.$	$f_1(x) = x_1$ $f_2(x) = g(x).h(f_1(x), g(x))$ $g(x) = 1 + \dfrac{9}{n-1} \sum_{i=2}^{n} x_i$ $h(f_1(x), g(x)) = 1 - \left(x_1 / g(x)\right)^2$	0

TABLE 20.1 Continued

S. No.	Function	No. of variables	Variable bunds	Objective functions (Minimize)	Constraint
9	ZDT3	30	$0 \leq x_i \leq 1,$ $i = 1, \ldots, 30.$	$f_1(x) = x_1$ $f_2(x) = g(x).h(f_1(x), g(x))$ $g(x) = 1 + \dfrac{9}{n-1} \sum_{i=2}^{n} x_i$ $h(f_1(x), g(x)) = 1 - \sqrt{\dfrac{x_1}{g(x)}} - \left(\dfrac{x_1}{g(x)} \right) \sin(10\pi x_1).$	0
10	ZDT4	10	$0 \leq x_1 \leq 1,$ $-5 \leq x_i \leq 5,$ $i = 2, \ldots, 10.$	$f_1(x) = x_1$ $f_2(x) = g(x).h(f_1(x), g(x))$ $g(x) = 1 + 10(n-1) + \sum_{i=2}^{n} \left(x_i^2 - 10\cos(4\pi x_i) \right)$ $h(f_1(x), g(x)) = 1 - \sqrt{\dfrac{x_1}{g(x)}}$	0

TABLE 20.1 Continued

S. No.	Function	No. of variables	Variable bounds	Objective functions (Minimize)	Constraint
11	ZDT6	10	$0 \leq x_i \leq 1,$ $i = 1, \ldots, 10.$	$f_1(x) = 1 - e^{(-4x_1)} \sin^6(6\pi x_1)$ $f_2(x) = g(x).h(f_1(x), g(x))$ $g(x) = 1 + 9\left(\left(\sum_{i=2}^{10} x_i\right)/9\right)^{0.25}$ $h(f_1(x), g(x)) = 1 - \left(\dfrac{f_1(x)}{g(x)}\right)^2$	0
12	BNH	2	$0 \leq x_1 \leq 5,$ $0 \leq x_2 \leq 3.$	$f_1(x) = 4x_1^2 + 4x_2^2$ $f_2(x) = (x_1 - 5)^2 + (x_2 - 5)^2$	$C_1(x) = (x_1 - 5)^2 + x_2^2 \leq 25$ $C_2(x) = \left. \begin{matrix} (x_1 - 8)^2 + \\ (x_2 + 3)^2 \geq 7.7 \end{matrix} \right\}$
13	SRN	2	$-20 \leq x_1 \leq 20,$ $-20 \leq x_2 \leq 20.$	$f_1(x) = 2 + (x_1 - 2)^2 + (x_2 - 1)^2$ $f_2(x) = 9x_1 - (x_2 - 1)^2$	$C_1(x) = x_1^2 + x_2^2 \leq 225$ $C_2(x) = x_1 - 3x_2 + 10 \leq 0$

TABLE 20.1 Continued

S. No.	Function	No. of variables	Variable bunds	Objective functions (Minimize)	Constraint
14	TNK	2	$0 \leq x_1 \leq \pi,$ $0 \leq x_2 \leq \pi.$	$f_1(x) = x_1$ $f_2(x) = x_2$	$C_1(x) \equiv \left\{ x_1^2 + x_2^2 - 1 \right. \left. -0.1\cos\left(16\arctan\left(\dfrac{x_1}{x_2}\right)\right) \right\} \geq 0$ $C_2(x) \equiv \left\{ (x_1 - 0.5)^2 + (x_2 - 0.5)^2 \right\} \leq 0.5$
15	CTP1	2	$0 \leq x_1 \leq 1,$ $0 \leq x_2 \leq 1.$	$f_1(x) = x_1$ $f_2(x) = g(x)e^{(-f_1(x)/g(x))}$ $g(x) = 1 + x_2$	$C_1(x) \equiv \left\{ \dfrac{f_2(x) - }{a_1 e^{(-b_1 f_1(x))}} \right\} \geq 0$ $C_2(x) \equiv \left\{ \dfrac{f_2(x)}{-a_2} \right. \left. e^{(-b_2 f_1(x))} \right\} \geq 0$ $a_1 = 0.858, b_1 = 0.541, a_2 = 0.728$ and $b_2 = 0.295$ (for two constraints)

TABLE 20.1 Continued

S. No.	Function	No. of variables	Variable bounds	Objective functions (Minimize)	Constraint
16	CTP2	2	$0 \leq x_1 \leq 1,$ $0 \leq x_2 \leq 1.$	$f_1(x) = x_1,$ $f_2(x) = g(x)\left(1 - \dfrac{f_1(x)}{g(x)}\right)$ $g(x) = 1 + x_2$	$C(x) \equiv \cos(\theta)\left[f_2(x) - e\right]$ $-\sin(\theta)f_1(x) \geq$ $a\left\|\sin\left\{b\pi.\right.\right.$ $\left.\left.\left[\sin(\theta)\left(f_2(x) - e\right)\right]^c\right.\right.$ $\left.\left.\left. + \cos(\theta)\left(f_1(x)\right)\right]^d\right\}\right\|$ $\theta = -0.2\pi,\ a = 0.2,\ b = 10,\ c = 1,\ d = 6 \text{ and}$ $e = 1$

TABLE 20.1 Continued

S. No.	Function	No. of variables	Variable bunds	Objective functions (Minimize)	Constraint
17	CTP5	2	$0 \le x_1 \le 1,$ $0 \le x_2 \le 1.$	$f_1(x) = x_1$ $f_2(x) = g(x)\left(1 - \dfrac{f_1(x)}{g(x)}\right)$ $g(x) = 1 + x_2$	$C(x) \equiv \cos(\theta)\big[f_2(x) - e\big]$ $-\sin(\theta)f_1(x) \ge$ $a\left|\sin\left\{ b\pi.\left[\sin(\theta)\big(f_2(x) - e\big) \atop + \cos(\theta)\big(f_1(x)\big) \right]^c \right\}\right|^d$ $\theta = -0.2\pi,\ a = 0.1,\ b = 10,\ c = 2,\ d = 0.5$ and $e = 1$

TABLE 20.1 Continued

S. No.	Function	No. of variables	Variable bounds	Objective functions (Minimize)	Constraint
18	CTP6	2	$0 \le x_1 \le 1,$ $0 \le x_2 \le 10.$	$f_1(x) = x_1$ $f_2(x) = g(x)\left(1 - \dfrac{f_1(x)}{g(x)}\right)$ $g(x) = 1 + x_2$	$C(x) \equiv \begin{cases} \cos(\theta)\left[f_2(x) - e\right] \\ -\sin(\theta)f_1(x) \ge \end{cases}$ $b\pi.$ $a\left|\sin\left\{\dfrac{\sin(\theta)\left(f_2(x) - e\right)}{+\cos(\theta)\left(f_1(x)\right)}\right]^c\right\}\right|^d$ $\theta = 0.1\pi, a = 40, b = 0.5, c = 1, d = 2 \text{ and } e = -2$
19	CTP7	2	$0 \le x_1 \le 1,$ $0 \le x_2 \le 10.$	$f_1(x) = x_1$ $f_2(x) = g(x)\left(1 - \dfrac{f_1(x)}{g(x)}\right)$ $g(x) = 1 + x_2$	$C(x) \equiv \begin{cases} \cos(\theta)\left[f_2(x) - e\right] \\ -\sin(\theta)f_1(x) \ge \end{cases}$ $b\pi.$ $a\left|\sin\left\{\dfrac{\sin(\theta)\left(f_2(x) - e\right)}{+\cos(\theta)\left(f_1(x)\right)}\right]^c\right\}\right|^d$ $\theta = -0.05\pi, a = 40, b = 5, c = 1, d = 6 \text{ and } e = 0$

TABLE 20.1 Continued

S. No.	Function	No. of variables	Variable bunds	Objective functions (Minimize)	Constraint
20	CTP8	2	$0 \leq x_1 \leq 1,$ $0 \leq x_2 \leq 10.$	$f_1(x) = x_1,$ $f_2(x) = g(x)\left(1 - \dfrac{f_1(x)}{g(x)}\right)$ $g(x) = 1 + x_2$	$C_1(x) \equiv \cos(\theta)\left[f_2(x) - e\right]$ $-\sin(\theta)f_1(x) \geq$ $a\left\|\sin\left\{b\pi\cdot\left[\dfrac{\sin(\theta)\left(f_2(x) - e\right)}{+\cos(\theta)\left(f_1(x)\right)}\right]^c\right\}\right\|^d$ $\theta = 0.1\pi,\ a = 40,\ b = 0.5,\ c = 1,\ d = 2$ and $e = -2$ (for constraint 1) $C_2(x) \equiv \cos(\theta)\left[f_2(x) - e\right]$ $-\sin(\theta)f_1(x) \geq$ $a\left\|\sin\left\{b\pi\cdot\left[\dfrac{\sin(\theta)\left(f_2(x) - e\right)}{+\cos(\theta)\left(f_1(x)\right)}\right]^c\right\}\right\|^d$ $\theta = -0.05\pi,\ a = 40,\ b = 2,\ c = 1,\ d = 6$ and $e = 0$ (for constraint 2)

solution set obtained by the algorithm (Q) from Pareto-optimal set of solutions (P*) [1, 9]. The GD is obtained from the following relation.

$$GD = \frac{\left(\sum_{i=1}^{|Q|} d_i^p\right)^{1/p}}{|Q|} \tag{17}$$

where d_i is the Euclidean distance and p and i are the constants. The value of p is equal to two. The d_i (in the objective space) is the distance between solutions $i \in Q$ and the nearest member of P*. The d_i is obtained from the following relation.

$$d_i = \min_{k=1}^{|P^*|} \sqrt{\sum_{m=1}^{M} \left(f_m^{(i)} - f_m^{*(k)}\right)^2} \tag{18}$$

where the $f_m^{*(k)}$ is the m-th objective function value of the k-th member of P* and $f_m^{(i)}$ is the m-th objective function value of the i-th member of Q. The GD of diversity metric is obtained from the following relation.

$$\Delta = \frac{\sum_{m=1}^{M} d_m^e + \sum_{i=1}^{|Q|} \left|d_i - \overline{d}\right|}{\sum_{m=1}^{M} d_m^e + |Q|\overline{d}} \tag{19}$$

where d_i is the distance measure between neighboring solutions, $\overline{d}$ is the mean value of these distances and d_m^e is the distance between the extreme solutions of P* and Q corresponding to m-th objective function.

The variance GD of convergence and diversity are computed, in each case, to find the performance of algorithm with SBX-A and SBX-LN crossover probability distributions. The probability of contracting crossover is more desirable than the expanding crossover. This increases the probability of creating offspring between the parents. This is more parent-centric for small value of η_c.

The use of mutation may destroy already found good information. It is suggested that GAs may work well with large p_c and small p_m values. Hence, the η_m is chosen with a small value. The value of p_m, for 20 multiobjective functions, is chosen as 0.01 instead of (1/N) where N is number of variables [9].

Five runs are made, for each function, with different random seeds. For all functions, the population size is 100, number of generations is 250, p_c is 0.8, p_m is 0.01, η_c is 0.05 and η_m is 0.5 [7, 9]. These results are shown in Table 20.2.

TABLE 20.2   Variance GD of Convergence and Diversity

S. No.	Function	NSGA-II (SBX-A)		NSGA-II (SBX-LN)		Remark
		Con.	Div.	Con.	Div.	
1	ZDT1	5.99E-05	**1.72E-02**	**4.55E-06**	4.64E-02	SBX-LN/SBX
2	ZDT2	6.84E-05	**7.90E-03**	**1.16E-05**	4.13E-02	SBX-LN/SBX
3	ZDT3	9.50E-05	**1.65E-02**	**1.87E-05**	4.97E-02	SBX-LN/SBX
4	ZDT4	2.97E-04	**1.72E-02**	**1.48E-04**	4.64E-02	SBX-LN/SBX
5	ZDT6	2.83E-04	3.43E-02	**1.81E-06**	**3.00E-02**	SBX-LN
6	SCH1	**2.18E-06**	**5.12E-02**	1.88E-05	5.30E-02	SBX
7	SCH2	**1.60E-05**	1.88E-02	1.90E-04	**1.25E-02**	SBX/ SBX-LN
8	POL	**3.72E-03**	**2.28E-02**	9.44E-03	3.55E-02	SBX
9	FON	**2.62E-05**	4.09E-02	8.14E-05	**3.90E-02**	SBX/SBX-LN
10	KUR	**2.86E-04**	**2.88E-02**	1.16E-03	3.30E-02	SBX
11	VNT	**4.75E-05**	**5.41E-02**	1.81E-04	8.65E-02	SBX
12	BNH	3.09E-03	**2.46E-02**	**2.54E-03**	2.71E-02	SBX-LN/SBX
13	SRN	**1.42E-02**	**2.73E-02**	1.47E-02	2.80E-02	SBX
14	CTP1	**3.12E-05**	2.88E-02	3.55E-05	**2.73E-02**	SBX/SBX-LN
15	CTP2	**5.45E-05**	**7.00E-03**	4.04E-04	2.64E-02	SBX
16	CTP5	**9.59E-05**	3.61E-02	7.40E-04	**2.79E-02**	SBX/SBX-LN
17	CTP6	**8.95E-05**	**1.51E-02**	1.73E-04	3.48E-02	SBX
18	CTP7	**6.46E-05**	**3.69E-03**	1.15E-04	5.56E-03	SBX
19	CTP8	**2.28E-04**	**4.66E-03**	2.85E-03	7.58E-02	SBX
20	TNK	1.24E-03	**4.29E-02**	**2.69E-04**	2.12E-01	SBX-LN/SBX

From Table 20.2, it is observed that the NSGA-II algorithm with SBX-LN crossover probability distribution found better optimum solutions for different types of functions. By comparing the variance GD of convergence and diversity metrics, it is observed that the NSGA-II algorithm with SBX-LN crossover probability distribution is having better convergence for seven functions and better diversity for five functions (9). Classification of these functions is shown in Table 20.3.

From Table 20.3, it is observed that the NSGA-II algorithm with SBX-A is having good convergence and better diversity for some functions. This is because the number of generations, for all the functions, is taken as 250. This is not acceptable for all the functions. Hence, a suitable number of generations, with sufficient number of function evaluations, are to be selected for each function to converge to the Pareto-optimal front (POF). From these

TABLE 20.3 Classification of Functions Outperformed by NSGA-II (SBX-LN)

S. No.	Parameter	NSGA-II (SBX-LN)	
		Convergence	Diversity
1	Continuous solutions	5 (ZDT1, ZDT2, ZDT4, ZDT6 and BNH)	3 (ZDT6, FON and CTP1)
2	Discontinuous solutions	2 (ZDT3 and TNK)	2 (SCH2 and CTP5)
3	Unconstrained functions	5 (ZDT1, ZDT2, ZDT3, ZDT4 and ZDT6)	3 (ZDT6, SCH2 and FON)
4	Constrained functions	2 (BNH and TNK)	2 (CTP1 and CTP5)
5	Unimodal functions	5 (ZDT1, ZDT2, ZDT3, ZDT4 and ZDT6)	3 (ZDT6, SCH2 and FON)
6	Multimodal functions	2 (ZDT3 and TNK)	2 (SCH2 and CTP5)
7	More epitasis	2 (ZDT4 and ZDT6)	1 (ZDT6)
8	One variable functions	0	1 (SCH2)
9	Two variable functions	2 (BNH and TNK)	2 (CTP1 and CTP5)
10	Five variable functions	0	1 (FON)
11	Ten variable functions	2 (ZDT4 and ZDT6)	1 (ZDT6)
12	Thirty variable functions	3 (ZDT1, ZDT2 and ZDT3)	0

results, it is observed that the performance of NSGA-II algorithm is improved by choosing a better parent centric crossover probability distribution.

20.7 SUMMARY

The NSGA-II algorithm with SBX-A and SBX-LN crossover probability distributions are used to test 20 multiobjective functions. The NSGA-II algorithm with SBX-LN crossover probability distribution found better optimal solutions with good diversity for different multiobjective functions. The major observations of this work are as follows.

1. The performance of NSGA-II algorithm, when used to find the optimal solutions of 20 multiobjective functions, is improved by the SBX-LN crossover probability distribution.
2. The NSGA-II algorithm with SBX-LN crossover probability distribution found better optimal solutions for various functions such as unconstrained and constrained functions, unimodal and multimodal functions, functions with different number of variables and functions having continuous and discontinuous solutions.

3. Two functions, ZDT4 and ZDT6, have more epitasis. NSGA-II algorithm with SBX-LN crossover probability distribution found the optimal solutions having better convergence for both functions and having better diversity for one function, ZDT6.

20.8   POINTS FOR FURTHER IMPROVEMENT

The performance of NSGA-II algorithm can be improved further by implementing the following points. Using a better crossover probability distribution. Other with SBX-LN crossover probability distributions will be improved by implementing the following points:

1. The number of generations, with sufficient number of function evaluations, should be selected for each function to convert to the POF.
2. The performance of NSGA-II algorithm will be improved further by using a better probability distribution of SBX-A operator.

KEYWORDS

- actual simulated binary crossover
- differential evolution
- evolutionary programming
- evolutionary strategies
- genetic algorithm
- genetic programming
- nondominated sorting algorithm

REFERENCES

1. Deb, K. *Multiobjective Optimization using Evolutionary Algorithms,* John Wiley & Sons Limited: Chichester, 2002.
2. Deb, K. *Optimization for Engineering Design,* Prentice-Hall of India Private Limited: New Delhi, 2004.
3. Rao, S. S. *Engineering Optimization,* John Wiley & sons: Inc., 1996.
4. Goldberg, D. E. *Genetic Algorithms in Search, Optimization and Machine Learning,* Pearson Education: New Delhi, 2006.

5. Xue, F., Sanderson, A. C., & Graves, R. J. Pareto-based Multiobjective Differential Evolution, *IEEE Congress on Evolutionary Computation, Canberra,* 2003, 2, 862–869.

6. Deb, K., Agarwal, S., Pratap. A., & Meyariven, T. A Fast and Elitist Multiobjective Genetic Algorithm: NSGA-II, *IEEE Transactions on Evolutionary Computation,* 2002, 6, 182–197.

7. Raghuwanshi, M. M., Singru, P. M., Kale, U., & Kakde, O. G. Simulated Binary Crossover with lognormal distribution, *Complexity International,* 12, 1–10.

8. Price, K. V., Storn, R. M., & Lampinen, J. A. *Differential Evolution – A Practical Approach to Global Optimization,* Springer: Verlag Berlin Heidelberg, 2005.

9. Prasad, K. V. R. B., & Singru, P. M. *Performance of Lognormal Probability Distribution in Crossover Operator of NSGA-II Algorithm,* Proceedings of Eighth International Conference on Simulated Evolution And Learning (SEAL-2010), Kanpur (Lucknow), India, December 01–04, 2010; Deb, K., Bhattacharya, A., Chakraborti, N., Chakroborty, P., Das, S., Dutta, J., & Gupta, S. K., Springer-Berlin: Heidelberg, 2010.

CHAPTER 21

EVOLUTIONARY ALGORITHMS FOR MALWARE DETECTION AND CLASSIFICATION

KAMRAN MOROVATI[1] and SANJAY S. KADAM[2]

[1]*Information Security Center of Excellence (ISCX), Faculty of Computer Science, University of New Brunswick, 550 Windsor St., Head Hall E128, Fredericton, NB, E3B 5A3, Canada*

[2]*Evolutionary Computing and Image Processing Group (ECIP), Center for Development of Advanced Computing (C-DAC), Savitribai Phule Pune University Campus, Ganeshkhind, Pune–411007, India*

CONTENTS

ABSTRACT

This chapter presents application of evolutionary algorithms for malware detection and classification based on the dissimilarity of op-code frequency patterns extracted from their source codes. To address this problem, the chapter first describes efforts to establish a correlation between the op-code frequencies and the pre-defined class types using the ANOVA test, and then elaborates on the use of Duncan multiple range test to extract effective op-codes (attribute reduction) for the classification purpose. It also elaborates on another attribute reduction technique based on ant colony optimization, rough sets, and InfoGain-Ranker in order to extract effective op-codes for recognition and classification purpose. Using a reduced set of attributes and randomly collected samples, the random forests classifier was built and used to classify the test samples into various malware classes.

21.1 INTRODUCTION

Malware, an abbreviation for "Malicious Software," is designed to disrupt normal execution of computer operations, gather sensitive information, or gain access to the computer systems. It can also be defined as hostile, intrusive, or annoying software that can appear in the form of code, scripts, and active content, which can infect a single computer, server, or an entire computer network. Malware is usually categorized into "families" (representing particular type of malware with unique characteristics) and "variants" (usually a different version of code in a particular family). Malware is injected into a system in order to harm or subvert the system for purposes other than those intended by its owner(s). Malware includes computer viruses, worms, Trojan horses, spyware, adware, rootkits, logic bombs, bots and other malicious programs.

The current anti-malware software solutions usually rely on the use of binary signatures to identify the corresponding malware. This approach cannot defend against new malware until unless the new malware samples are obtained and analyzed, and their unique signatures are extracted. Therefore, the signature-based approaches are not effective against new and unknown malware. Certain anti-malware solutions use other malware detection techniques such as the "abnormal behavior detection" or "heuristic analysis." These methods also have some disadvantages like high false alarm rate or are costly to implement.

In this chapter we present a new malware detection technique based on op-code frequency patterns. This work is an extension to the research conducted by Bilar [1], showing the effectiveness of op-codes as malware predictor. We first establish a correlation between op-code frequencies and the pre-defined class types using the ANOVA test. We then use Duncan multiple range test for extracting the effective op-codes in order to increase the accuracy of the classification. To the best of our knowledge, the use of the Duncan test is a novel feature selection method used for malware detection. The classifiers used in our research work include decision trees, Artificial Neural Networks, naïve Bayes,' and support vector machines. The classification has been done in two phases: identifying of malware against benign files (two-class problem), and use of six-class labels to further determine sample file types (virus, Trojan, worm, adware, etc.).

In addition to the Duncan test, we have also employed another technique to reduce the number of op-codes and extract the effective ones for classification of malware. This technique comprises a feature selection method known as ant rough set attribute reduction (ARSAR), which is based on ACO, and uses the rough set dependency measure. It then employs *infogain* and a ranker method to rank the selected op-codes. The ranking was based on the "Information Gain" value of the op-codes. We then apply the random forests classifier several times to perform multi-class classification (with class labels as benign, virus, Trojan, worm, adware and rootkit) using the ranked op-codes; adding one op-code at a time to the previous op-code set after completion of every iteration. In this process we could select the best 54 op-codes that gave the maximum classification accuracy.

The chapter is organized as follows. Section 21.2 describes method to extract the op-codes and calculate their frequencies in each malware. Section 21.3 describes the ANOVA test to establish a correlation between the op-code frequencies and the pre-defined class types. It describes the Duncan multiple range test used for extracting the effective op-codes. Section 21.4 describes techniques for classifying malware against the benign files, and methods to identify classes within the malware types (virus, Trojan, worm, adware, etc.). Section 21.5 describes methods to improve classification accuracy using ant colony optimization and rough sets. The chapter concludes with a summary of the research work.

21.2   CALCULATING MALWARE OP-CODE FREQUENCIES

The malware samples were collected from various online virus repositories such as the VX-Heaven[1] and the Virus Sign websites[2]. Out of thousands of malicious files, a total of 200 malware from different families were selected randomly. The normal files representing the "Benign class" comprised Portable Executable (PE) files randomly selected from Cygwin software folders (a collection of tools that provide a Linux look and feel environment for Windows) and "System32" folder of MS-Windows 7 Ultimate version. Malware developers use packing techniques to obfuscate malware source code or to compress the executables. Appropriate unpacking tools available on the web (e.g., www.woodmann.com) were therefore used to acquire the original content of the malware and the normal files. The unpacked files were loaded into the de-facto industry standard disassembler, IDA Pro (6.1) [2], which translates binary content and generates assembly language source code from the machine-executable file format.

After loading each malware sample into IDA Pro and running the InstructionCounter [26], a modified plugin of the IDA disassembler, we extracted the assembly function statistics. For each malware, one text file

1 http://www.vxheaven.org
2 http://www.virussign.com

was generated, which contained the frequency of used assembly functions in the corresponding binary file of the malware. Subsequently, these text files were imported to an MS-Excel spreadsheet and were augmented with the complete list of x86 instruction set available in Ref. [7], totaling 681 assembly functions or op-codes. The total frequency count of the frequency of each of the 681 op-codes across all the 100 benign sample files was around 2 million, and these op-codes were spread across 130 different categories of the assembly functions.

Amongst these op-codes, 85 op-codes accounted for more than 99.8% (ratio of the sum of their frequency count to the total frequency count, expressed in percentage) over the total op-codes found, 14 op-codes accounted for more than 91%, and the top 5 op-codes accounted for 67.7% over the total op-codes extracted. Similarly, the total of the frequency count of each of the 681 op-codes across all the 200-malware sample files was around 7 million, and these op-codes were spread across 163 different categories of the assembly functions. Table 21.1, shows some descriptive statistics of the samples.

TABLE 21.1   Descriptive Statistics of Samples

File Type	Total Op-codes	Op-codes found	Top 14 Op-codes found (Samples' Average)
Goodware	19798076	>130	mov(48%), call(7%), jmp(4%), cmp(4%), jz(3%), test(3%), lea(3%), push(3%), add(3%), pop(3%), sub(2%), jnz(2%), movzx(1%), retn(1%)
Viruses	932016	>141	mov(23%), push(20%), call(10%), pop(10%), cmp(4%), lea(4%), jz(4%), jmp(3%), test(3%), jnz(3%), add(3%), xor(2%), retn(2%), sub(1%)
Trojans	3083467	>146	mov(28%), push(16%), call(9%), pop(6%), lea(4%), add(4%), cmp(4%), jmp(3%), jz(3%), xor(3%), sub(2%), test(2%), retn(2%), jnz(2%)
Worms	1043123	>133	mov(26%), push(21%), call(12%), lea(8%), pop(4%), cmp(4%), add(3%), jz(3%), jmp(3%), test(2%), jnz(2%), xor(2%), retn(2%), sub(1%)
Ad-ware	1614391	>163	mov(24%), push(17%), call(9%), pop(6%), cmp(6%), jz(4%), lea(4%), test(3%), jnz(3%), add(3%), jmp(3%), xor(2%), retn(2%), sub(1%)
Rootkits	.20072	~100	mov(26%), push(19%), call(9%), pop(6%), nop(4%), lea(4%), cmp(3%), add(3%), jz(3%), xor(2%), jmp(2%), retn(2%), test(2%), jnz(2%)

21.3　ANOVA AND DUNCAN MULTIPLE RANGE TESTS

According to Ref. [7], a total of 681 assembly op-codes are defined to date, but we observed that less than 200 op-codes are relevant based on the samples we collected for our research study. In a Microsoft Excel spread sheet, a 300 X 681 contingency table was designed (rows represented samples of different classes; columns represented the op-code frequencies). In this phase our aim was to detect the effective op-codes and reduce the number of assembly functions that we were going to be used as inputs to our classifier. We also wanted to know if there was a statistically significant difference in op-code frequencies between the classes for which we used the ANOVA (Analysis of variance) and Duncan Multiple Range tests.

The result of the ANOVA test confirms that there are some associations between the op-codes and the corresponding classes. For example, the variance of the frequency pattern of the 681 op-codes for a Virus is different from the variance of similar such pattern for a Trojan or any other malware or non-malware. This means that the results of the ANOVA test imply that the frequency patterns of the 681 op-codes for different samples are useful in classification of the samples into different malware types and the benign class. After observing the results of the ANOVA test, we run Duncan test to know the most significant and the least significant op-code clusters. Generally, the Duncan test is based on the idea that the means must be compared according to the variable range. In our experiment, the Duncan test clustered the op-codes into 27 segments based on their significance.

Table 21.2, summarizes the segments created by the Duncan multiple range test. Means for groups in homogeneous subsets are displayed. For example, segment 27 has an op-code (Mov) selected from the 681 op-codes. Its 'Significant' value is 1, which is based on the mean calculated for the "Mov" op-code. This mean is computed by accumulating the frequency of the "Mov" op-code across each of the 300 samples and taking the mean with respect to the total frequency count of all the 681 op-codes across each of the 300 samples. The entries in Table 21.2, help us to establish that the op-codes are good malware predictors. In addition, Duncan test also helps us to segment the op-codes. In this segmentation, op-code functions belonging to each group have same significance, hence, practically only one member function (op-code) is enough to represent the entire segment/cluster. This significantly reduces the number of the effective op-codes from 681 to 27. Table 21.3 shows a list of selected op-codes by the above-mentioned procedure. For example, "neg" op-code (Segment 10) in Table 21.3 represents the entire cluster 10 depicted in Table 21.2.

TABLE 21.2 Duncan Multiple Test Results

Seg.No.	Op-code (Mean)	Significant
27	Mov(0.336307018)	1
26	Push(0.129956140)	1
25	Call(0.092000351)	1
24	Pop(0.045806667), Lea(0.044997544)	0.142
23	Cmp(0.041374737)	1
22	Jz(0.034317895), Jmp(0.033304211)	0.066
21	Add(0.032072281)	1
20	Test(0.029306316)	1
19	Jnz(0.022636842)	1
18	Sub(0.019944211), Xor(0.019635439)	0.576
17	Retn(0.018785614), Xor(0.019635439)	0.123
16	Inc(0.009570526), Movzx(0.008841053)	0.186
15	Andnpd(0.008094737), Movzx(0.008841053)	0.176
14	Or(0.006754035), Dec(0.006129474)	0.257
13	Jge(0.001707719),fstp(0.001731228), jnb(0.001824912), jg(0.001862456), jl(0.001952281), ja(0.002014035), jbe(0.002148421), leave(0.002558947), shr(0.002677895), shl(0.002889123), jle(0.002902456), jb(0.002942807), nop(0.003014035)	0.051
12	Imul(0.001677895), jge (0.001707719), fstp(0.001731228), jnb(0.001824912), jg(0.001862456), jl(0.001952281), ja(0.002014035), jbe(0.002148421), leave(0.002558947), shr(0.002677895), shl(0.002889123),jle(0.002902456), jb(0.002942807)	0.060
11	Sar(0.001381404), movs(0.001405263), movsx(0.001540000), Fld(0.001556140),Imul(0.001677895), jge(0.001707719), fstp(0.001731228), jnb(0.001824912), jg(0.001862456), jl(0.001952281), ja(0.002014035), jbe(0.002148421), leave(0.002558947), shr(0.002677895)	0.054
10	sbb(0.000819649), stos(0.000825614), not(0.000835088), rol(0.000835088), ror(0.000868772), xchg(0.000956140), neg(0.000971930), sar(0.001381404), movs(0.001405263), movsx(0.001540000), fld(0.001556140), imul(0.001677895), jge(0.001707719), fstp(0.001731228), jnb(0.001824912), jg(0.001862456), jl(0.001952281), ja(0.002014035),jbe(0.002148421)	0.054
9	setnz(0.000749123), punpcklwd(0.000763860), monitor(0.000768070), setz(0.000769825), js(0.000778246), sbb(0.000819649), stos(0.000825614), not(0.000835088), rol(0.000835088), ror(0.000868772), xchg(0.000956140), neg(0.000971930), sar(0.001381404), movs(0.001405263), movsx(0.001540000), fld(0.001556140), imul(0.001677895), jge(0.001707719), fstp(0.001731228), jnb(0.001824912), jg(0.001862456), jl(0.001952281), ja(0.002014035)	0.071

TABLE 21.2 continued

Seg.No.	Op-code (Mean)	Significant
8	jns(0.000595789), setnz(0.000749123), punpcklwd(0.000763860), monitor(0.000768070), setz(0.000769825), js(0.000778246), sbb(0.000819649), stos(0.000825614), not(0.000835088), rol(0.000835088), ror(0.000868772), xchg(0.000956140), neg(0.000971930), sar(0.001381404), movs(0.001405263), movsx(0.001540000), fld(0.001556140), imul(0.001677895), jge(0.001707719), fstp(0.001731228), jnb(0.001824912), jg(0.001862456), jl(0.001952281)	0.051
7	cdq(0.000472281), jns(0.000595789), setnz(0.000749123), punpcklwd(0.000763860), monitor(0.000768070), setz(0.000769825), js(0.000778246), sbb(0.000819649), stos(0.000825614), not(0.000835088), rol(0.000835088), ror(0.000868772), xchg(0.000956140), neg(0.000971930), sar(0.001381404), movs(0.001405263), movsx(0.001540000), fld(0.001556140), imul(0.001677895), jge(0.001707719), fstp(0.001731228), jnb(0.001824912), jl(0.001952281)	0.051
6	fxch(0.000347368), bts(0.000351930), pusha(0.000355088), fmul(0.000363509), fild(0.000364211), cmovz(0.000368421), div(0.000376842), cmps(0.000392281), adc(0.000443860), cdq(0.000472281), jns(0.000595789), setnz(0.000749123), punpcklwd(0.000763860), monitor(0.000768070), setz(0.000769825), js(0.000778246), sbb(0.000819649), stos(0.000825614), not(0.000835088), rol(0.000835088), ror(0.000868772), xchg(0.000956140), neg(0.000971930), sar(0.001381404), movs(0.001405263), movsx(0.001540000), fld(0.001556140), imul(0.001677895), jge(0.001707719), fstp(0.001731228)	0.050
5	fnstcw(0.000154737), pmvzb(0.000157544), mul(0.000159649), fldz(0.000160351), fadd(0.000161404), fucomip(0.000164561), ldmxcsr(0.000177895), fistp(0.000179649), wait(0.000197193), setnle(0.000203509), scas(0.000219649), movdqa(0.000252982), idiv(0.000280000), xadd(0.000280000), popa(0.000283509), fldcw(0.000315088), cld(0.000332982), fxch(0.000347368), bts(0.000351930), pusha(0.000355088), fmul(0.000363509), fild(0.000364211), cmovz(0.000368421), cdq(0.000472281), jns(0.000595789), setnz(0.000749123), punpcklwd(0.000763860), monitor(0.000768070), setz(0.000769825), js(0.000778246), sbb(0.000819649), stos(0.000825614), not(0.000835088), rol(0.000835088), ror(0.000868772), xchg(0.000956140), neg(0.000971930), sar(0.001381404), movs(0.001405263), movsx(0.001540000), fld(0.001556140)	0.052
4	fldcw(0.000315088), cld(0.000332982), fxch(0.000347368), bts(0.000351930), pusha(0.000355088), fmul(0.000363509), fild(0.000364211), cmovz(0.000368421), div(0.000376842),	0.054

TABLE 21.2 continued

Seg.No.	Op-code (Mean)	Significant
	cmps(0.000392281), adc(0.000443860), cdq(0.000472281), jns(0.000595789), setnz(0.000749123), punpcklwd(0.000763860), monitor(0.000768070), setz(0.000769825), js(0.000778246), sbb(0.000819649), stos(0.000825614), not(0.000835088), rol(0.000835088), ror(0.000868772), xchg(0.000956140), neg(0.000971930), sar(0.001381404), movs(0.001405263), movsx(0.001540000), fld(0.001556140), imul(0.001677895), monitor(0.000768070), setz(0.000769825), js(0.000778246), sbb(0.000819649), stos(0.000825614), not(0.000835088), rol(0.000835088), ror(0.000868772), xchg(0.000956140), neg(0.000971930), sar(0.001381404), movs(0.001405263), movsx(0.001540000), fld(0.001556140)	
3	fnstcw(0.000154737), pmvzb(0.000157544), mul(0.000159649), fldz(0.000160351), fadd(0.000161404), fucomip(0.000164561), ldmxcsr(0.000177895), fistp(0.000179649), wait(0.000197193), setnle(0.000203509), scas(0.000219649), movdqa(0.000252982), idiv(0.000280000), xadd(0.000280000), popa(0.000283509), fldcw(0.000315088), cld(0.000332982), fxch(0.000347368), bts(0.000351930), pusha(0.000355088), fmul(0.000363509), fild(0.000364211), cmovz(0.000368421), div(0.000376842), cmps(0.000392281), adc(0.000443860), cdq(0.000472281), jns(0.000595789), setnz(0.000749123), punpcklwd(0.000763860), monitor(0.000768070), setz(0.000769825), js(0.000778246), sbb(0.000819649), stos(0.000825614), not(0.000835088), rol(0.000835088), ror(0.000868772), xchg(0.000956140),n eg(0.000971930), sar(0.001381404), movs(0.001405263), movsx(0.001540000), fld(0.001556140)	0.052
2	lods(0.000128772), std(0.000135789), fnstcw(0.000154737), pmvzb(0.000157544), mul(0.000159649), fldz(0.000160351), fadd(0.000161404), fucomip(0.000164561), ldmxcsr(0.000177895), fistp(0.000179649), wait(0.000197193), setnle(0.000203509), scas(0.000219649),	
	movdqa(0.000252982), idiv(0.000280000), xadd(0.000280000), popa(0.000283509), fldcw(0.000315088), cld(0.000332982), fxch(0.000347368), bts(0.000351930), pusha(0.000355088), fmul(0.000363509), fild(0.000364211), cmovz(0.000368421), div(0.000376842), cmps(0.000392281), adc(0.000443860), cdq(0.000472281), jns(0.000595789), setnz(0.000749123), punpcklwd(0.000763860), monitor(0.000768070), setz(0.000769825), js(0.000778246), sbb(0.000819649), stos(0.000825614), not(0.000835088), rol(0.000835088), ror(0.000868772), xchg(0.000956140), neg(0.000971930), sar(0.001381404), movs(0.001405263), movsx(0.001540000)	0.050
1	Op-codes with the mean value <= 0.001405263	0.069

TABLE 21.3 Selected Op-Codes From the Resulting Clusters of Duncan Test

			Selected op-codes according to the Duncan multiple range test		
Segment #	Selected Op-code	Segment #	Selected Op-code	Segment #	Selected Op-code
1	cmovnz(0.00012316)	10	neg(0.000971930)	19	jnz(0.022636842)
2	std(0.000135789)	11	fld(0.001556140)	20	test(0.0293063)
3	div(0.000376842)	12	imul(0.0016677895)	21	add(0.03207229)
4	fldcw(0.0003315088)	13	nop(0.003014035)	22	jz(0.034317895)
5	cld(0.0003332982)	14	or(0.006754035)	23	cmp(0.04137473)
6	adc(0.000443860)	15	movzx(0.00884105)	24	pop(0.04580667)
7	cdq(0.0004472281)	16	inc(0.009570526)	25	call(0.092000351)
8	jns(0.000595789)	17	retn(0.018785614)	26	push(0.12995614)
9	js(0.0000778246)	18	sub(0.019944211)	27	mov(0.33630701)

21.4 CLASSIFICATION OF SAMPLES

After limiting the op-codes to 27, the next task is to apply different classifiers to classify the samples in two different ways, namely, the binary classification (i.e., malware v/s non-malware) and multi-class classification (i.e., different malware classes and the benign class). In our research work, four different classifiers were tested and compared.

21.4.1 BINARY CLASSIFICATION

21.4.1.1 Decision Tree Classification

To classify our samples using decision trees, we converted our excel database into the CSV (comma-separated values) format and then we imported this data file into the WEKA software [6]. Weka is a collection of machine learning algorithms for data mining tasks and it contains tools for data preprocessing, classification, regression, clustering, association rules, and visualization. It is also well-suited for developing new machine learning schemes. We tested various decision tree classification algorithms, which are available in the tree classifier section of the Weka software. These algorithms include J48 [25], J48-Graft [11], Best-First tree (BFTree) [13], Functional Tree (FT) [14], Multiclass alternating decision trees (LADTree) [10], Logistic Model Trees (LMT) [20], Naive Bayes Tree (NBTree) [24], Random Forests [16], Fast decision tree learner (REPTree) [4], Classification and Regression Trees (SimpleCart) [17].

The results of classification with two class labels (Malware and Benign) using the decision tree classifiers are depicted in Table 21.4. According to our observation, the NB Tree and Random Forest methods achieved the best accuracy with more than 98% classification success rate. Random forests are an ensemble learning method for classification that operate by constructing a multitude of decision trees at training time and output the class that is the mode of the classes output by individual trees [16]. The method combines the "bagging" [15] idea and the random selection of features, in order to construct a collection of decision trees with controlled variation. The detailed procedure of the Random Forests algorithm is explained in [16]. Tables 21.4–21.6 presents more details regarding these two methods.

TABLE 21.4 Decision Tree Classification Summary (2 Class Labels)

	J48	J48 Graft	BF Tree	FT	LAD Tree	LMT	NB Tree	Random Forest	Random Tree	REP Tree	Simple Cart	Decision Stump
Total Number of Instances	300	300	300	300	300	300	300	300	300	300	300	300
Correctly Classified Instances	283	287	290	289	291	291	295	295	292	290	290	291
(%)	94.33%	95.66%	96.66%	96.33%	97%	97%	98.33%	98.33%	97.33%	96.66%	96.66%	97%
Incorrectly Classified Instances	17	13	10	11	9	9	5	5	8	10	10	9
(%)	5.66%	4.33%	3.33%	3.66%	3%	3%	1.66%	1.66%	2.66%	3.33%	3.33%	3%
Kappa statistic	0.8709	0.9018	0.9246	0.9169	0.932	0.9327	0.9622	0.9622	0.9394	0.9246	0.9246	0.9327
Mean absolute error	0.0673	0.0542	0.047	0.0416	0.034	0.075	0.0192	0.0399	0.0267	0.0526	0.0502	0.051
Root mean squared error	0.2368	0.2074	0.1733	0.1715	0.15	0.1601	0.1273	0.1328	0.1633	0.1801	0.1723	0.17
Relative absolute error	15.21%	12.23%	10.62%	9.40%	7.68%	16.94%	4.34%	9.01%	6.02%	11.89%	11.34%	11.51%
Root relative squared error	50.37%	44.12%	36.87%	36.47%	31.90%	34.05%	27.08%	28.24%	34.73%	38.31%	36.64%	36.15%

TABLE 21.5 Evaluation of NBTree and Random Forest Classifiers

Class Label	TP Rate	FP Rate	Precision	Recall	F-Measure	ROC Area
Malware	0.99	0.03	0.985	0.99	0.988	0.991
Benign	0.97	0.01	0.98	0.97	0.975	0.991
Weighted Avg.	0.983	0.024	0.983	0.983	0.983	0.991

TABLE 21.6 Confusion Matrix of NBTree and Random Forest Classifier

Classified As →	Malware	Benign
Malware	199	2
Benign	3	96

21.4.1.2 ANN Classification

The Neurosolution™ [3] software was used for creating an appropriate ANN model, training the model, and testing and performing the classification. In our experiment, we have used a three-layer (input-hidden-output) feed forward ANN model with Levenberg–Marquardt back propagation learning algorithm and sigmoid transfer function (Figure 21.1). For classification with two class labels, the malware and benign labels were tagged as desired outputs, and the 27 op-code frequencies, selected through the Duncan test, were tagged as inputs. Out of the 300 samples collected, 50% of these samples were used for training the neural network, 15% samples were used for cross-validation, and the remaining 35% samples were used for testing. Cross validation set was used to prevent over training and ensure generalization.

The network trained itself for 500 iterations, just before the error for the cross validation set started increasing (which constituted the stopping criteria). The observed MSE versus the epoch values is shown in Figure 21.2. The accuracy for binary classification using ANN was around 97% (for the test samples) as shown in Table 21.7. After training the network, we test the network performance on data that the network was not trained with (test data). The resultant confusion matrix and other the observed results are summarized in Table 21.7.

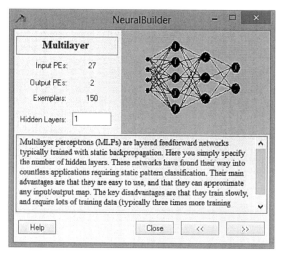

FIGURE 21.1 ANN architecture.

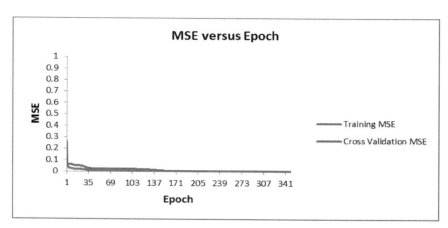

FIGURE 21.2 MSE versus Epoch diagram.

TABLE 21.7 Confusion Matrix (Binary Classification)

Actual	Target					
	Training		Cross Validation		Testing	
	Malware	Benign	Malware	Benign	Malware	Benign
Malware	115(69.7%)	1(0.61%)	19(63.33%)	0(0%)	65(61.90%)	1(0.95%)
Benign	0(0%)	49(29.70%)	0(0%)	11(33.67%)	2(1.90%)	37(35.24%)

ROC is a matrix used to show how changing the detection threshold affects detection versus false positives. If the threshold is set too high, the network would miss several detections. On the other hand if the threshold is set too low, there would be many false positive. In the following ROC chart, the True Positive Rate is placed on Y axis and the False Positive Rate on the X axis. Area under curve (ROC) is equal to 0.984681854 (Figure 21.3). Since the ROC value is close to 1, we can infer that the performance of the ANN classifier is satisfactory.

21.4.1.3   Naïve Bays Classification

The Bayesian Classifier is capable of calculating the most probable output depending on the input. Naïve Bayes classifier estimates the class conditional probability by assuming that the attributes are conditionally independent, given the class label. Bayesian classification provides practical learning algorithms and prior knowledge and observed data can be combined. Bayesian Classification provides a useful perspective for understanding and evaluating many learning algorithms. It calculates explicit probabilities for hypothesis and it is robust to noise in the input data. Naïve Bayes classifier is a probabilistic learning method based on Bayes' theorem. It combines evidence *e* from multiple sources of data to estimate the probability of a hypothesis *h*:

$$P(h/e) = \frac{P(h)P(e/h)}{P(e)} \qquad (1)$$

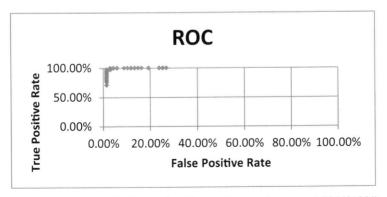

FIGURE 21.3   ROC curve for binary classification (area under curve=0.984681854).

In our experiment, an implementation of the Naïve Bayes Classifier in WEKA software was used. According to WEKA's manual, *weka.classifiers. bayes.NaiveBayes* class is a Naive Bayes classifier, which uses estimator classes. Numeric estimator precision values are selected based on the analysis of the training data. For testing the classifier, we used the 10-fold cross validation method. Tables 21.8 and 21.9 show the Naïve Bays binary classification results. Table 21.10 summarizes the classifier evaluation results.

21.4.1.4 Support Vector Machine Classification

The support vector machine is one of the most popular and accurate classifiers. A classification task usually involves dividing the data into training and testing sets. Each instance in the training set contains one "target value"

TABLE 21.8 Naïve Bays Binary Classification Stratified Cross-Validation Results

	Naïve Bayes Classifier
Total Number of Instances	300
Correctly Classified Instances	294 (98%)
Incorrectly Classified Instances	6 (2%)
Kappa statistic	0.9545
Mean absolute error	0.0179
Root mean squared error	0.1279
Relative absolute error	4.03%
Root relative squared error	27.20%

TABLE 21.9 Naïve Bays Binary Classification Confusion Matrix

Classified As ➜	Malware	Benign
Malware	199	2
Benign	4	95

TABLE 21.10 Naïve Bays Binary Classification Evaluation Results

Class Label	TP Rate	FP Rate	Precision	Recall	F-Measure	ROC Area
Malware	0.99	0.03	0.985	0.99	0.988	0.991
Benign	0.97	0.01	0.98	0.97	0.975	0.991
Weighted Avg.	0.983	0.024	0.983	0.983	0.983	0.991

(known as class label) and several "attributes" (i.e., the features or observed variables). The goal of SVM is to produce a model (based on the training data), which predicts the target values of the test data given only the test data attributes. Given a training set, the SVM requires the solution of the following optimization problem [9]:

$$\min_{w,b,s} \frac{1}{2} w^T w + C \sum_{i=1}^{l} \varepsilon_i \tag{2}$$

subject to $\quad y_i \left(w^T \varphi(x_i) + b \right) \geq 1 - \varepsilon_i$, where $\varepsilon_i \geq 0$

Here, the training vectors x_i are mapped onto a higher (may be infinite) dimensional space by the function φ. SVM finds a linear separating hyper plane with the maximal margin in this higher dimensional space. The variable $C > 0$ is the penalty parameter of the error term. Furthermore, $K(x_i, y_i) \equiv \varphi(x_i)^T \varphi(x_j)$ is called the kernel function.

To classify our samples using SVM, the LIBSVM [8] package was used. We used the RapidMiner[3] software, which has implemented the SVM classifier by using the LIBSVM package. Figure 21.4 illustrates the model structure in RapidMiner. To evaluate the SVM classifier, X-Validation operator was used to implement a 10-fold cross validation. It is a nested operator. It has two sub-processes: a training sub-process and a testing sub-process. The training sub-process is used for training a model. The trained model is then applied in the testing sub-process. The performance of the model is also measured during the testing phase.

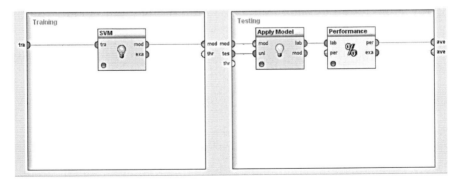

FIGURE 21.4   SVM classifier in RapidMiner.

---

3 http://rapid-i.com

The input set is partitioned into k subsets of equal size. Of the k subsets, a single subset is retained as the testing dataset and the remaining $k - 1$ subsets are used as training data. The cross-validation process is then repeated k times, with each of the k subsets used exactly once as the testing data. The k results from the k iterations can then be averaged (or otherwise combined) to produce a single estimation. The value k can be adjusted using a number of validation parameters. The learning process usually optimizes the model to fit the training data as best as possible. Setting the optimal values for different parameters in a particular SVM kernel results in better classification and more accuracy. By using the "Optimize Parameters (Grid)" component in RapidMiner, we tested different parameter values to see which combination achieves the best accuracy. This operator finds the optimal values of the selected parameters (Figure 21.5).

In our experiment we classified our samples using all four basic SVM kernels. Table 21.11 shows the obtained results. The confusion matrixes and ROC values of different SVM kernels are listed in Table 21.12. As can be seen from Tables 21.11 and 21.12, the SVM classifier gives relatively better results for binary classification in comparision to other classifiers (i.e., Decision tree, ANN, and Naïve Bayes).

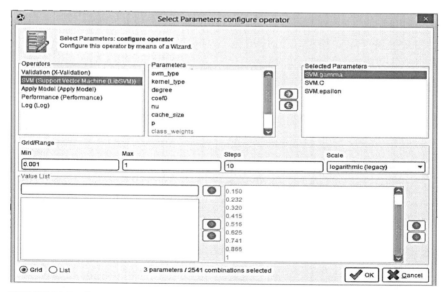

FIGURE 21.5   Parameter configuration operator to find optimal combination for RBF kernel.

TABLE 21.11 SVM Binary Classification Results

SVM Kernel Type	Accuracy	Precision	Recall	Gamma	Cost	Epsilon	Coefficient	Degree
RBF	98.00%	97.27%	97%	0.1009	0.0	1.0	-	-
Linear	99.00%	98.18%	99.00%	-	790.21	1	-	-
Polynomial	99.00%	98.09%	99.00%	7.90021	300.0	0.50095	0.0	1
Sigmoid	99.00%	98.33%	99.00%	1.7	820.18	0.001	-	-

TABLE 21.12 SVM Binary Classification ROC Values and Confusion Matrixes

	RBF ROC = 0.986		Linear ROC= 0.986		Polynomial ROC= 0.986		Sigmoid ROC= 0.987	
	True Malware	True Benign	True Malware	True Benign	True Malware	True Benign	True Malware	True Benign
Predicted Malware	196	3	199	1	199	1	199	1
Predicted Benign	3	96	2	98	2	98	2	98

21.4.2   MULTI-CLASS CLASSIFICATION

21.4.2.1   Decision Tree Classification

Table 21.13, summarizes the experimental results of the classification with six class labels representing: Virus, Trojan, Adware, Worm, Rootkit and Benign. In this case, again the Random Forest algorithm yields the best accuracy with 82.33% success rate. Table 21.14, recapitulates the classifier evaluation information and Table 21.15 shows the respective confusion matrix.

21.4.2.2   ANN Classification

Classification with six class labels was done using the Matlab software. The ANN architecture is similar as in the previous study except that this time the desired outputs have six class labels instead of two. The dataset was divided as follow: 50% of the total 300 samples were marked as training data, 15% as cross validation set, and 35% marked for testing purpose. Figure 21.6 shows the network's performance as it improved during the training process. Performance is measured in terms of the mean squared error and is shown in log scale. It decreased rapidly as the network was trained. Performance is shown for each of the training, validation and test datasets.

A good measure of how well the neural network has fitted the data is the confusion matrix plot. Here, the confusion matrix is plotted across all the samples. The confusion matrix shows the percentage of correct and incorrect classifications. Correct classifications are the green squares on the matrix diagonal. Incorrect classifications are represented by the red squares. If the network has learned to classify correctly, the percentages in the red squares should be very small, indicating few misclassifications. If this is not the case then further training or training a network with more hidden neurons could improve the accuracy. Figure 21.7 shows the overall percentages of correct and incorrect classification. The accuracy of the testing phase is 70.5% and overall accuracy is 73.7%.

Another measure of how well the neural network has fitted the data is the ROC (receiver operating characteristic) plot. This plot shows how the false positive and true positive rates relate as the thresholding of outputs is varied from 0 to 1. The nearer the line is to the left as well as to the top; the fewer are false positives that need to be accepted in order to get a high true positive rate. The best classifiers will have a line whose one end goes from

TABLE 21.13 Decision Tree Classification Summary (6 Class Labels)

	J48	J48 Graft	BF Tree	FT	LAD Tree	LMT	NB Tree	Random Forests	Random Tree	REP Tree	Simple Cart	Decision Stump
Total Number of Instances	300	300	300	300	300	300	300	300	300	300	300	300
Correctly Classified Instances	220	222	219	225	213	217	236	247	233	210	218	169
(%)	73.33%	74%	73%	75%	71%	72.33%	78.66%	82.33%	77.66%	70%	72.66%	56.33%
Incorrectly Classified Instances	80	78	81	75	87	83	64	53	67	90	82	131
(%)	26.6%	26%	27%	25%	29%	27.66%	21.33%	17.66%	22.33%	30%	27.33%	43.66%
Kappa statistic	0.6585	0.6665	0.651	0.6809	0.6263	0.6467	0.725	0.7712	0.7133	0.6098	0.6471	0.394
Mean absolute error	0.0946	0.0929	0.1022	0.0931	0.1251	0.1056	0.0941	0.0976	0.0767	0.1291	0.1061	0.1786
Root mean squared error	0.2805	0.2784	0.2718	0.2789	0.2697	0.2739	0.2324	0.2097	0.2735	0.2742	0.276	0.3002
Relative absolute error	36.38%	35.74%	39.30%	35.80%	48.14%	40.65%	36.22%	37.54%	29.49%	49.67%	40.83%	68.70%
Root relative squared error	77.85%	77.27%	75.44%	77.41%	74.87%	76.02%	64.52%	58.20%	75.92%	76.10%	76.60%	83.32%

TABLE 21.14   Evaluation of Random Forest Classifier (6 Class Labels)

Class Label	TP Rate	FP Rate	Precision	Recall	F-Measure	ROC Area
Virus	0.606	0.045	0.625	0.606	0.615	0.888
Adware	0.688	0.034	0.71	0.688	0.698	0.919
Worm	0.622	0.039	0.737	0.622	0.675	0.864
Benign	0.98	0.035	0.933	0.98	0.956	0.984
Trojan	0.776	0.094	0.738	0.776	0.756	0.932
Rootkit	0.867	0.007	0.867	0.867	0.867	0.923
Weighted Avg.	0.797	0.05	0.793	0.797	0.794	0.932

TABLE 21.15   Confusion Matrix for Random Forest Classifier (6 Class Labels)

Classified As →	Virus	Adware	Worm	Benign	Trojan	Rootkit
Virus	20	3	5	0	5	0
Adware	3	22	0	0	7	0
Worm	5	2	28	4	6	0
Benign	0	0	1	97	1	0
Trojan	4	4	4	3	59	2
Rootkit	0	0	0	0	2	13

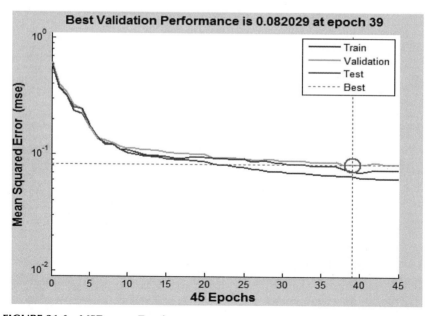

FIGURE 21.6   MSE versus Epoch.

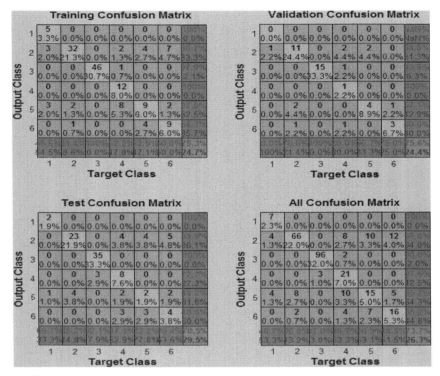

FIGURE 21.7   Confusion matrix.

the bottom left corner to the top left corner and the other end goes to the top right corner, or closer to that. Figure 21.8 shows the ROC plot for training, cross validation and testing sets. From Figure 21.8, it can be observered that the multi-class classification using ANN was not as expected.

21.4.2.3   Naïve Bays Classifier Classification

For multiclass classification, the Bayes' theorem provides the probability of each hypothesis being true given the evidence. During the training step, a Naïve Bayes classifier uses the training data to estimate the parameters of a probability distribution (on the right-hand side of the Eq. 1). Again, a 10-fold cross validation was used to test the model. During the testing phase, a naïve Bayes classifier uses these estimations to compute the posterior probability (the left-hand side of the Eq. 1) given a test sample. Each test sample is then classified using the hypothesis with the largest posterior probability. Tables 21.16 and 21.17 show the Naïve Bays multi-class classification results. Table 21.18 summarizes the classifier evaluation results.

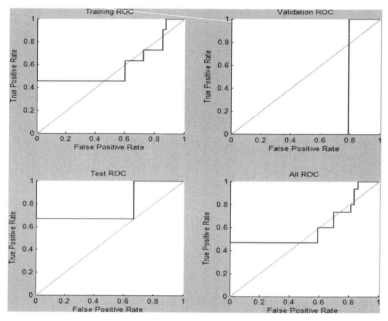

FIGURE 21.8　ANN multiclass classification ROC plot.

TABLE 21.16　Naïve Bays Multiclass Classification Stratified Cross-Validation Results

	Naïve Bayes Classifier
Total Number of Instances	300
Correctly Classified Instances	190 (63.33%)
Incorrectly Classified Instances	110 (36.66%)
Kappa statistic	0.5408
Mean absolute error	0.1235
Root mean squared error	0.3362
Relative absolute error	47.50%
Root relative squared error	93.30%

TABLE 21.17　Naïve Bays Multiclass Classification Confusion Matrix

Classified As ➔	Virus	Adware	Worm	Benign	Trojan	Rootkit
Virus	22	3	4	0	2	2
Adware	11	19	0	0	2	0
Worm	14	7	20	2	2	0
Benign	0	0	2	96	1	0
Trojan	38	7	4	1	26	0
Rootkit	3	2	0	0	3	7

TABLE 21.18   Naïve Bays Multiclass Classification Evaluation Results

Class Label	TP Rate	FP Rate	Precision	Recall	F-Measure	ROC Area
Virus	0.607	0.247	0.25	0.667	0.364	0.789
Adware	0.594	0.071	0.5	0.594	0.543	0.866
Worm	0.444	0.039	0.667	0.444	0.533	0.82
Benign	0.97	0.015	0.97	0.97	0.97	0.981
Trojan	0.342	0.045	0.722	0.342	0.464	0.896
Rootkit	0.467	0.007	0.778	0.467	0.583	0.913
Weighted Avg.	0.633	0.057	0.723	0.633	0.645	0.898

21.4.2.4   Support Vector Machine Classification

To test the performance of the SVM classifier in the multiclass case, we first determined the optimal parameters for each SVM kernel using "Optimize Parameters (Grid)" component in the RapidMiner software. We then performed the classification using the LIBSVM wrapper available in the WEKA software. The reason for selecting the WEKA LIBSVM wrapper is its ability to provide more detailed information (for the multi-class case) compared to other data mining software (such as R, LionSolver, etc.). Tables 21.19–21.21 show the multi-class classification results obtained by the SVM classifier. The results are stratified for each SVM kernel.

Table 21.22 summarizes the experiment results of the Binary classification. As seen from Table 21.22, for binary classification with Malware and Benign class labels, the SVM classifier (Linear, Polynomial and Sigmoid kernels) performed better than others.

21.5   IMPROVING MALWARE DETECTION USING ACO AND ROUGH SET

In this section we attempt to further improve the classification accuracy of the multi-class classification of the malware. In order to perform this task, we first apply a feature selection method based on the ant colony optimization algorithm to get a reduced feature set comprising the op-codes. We then apply an appropriate classifier on this reduced feature set to classify the samples into various classes (i.e., into six class labels as discussed in Section 21.3). For feature selection, the Ant Rough Set Attribute Reduction algorithm (ARSAR) [23] has been used, which is based on the Rough Set Dependency Measure and Ant Colony Optimization (ACO). In our research work, since the Random Forests

TABLE 21.19   SVM Multiclass Classification

SVM Kernel Type	RBF	Linear	Polynomial	Sigmoid
Accuracy	72.66%	72%	65.67%	33%
Total Number of Instances	300	300	300	300
Correctly Classified Instances	218	216	197	99
Incorrectly Classified Instances	82	84	103	201
Kappa statistic	0.6452	0.6372	0.5358	0
Mean absolute error	0.0911	0.0933	0.1144	0.2233
Root mean squared error	0.3018	0.3055	0.3383	0.4726
Relative absolute error	35.05%	35.90%	44.02%	85.92%
Root relative squared error	83.78%	84.79%	93.90%	131.17%
Optimal Kernel Parameters				
Gamma	9.60004	880.12	2.0008	0.001
Cost	50	-	10	1
Epsilon	1	-	10	1
Coefficient	-	-	0	-
Degree	-	-	1	-

TABLE 21.20   SVM Multiclass Classification Evaluation

SVM Kernel Type	Class Label	TP Rate	FP Rate	Precision	Recall	F-Measure	ROC Area
Linear	Virus	0.242	0.019	0.615	0.242	0.348	0.612
	Adware	0.719	0.149	0.365	0.719	0.484	0.785
	Worm	0.467	0.027	0.75	0.467	0.575	0.72
	Benign	0.99	0.025	0.951	0.99	0.97	0.983
	Trojan	0.776	0.121	0.686	0.776	0.728	0.828
	Rootkit	0.467	0	1	0.467	0.636	0.733
	Weighted Avg.	0.72	0.061	0.757	0.72	0.713	0.83
RBF	Virus	0.303	0.019	0.667	0.303	0.417	0.642
	Adware	0.719	0.108	0.442	0.719	0.548	0.805
	Worm	0.489	0.059	0.595	0.489	0.537	0.715
	Benign	0.98	0.03	0.942	0.98	0.96	0.975
	Trojan	0.776	0.121	0.686	0.776	0.728	0.828
	Rootkit	0.467	0	1	0.467	0.636	0.733
	Weighted Avg.	0.727	0.063	0.744	0.727	0.718	0.832

TABLE 21.20   continued

SVM Kernel Type	Class Label	TP Rate	FP Rate	Precision	Recall	F-Measure	ROC Area
Polynomial	Virus	0.212	0.034	0.438	0.212	0.286	0.589
	Adware	0	0	0	0	0	0.5
	Worm	0.4	0008	0.9	0.4	0.554	0.696
	Benign	0.98	0.03	0.942	0.98	0.96	0.975
	Trojan	0.895	0.384	0.442	0.895	0.591	0.755
	Rootkit	0.467	0	1	0.467	0.636	0.733
	Weighted Avg.	0.657	0.112	0.656	0.657	0.613	0.772
Sigmoid	Virus	0	0	0	0	0	0.5
	Adware	0	0	0	0	0	0.5
	Worm	0	0	0	0	0	0.5
	Benign	1	1	0.33	1	0.496	0.5
	Trojan	0	0	0	0	0	0.5
	Rootkit	0	0	0	0	0	0.5
	Weighted Avg.	0.33	0.33	0.109	0.33	0.164	0.5

TABLE 21.21   SVM Multiclass Classification Confusion Matrices

SVM Kernel Type	Classified as →	Virus	Adware	Worm	Benign	Trojan	Rootkit
Linear	Virus	8	12	3	0	10	0
	Adware	2	23	0	0	7	0
	Worm	1	14	21	3	6	0
	Benign	0	0	1	98	0	0
	Trojan	2	12	1	2	59	0
	Rootkit	0	2	2	0	4	7
RBF	Virus	10	11	2	0	10	0
	Adware	2	23	0	0	7	0
	Worm	3	11	22	4	5	0
	Benign	0	0	1	97	1	0
	Trojan	0	7	8	2	59	0
	Rootkit	0	0	4	0	4	7

TABLE 21.21   continued

SVM Kernel Type	Classified as →	Virus	Adware	Worm	Benign	Trojan	Rootkit
Polynomial	Virus	7	0	0	0	26	0
	Adware	3	0	0	0	29	0
	Worm	3	0	18	2	22	0
	Benign	0	0	1	97	1	0
	Trojan	3	0	1	4	68	0
	Rootkit	0	0	0	0	8	7
Sigmoid	Virus	0	0	0	33	0	0
	Adware	0	0	0	32	0	0
	Worm	0	0	0	45	0	0
	Benign	0	0	0	99	0	0
	Trojan	0	0	0	76	0	0
	Rootkit	0	0	0	15	0	0

TABLE 21.22   Classification Summary

Classifier	Classification Accuracy (Multiclass)	Classification Accuracy (Binary)
Decision Tree (Random Forest)	82.33%	98.33%
ANN	73.7%	97.36%
Naïve Bayes	63.33%	98%
SVM	72.66%	99%

classifier performed relatively better than the other classifiers, it has been used to perform the multi-class classification using the reduced feature set.

The procedure to improve the accuracy of the multi-class classification process can be summarized as follows:

1) Apply the ARSAR algorithm several times on the op-code frequency dataset to reduce the number of features and find the optimal subset of these op-codes.
2) Rank the reduced op-codes from the optimal set based on their ability of partitioning the dataset ("Information gain" value).
3) Sort the op-codes according to their "Infogain" value, and iteratively perform multi-class classification starting with only one op-code with the highest Infogain value in the first iteration, followed by

multi-class classification using the next higher-ranked op-code added to the previous op-code list during each iteration.

4) Plot the observed accuracies for each op-code set (of each iteration) to determine the global maximum (max. accuracy).

The rest of the discussion elaborates on the algorithm used for feature selection, the classification process, and the results.

21.5.1 ANT COLONY OPTIMIZATION

Real ants are able to find the shortest path between their nest and the food sources due to deposition of a chemical substance called as pheromone on their paths. The pheromone evaporates over time. Hence, the shorter paths will contain more pheromone (as the rate of pheromone deposition is relatively greater than the rate of evaporation for such paths) and will subsequently attract a greater number of ants in comparison to the longer paths (which would be taken by only fewer ants). ACO, loosely inspired by the behavior of the real ants, was initially proposed by Colorni, Dorigo and Maniezzo [18, 19]. The main underlying idea is to use ants to find parallel solutions to a given problem based on the local problem data and on a dynamic memory structure containing information (e.g., amount of pheromone depositions, path length, etc.) on the quality of previously obtained result. The collective behavior emerging from the interaction of the different search processes has proved effective in solving combinatorial optimization problems.

21.5.2 DATA REDUCTION WITH ROUGH SET THEORY

Using the rough set theory, the reduction of attributes is achieved by comparing equivalence relations generated by the sets of attributes. Attributes are removed so that the reduced set provides the same predictive capability of the decision feature as the original. A_{reduct} is defined as a subset of minimal cardinality R_{min} of the conditional attribute set C such that $\gamma_R(D) = \gamma_C(D)$, where γ is the attribute dependency measure [5].

$$R = \{X : X \subseteq C, \gamma_X(D) = \gamma_C(D)\} \qquad (3)$$

$$R_{min} = \{X : X \in R, \forall Y \in R, |X| \leq |Y|\} \qquad (4)$$

The intersection of all the sets in R_{min} is called the core, the elements of which are those attributes that cannot be eliminated without introducing more contradictions to the dataset. Using Rough Set Attribute Reduction (RSAR), a subset with minimum cardinality is searched. The problem of finding a *reduct* of an information system has been addressed by numerous researchers. The most basic solution to locating such a subset is to simply generate all possible subsets and retrieve those with a maximum rough set dependency degree. Obviously, this is an expensive solution to the problem and is only practical for very simple datasets. Most of the time only one *reduct* is required as typically only one subset of features is used to reduce a dataset, so all the calculations involved in discovering the rest are pointless. To improve the performance of the above method an element of pruning can be introduced. By noting the cardinality of any pre-discovered reducts, the current possible subset can be ignored if it contains more elements. However, a better approach is needed one that will avoid computational effort [23].

The QuickReduct algorithm listed below (adapted from Ref. [27]) attempts to calculate a *reduct* without exhaustively generating all possible subsets. It starts off with an empty set and adds in turn, one at a time, those attributes that result in the greatest increase in the rough set dependency metric, until this produces the maximum possible value for the dataset. When there is no further increase in the dependency measure, this could be considered as a termination criterion. This will produce exactly the same path to a *reduct* due to the monotonicity of the measure without the computational overhead of calculating the dataset consistency.

$$QUICKREDUCT(C, D)$$

C = the set of all conditional features;

D = the set of decision features

$R \leftarrow \{ \ \}$

Do

$\quad T \leftarrow R$

$\quad \forall x \in (C - R)$

$\quad if \ \gamma_{R \cup \{x\}}(D) > \gamma_T(D)$

$\quad T \leftarrow R \cup \{x\}$

$\quad R \leftarrow T$

$\quad until \ \gamma_R(D) == \gamma_C(D)$

Return R

The QuickReduct algorithm, however, is not guaranteed to find a minimal subset as has been shown in [12]. Using the dependency function to discriminate between candidates may lead the search down a non-minimal path. It is impossible to predict which combinations of the attributes will lead to an optimal *reduct* based on the changes in dependency with the addition or deletion of single attributes. It does result in a close-to-minimal subset, though, which is still useful in greatly reducing dataset dimensionality. However, when maximal data reductions are required, other search mechanisms must be employed [23]. In our research work, we employ the QuickReduct algorithm by building the *reduct* iteratively, where, during each iteration, one additional op-code gets added to the previous *reduct* set.

21.5.3 ACO FRAMEWORK

ACO based feature selection method is attractive for feature selection as there seems to be no heuristic that can guide search to the optimal minimal subset (of features). Additionally, the ants may discover the best feature combinations as they proceed throughout the search space. On the other hand, Rough set theory [28] has successfully been used as a selection tool to discover data dependencies and reduce the number of attributes contained in a structural dataset. In general, an ACO algorithm can be applied to any combinatorial problem which satisfies the following requirements [22]:

- **Appropriate problem representation.** A description of the problem as a graph with a set of nodes and edges between nodes.
- **Heuristic desirability (η) of edges.** A suitable heuristic measure of the "goodness" of paths from one node to every other connected node in the graph.
- **Construction of feasible solutions.** A mechanism to efficiently create possible solutions.
- **Pheromone updating rule.** A suitable method of updating the pheromone levels on edges with a corresponding evaporation rule. Typical methods involve selecting the nbest ants and updating the paths they chose.
- **Probabilistic transition rule.** A mechanism for the determination of the probability of an ant traversing from one node in the graph to the next.

Each ant in the artificial colony maintains a memory of its history, remembering the path it has chosen so far in constructing a solution. This history can be used in the evaluation of the resulting solution and may also contribute to the decision process at each stage of the solution construction.

Two types of information are available to ants during their graph traversal, local and global, controlled by the parameters α and β, respectively. Local information is obtained through a problem-specific heuristic measure.

The extent to which the measure influences an ant's decision to traverse an edge is controlled by β parameter. This will guide ants towards paths that are likely to result in good solutions. Global knowledge is also available to ants through the deposition of artificial pheromone on the graph edges by their predecessors over time. The impact of this knowledge on an ant's traversal decision is determined by the α parameter. Good paths discovered by past ants will have a higher amount of associated pheromone. How much pheromone is deposited, and when, is dependent on the characteristics of the problem. No other local or global knowledge is available to the ants in the standard ACO model.

21.5.3.1 Feature Selection with ACO

Feature selection task in our research work can be formulated as an ACO problem. An ACO method requires a problem to be represented as a graph, where nodes represent features (op-codes in our case) and edges denote the choice of the next feature. The search for the optimal feature subset is then an ant traversal through the graph, where a minimum number of nodes (representing a subset of the best features selected by the ants) are visited such that they satisfy a traversal stopping criterion. The stopping criterion could simply be the classification accuracy, which must be achieved with this subset, assuming that the selected features are used to classify certain objects. The ants terminate their traversal and output this best feature subset (optimal op-code set in our case) as a candidate for data reduction.

A suitable heuristic desirability function that enables traversing between the features on the problem graph and the subset evaluation function (i.e., the stopping criteria) can be performed by using different techniques such as entropy-based measure [21], the rough set dependency measure, and several others. Optimality is defined as criteria to select the best path among several paths undertaken by the corresponding ants at the end of each iteration. Depending on how optimality is defined for a particular application, the pheromone may be updated accordingly. For instance, subset minimality and "goodness" are two key factors (or optimality criteria), where the pheromone updates is proportional to the "goodness" and inversely proportional to the size of the path traversed. How "goodness" is determined will also depend on the application. In some cases, there may be a heuristic evaluation of the subset, in others it may be based on the resulting classification accuracy of a classifier produced using the subset [23].

In our research work, we have used entropy-based measure [21] as a heuristic, rough set dependency measure as a subset evaluator (or stopping criteria), and minimality of the paths as optimality criteria. By minimality of the paths we mean the minimum number of nodes or features (op-codes) traversed by the ants. The heuristic desirability function and pheromone factors are combined to form the so-called probabilistic transition rule, denoting the probability of an ant k at feature i choosing to move to feature j at time t:

$$P_{ij}^k (t) = \frac{\left[\tau_{ij} (t) \right]^\alpha \left[\eta_{ij} \right]^\beta}{\sum_{l \in j_i^k} \left[\tau_{ij} (t) \right]^\alpha \left[\eta_{ij} \right]^\beta} \tag{5}$$

where, j_i^k is the set of ant k's unvisited features, η_{ij} is the heuristic desirability of selecting feature j (when at feature i), and $\tau_{ij}(t)$ is the amount of virtual pheromone on edge (i, j). The choice of α and β is determined experimentally.

Figure 9 (adapted from [23]), illustrates the attribute selection process with ACO. It begins by generating a number of ants k, which are then placed randomly on the graph (i.e., each ant starts with one random feature). The number of ants could be equal to the number of features.

From these initial positions, the ants traverse edges probabilistically until a traversal stopping criterion is satisfied. The resulting feature subsets are gathered and then evaluated. If an optimal subset has been found or the algorithm has executed a certain number of iterations, then the process halts and outputs the best feature subset encountered. If neither condition holds, then the pheromone is updated, a new set of ants is created and the process iterates again [23].

21.5.3.2 Pheromone Update

As discussed in the earlier section, the pheromone may be updated based on the optimality criteria specified for the particular application. The use of the rough set dependency measure to find the rough set *reduct* could be

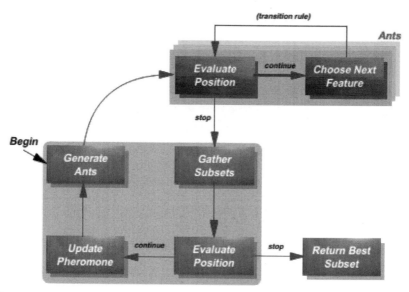

FIGURE 21.9   Feature selection with ACO.

used as the stopping criterion. This means that an ant will stop building its feature subset when the dependency of the subset reaches the maximum for the dataset (a value 1 for consistent datasets). The pheromone on each edge is updated according to the following equation:

$$\tau_{ij}(t+1) = (1-\rho)\tau_{ij}(t) + \Delta\tau_{ij}(t) \tag{6}$$

where,

$$\Delta\tau_{ij}(t) = \sum_{k=1}^{n}\left(\gamma'\left(S^{k}\right)/\left|S^{k}\right|\right) \tag{7}$$

The above equation is used to update the pheromone if an edge *(i, j)* has been traversed, otherwise, The value ρ is decay constant used to simulate the evaporation of the pheromone, S^{k} is the feature subset found by an ant k. The pheromone is updated according to both the rough measure of the "goodness" of the ant's feature subset (γ') and the size of the subset itself. By this definition, all ants update the pheromone. Alternative strategies may be used for this, such as allowing only those ants with the most current best feature subsets to proportionally increase the pheromone [23].

21.5.3.3   Experiment Setup

The java version of the ARSAR, which is ported to the WEKA by Richard Jensen[4], was used to reduce the number of attributes (op-codes). In our experiment, the pre-computed heuristic desirability of edge traversal was the entropy measure; with the subset evaluation performed using the rough set dependency heuristic (weka.attributeSelection.InfoGainAttributeEval class). In each iteration of applying the ARSAR, the number of ants used was set to the number of features, with each ant starting on a different feature. The ants continue to construct possible solutions until they reach a rough set *reduct*. To avoid unproductive searches, the size of the best *reduct* set was used to reject those subsets whose cardinality exceeded this value. The pheromone levels were set at 0.5 with a small random variation added. The pheromone levels were increased by only those ants, which achieved the *reduct* set. The parameter α was set to 1 and β was set to 0.5.

For classification, the Random Forests algorithm with 100 trees was used. Table 21.23 summarizes the feature selection and classification results using the ARSAR and Random Forests classifier. As can be seen in Table 21.23 while the size of the feature set decreases from 681 to 46, the accuracy also decreases. The best accuracy was achieved by using the first optimal subset, which contains 116 op-codes, that is, the six-label multi-class classification was performed using 116 op-codes selected by the ants.

21.5.3.4   Entropy Based Ranking of the Op-Codes

After getting the optimal feature set (116 op-codes in our problem), in order to further increase the accuracy of the multi-class classification, we ranked each feature in the optimal feature set. We used an entropy measure to rank the op-codes. We ran the InfoGain + Ranker algorithms in the feature selection tab of the Weka software in order to determine the significance of each individual variable that contributed to the classification results. The InfoGainAttributeEval class, evaluates the worth of an attribute by measuring the information gain with respect to the class.

$$\text{InfoGain(Class, Attribute)} = H(\text{Class}) - H(\text{Class} \mid \text{Attribute})$$

---

4 Download URL: http://users.aber.ac.uk/rkj/book/wekafull.jar

TABLE 21.23 Applying ASRAR Algorithm Multiple Times

Iteration	Initial Set Size	Reduced Subset size	Classification Accuracy
Round 1	681	116	83.67%
Round 2	116	77	83%
Round 3	77	54	82.33%
Round 4	54	46	81.67%

where, **H** is the information entropy. The entropy function for a given set is calculated based on the class distribution of the tuples in the set. Let **D** consist of data tuples defined by a set of attributes and a class-label attribute. The class-label attribute provides the class information per tuple.

The entropy function for a given set $(\mathbf{D_1})$ is calculated based on the class distribution of the tuples in the set. For example, given m classes, C_1, C_2, ..., C_m, the entropy of $\mathbf{D}_1$ is:

$$Entropy\ (D_1) = -\sum_{i=1}^{m} p_i \log_2 (p_i) \qquad (8)$$

where, p_i is the probability of class C_i in D_1, determined by dividing the number of tuples of class C_i in D_1 by $|D_1|$, the total number of tuples in D_1. After applying ARSAR multiple times and selecting the best subset, we ranked all the op-codes in the selected optimal subset. Table 21.24 shows the corresponding output.

21.5.4 FINDING THE MAXIMUM ACCURACY

After ranking the op-codes and using the Random Forests algorithm, the six-label multi-class classification was performed multiple times. Starting with only one op-code with the highest rank (i.e., push) and each time adding the next higher ranked op-code from the Table 21.24, we repeated the classification. The resulting plot is illustrated in Figure 21.10. The best accuracy (84.33%) was observed with the first 50 four ranked op-codes from Table 21.24.

Table 21.25 summarizes the Random Forests multi-class classification results, which are obtained by using the first 54 highest ranked op-codes from Table 21.24. Tables 21.26 and 21.27 show the confusion matrix and the evaluation results, respectively. To sum up, the combination of RoughSet-ACO

TABLE 21.24 Ranked Op-Codes

Information Gain Ranking Filter								
Rank	InfoGain	Op-code	Rank	InfoGain	Op-code	Rank	InfoGain	Op-code
1	0.8709	push	40	0.2195	pushf	79	0.1199	fucomi
2	0.8176	dec	41	0.2142	fxch	80	0.1138	cmovg
3	0.7802	mov	42	0.2126	wait	81	0.1124	cld
4	0.6791	test	43	0.2124	setb	82	0.103	jnb
5	0.6722	nop	44	0.2089	popa	83	0.1007	fst
6	0.6447	pop	45	0.2082	setbe	84	0.1001	fsubp
7	0.6374	movs	46	0.2036	cmovbe	85	0.088	shr
8	0.6361	retn	47	0.2023	setz	86	0.088	add
9	0.6347	leave	48	0.1969	fldcw	87	0.0877	setns
10	0.6245	inc	49	0.1855	cpuid	88	0.0709	fsub
11	0.5754	jle	50	0.1835	cmova	89	0.0611	rcl
12	0.5681	xor	51	0.1835	ja	90	0.0556	xadd
13	0.5479	or	52	0.177	cmovns	91	0	cmovnc
14	0.5369	stos	53	0.1768	bts	92	0	psrlw
15	0.5284	sub	54	0.1732	faddp	93	0	popfq
16	0.4969	jl	55	0.1711	movapd	94	0	andnps
17	0.4917	js	56	0.1684	rcr	95	0	aesdec
18	0.4536	jmp	57	0.1621	not	96	0	aesimc
19	0.4531	movzx	58	0.162	jbe	97	0	setg
20	0.4407	jnz	59	0.1596	fnclex	98	0	vfmaddsd

TABLE 21.24   continued

			Information Gain Ranking Filter					
Rank	InfoGain	Op-code	Rank	InfoGain	Op-code	Rank	InfoGain	Op-code
21	0.3949	setnbe	60	0.159	div	99	0	sysexit
22	0.3891	cdq	61	0.1579	popf	100	0	movnti
23	0.3759	cmp	62	0.1575	cmovle	101	0	fdecstp
24	0.3691	call	63	0.1575	cmovs	102	0	pcmpeqq
25	0.3617	xchg	64	0.1553	sar	103	0	out
26	0.3288	jge	65	0.1547	sbb	104	0	imul
27	0.3224	cmovz	66	0.1546	fcomp	105	0	fsetpm
28	0.3199	and	67	0.1542	scas	106	0	ficomp
29	0.3067	cmps	68	0.1521	bt	107	0	fldln2
30	0.3014	jb	69	0.1463	pxor	108	0	cmovpo
31	0.2893	cmovnz	70	0.1387	lea	109	0	pmovsxbq
32	0.2875	std	71	0.1385	fucomip	110	0	popad
33	0.27	lods	72	0.1385	cmovge	111	0	pmvzb
34	0.2683	movsx	73	0.1337	loop	112	0	cmpsd
35	0.2602	setnz	74	0.1323	cmovl	113	0	pcmpgtq
36	0.2601	neg	75	0.1314	jz	114	0	pmaxsd
37	0.2499	fnstcw	76	0.121	retf	115	0	pf2iw
38	0.2403	movdqa	77	0.1206	fld	116	0	pfpnacc
39	0.22	pusha	78	0.1204	jg			

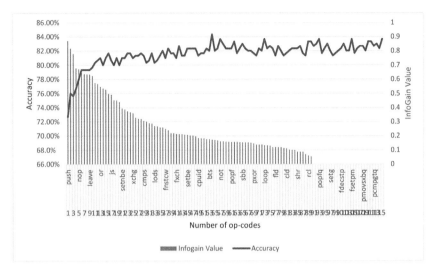

FIGURE 21.10 Accuracy Plot.

TABLE 21.25 Random Forests Multiclass Classification Stratified Cross-Validation Results

	Random Forests Classifier
Total Number of Instances	300
Correctly Classified Instances	253 (84.33%)
Incorrectly Classified Instances	47 (15.67%)
Kappa statistic	0.7966
Mean absolute error	0.0935
Root mean squared error	0.2032
Relative absolute error	36.0262%
Root relative squared error	56.427%

TABLE 21.26 Random Forests Multiclass Classification Confusion Matrix

Classified As →	Virus	Adware	Worm	Benign	Trojan	Rootkit
Virus	22	2	1	1	7	0
Adware	4	22	1	1	2	1
Worm	4	1	30	3	7	0
Benign	0	0	1	98	0	0
Trojan	2	3	2	2	68	0
Rootkit	0	1	0	0	1	13

TABLE 21.27   Random Forests Multiclass Classification Evaluation Results

Class Label	TP Rate	FP Rate	Precision	Recall	F-Measure	ROC Area
Virus	0.667	0.037	0.688	0.667	0.677	0.93
Adware	0.71	0.026	0.759	0.71	0.733	0.926
Worm	0.667	0.02	0.857	0.667	0.75	0.918
Benign	0.99	0.035	0.933	0.99	0.961	0.995
Trojan	0.883	0.076	0.8	0.883	0.84	0.96
Rootkit	0.867	0.004	0.929	0.867	0.897	0.958
Weighted Avg.	0.843	0.041	0.842	0.843	0.84	0.958

and InfoGain-Ranker methods, resulted in a better op-code selection. The multi-class classification with this subset produced better results compared to the Duncan test (i.e., 2% accuracy improvement).

21.6   CONCLUSION

This chapter presented application of evolutionary algorithms for malware detection and classification based on the dissimilarity of op-code frequency patterns extracted from their source codes. In binary classification, a new method of malware detection has been proposed. Initially, by using the ANOVA test we showed that, statistically, the op-code frequency patterns of our classes differ significantly, which aided their classification. The op-codes were reduced by the Duncan multiple range test. Finally, we performed the classification using multiple classifiers, including decision trees, ANN, Naïve Bayes and SVM. According to Table 21.22, for binary classification with Malware and Benign class labels, the SVM classifier (Linear, Polynomial and Sigmoid kernels) performed better than others. In multi-class classification, the Random Forest classifier yielded betters result in comparison to other classifiers. The goal of this approach is to detect malicious files (binary classification). In this case, all the classifiers successfully achieved a very high accuracy, which confirms the strength and efficiency of the suggested malware detection method. If we consider ±2% as an accuracy tolerance to compare classifiers together, this can be concluded all techniques presented here have a good potential to be used as the malware detector.

In ACO and Rough Set based Classification, we attempted to improve the classification accuracy of the multi-class classification process. We applied a feature selection method known as Ant Rough Set Attribute Reduction (ARSAR), which is based on ACO and uses the Rough Set Dependency Measure, to get a reduced feature set comprising the op-codes. In the next step, we employed the InfoGain+Ranker method to rank the selected features. The ranking was based on the InfoGain value of the op-codes. We then applied the Random Forests classifier several times to perform multi-class classification using the ranked op-codes, adding one op-code to previous op-code set after each iteration. The best accuracy was observed using first 54 ranked op-codes. The multi-class classification with this subset produced better results in comparison to the use of the Duncan test (i.e., 2% accuracy improvement).

KEYWORDS

- **ANN and SVM classifier**
- **ANOVA test**
- **Ant Rough Set Attribute Reduction**
- **decision trees classifier**
- **Duncan multiple range**
- **malware detection and classification**
- **Naïve Bayes classifier**
- **random forests classifier**

REFERENCES

1. Bilar, D. Opcodes as predictor for malware, *International Journal of Electronic Security and Digital Forensics.* 2007, 1–2, 156–168.
2. Hex-Rays, IDA Pro Dissasember, An Advanced Interactive Multiprocessor Disassembler, http://www.hex-rays.com (Accessed Oct, 2012).
3. NeuroSolutions for Excel Product Summary, http://www.neurosolutions.com/products/nsexcel/ (Accessed Jun, 2012).
4. Weka, REPTree Algorithm, http://weka.sourceforge.net/doc/weka/classifiers/trees/REPTree.html (Accessed Aug, 2012).
5. Wikipedia, Rough set, http://en.wikipedia.org/wiki/Rough_set (Accessed Oct 05, 2013).
6. WEKA, The University of Waikato, http://www.cs.waikato.ac.nz/~ml/weka/ (Accessed Oct, 2012).
7. Wikipedia, x86 instruction listings, http://en.wikipedia.org/wiki/X86_instruction_listings (Accessed May, 2012).

8. Chang, C-C., & Lin, C-J., LIBSVM: A Library for Support Vector Machines, *ACM Transactions on Intelligent Systems and Technology (TIST)* 2011, 2–3, 1–2.

9. Hsu, C. W., Chang, C. C., & Lin, C. J. *A Practical Guide to Support Vector Classification*, Department of Computer Science, National Taiwan University, Taipei, Taiwan, 2010.

10. Holmes, G., Pfahringer, B., Kirkby, R., Frank, E., & Hall, M. Multiclass Alternating Decision Trees, *Lecture Notes in Computer Science*, 2002, 2430, 161–172.

11. Webb, G., *Decision Tree Grafting From the All-Tests-But-One Partition*, 1999.

12. Chouchoulas, A., Halliwell, J., & Shen, Q. On the Implementation of Rough Set Attribute Reduction, *Proceedings of the 2002 UK Workshop on Computational Intelligence*, 2002.

13. Shi, H. *Best-first decision tree learning*, University of Waikato: Hamilton, NZ,. 2007.

14. Gama, J. Functional Trees, *Machine Learning* 2004, 55, 219–250.

15. Breiman, L. Bagging predictors, *Machine Learning* 1996, 24–2, 123–140.

16. Breiman, L. Random Forests, *Machine Learning* 2001, 45–1, 5–32.

17. Breiman, L., Friedman, J. H., Olshen, R. A., & Stone, C. J. *Classification and Regression Trees*, Wadsworth International Group: Belmont, California, 1984.

18. Colorni, A., Dorigo, M., & Maniezzo, V. Distributed Optimization by Ant Colonies, Proceedings of ECAL'91, European Conference on Artificial Life, Paris, France, 134–142, Elsevier Publishing, 1991.

19. Dorigo, M. Optimization, learning and natural algorithms, PhD Thesis, Polite cnico di Milano, 1992.

20. Landwehr, N., Hall, M., & Frank, E. Logistic Model Trees, *Machine Learning* 2005, 95, 161–205.

21. JR, Q. C4.5: *Programs for Machine Learning*, San Mateo; Morgan Kaufmann Publishers: CA, 1993.

22. Jensen, R., & Shen, Q. Computational Intelligence and Feature Selection, IEEE Ed., Hoboken; Wiley Publication: New Jersy, 2008.

23. Jensen, R. Performing Feature Selection with ACO, *Swarm Intelligence and Data Mining*, Springer-Verlag, 2006, 45–73.

24. Kohavi, R. Scaling Up the Accuracy of Naive-Bayes Classifiers: A Decision-Tree Hybrid, Proceedings of the Second International Conference on Knowledge Discovery and Data Mining, 1996, 202–207.

25. Quinlan, R. *C4.5: Programs for Machine Learning*; Morgan Kaufmann Publishers: San Mateo, CA, 1993.

26. Porst, S. InstructionCounter v.1.02 (IDA Plugin), *http://www.the-interweb.com/serendipity/index.php?/archives/62-IDA-InstructionCounter-plugin-1.02.html* (Accessed 2012).

27. Chouchoulas, A., & Shen, Q. Rough Set-Aided Keyword Reduction for Text Categorisation, *Applied Artificial Intelligence*, 2001, 15–9, 843–873.

28. Zheng, S. Y., & Liu, J. J. A Secure Confidential Document Model and Its Application, International Conference on Multimedia Information Networking and Security, Nov 4–6, 2010, 516–519.

INDEX

N